International Textbooks in Chemical Engineering

Consulting Editor

PAUL W. MURRILL

Professor of Chemical Engineering
Louisiana State University

Formulation and Optimization of Mathematical Models

Formulation and Optimization of Mathematical Models

CECIL L. SMITH
RALPH W. PIKE
PAUL W. MURRILL

Department of Chemical Engineering
Louisiana State University

INTERNATIONAL TEXTBOOK COMPANY
An Intext *Publisher*
Scranton, Pennsylvania 18515

ISBN O-7002-2245-6

Library of Congress Catalog Card Number 70-105073

Copyright © 1970 by International Textbook Company

to our wives
CHARLOTTE, PAT, and NANCY

Preface

As evidenced by its tremendous impact on our present-day society, the development of the digital computer must rank as one of the more significant technological innovations of this century. Although not so well-known outside of engineering fields, the analog computer has also contributed significantly to the development and application of scientific knowledge. Defining the practice of engineering as "obtaining numerical answers to numerical problems," these two types of computing machines have transformed the practice of engineering from one of obtaining approximate answers utilizing simple and approximate models to one of analyzing the detailed mechanisms of practical problems to obtain a better understanding and a more exact answer.

Taking advantage of the capabilities of these machines requires that the pertinent variables in the system be related by mathematical equations, the collection of which has come to be known as the mathematical model. A study of the development and utilization of these models is the central theme of this book. Since the process of "obtaining a numerical answer to a numerical problem" often involves obtaining a solution which is *best* in some sense, optimization methods are frequently involved. We have thus chosen to incorporate material on development of models, on methods to solve the equations involved, and on optimization techniques into one volume.

In this text, the formulation of both ordinary and partial differential equations from the basic laws of conservation, rate expressions, and other fundamental equations are presented. To solve these equations, numerical and analog techniques are emphasized over the classical analytical methods. Associated topics incorporated into the text include development of models for control systems, derivation of changes of variables, use of analog models to show the similarity of apparently diverse systems, matrix formulation of models, solution of nonlinear algebraic equations describing large-scale systems, dimensional analysis and statistical techniques. The applications of the most widely used optimization techniques of analytical and numerical optimum seeking methods, including linear and dynamic programming, are discussed in the context of applying these methods to problems involving mathematical models.

The authors have presented the material in this text to a variety of audiences. As originally developed, the material in the first part of the book was used mainly in a course at the senior level composed primarily of chemical engineering students but with some engineering students from other disciplines. The entire text has been used in various extension and continuing-education courses for practicing engineers. By selecting appropriate topics from Chapters 1, 2, 3, 5, 6, 7, 8, and 11, we have used this text in a sophomore-level course on applications of mathematical methods to engineering problems. In preparing the manuscript an effort was made to keep the chapters reasonably independent to give the instructor considerable freedom in matching subjects in the text to his course outline.

The preparation of this material has represented a large secretarial and reproduction effort. We would like to give a special word of thanks to Louisiana State University's Department of Chemical Engineering, the Division of Engineering Research, and The Continuing Education Division for their able assistance. We would especially like to thank Dr. Philip A. Bryant for his contributions in Chapter 15. We would also like to acknowledge the assistance of Dr. Alfredo López, Dr. Abel DeSouza, Dr. Gary C. April, Dr. John A. Miller, Dr. Charles F. Moore, Mr. George P. Burdell, and a number of others who have given helpful suggestions.

Cecil L. Smith
Ralph W. Pike
Paul W. Murrill

Baton Rouge, Louisiana
February, 1970

Contents

chapter 3   LUMPED-PARAMETER SYSTEMS . . . 89

chapter 4   SIMPLE CONTROL-SYSTEM MODELS . . . 124

chapter 5   ANALOG-COMPUTER SOLUTION TECHNIQUES . . . 155

chapter 6 NUMERICAL SOLUTIONS OF ORDINARY DIFFERENTIAL
EQUATIONS . . . 185

chapter 7 PROBLEMS REQUIRING PARTIAL DIFFERENTIAL
EQUATIONS . . . 211

chapter 8   SOLUTION OF PARTIAL DIFFERENTIAL
            EQUATIONS . . . 241

chapter 9   TRANSFORMATIONS . . . 262

chapter 10   PHYSICAL ANALOGIES . . . 284

chapter 11   LARGE-SCALE SYSTEMS . . . 323

chapter 12   MATRIX MODELS . . . 360

chapter 13   SYSTEMS THEORY . . . 396

chapter 14   DIMENSIONAL ANALYSIS . . .   425

chapter 15   STATISTICAL TECHNIQUES. . .   441

chapter 1

An Introduction to Math Models

Although the physical systems and processes encountered in engineering practice differ widely, the basic approach to deriving the equations describing them varies only slightly. A number of examples will be presented to illustrate the different techniques.

1-1. DEFINING THE SYSTEM [1,2,3,4][1]

Before proceeding with an analysis it is desirable to explicitly define the object of attention. In order to write the descriptive equations for some physical entity, the system must be defined and this is often accomplished by specifying the boundaries. For example, an abstract system may be defined as the cross-hatched area enclosed by the line in Fig. 1-1a. For the well-mixed tank in Fig. 1-1b, the system could be the contents of the tank. For the steam-jacketed pipe of Fig. 1-1c, the system could be a small entity of infinitesimal thickness Δz located a distance z from the inlet. Why is this different from the infinitesimal element in the last case? Consider the expressions *the temperature of the liquid in the well-mixed tank* and *the temperature of the liquid in the steam jacketed exchanger.* One of these is meaningful and one is meaningless. If the tank is agitated very well, the contents should be reasonably uniform. Thus the liquid is all at the same temperature, and it is reasonable to talk about the temperature of the liquid in the tank. But if asked about the temperature of the liquid in the exchanger, the logical response would be to inquire at what point in the exchanger is the liquid temperature desired. As the temperature varies continuously from inlet to outlet, it is completely ambiguous to talk about the temperature of the liquid in the exchanger. Instead, a temperature profile —i.e., the temperature as a function of position, or $T(z)$— is more reasonable.

When writing equations to describe a system it is desirable that the contents of the system be uniform. Certainly the tank presents no difficulties in this respect but, superficially, the exchanger does present problems. Even though the temperature varies significantly from inlet to outlet, however, what about over

---

[1] Bracketed numbers refer to references at the end of the chapter.

an infinitesimal distance Δz?   Certainly the change would be small, and as justified later, it is indeed negligible.   Thus defining the system as an infinitesimal length of the exchanger permits equations to be written for a system whose contents are uniform.   The terminology attached to such a system is an *infinitesimal element* or *control volume.*

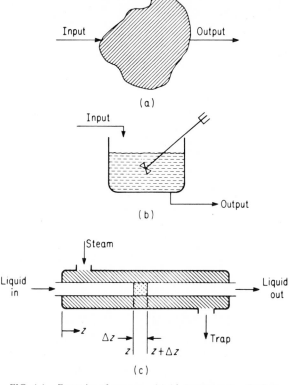

FIG. 1-1.   Examples of systems.   (a) Abstract system.   (b) Well-stirred tank.   (c) Steam-jacketed pipe.

In many books on formulation of mathematical models the infinitesimal length of the system is denoted by the differential *dz,* but it will be denoted by Δz in this book.   While these two symbols have different connotations, either may be used as long as the user is aware that he is denoting an infinitesimal or differential length in the direction of change.

Hereafter, the terminology for distinguishing between systems will be *lumped-parameter* vs. *distributed-parameter* systems.   The properties of the liquid in the well-mixed tank can be lumped, allowing one to talk about the heat content of the liquid in the tank.   Thus the term *lumped-parameter system* is applied.   On the other hand, knowing the heat content (in Btu) of all the liquid in the exchanger is of little value; instead, the heat content per unit length (in Btu/ft) is more appropriate. As the properties must remain distributed throughout the system, the term *distributed-parameter system* is applied.

Another method to distinguish these types of systems is to specify their dimensionality. As the temperature of the liquid in the tank is independent of position in the tank, this is a zero-dimensional problem. On the other hand, the temperature of the liquid in the exchanger is a function only of the distance from the inlet (ignoring axial and rotational effects), and this is a one-dimensional problem. Higher-dimensional problems are postponed until Chapter 7.

1-2. QUANTITY CONSERVED

As the boundaries of the system have now been defined, the next logical step might be to determine what crosses these boundaries. That is, the inputs and outputs as shown schematically in Fig. 1-1a must be determined. But input or output of *what?* Does one need to know the heat going in, the total mass flow, the input of a specific component, and the like? This is the quantity conserved, and developing a mathematical model usually reduces to making a mass balance, heat balance, force balance, an economic balance, or applying some law of conservation.

Knowing the inputs and outputs of a given quantity conserved, the total input and the total output can be determined. What goes in (including generation) must either come out (including consumption) or stay within the system. This latter term is the accumulation, and the relationship can be expressed in equation form as follows:

$$\begin{bmatrix} \text{Input of quantity} \\ \text{conserved} \end{bmatrix} - \begin{bmatrix} \text{Output of quantity} \\ \text{conserved} \end{bmatrix} = \begin{bmatrix} \text{Accumulation of} \\ \text{quantity conserved} \end{bmatrix} \quad (1\text{-}1)$$

This equation is convenient for nonflow processes, but the following time derivative of Eq. 1-1 is more convenient for flow processes:

$$\begin{bmatrix} \text{Rate of input of} \\ \text{quantity conserved} \end{bmatrix} - \begin{bmatrix} \text{Rate of output} \\ \text{of quantity conserved} \end{bmatrix} = \begin{bmatrix} \text{Rate of accumulation} \\ \text{of quantity conserved} \end{bmatrix}$$

$$(1\text{-}2)$$

If the quantity conserved is mass, the above equations become the law of conservation of mass. For convenience, the above equations are shortened to

$$\text{In} - \text{Out} = \text{Accumulation (Acc)} \quad (1\text{-}3)$$

Although it is possible to use depletion instead of accumulation in the above equations, the use of accumulation is more common. As depletion is simply a negative accumulation, no generality is lost.

Under some conditions the accumulation term will be zero. Physically, this means that the system is in a static condition, i.e., unchanging with time. For example, if all the heat that enters the system appears in the output, the temperature of the system cannot change. Thus the distinguishing factor between steady-state and unsteady-state conditions is that the accumulation term is zero under the former conditions. The previous equations reduce to

$$\text{In} - \text{Out} = 0 \qquad\qquad (1\text{-}4)$$

The best way to learn how to develop models is to study specific examples. Any further discussion of general concepts and ideas concerning model development will be done in the context of such examples. The first example will be very simple, and each subsequent example will introduce additional points. In this chapter and the next the examples will consist of one-dimensional, steady-state problems. Zero-dimensional, unsteady-state problems will be considered in Chapter 3, and subsequent chapters will treat more complex problems.

1-3. HEAT CONDUCTION IN A ROD [5]

Consider a steam-line hanger, such as shown in Fig. 1-2, connecting two bodies at different temperatures. The temperature varies continuously along the bar, and the objective of this example is to derive an expression for the temperature profile. Let the temperature of the ceiling and steam line be T_a and T_s, respectively, and let the support be a square bar of width W and length L. The heat loss by convection to the air at temperature T_a is given by Newton's law, the heat-transfer coefficient being h.

For simplicity, the temperature is assumed to vary only in the z-direction, i.e., it is considered to be constant across a cross-section of the bar. Although this latter assumption is valid when $h = 0$, it is not exactly true when heat is lost from the surface of the bar. However, the errors incurred in neglecting tem-

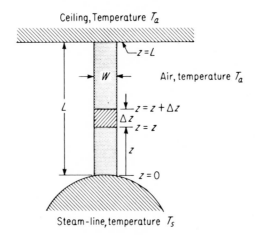

FIG. 1-2. Steam-line hanger.

perature variations across a cross section are usually not serious for this type of problem, as the heat flux in the cross-sectional direction is small compared to the heat flux in the longitudinal direction. In Chapter 7 the techniques are presented for analyzing cross-sectional variations in temperature.

System: Since this is a one-dimensional problem the system is defined as an infinitesimal element of length Δz in the z-direction as shown in Fig. 1-2.

Quantity Conserved: Thermal energy or heat.

Independent Variable: Length z.

Dependent Variables: Temperature $T(z)$ and heat flux $q(z)$. The heat flux is considered positive in the positive z-direction. *As this convention of positive flows in the positive z-direction will be adhered to strictly in subsequent examples, no further mention will be made.*

Input: The heat flows in at face z at a rate of $W^2 q(z)$.

Output: There are two output terms. First, the heat flows out at face $z + \Delta z$ at rate $W^2 q(z + \Delta z)$. Second, the rate of heat loss from the surface of the control volume is given by Newton's law:

$$4 Wh[\, T(z) - T_a \,] \Delta z$$

where $4W\Delta z$ is the surface area for convective heat transfer. This term is treated as an output or loss term. Alternatively, it could be treated as an input term, in which case it would be given by $4Wh[\, T_a - T(z)] \Delta z$. The student should satisfy himself that the final result would be the same for both ways.

Substituting these terms into Eq..1-3 gives the heat balance

$$W^2 q(z) - W^2 q(z + \Delta z) - 4\, Wh\, [\, T(z) - T_a]\ \Delta z = 0 \tag{1-5}$$

From this point, several superficially different routes can be taken to achieve the identical result.

One common approach is to divide Eq. 1-5 by Δz to obtain

$$\frac{q(z) - q(z + \Delta z)}{\Delta z} = \frac{4\, h}{W} [\, T(z) - T_a]$$

Recall the definition of the derivative:

$$\frac{df(x)}{dx} = \lim_{\Delta x \to 0} \frac{f(x + \Delta x) - f(x)}{\Delta x}$$

Thus, taking the limit of both sides of the previous equation as Δz approaches zero yields

$$\frac{dq(z)}{dz} = -\frac{4\, h}{W} [\, T(z) - T_a] \tag{1-6}$$

Alternatively, the Taylor series expansion can be used. Expanding $q(z + \Delta z)$ about z yields

$$q(z + \Delta z) = q(z) + \frac{dq(z)}{dz} \Delta z + \frac{d^2 q(z)}{dz^2} \frac{\Delta z^2}{2} + \cdots$$

$$= q(z) + \frac{dq(z)}{dz} \Delta z + O(\Delta z^2) \tag{1-7}$$

Substituting this expression into Eq. 1-5 gives

$$q(z) - \left[q(z) + \frac{dq(z)}{dz} \Delta z + O(\Delta z^2) \right] = \frac{4h}{W} \left[T(z) - T_a \right] \Delta z$$

Canceling $q(z)$ and recalling that higher-order differentials can be neglected in comparison to first-order differentials, the equation reduces to

$$\frac{dq(z)}{dz} = -\frac{4h}{W} [T(z) - T_a] \tag{1-6}$$

As the higher-order differentials will always disappear, the Taylor series will be truncated after two terms in future examples.

It is somewhat pointless to debate which of these approaches is preferable. Both will always give the same result, and a student is well advised to become proficient enough at both to recognize each when encountered. The authors prefer the Taylor series expansion, but will use both approaches in subsequent examples.

Fourier's Law: As Eq. 1-6 contains two dependent variables, $q(z)$ and $T(z)$, one of these must be expressed in terms of the other before a solution can be obtained. Recall that the heat flux is related to the temperature gradient by Fourier's law:

$$q(z) = -k \frac{dT(z)}{dz} \tag{1-8}$$

Substituting into Eq. 1-6, assuming k is constant, yields

$$\frac{d^2 T(z)}{dz^2} = \frac{4h}{Wk} [T(z) - T_a] \tag{1-9}$$

Boundary Conditions: As Eq. 1-7 is second-order, the two boundary conditions required to specify a solution are

$$T(0) = T_s \tag{1-10}$$

$$T(L) = T_a \tag{1-11}$$

Solution: Equation 1-7 is a second-order, linear, nonhomogeneous differential equation, and can be solved analytically to yield

$$\frac{T(z) - T_a}{T_s - T_a} = \frac{\sinh[b(L - z)]}{\sinh[bL]} \tag{1-12}$$

where $b = \sqrt{\dfrac{4h}{Wk}}$. The solution for a specific case is given in Fig. 1-3.

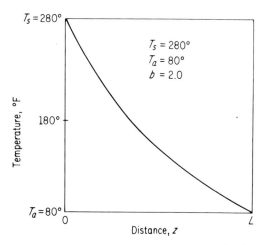

FIG. 1-3.   Temperature profile in an uninsulated rod.

Differentiating Eq. 1-12 gives

$$\frac{dT(z)}{dz} = -\frac{(T_s - T_a)b}{\sinh(bL)} \cosh|b(L-z)|$$

Substituting into Fourier's law gives an expression for $q(z)$:

$$q(z) = \frac{kb(T_s - T_a)}{\sinh(bL)} \cosh b(L - z) \qquad (1\text{-}13)$$

1-4.   INTRODUCTION TO CHANGES OF VARIABLE [7]

In many problems a change of variable greatly simplifies the form of the equation and also makes the equation easier to solve. In most cases the new variables are both dimensionless and normalized, i.e., they vary from 0 to 1. A dimensionless, normalized variable can often be obtained from the original variable via the following equation:

$$\begin{bmatrix} \text{Dimensionless,} \\ \text{normalized} \\ \text{variable} \end{bmatrix} = \frac{\left[\begin{pmatrix} \text{Original} \\ \text{Variable} \end{pmatrix} - \begin{pmatrix} \text{Reference or characteristic} \\ \text{value of original variable} \end{pmatrix} \right]}{\left[\begin{matrix} \text{Characteristic or reference length,} \\ \text{difference, driving force, etc.} \end{matrix} \right]} \qquad (1\text{-}14)$$

In many cases the new variables can be obtained upon an inspection of the problem. In the case of the bar connecting the steam line and the ceiling in

Sec.1-3, the temperature varies from T_a to T_s. Therefore, a normalized, dimen-sionless temperature θ is defined as

$$\theta = \frac{T(z) - T_a}{T_s - T_a}$$

In this equation T_a corresponds to the reference value for the temperature, and $T_s - T_a$ is the characteristic temperature difference. In addition, the independent variable z varies from 0 to L, and a normalized length ξ may be defined as

$$\xi = \frac{z}{L}$$

To introduce these new variables into the original equation, $\frac{d^2 T(z)}{dz^2}$ and $T(z)$ must be expressed in terms of θ and ξ. These quantities can be obtained as follows:

$$T(z) = (T_s - T_a)\theta + T_a$$

$$\frac{d^2 T(z)}{dz^2} = (T_s - T_a)\frac{d^2\theta}{dz^2} = \frac{(T_s - T_a)}{L^2}\frac{d^2\theta}{d\xi^2}$$

Substituting into Eq.1-9 yields:

$$\frac{(T_s - T_a)}{L^2}\frac{d^2\theta}{d\xi^2} - \frac{4h}{Wk}\left[(T_s - T_a)\theta + T_a - T_a\right] = 0$$

or:

$$\frac{d^2\theta}{d\xi^2} - b'^2\theta = 0 \qquad\qquad (1\text{-}15)$$

where $b' = \sqrt{4hL^2/Wk}$

The change of variable must also be introduced into the boundary conditions. Consider the boundary condition

$$T(0) = T_s \qquad\qquad [1\text{-}10]$$

At $z = 0$,

$$\xi = \frac{0}{L} = 0$$

At $T(z) = T_s$,

$$\theta = \frac{T_s - T_a}{T_s - T_a} = 1$$

Since θ is a function of ξ, it may be denoted by $\theta(\xi)$. In this notation, the first boundary condition becomes

$$\theta(0) = 1 \qquad (1\text{-}16)$$

By similar reasoning, the second boundary condition, Eq. 1-11, yields

$$\theta(1) = 0 \qquad (1\text{-}17)$$

By making the change of variable, the differential equation in Eq. 1-9 became a homogeneous differential equation, which is easier to solve than the original nonhomogeneous equation. The solution of Eq. 1-15 and its boundary conditions is

$$\theta(\xi) = \frac{\sinh |b'(1-\xi)|}{\sinh(b')}$$

The main advantage of changes of variable of this type is that the solution to problems can be made more general. Note that the only variables and/or parameters appearing in Eq. 1-15 and its boundary conditions are, θ, ξ, and b'. This permits the construction of graphs such as the one in Fig. 1-4 that display the solution of the original problem for essentially all possible numerical values of the parameters. Note that the line corresponding to $b' = 0$ is the solution for an insulated bar.

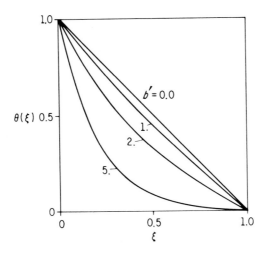

FIG. 1-4. Plot of θ vs. ξ for various values of b'.

A different change of variable for this equation is suggested in the problems at the end of this chapter, and the subject of change of variables is treated in more detail in Chapter 9.

1-5. TEMPERATURE DISTRIBUTION
IN CYLINDRICAL COORDINATES

The formulation of many engineering problems is much easier when using coordinate systems other than rectangular, e.g., the cylindrical coordinate system. The approach to the formulation of the mathematical model is essentially the same regardless of the coordinate system used, but illustrating the techniques in another coordinate system should be enlightening. The following example illustrates the use of cylindrical coordinate systems along with helpful principles in formulating boundary conditions.

The transportation of electric current over large distances is accomplished with the use of long wires of copper, aluminum, or other conducting metal. Although the power loss due to the resistance of the wire to a flow of current is minimized by use of high voltage, it is still present. In such long wires the conduction of heat in the axial direction is negligibly small, and all heat generated must be dissipated by convection to the air at temperature T_a. Let the radius of the wire be R, and assume that the resistance does not vary appreciably with temperature. Under these conditions the heat generated per unit volume $\dot{q}$ is essentially constant throughout the wire. The heat-transfer coefficient between the wire and the air is h, and the termal conductivity of the wire is k. The curvature of the wire between supports is neglected.

The objective of the analysis is to determine the steady-state temperature profile over the cross section of the wire shown in Fig. 1-5.

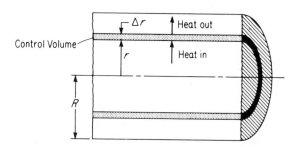

FIG. 1-5. Control volume for cylindrical conductor.

Quantity Conserved: Heat.

Control Volume: In this case the system must again be of infinitesimal thickness in the direction of change of the independent variable. As the temperature is a function of the distance r from the center of the wire, the infinitesimal element is an annular ring of inner radius r and thickness Δr as shown in Fig. 1-5. The length of the element in the axial direction is taken as a unit length for simplicity.

Independent Variable: Distance from center r.

Dependent Variables: Temperature T and conductive heat flow Q.

Input terms: In this problem there are two input terms. First, heat is conducted into the element at the inner face at rate $Q(r)$. The second input is the heat being generated within the element at rate $(2\pi r \Delta r)q$, the volume of the element being $2\pi r \Delta r$.

Output terms: The only output term is the heat of conduction at $r + \Delta r$, namely, $Q(r + \Delta r)$.

Formulation of the Equation:

$$\text{In} - \text{Out} = 0$$

$$Q(r) + 2\pi \dot{q} r \Delta r - Q(r + \Delta r) = 0$$

Using the Taylor series expansion for $Q(r + \Delta r)$ gives

$$\frac{dQ(r)}{dr} = (2\pi \dot{q})r \qquad (1\text{-}18)$$

One boundary condition for $Q(r)$ is required to obtain a unique solution.

To obtain a differential equation in terms of $T(r)$ instead of $Q(r)$, substitute Fourier's law, which is

$$Q(r) = -kA\frac{dT(r)}{dr} = -k(2\pi r)\frac{dT(r)}{dr}$$

Substituting into Eq. 1-18 and assuming constant thermal conductivity,

$$-\frac{d}{dr}\left[r\frac{dT(r)}{dr}\right] = \frac{\dot{q}}{k}r$$

or

$$\frac{d^2 T(r)}{dr^2} + \frac{1}{r}\frac{dT(r)}{dr} = -\frac{\dot{q}}{k} \qquad (1\text{-}19)$$

Since this is a second-order differential equation, two boundary conditions on T are required.

Boundary Conditions: In this problem the boundary conditions are not as obvious as in earlier problems. Although an experienced person could write directly the boundary conditions for this problem, many problems of considerable complexity require a special approach, which will be illustrated in deriving the boundary conditions for this problem. Basically, the technique is nothing new, and the approach is to make balances for special control volumes at the boundaries of the system just as has been done for more general control volumes within the system.

To derive the first boundary condition, consider a control volume at the center of the wire as shown in Fig. 1-6. Since $Q(r)$ is positive in the positive r-direction, no heat enters by conduction.

The only input term is the heat-generation term $\pi(\Delta r)^2 \dot{q}$. The output term is the rate of heat conduction at distance Δr from the center, namely $Q(0 + \Delta r)$.

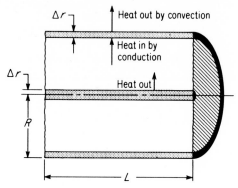

FIG. 1-6.   Special control volume for deriving boundary conditions.

The equation describing this special control volume is formulated exactly like the equation describing a general control volume:

$$\text{In} - \text{Out} = 0$$

$$\pi(\Delta r)^2 \dot{q} - Q(0 + \Delta r) = 0 \tag{1-20}$$

Taking the limit as $\Delta r \to 0$ gives

$$Q(0) = 0 \tag{1-21}$$

Alternatively, by Fourier's law

$$Q(0 + \Delta r) = -k\pi\Delta r \frac{dT(0 + \Delta r)}{dr}$$

Substituting this expression into Eq. 1-20 gives

$$\pi(\Delta r)^2 \dot{q} + k\pi\Delta r \frac{dT(0 + \Delta r)}{dr} = 0$$

Dividing by Δr and then taking the limit as $\Delta r \to 0$ yields

$$\frac{dT(0)}{dr} = 0 \tag{1-22}$$

To formulate the second boundary condition, a control volume at the surface must be examined as shown in Fig. 1-6. The rate of heat conduction into this control volume is $Q(R - \Delta r)$, i.e., the heat flow across the inner surface. The rate of heat generation within this control volume is $2\pi R(\Delta r)\dot{q}$. The output term for this control volume is the rate of heat transfer by convection at the surface Q_v:

$$Q_v = h(2\pi R)[T(R) - T_a]$$

Again the equation is formulated exactly as for a more general control volume:

$$\text{In} - \text{Out} = 0$$

$$Q(R - \Delta r) + 2\pi R\dot{q}\Delta r - Q_v = 0$$

Substituting Fourier's law as well as the above relationship for Q_v yields

$$- k[2\pi(R - \Delta r)]\frac{dT(R - \Delta r)}{dr} + 2\pi R(\Delta r)\dot{q}$$

$$- h(2\pi R)[T(R) - T_a] = 0$$

Taking the limit as $\Delta r \to 0$ and simplifying

$$\frac{dT(R)}{dr} = -\frac{h}{k}[T(R) - T_a] \qquad (1\text{-}23)$$

This equation could be written directly in this case by noting that the rate of heat conduction to the surface equals the rate of heat convection at the surface.

Solution: In terms of the temperature (the desired result), the differential equation and boundary conditions are as follows:

$$\frac{d^2T(r)}{dr^2} + \frac{1}{r}\frac{dT(r)}{dr} = -\frac{\dot{q}}{k} \qquad [1\text{-}19]$$

$$\frac{dT(0)}{dr} = 0 \qquad [1\text{-}22]$$

$$\frac{dT(R)}{dr} = -\frac{h}{k}[T(R) - T_a] \qquad [1\text{-}23]$$

Alternatively, the following equations can be solved for $Q(r)$:

$$\frac{dQ(r)}{dr} = (2\pi\dot{q})r \qquad [1\text{-}18]$$

$$Q(0) = 0 \qquad [1\text{-}21]$$

Integrating Eq. 1-19 and evaluating the constant of integration from the boundary condition yields

$$Q(r) = \pi\dot{q}r^2$$

Either substituting Fourier's law into this equation or, alternatively, solving the original differential equation yields the following equation for the temperature profile:

$$T(r) - T_a = \frac{\dot{q}}{4k}(R^2 - r^2) + \frac{\dot{q}R}{2h}$$

The results for a specific case are shown graphically in Fig. 1-7.

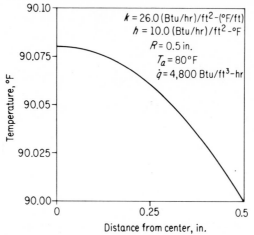

FIG. 1-7.   Specific solution for temperature profile in
the wire.

1-6.   SYSTEM WITH BULK FLOW OF MASS [1]

In the previous problems the input and output quantities have been heat transfer by conduction, heat transfer by convection, or heat-generation terms. Another input or output term commonly encountered in flow processes is the *thermal energy* associated with the bulk flow.   Of course, the thermal energy of matter cannot be measured in absolute terms, but only in relative terms.   Therefore, to account for the input of thermal energy associated with a flowing stream, the enthalpy of this stream must be included.   To calculate the enthalpy of the stream, a set of reference conditions must be stated.   This usually reduces to defining a reference temperature and a reference state at which the enthalpy of the material is considered to be zero.   There is a high degree of flexibility in defining reference conditions, although some are usually more convenient than others. The main restriction is that the reference conditions must be consistent throughout the problem.   For example, if the enthalpy of steam is required in the problem, the reference state for water used in the problem must be the same as that used in the steam table from which the enthalpy of the steam is obtained.   In most problems the numerical values for the reference conditions are not specified until the equations are completely formulated and a solution is undertaken.   The principles involved in this type of problem are illustrated in the following example [8]:

In a vapor-liquid heat exchanger operating at steady-state, the temperature on the vapor side is constant.   Since the liquid temperature varies from inlet to outlet, the temperature difference to be used in calculating the rate of heat transfer is not constant. Considering the case in which the overall heat-transfer coefficient U_i (based on inside area of pipe) and steam temperature T_s are constant, a differential equation can be derived and solved for the temperature of the liquid as a function of position along the pipe. The liquid is assumed to be flowing at a mass velocity w (lb/hr), and heat conduction in the axial direction is neglected. The liquid inlet temperature is assumed to be T_f, and the inside tube radius is R. The liquid heat capacity is c_p (Btu/lb-°F). For convenience in discussing the problem,[2] consider the liquid to be acetone.

Quantity Conserved: Heat.

Control Volume: Infinitesimal element of thickness Δz as shown in Fig. 1-8.

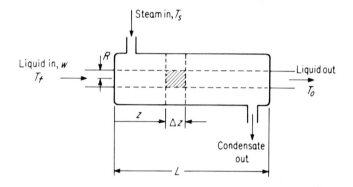

FIG. 1-8. Single-tube vapor-liquid heat exchanger.

Independent Variable: Distance z from the inlet.

Dependent Variable: Temperature T.

Input: Assuming the temperature T_s in the steam chest is higher than the temperature of the acetone in the tube, the heat transferred from the steam to the acetone in the control volume is

$$q_v(z) = U_i A_i (T_s - T) = U_i (2\pi R \Delta z)[T_s - T(z)]$$

The rate at which thermal energy enters the control volume by bulk flow is equal to the rate at which acetone enters the control volume multiplied by the enthalpy of the liquid. The reference conditions for calculating the enthalpy of acetone can be assumed to be liquid acetone at temperature T_R. The reference state could be solid acetone or acetone vapor at some temperature, but the change in enthalpy associated with a change in state would have to be included in calculating the enthalpy. To avoid these obvious inconveniences the reference state used is the

---

[2] In this and subsequent problems the arguments of the dependent variable are used only when convenient, e.g., in starting boundary conditions. Where temperature as a function of x has been denoted by $T(x)$, this notation will frequently be shortened to simply T except when stating boundary conditions.

liquid state. The rate of input of thermal energy with the bulk stream q_b is therefore given by

$$q_b(z) = wc_p [T(z) - T_R]$$

Output: The output of thermal energy across the face at $z + \Delta z$ is $q_b(z + \Delta z)$.

Formulation of the Equation:

$$\text{In} - \text{Out} = 0$$

$$q_b(z) + q_v(z) - q_b(z + \Delta z) = 0$$

Expanding $q_b(z + \Delta z)$ in a Taylor series, substituting the above relationships for $q_b(z)$ and $q_v(z)$, and simplifying yields

$$\frac{dT(z)}{dz} + \frac{2\pi RU_i}{wc_p} [T(z) - T_s] = 0 \tag{1-24}$$

Next, the following change of variable is made:

$$\varphi(z) = \frac{T(z) - T_s}{T_f - T_s}$$

Substituting into Eq. 1-24 and simplifying yields

$$\frac{d\varphi(z)}{dz} + \frac{2\pi RU_i}{wc_p} \varphi(z) = 0 \tag{1-25}$$

Boundary Condition: Since the inlet temperature is T_f, the boundary condition for Eq. 1-24 is

$$T(0) = T_f$$

In terms of φ, this boundary condition is

$$\varphi(0) = 1 \tag{1-26}$$

Solution: The solution to Eq. 1-25 and its boundary condition is

$$\varphi(z) = \frac{T(z) - T_s}{T_f - T_s} = \exp\left(- \frac{2\pi RU_i}{wc_p} z \right)$$

The temperature profile for a specific example is given in Fig. 1-9.

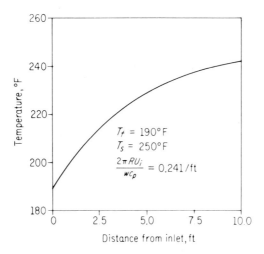

FIG. 1-9. Temperature profile in the exchanger of
Fig. 1-8.

1-7. MASS-TRANSFER APPLICATIONS

Mass transfer, the tendency of a component to travel from a region of high concentration to one of low concentration, is analogous to the flow of heat from a region of high temperature to a region of low temperature. Mass transfer may occur by two mechanisms: first, mass transfer may be by molecular diffusion, which is analogous to heat transfer by conduction; and second, mass transfer may be accomplished by bulk flow of a medium, analogous to heat transfer by convection.

The equations describing the rate of mass transfer by diffusion are variations of Fick's first law, whose fundamental form for binary mixtures is [3,4]

$$J_a = - D_{AB} \frac{dC_A}{dz} = - D_{AB} \frac{d(\rho_m x_A)}{dz} \tag{1-27}$$

where C_A = molar concentration of component A, lb-mole/ft^3

x_A = mole fraction of A

z = direction of change, ft

ρ_m = molar density, lb-mole/ft^3

D_{AB} = molecular diffusivity of A through B, ft^2/hr

J_A = molar flux, lb-mole/hr-ft^2

This equation is analogous to Fourier's law for heat conduction, and states that the rate of mass transfer is proportional to the negative of the concentration gradient. In Eq. 1-27 the molar flux J_A is superimposed upon any bulk flow, i.e., J_A is relative to any motion of the mixture. A more convenient form for many problems is Fick's law in terms of N_A, the molar flux of component A relative to fixed coordinates in space

$$N_A = x_A(N_A + N_B) + J_A \qquad (1\text{-}28)$$

where N_B = molar flux of B relative to fixed coordinates, lb-mole/hr-ft². The first term on the right is the molar flux of A resulting from bulk flow of the mixture, and the second term results from the diffusion superimposed on the bulk flow.

The two cases usually distinguished in books on mass transfer are

1. Equimolal counterdiffusion, where the moles of the component diffusing in one direction exactly equals the moles of the component diffusing in the opposite direction. For this case,

$$N_A = -N_B$$

and the first term on the right-hand side of Eq. 1-28 disappears. Also, it turns out that

$$J_A = N_A$$

2. Component A diffusing through stagnant B, where N_B is zero. An interesting point here is that although N_B (molar flux of B with respect to fixed coordinates) is zero, J_B (flux with respect to fluid motion) is not zero. As D_{AB} is approximately equal to D_{BA}, then

$$J_B = -D_{AB}\frac{d(\rho_m x_B)}{dz} = D_{AB}\rho_m \frac{dx_A}{dz}$$

for constant ρ_m. As A is diffusing, dx_A/dz is nonzero, and consequently J_B is nonzero. Writing Eq. 1-28 for N_B gives

$$N_B = x_B(N_A + N_B) + J_B$$

As N_B is zero, this reduces to

$$J_B = -N_A x_B = -N_A(1 - x_A)$$

This equation says that the molar flux of B by diffusion is offset by the B moving with the bulk motion in the opposite direction.

In problems in which the main contribution to mass transfer is the turbulent motion of the fluid in which mass transfer is occurring, the following equation, analogous to Newton's law for heat transfer by convection, applies:

$$N_A = k_m \Delta C_A \qquad (1\text{-}29)$$

where k_m = mass transfer coefficient, mole/hr-ft² (concentration difference)
ΔC_A = concentration difference

Mass-transfer coefficients are common for driving forces in terms of mole fraction, mass fraction, or partial pressure.

The following example illustrates the techniques for formulating problems involving mass transfer [4].

An idealized form of an apparatus used to measure the diffusion of a vapor through a gas is shown in Fig. 1-10 For ease in discussing this example the

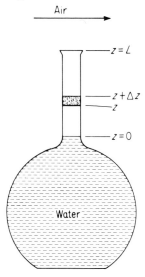

FIG. 1-10. Simple apparatus
for measurement of diffusivities.

vapor and gas are considered to be water and air, respectively. Air containing a known mole fraction y_{wo} of water vapor is passed slowly over the top of the flask, and the water vapor diffuses through a stagnant column of air to the air stream. The entire apparatus is maintained at a constant temperature and pressure, and the rate of change in the liquid level is measured. Although the process is actually in an unsteady-state condition, the rate of change is so slow that steady state conditions may be assumed at any instant. From a measurement of the rate of change in liquid height, the mass flux of water vapor can be calculated. The objective of this example is to develop an equation relating the diffusivity (the unknown) to the mass flux, concentration of water vapor in the air stream, and the concentration of water vapor at the surface of the liquid.

In deriving this expression, the following assumptions are made:
1. Air-water vapor mixtures are ideal.
2. No turbulence occurs in the neck of the flask, i.e., mass transfer is by molecular diffusion only.

3. The ratio of the change in liquid level to the length of the neck of the flask is small.
4. The vapor pressure of the water vapor is P_v, and the total pressure is P_t.
5. Pressure variations are neglected.
6. The concentration of water vapor at the surface of the liquid is its equilibrium concentration. This has been shown to be experimentally true unless the mass transfer rate is very high.
7. The variation of the molar density ρ_m of the air-water vapor mixture with concentration is neglected.

The following nomenclature is used:

A_c = Area of a cross section of the neck of the flask, ft^2
y^*_w = equilibrium mole fraction of water vapor at the liquid surface
L = distance from liquid surface to top of the flask, ft
N_w = molar flux of water vapor, lb-mole/hr
N_a = molar flux of air, lb-mole/hr
y_w = mole fraction of water vapor in the mixture
D_{wa} = Diffusivity of water vapor through air, ft^2/hr

Quantity Conserved: Moles of water vapor.

Control Volume: Section of the neck of the flask of thickness Δz as shown in Fig. 1-10.

Independent Variable: Distance z from the liquid surface.

Dependent Variable: Molar flux N_w or mole fraction y_w.

Molar Flux at z: Since the air in the neck of the flask is stagnant, its molar flux N_a is zero. In terms of the nomenclature used in this example, the form of Fick's law in Eq. 1-28 becomes

$$N_w(z) = y_w(z)N_w(z) - D_{wa}\rho_m \frac{dy_w(z)}{dz}$$

or

$$N_w(z) = -\frac{D_{wa}\rho_m}{1 - y_w(z)} \frac{dy_w(z)}{dz} \qquad (1\text{-}30)$$

However, the differential equation will be formulated in terms of $N_w(z)$ and the above expression will be substituted.

Molar Flux at z + \Delta z: In terms of the molar flux at z, the flux at $z + \Delta z$ is given by

$$N_w(z + \Delta z) = N_w(z) + \frac{dN_w(z)}{dz} \Delta z$$

Formulation of the Equation:

$$\text{In} - \text{Out} = 0$$

$$N_w(z) - \left\{ N_w(z) + \frac{dN_w(z)}{dz} \Delta z \right\} = 0$$

or

$$\frac{dN_w(z)}{dz} = 0$$

Although this equation can be integrated, no boundary conditions on $N_w(z)$ are available. Thus, Eq. 1-30 is substituted to yield, after simplification, an equation involving $y_w(z)$:

$$\frac{d}{dz}\left[\frac{1}{1 - y_w(z)}\frac{dy_w(z)}{dz}\right] = 0 \qquad (1\text{-}31)$$

Boundary Conditions: Since the mole fraction of the water vapor in the air stream is known, the first boundary condition is

$$y_w(L) = y_{wo}$$

Assuming ideal behavior, the mole fraction of water vapor in equilibrium with the liquid is

$$y_w(0) = \frac{P_v}{P_t} = y_w^*$$

Solution: Equation 1-31 and its boundary conditions can be solved to yield

$$\left(\frac{1 - y_w(z)}{1 - y_w^*}\right) = \left(\frac{1 - y_{wo}}{1 - y_w^*}\right)^{\frac{z}{L}}$$

The concentration profile for a typical case is shown in Fig. 1-11.

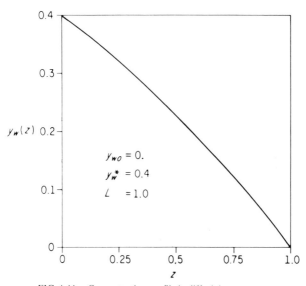

FIG. 1-11. Concentration profile in diffusivity apparatus.

The equation for calculating the diffusivity from the experimental data can be obtained by substituting this expression for $y_w(z)$ into Eq. 1-30 and solving for D_{wa} to obtain

$$D_{wa} = \frac{N_w(0)L}{\rho_m \ln \left(\dfrac{1 - y_{wo}}{1 - y_w^*} \right)}$$

The numerical value for $N_w(0)$ is determined from the change in liquid level in the flask.

Further examples illustrating problems involving molecular diffusion are presented in the problems at the end of this chapter.

As another example of the use of mass balances, consider the following problem associated with distillation-column tray efficiency data [9]. On a tray in a distillation column the liquid flows onto the tray, contacts the vapor flowing up through the tray, and leaves the tray on the opposite side. A schematic diagram of a simple tray illustrating the nomenclature is shown in Fig. 1-12. The flow patterns and arrangements of trays are varied and complex, and no exact analysis

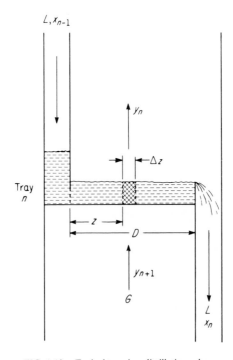

FIG. 1-12.   Typical tray in a distillation column.

has been achieved. However, two limiting cases have been analyzed, and are as follows:

1. The liquid on the tray is perfectly mixed in all directions, and there are no concentration gradients on the tray. This is the so-called perfectly mixed case.

2. No mixing occurs in the direction of flow as the liquid flows across the plate, although it is assumed that there are no concentration gradients in the vertical direction. This is the plug-flow case.

The efficiency for an actual tray is intermediate between these two extremes; therefore, predicting the tray efficiencies for these two cases brackets the efficiency for the actual tray.

The point efficiency, i.e., the efficiency at a point on the tray, is easier to correlate than the overall tray efficiency. Although there are numerous definitions of efficiencies based on either the liquid or vapor phases, the following two are probably most common:

Point efficiency for the vapor phase:

$$E_{OG} = \frac{y(z) - y_{n+1}}{y^*(z) - y_{n+1}}$$

where $y(z) = $ mole fraction light component in vapor leaving tray at point z;

$y^*(z) = $ mole fraction light component in equilibria with the liquid at point z

Murphree tray efficiency based on vapor phase:

$$E_{MV} = \frac{y_n - y_{n+1}}{y_n^* - y_{n+1}}$$

where $y_n^* = $ mole fraction vapor in equilibrium with liquid of composition x_n.

In the following example, an expression for the Murphree tray efficiency in terms of the point efficiency is developed for the plug-flow case. The consideration of the perfectly mixed case and intermediate cases are reserved for a later chapter.

For simplicity, a square tray (often used in experimental work) is considered. As is the customary nomenclature, the liquid flow rate in the column is L (mole/hr), and the vapor flow rate is G (mole/hr). The vapor entering the tray is considered to be of uniform composition, and equimolal overflow is assumed. The value of E_{OG} is assumed to be constant across the tray, and the vapor-liquid equilibrium data are represented by:

$$y^*(z) = mx(z) + b$$

where m and b are constants.

Quantity Conserved: Moles light component.

Control Volume: Cross section of tray of infinitesimal length Δz as shown in Fig. 1-12.

Independent Variable: Distance z from inlet.

Dependent Variables: Liquid composition $x(z)$ and vapor composition $y(z)$ at a point on the tray. Note that the equilibrium composition $y^*(z)$ is also a function of position.

Input Terms: In a component material balance there are two input terms for the control volume. First, the moles of light component entering with the liquid is $Lx(z)$. The moles of light component entering in the vapor from the tray below is given by

$$\frac{Gy_{n+1}}{D} \Delta z$$

Output Terms: The moles of light component leaving the control volume with the liquid at $z + \Delta z$ is $Lx(z + \Delta z)$. The moles leaving in the vapor is

$$\frac{Gy(z)}{D} \Delta z$$

Formulation of Equation:

$$\text{In} - \text{Out} = 0$$

$$Lx(z) + \frac{Gy_{n+1}}{D} \Delta z - Lx(z + \Delta z) - \frac{Gy(z)}{D} \Delta z = 0$$

Simplifying:

$$\frac{dx(z)}{dz} + \frac{G}{LD} \lceil y(z) - y_{n+1} \rceil = 0 \tag{1-32}$$

Boundary Condition:

The boundary condition to be used is the composition of the liquid leaving the tray, which is

$$x(D) = x_n$$

Solution: In order to solve this equation, $y(z)$ must be replaced by an expression in terms of $x(z)$, or vice versa. First, the definition of the point efficiency is substituted into Eq. 1-32 to obtain.

$$\frac{dx(z)}{dz} + \frac{GE_{OG}}{LD} \lceil y^*(z) - y_{n+1} \rceil = 0$$

Now $y^*(z)$ is replaced by the equilibrium relationship $y^*(z) = mx(z) + b$ to yield an equation involving only $x(z)$. Integrating and evaluating the constant of integration from the boundary condition yields the concentration profile across the plate:

$$\frac{mx(z) + b - y_{n+1}}{mx_n + b - y_{n+1}} = e^{\lambda E_{OG}(1 - z/D)}$$

where $\lambda = \dfrac{mG}{L}$

Substituting the equilibrium relationship and the definition of the point efficiency yields

$$\frac{y(z) - y_{n+1}}{E_{OG}(y_n^* - y_{n+1})} = e^{\lambda E_{OG}(1 - z/D)}$$

Integrating across the plate (from 0 to D) gives

$$\frac{y_n - y_{n+1}}{y_n^* - y_{n+1}} = \frac{1}{\lambda}\ (e^{\lambda E_{OG}} - 1)$$

As the left-hand side is the Murphree tray efficiency, the equation becomes

$$E_{MV} = \frac{1}{\lambda}\ (e^{\lambda E_{OG}} - 1)$$

Thus the point efficiency is related to the overall tray efficiency.

1-8. CHEMICAL REACTIONS

Previous examples in this chapter have treated heat and mass transfer in one-dimensional, steady-state situations. With the exception of heat generation due to electrical resistance, all input and output terms corresponded to movements of the quantity conserved across the boundaries of the system. The appearance or disappearance of a component due to chemical reaction is similar to the heat generation in that it does not represent the movement of the quantity conserved across the system boundaries.

To define several of the terms used in kinetics, consider the following irreversible reaction:

$$aA + bB + \cdots + dD \rightarrow rR + sS + \cdots + qQ \tag{1-33}$$

where A, B, ..., D are the reactants and R, S, ..., Q are the products. The coefficients a, b, ..., d, r, s, ..., q are the soichiometric ratios in which the reactants and products appear or disappear as the reaction proceeds.

The reaction rate r_s is defined as the rate of formation of component s by reaction. As the stoichiometric ratios of each component in the reaction are known, the rates of appearance or disappearance of the other components in the reaction can be related to the known reaction rate. For the reaction in Eq. 1-33 the rate of disappearance of A equals $(a/s)r_s$. It is also customary to define rates of disappearance as negative rates of appearance. Thus r_i will always be used in this book as the *rate of appearance of component i by reaction*. We can therefore talk about rates of appearance for reactants, but these numerical rates will be negative. For example, the rate of appearance of A in Eq. 1-33 is

$$r_A = -(a/s)r_s$$

This is nothing more than a convention of signs.

The reaction rate is typically expressed on a convenient basis for the specific reacting medium. For homogeneous media the reaction rate is based on a unit volume of reacting fluid, making its units (lb-mole/hr-ft³). For surface reactions a more convenient basis is a unit surface area, the units being (lb-mole/hr-ft²-surface area). For solid-fluid systems the convenient basis might be a unit mass of solid with the units (lb-mole/hr-lb-solid). Other units are certainly possible.

This book by no means intends to delve deeply into the subject of reaction kinetics. Texts [10] devoted to kinetics consider mechanisms of various reactions and ways to relate the reaction rate to the variables upon which it depends. In each case in this book it will be assumed that this relationship is either known beforehand or can be obtained. This is not to imply, however, that obtaining rate expressions is easy. The following paragraphs review only the points pertinent to problems that follow.

To take into account the variation of the reaction rate with concentration, the rate of formation of component A by the reaction in Eq. 1-33 is given by (for irreversible reactions, the product concentrations do not affect the reaction rate):

$$r_A = -kC_A^{n_A} C_B^{n_B} \cdots C_D^{n_D} \tag{1-34}$$

where k is the rate constant; C_A, C_B, ..., C_D are the concentrations of components A, B, ..., D; and n_A, n_B ..., n_D are experimentally determined exponents. The rate constant k is not a function of concentration, but is a function of temperature and other variables. Except for simple elementary reactions, Eq. 1-34 is only approximate, but it is nevertheless so convenient that it has enjoyed widespread use.

The order of the reaction with respect to a component is the value of the exponent in the expression for the reaction rate. The reaction in Eq. 1-33, whose rate is given by Eq. 1-34, is

n_Ath order with respect to A

n_Bth order with respect to B

.
.
.

n_Dth order with respect to D

The overall reaction order is the sum of the orders with respect to the various components. If we define n as

$$n = n_A + n_B + \cdots + n_D$$

the overall order of the reaction in Eq. 1-33 is n.

For simple elementary reactions the exponents in the rate equations are integer numbers. For typical complex reactions, the exponents are not usually integer numbers, since the exponents in the rate expression in Eq. 1-34 are determined so as to give the best approximation to experimental rate data. In addition, the exponents n_A, n_B, ..., n_D are not necessarily related to the stoichiometric coefficients a, b, ..., d in Eq. 1-33. That is, there is no relationship between molecularity and order except for very elementary reactions.

Prediction of numerical values for reaction rates is very difficult, primarily because of the numerous variables which affect the rate. The typical variables considered for homogeneous reactions are concentration of reactants, temperature, and pressure (gas), although other variables may also enter. For heterogeneous reactions, additional variables such as catalyst concentration, catalyst activity, etc., may be significant.

1-9. FORMULATION OF EQUATIONS FOR REACTING SYSTEMS

To derive the descriptive equations for a reacting system, the balance given in Eq. 1-4 is used. For a reacting system the quantity conserved is the moles of a component, and Eq. 1-4 becomes

$$
\begin{bmatrix} \text{Rate of input} \\ \text{of component} \\ \text{to the system} \end{bmatrix}
-
\begin{bmatrix} \text{Rate of output} \\ \text{of component} \\ \text{from the system} \end{bmatrix}
=
\begin{bmatrix} \text{Rate of accumulation} \\ \text{of component within} \\ \text{the system} \end{bmatrix}
$$

If the component is a reactant, one of the output terms is the rate of disappearance of the component by reaction. On the other hand, if the component is a product, one of the input terms is the rate of formation of the component by reaction.

If the equation is always developed in terms of the rate of appearance r_A of a component regardless of whether it is being formed or depleted, the signs will be taken care of by the definition of r_A. Thus even for the reactants the equations will be developed as if they are appearing, and the correct sign will be assured in the final expression by the definition of r_A. In this regard it may be more convenient to modify the above equation as follows:

$$
\begin{bmatrix} \text{Rate of input of} \\ \text{component to system} \end{bmatrix}
+
\begin{bmatrix} \text{Rate of appearance of} \\ \text{component by reaction} \end{bmatrix}
$$

$$
-
\begin{bmatrix} \text{Rate of output of} \\ \text{component from system} \end{bmatrix}
=
\begin{bmatrix} \text{Rate of accumulation of} \\ \text{component within system} \end{bmatrix}
$$

This convention of signs for r_A will become more important when the reacting systems become very complex, especially when parallel, series, and reversible reactions exist.

As the expression for the reaction rate normally involves concentrations of the various components, the concentration, e.g., $C_A(z)$, seems a natural choice for the dependent variable in the mass balance. Indeed, this is frequently the case, but two other quantities are sometimes more convenient. One of these is the molar flow rate $N_A(z)$, which is related to $C_A(z)$ by the equation

$$N_A(z) = vC_A(z)$$

where v is the volumetric flow rate (may also be a function of z). The second is the fraction conversion $X_A(z)$, which is related to $N_A(z)$ as follows:

$$X_A(z) = \frac{N_A(0) - N_A(z)}{N_A(0)}$$

The fraction conversion is simply the fraction of A in the feed that has been consumed by reaction.

Now to consider a reactor that is described by one-dimensional steady-state equations. In Fig. 1-13a is a diagram of a plug-flow tubular reactor. In this type of reactor, there is no mixing in the direction of flow, although there may be some mixing in the direction perpendicular to flow. Furthermore, all elements of the fluid are moving with the same velocity, and consequently the residence time (time a given particle of fluid remains in the reactor) is the same for each element of fluid.

Since the concentration in such a reactor varies continuously from input to output, the descriptive equation must be developed by examining an element of infinitesimal thickness Δz in the axial direction, as shown in Fig. 1-13b. The total volume of the reactor is V and its length is L. The feed enters at concen-

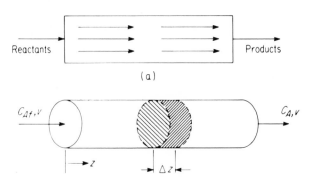

FIG. 1-13. Aspects of a tubular reactor. (a) Schematic representation of plug flow. (b) Infinitesimal element within a tubular reactor.

tration C_{Af} and volumetric flow rate v.   Suppose component A reacts by the following reaction:

$$A \rightarrow B$$

The rate constant is k, and the order is n (i.e., $r_A = -kC_A^n$).   The reactor is operated isothermally, and there is no change of volume upon reaction.

Tubular reactors are normally operated under steady-state conditions, and the element in Fig. 1-13b will be examined in order to derive an expression for the concentration C_A (dependent variable) as a function of z (independent variable). The flow rate of component A into the element is vC_A, and the flow rate out is

$$vC_A(z + \Delta z) = vC_A + \frac{d}{dz}(vC_A)\Delta z$$

Since the volume of fluid within the element is $\dfrac{V\Delta z}{L}$, the rate of appearance of A by reaction within this element is

$$\frac{V\Delta z}{L} r_A$$

Since there is no accumulation term for steady-state operation, the above terms yield

$$vC_A + \frac{V\Delta z}{L} r_A - \left[vC_A + \frac{d}{dz}(VC_A)\Delta z\right] = 0$$

or

$$\frac{V}{L} r_A = v\frac{dC_A}{dz} \tag{1-35}$$

Since there is no volume change upon reaction, the volumetric flow rate v is not a function of z.   The boundary condition is

$$C_A(0) = C_{Af}$$

With this boundary condition Eq. 4-16 can be separated and integrated to obtain $(r_A = -kC_A^n)$

$$\frac{C_A(z)}{C_{Af}} = \left[(n-1)\frac{kV}{vL} zC_{Af}^{n-1} + 1\right]^{-\frac{1}{n-1}}, \quad n \neq 1$$

or

$$\frac{C_A(z)}{C_{Af}} = \exp\left(-\frac{kV}{vL} z\right), \quad n = 1$$

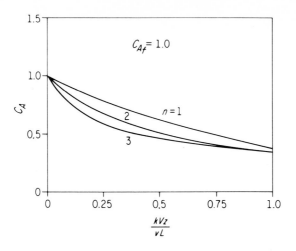

FIG. 1-14.   Concentration profiles in a tubular reactor.

In Fig. 1-14 is shown the concentration profiles in a tubular reactor for various values of n.

1-10.   MOMENTUM TRANSFER[3,4]

Until now all the emphasis has been placed on problems involving heat and mass balances, which are comparatively straightforward. Although there is nothing inherently difficult about momentum transfer, heat and mass transfer problems are usually easier to visualize. Engineering problems typically involve heat, mass, and momentum transfer; consequently consideration of momentum transfer is essential, especially for problems involving flow of fluids. In the ensuing discussion it is assumed that the reader is familiar with the elementary concepts of fluid dynamics, although several of these concepts will be reviewed briefly.

Equations describing momentum transfer can be formulated in at least two different ways. First, a force balance can be made over an infinitesimal element of the fluid in much the same manner as a force balance is made over a structural beam or other rigid body. Second, a momentum balance similar to the heat and mass balances in the previous sections can be made over an infinitesimal element of the fluid, but it is convenient to retain a sum of forces term in the balance to account for such terms as gravitational attraction or centrifugal forces. Consequently, Eq. 1-3 is modified as follows for momentum transport:

$$\begin{bmatrix} \text{Rate of} \\ \text{momentum in} \end{bmatrix} - \begin{bmatrix} \text{Rate of} \\ \text{momentum out} \end{bmatrix} + \begin{bmatrix} \text{Sum of} \\ \text{forces acting} \\ \text{on system} \end{bmatrix}$$

$$= \begin{bmatrix} \text{Product of} \\ \text{acceleration} \\ \text{and mass with-} \\ \text{in the system} \end{bmatrix}$$

The acceleration term is equivalent to the rate of change (derivative with respect to time) of the momentum of the mass in the system. For steady-state problems this term is zero, reducing the above equation to

$$\begin{bmatrix} \text{Rate of} \\ \text{momentum in} \end{bmatrix} - \begin{bmatrix} \text{Rate of} \\ \text{momentum out} \end{bmatrix} + \begin{bmatrix} \text{Sum of forces} \\ \text{acting on system} \end{bmatrix} = 0 \qquad (1\text{-}36)$$

The momentum balance approach to the formulation of momentum transfer problems will be used in the examples since it is very similar to the heat and mass balances discussed previously.

Perhaps the most perplexing aspect of fluid-flow problems is the nomenclature, especially that for the shear stresses. For example, consider the layer of fluid between two large parallel plates separated by a small distance δ as shown in Fig. 1-15. If the lower plate is moving with velocity v_0 with respect to the upper plate, the fluid between the plates is subjected to a shear stress. The shear stress exerted in the x-direction on a fluid surface of constant y by the fluid in the region of lesser y is denoted as τ_{yx}, with the first subscript denoting the face upon which the shear occurs and the second denoting the direction of the shear. This nomenclature is extended to include τ_{zx}, τ_{xy}, etc. in rectangular coordinates, and a similar

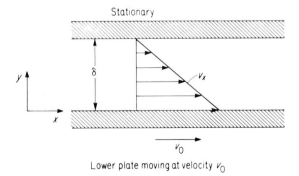

FIG. 1-15. Fluid between two parallel plates.

notation is used for shear stresses on elements in other coordinate systems. Fig. 1-16 illustrates this for rectangular and cylindrical coordinates.

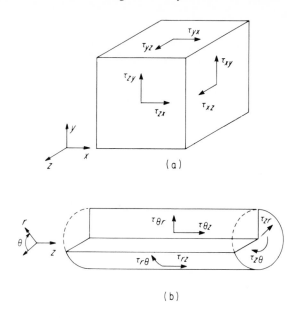

(a)

(b)

FIG. 1-16.   Shear stresses on three-dimensional elements.   (a) Shear stresses on a rectangular element.   (b) Shear stresses on a cylindrical element.

The velocity in the x-direction is designated as v_x, with the subscript denoting direction and *not* partial differentiation.   In fluid dynamics problems the direction of stresses and velocities is very important, and necessitates much of the complex nomenclature.   In Fig. 1-15 the velocity gradient is dv_x/dy, that is, rate of change of x-velocity in the y-direction.   The shear stress is related to the velocity gradient by Newton's law of viscosity:

$$\tau_{yx} = -\frac{\mu}{g_c}\frac{dv_x}{dy} \tag{1-37}$$

where the proportionality constant μ is the viscosity of the fluid.   This relationship states that the shear stress is proportional to the negative of the velocity gradient, which is analogous to Fourier's law for heat conduction and Fick's law for diffusion.   Essentially all gases and many liquids obey the above relationship and are termed Newtonian fluids.   Fluids, such as polymer solutions and melts, that do not obey this relationship are referred to as non-Newtonian fluids.   For such fluids Eq. 1-37 is usually employed but the proportionality constant is a function of the velocity gradient (rate of shear).

Since the shear stress is simply the force per unit area, it may be easily incorporated into a force balance for an element of fluid. However, it may be just as easily incorporated into a momentum balance. In Fig. 1-15 the fluid at the surface $y = 0$ acquires a certain amount of x-momentum from the lower plate. This fluid imparts some of its momentum to the adjacent layer of fluid, which in turn imparts momentum to the next layer. Thus, x-momentum is transmitted from a region of high velocity to a region of low velocity. In Fig. 1-15 momentum is transmitted in the y-direction, and τ_{yx} can be interpreted as the flux (rate of transfer per unit area) of x-momentum in the y-direction. Thus the shear stress has been shown to be equivalent to force per unit area and to momentum flux, and these two are indeed equivalent.

In addition, the units of the stress and momentum flux can be examined to give further insight. In the above paragraph it has been shown that stress is equivalent to force per unit area. Now the units of the momentum flux are examined. From physics we know that momentum equals mass × velocity, for which the units are lb_m-ft/sec in the American engineering set of units. Consequently, the units of momentum flux are $lb_m /(ft\text{-}sec^2)$. If the momentum flux is divided by the gravitational constant g_c (ft-lb_m/lb_f-sec^2), the units of the resulting expression are lb_f/ft^2, or force per unit area.

The boundary conditions for fluid dynamics problems typically describe the shear stress or velocity at a solid-fluid, liquid-gas, or liquid-liquid interface. At a liquid-gas interface the momentum flux in the liquid phase can be assumed zero in most cases. At a solid-liquid or solid-gas interface the fluid velocity equals the velocity of the surface; i.e., there is normally assumed to be no slip at the surface. At a liquid-liquid interface the velocities of the liquid in the two phases are equal, and the momentum flux perpendicular to the interface is the same in each phase. In some cases the symmetry of the problem will permit the use of other types of boundary conditions, and in other cases a balance over a specific infinitesimal element must be made as done in Sec. 1-6 for heat transfer.

1-11. LAMINAR FLOW IN CIRCULAR TUBES

Fluids are typically transported from one vessel to another or from one location to another via circular tubes or pipes. When the fluid is in laminar flow the momentum balance can be used to formulate the descriptive equations for flow in a very long tube. At the ends of a tube, entrance and exit effects cause deviations from normal flow patterns. However, for a long tube these effects on such quantities as the overall pressure drop are minimized. The objective of the following discussion is to develop an equation for the velocity profile in a liquid flowing through a pipe.

Control Volume: A cylindrical shell of radius r, thickness Δr, and length L as shown in Fig. 1-17.

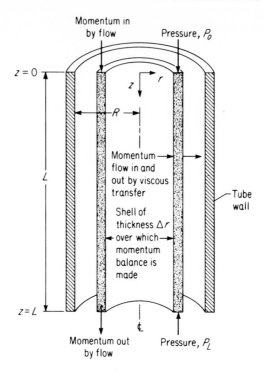

FIG. 1-17. Cylindrical infinitesimal element in pipe.
(Reprinted by permission from R. B. Bird, W. E. Stewart,
and E. N. Lightfoot, *Transport Phenomena,* John Wiley
& Sons, Inc., New York, 1960.)

Quantity Conserved: Momentum.
Independent Variable: r.
Dependent Variables: $\tau_{rz}(z)$, $v_z(z)$.
Viscous Terms: In formulating momentum balances the "in" and "out"
directions are always taken in the direction of the positive r-and z-axes. If the
transfer is actually in the opposite direction, the signs of the shear stresses and
velocities will be negative. Therefore, the rate of momentum transfer in across
the face at r is

$$2\pi r L \tau_{rz}(r)$$

The rate of momentum transfer out across the face at $r + \Delta r$ is

$$2\pi r L \tau_{rz}(r) + \frac{d}{dr}\left[2\pi r L \tau_{rz}(r)\right]\Delta r$$

Bulk-Flow Terms: Since the fluid entering and leaving the element has a velocity, there is input and output of momentum at the faces $z = 0$ and $z = L$, respectively. At $z = 0$ the input term is

$$\{[2\pi r\Delta\, r\rho v_z(r)]v_z(r)\}_{z=0}$$

At $z = L$ the output term is

$$\{[2\pi r\Delta\, r\rho v_z(r)]v_z(r)\}_{z=L}$$

Pressure Forces: At the faces $z = 0$ and $z = L$ pressure forces are acting on the element. At $z = 0$ the force is the product of the pressure and the area:

$$(2\pi r\Delta\, r)P_0$$

Note that this force is in the positive z-direction and is taken as a positive term in the sum of forces. At $z = L$ the force is

$$-(2\pi r\Delta\, r)\, P_L$$

This force is in the negative z-direction, and thus the origin of the minus sign.

Gravitational Forces: The force exerted by gravity on the fluid in the control volume is

$$(2\pi r\Delta\, rL)\rho g/g_c$$

where g is the local acceleration due to gravity. This term is acting in the positive z-direction and is taken as a positive term in the sum of forces.

Formulation of Equation: Substituting the above terms into Eq. 1-36 and simplifying yields

$$\{2\pi r\Delta\, rv_z(r)\,[\rho v_z(r)]\}_{z=0} - \{2\pi r\Delta\, rv_z(z)\,[\rho v_z(r)]\}_{z=L}$$

$$-\frac{d}{dr}\,[2\pi rL\tau_{rz}(r)]\,\Delta r + 2\pi r\Delta r(P_0 - P_L)$$

$$+\frac{2\pi r\Delta\, rL\rho g}{g_c} = 0 \tag{1-38}$$

For an incompressible fluid the densities at $z = 0$ and $z = L$ are equal. Furthermore, a total mass balance on the element indicates

$$(v_z)_{z=0} = (v_z)_{z=L}$$

Substituting into Eq. 1-38, the first two terms cancel. Canceling $2\pi\Delta r$ from the remaining terms gives

$$\frac{d}{dr}[rL\tau_{rz}(r)] = r(P_0 - P_L) + \frac{r(\rho gL)}{g_c}$$

$$= r\left[(P_0 + \frac{\rho gL}{g_c}) - P_L\right]$$

$$= r\Delta P_t \tag{1-39}$$

where ΔP_t is the total head loss including the static head.

Boundary Conditions: At the tube wall ($r = R$), the velocity in the z-direction is zero:

$$v_z(R) = 0 \qquad\qquad (1\text{-}40)$$

At the center of the pipe the boundary condition is derived by making a balance over an element of radius Δr, yielding

$$\tau_{rz}(0) = 0 \qquad\qquad (1\text{-}41)$$

Solution: Since ΔP_t is not a function of r (pressure is constant over a cross section of the pipe), Eq. 1-39 can be integrated with respect to r.   Evaluating the constant of integration via Eq. 1-41, we obtain

$$\tau_{rz}(r) = \frac{r\Delta P_t}{2L}$$

Substituting Newton's law of viscosity in cylindrical coordinates and simplifying gives

$$\frac{dv_z(r)}{dr} = -\left(\frac{g_c\Delta P_t}{2\mu L}\right) r$$

Integrating and evaluating the constant of integration via Eq. 1-40, we obtain

$$v_z(r) = g_c\frac{\Delta P_t R^2}{4\mu L}\left[1 - \left(\frac{r}{R}\right)^2\right]$$

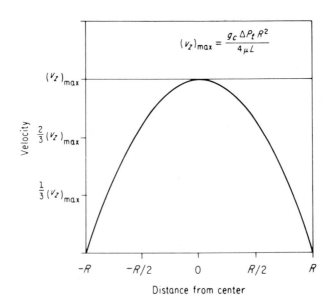

FIG. 1-18.   Parabolic velocity profile in a circular pipe.

This indicates that the velocity profile in a circular pipe is parabolic as shown in Fig. 1-18. This equation can be integrated over a cross section of the pipe to obtain the Hagen-Poiseuille law.

1-12. REVIEW OF GENERAL PROCEDURE

Although the procedures of this chapter have been illustrated only for a few systems to this point, the principles hold for other processes as well. Beginning with the next chapter, these concepts will be applied to more complex systems. Since the general procedure is the same, an outline is helpful.
1. Define the quantity conserved, the control volume for which the equations are to be written, and select the independent and dependent variables. If only one independent variable appears, an ordinary differential equation will result.
2. Apply the physical laws that yield the input and output expressions for the quantity conserved.
3. Substitute these expressions into the appropriate terms in the equation

$$\text{Input } - \text{Output} = \text{Accumulation}$$

4. Develop the necessary boundary conditions.
5. Solve the resulting differential equation and boundary conditions. The solution technique may be analytic, numerical, or otherwise.

1-13. SUMMARY

In this chapter the introduction to the formulation of equations for one-dimensional steady-state systems was presented. The basic principles are the same for all systems, and in the following chapters these principles will be extended to higher-dimensional systems, unsteady-state problems, lumped-parameter models, and a variety of other situations.

REFERENCES

1. H. S. Mickley, T. K. Sherwood, and C. E. Reed, *Applied Mathematics in Chemical Engineering,* McGraw-Hill, New York, 1957.
2. B. G. Jenson and G. V. Jeffreys, *Mathematical Methods in Chemical Engineering,* Academic Press, New York, 1963.
3. R. B. Bird, W. E. Stewart and E. N. Lightfoot, *Transport Phenomena,* Wiley, New York, 1960.
4. C. O. Bennett and J. E. Myers, *Momentum, Heat, and Mass Transfer,* McGraw-Hill, New York, 1962.

5. F. Kreith, *Principles of Heat Transfer*, 2d ed., International Textbook, Scranton, Pa., 1965.

6. R. V. Churchill, *Operational Mathematics*, McGraw-Hill, New York, 1958.

7. J. D. Hellums and S. W. Churchill, "Simplification of the Mathematical Description of Boundary and Initial Value Problems," *AIChE Journal*, Vol. 10, No. 1 (January 1964), pp. 110-114.

8. D. Q. Kern, *Process Heat Transfer*, McGraw-Hill, New York, 1950.

9. *Bubble-Tray Design Manual*, A.I.Ch.E., New York, 1958.

10. O. Levenspiel, *Chemical Reaction Engineering*, Wiley, New York, 1962.

PROBLEMS

1-1. Develop an expression for the volume of the spherical shell of radius r and infinitesimal thickness Δr as shown.

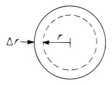

PROB. 1-1

1-2. The inside surface of a circular pipe is maintained at temperature T_i and the outside surface at T_0. The inner and outer radii are r_i and r_o, respectively, and the thermal conductivity k of the metal in the pipe walls is independent of temperature. (a) Develop an expression for the temperature profile in the pipe. (b) What is Q, the rate of heat loss per foot of pipe? (c) If the rate of heat transfer is to be given by $Q = kA_m (T_i - T_0)/(r_0 - r_i)$, what is the correct expression for A_m?

1-3. Consider a spherical container holding a liquid boiling at T_i. The liquid is insulated from the outside air by insulation of thickness x. (a) If the radius of the container is r_i and the thermal conductivity of the insulation is k, determine the temperature profile in the insulation assuming the temperature at the outer surface is T_o. (b) Derive an expression for the rate of heat loss. (c) If the rate of heat transfer is to be given by the expression $Q = kA_m (T_o - T_i)/x$, what is the appropriate expression for A_m?

1-4. The film theory for mass transfer in turbulent fluids contends that the major resistance occurs in a thin laminar film in the fluid adjacent to the interface. Consider the case in which component A is diffusing through stagnant B. Assuming that the diffusivity of A through B is D_{AB}, the molar density is ρ, and the film thickness is δ; relate the rate of mass transfer N_A to the interface concentration x_{A_i} and bulk phase concentration x_{Ab}. Repeat for A and B counterdiffusing.

1-5. Consider a mothball (composed primarily of naphthalene) surrounded by an infinite mass of stagnant air. The rate of diffusion is so slow that the radius R_o of the mothball (a sphere) can be considered constant. Furthermore, the partial pressure of naphthalene at the surface equals its vapor pressure. The concentration of naphthalene in the air is so low that its influence upon the air density is negligible. Develop an equation, along with the necessary boundary conditions, to give the concentration of naphthalene x_A as a function of distance from the center of the mothball. Solve these equations, and develop an expression for the rate of loss of naphthalene from the mothball.

1-6. Consider a wire of length L submerged in a fluid at constant temperature T_f. The radius of the wire is R, and its thermal conductivity k is independent of temperature. The convective heat-transfer coefficient at the surface of the wire is h. If the ends of the wire are held at constant temperature T_e, determine the equations and boundary conditions describing the temperature profile in the wire. Neglect temperature variations over a cross section.

1-7. Repeat the above problem including a heat-generation term $\dot{q}$ per unit volume of the wire. What changes are necessary if the ends of the wire are insulated?

1-8. Consider a stagnant spherical film of inner radius R_i and outer radius R_o such as may occur around a spherical particle in a liquid or a drop of liquid in air. The concentration on the inside of the film is x_i and on the outside is x_o. If mass transfer occurs by diffusion of one component through a stagnant mixture (diffusivity $= D$), determine the concentration profile and rate of mass transfer through the film. The density is independent of concentration.

1-9. Consider a wire of radius r_i coated with a material of thermal conductivity k. The temperature of the wire (and the inside surface of the coating) is T_i, and the surrounding air is at temperature T_a. The convective heat-transfer coefficient between the coating and the air is h, a constant. Determine the radius r_o of the surface of the coating for which the heat loss from the wire to the air is minimized.

1-10. Liquids boiling at low temperatures —oxygen and hydrogen, for example— are usually kept in spherical containers insulated on the outside. Assume that the inside surface (radius r_i) of the insulation is at the boiling point T_b of the liquid. The outer radius of the insulation is r_o, and the convective heat-transfer coefficient between the insulation and the air at temperature T_a is h. The thermal conductivity of the insulation is k. Determine the thickness (outer radius) of the insulation for which the heat loss is a minimum.

1-11. Consider a round rod, such as a thermocouple, projecting into a fluid of constant temperature. For simplicity, the rod (radius $= R$, length $= L$) is assumed to be a solid composed of a material of thermal conductivity k. The base of the rod is maintained at temperature T_b, and the surface is exposed to a fluid of constant temperature T_f. The convective heat-transfer coefficient between the rod and the fluid is h. Neglecting temperature variations in the radial direction and neglecting heat transfer at the end of the rod, determine the equation and boundary conditions describing the temperature in the rod as a function of position. What changes are necessary to include heat transfer at the end of the rod?

1-12. One technique for increasing the rate of heat transfer between a metallic surface and a fluid is to increase the area of contact by using extended surfaces or fins. Consider a fin whose cross section is shown in the figure. The base is maintained at T_b and the fluid at temperature T_f. The convective heat-transfer coefficient between the fin and the fluid is h_v, and the thermal conductivity is k. Determine the temperature profile in the fin and an expression for the rate of heat transfer between the fin and the fluid. Neglect temperature variations on a cross-sectional surface.

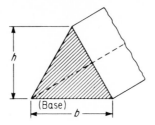

PROB. 1-12.   Tapered fin.

1-13. One common type of fin for tubes is the circular fin shown in the figure. The outer radius of the fin is R_o, and the outer radius of the tube wall is R_i. The temperature at R_i is assumed to be constant at T_b, and the fluid temperature is T_f. The thickness of the fin is δ, and the thermal conductivity k is constant. The convective heat-transfer coefficient between the surface of the fin and the fluid is h. Temperature variations over a cross section are neglected. Neglecting heat transfer at the outer rim, develop the equations and the necessary boundary conditions to describe the temperature as a function of the radius. What changes are required to account for heat transfer at the outer rim?

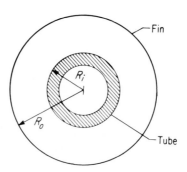

PROB. 1-13.

1-14. Consider the wetted-wall column shown. The gas enters the column at mole fraction x_{A_o} of component A. The convective mass-transfer coefficient between the liquid and the gas is k. Neglecting the change in molar flow rate w with height and neglecting concentration variations across a cross section, develop an equation for x_A as a function of z. Thickness of the liquid film is very small in comparison to the diameter of the column. Neglect diffusion in axial direction. Let the mole fraction of the vapor in equilibrium with the liquid be x_{A_i}.

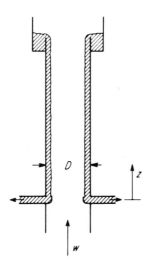

PROB. 1-14. Wetted-wall
column.

1-15. Consider bubbles rising through a fluid at constant temperature T_b. The average number of bubbles passing a given level per second is N, and the velocity of rise is v. The average mass of fluid in each bubble is m, and the bubbles enter at temperature T_i. The gas within the bubbles is well mixed, and the heat transfer from the fluid to the bubbles is given by Newton's Law (heat-transfer coefficient is constant a h). The gas a within the bubbles obeys the ideal gas law (molecular weight $= M$), and thus the radius of the bubbles varies with temperature. Develop a differential equation that can be solved for the temperature of the bubbles as a function of height within the liquid. The density of the liquid is ρ_L, and the depth at which the bubbles originate is L. Let the bubbles be spherical.

1-16. Pyrex tubing has been shown to be more permeable to helium than to the other gases normally accompanying helium, and is thus the potential basis of a separation process for helium. Consider a tube of inside radius R_i, outside radious R_o, and length L. The diffusivity of helium through pyrex is D. Assume the molar density of the gas is independent of the concentration of helium. The mole fraction of helium in the gas outside the tube is x_o, and the gas entering the tube at flow rate w is of mole fraction x_i of helium. (a) Consider a cross section of the tube wall in order to develop an expression for the rate of transfer of helium across the tube. (b) Use the result obtained in (a) to obtain the solution for the concentration profile down the tube. Neglect diffusion in the gas stream, and neglect concentration variations over a cross section. Neglect variation of total flow down the tube caused by diffusion of helium through the wall, i.e., assume w is constant.

1-17. In many situations fluid flows in an annular space between two concentric pipes. Consider such an annular space of inner radius R_i and outer radius R_o. Determine the velocity profile when a fluid of density ρ and viscosity μ flows through such a horizontal annulus for which the pressure gradient is constant. Assume laminar flow.

1-18. Determine the velocity profile when a fluid of viscosity μ flows down a slit of width h under the influence of gravity.   Neglect acceleration and assume laminar flow.

1-19. Consider a fluid of viscosity μ and density ρ flowing down an inclined surface. The thickness of the fluid film is δ.   Assuming laminar flow, determine the velocity profile.

1-20. Two immiscible fluids are flowing along a horizontal slit under the influence of a constant pressure gradient as shown.   The viscosities of the fluids are μ_1 and μ_2.   Determine the velocity profiles in each of the fluids if the flow in each zone is laminar.

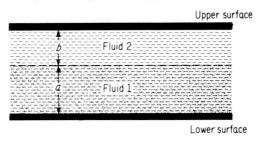

PROB. 1-20.   Flow of immiscible fluids.

1-21. Fluid is flowing up a circular tube and then down around the outside as shown. Determine the velocity profile in the film on the outside of the tube if the radii of the tube and outer film are R_1 and R_2 respectively.   Assume laminar flow.

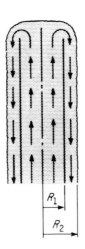

PROB. 1-21.
Falling film.

1-22. Consider fluid in an annular space in which the center tube is moving as shown. Such a phenomenon occurs in wire coating devices and similar operations.   The pressure gradient in such apparatus is usually negligible, the fluid motion being caused only by the moving center tube.   Determine the velocity profile and volumetric flow rate through such a space if the center tube moves with velocity V.

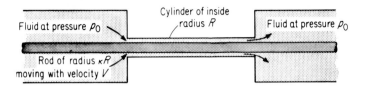

Cylinder of inside radius R

Fluid at pressure p_0

Fluid at pressure p_0

Rod of radius κR moving with velocity V

PROB. 1-22. Wire coater. (Reprinted by permission from R. B. Bird, W. E. Stewart, and E. N. Lightfoot, *Transport Phenomena*, John Wiley & Sons, Inc., New York, 1960.)

1-23. A "power-law fluid" is one that obeys the following shear stress-shear rate relationship:

$$\tau_{rz} = m\left(-\frac{dv_z}{dr}\right)^n$$

The coefficient m is referred to as the apparent viscosity. For $n < 1$ the behavior is referred to as *pseudoplastic* and for $n > 1$, *dilatant*. Determine the velocity profile for these fluids in the circular pipe in Sec. 1-11. Sketch the velocity profiles for $n < 1$ and $n > 1$ and compare to the Newtonian case ($n = 1$).

1-24. For cases in which the analysis must include heat generated by viscous friction in fluids in laminar flow, it is convenient to have an expression for the heat generation per unit volume. Consider the rectangular element shown in the figure in a one-dimensional flow pattern (unit length perpendicular to plane of paper).

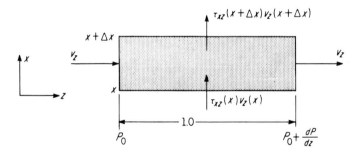

PROB. 1-24

Use the equation

$$\text{Friction} = \text{Work in} - \text{Work out}$$

to obtain the expression $\mu\left(\dfrac{dv_z}{dz}\right)^2 \Delta x$ for the viscous dissipation in the above element.

1-25. Consider the steady laminar flow of a Newtonian fluid with constant properties between two flat plates as shown, each maintained at temperature T_0. Including viscous heating due to the flow, derive the equation for the temperature profile far from the ends of the plate. The velocity profile is given by

$$v_z = -\frac{B^2}{2\mu}\frac{dP}{dz}\left[1-\left(\frac{x}{B}\right)^2\right]$$

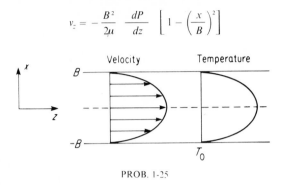

PROB. 1-25

1-26. Consider the web formation in the paper-marking process as illustrated. The fresh stock of about 0.25% solids ("consistency") by weight and flow rate w is sent to the "headbox" where it is uniformly distributed over a wire screen of width W and moving with velocity V. The pressure in the headbox is regulated so that the velocity of the liquid from the slice is approximately equal to the velocity of the wire. On the table roll section the liquid drains through the wire, depositing some of the suspended fiber on the wire. The suction boxes on the latter section of the wire apply a vacuum to promote drainage over that section. The following terms are frequently used in the paper industry:

B = Basis weight, the lb/ft² of fiber in the sheet

R = Retention, the fraction of the fiber impinging on the wire that is retained. The fraction $1 - R$ passes through. The retention at a point on the wire is a function of B at that point.

Let $z = 0$ be at the slice and $z = L$ be at the couch roll. Let $f(B,z)$ be the mass flow rate in lb/hr-ft² of liquid and fiber that impinges on the screen at point z on the wire. Determine the necessary equations to give $B(z)$, the basis weight as a function of position on the wire. Include equations that give the point on the wire at which all liquid has drained through.

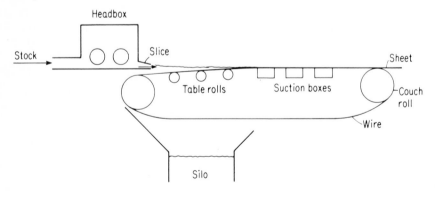

PROB. 1-26

1-27. Derive the equations for the tubular reactor in Sec. 1-9 if the following reaction occurs under isothermal conditions in the liquid phase (negligible volume change):

$$2A + B \rightarrow 2C$$

The reaction rate is second-order with respect to A and first-order with respect to B. The feed is of concentration C_{A_0} of A and C_{B_0} of B.

1-28. Consider the reversible reaction

$$A \rightleftharpoons B$$

This reaction can be viewed as two irreversible reactions, namely, the "forward reaction":

$$A \rightarrow B$$

and the "reverse reaction":

$$B \rightarrow A$$

The expressions for reaction rates for these two reactions are derived in the usual manner. Suppose both reactions are first-order, the rate constants being k_1 and k_2 for the forward and reverse reactions, respectively. The rate of formation of A by both reactions is then given by

$$r_A = k_2 C_B - k_1 C_A$$

For the case in which this reaction is carried out in the tubular reactor in Sec. 1-9, derive the differential equation and boundary conditions that can be solved for the concentration profile.

1-29. Consider the first-order irreversible reaction $A \rightarrow 2B$. If this reaction occurs in the gas phase, there is obviously a change in volume upon reaction. One convenient way to take this into account is with a "coefficient of volumetric expansion" ϵ_A defined as

$$\epsilon_A = \frac{V_{X_A=1} - V_{X_A=0}}{V_{X_A=0}}$$

where X_A is the fraction conversion. The volume V is related to X_A as follows:

$$V = V_0(1 + \epsilon_A X_A)$$

For the gas-phase reaction $A \rightarrow 2B$ at constant pressure, 1.0 ft³ of A would react to form 2.0 ft³ of B. Thus,

$$\epsilon_A = \frac{2.0 \text{ ft}^3 - 1.0 \text{ ft}^3}{1.0 \text{ ft}^3} = 1$$

If this reaction is carried out in the tubular reactor considered in Sec. 1-9, determine the differential equation and boundary conditions that can be solved for the concentration (or conversion) profile. Let the inlet volumetric flow rate be v_0 ft³/hr of pure A.

1-30. Consider the tubular reactor in Sec. 1-9 operating under the same conditions except that the diffusivity D_{AB} is significant.

1. Derive the differential equation for $C_A(t)$.

2. As usual the boundary conditions arise from physical considerations at the extremities. For such reactions, it is generally said that there is no diffusion into or out of the reactor at the ends. From these considerations and the fact that the inlet feed is at concentration C_{A_0}, the following three boundary conditions could be postulated:

$$(1) \qquad C_A(0) = C_{A_0}$$

$$(2) \qquad \frac{dC_A(0)}{dz} = 0$$

$$(3) \qquad \frac{dC_A(L)}{dz} = 0$$

Of course, only two of these are needed.

3. Show that $\lim_{z \to \infty} |C_A(z)| = \infty$ when the first two boundary conditions are used. Thus this combination can be dismissed.

4. Note that the last two boundary conditions give the trivial solution $C_A(z) = 0$. Thus this combination is unreasonable.

5. Note that the remaining combination

$$C_A(0) = C_{A_0}$$

$$\frac{dC_A(L)}{dz} = 0$$

give a solution that indicates diffusion at $z = 0$.

The difficulty with formulating these boundary conditions is the physical discontinuities at the inlet and outlet of the reactor. Although these give serious problems analytically, they can be handled numerically (see Prob 3-25).

1-31. Consider the heat exchanger in Sec. 1-6 operating under the same conditions except that the thermal conductivity of the liquid is significant.

1. Derive the differential equation for $T(z)$.

2. It seems reasonable to make the restriction that no heat is conducted into or out of the ends of the exchanger. Show that this leads to the same problems with boundary conditions as encountered in the previous problem. It will be convenient to make the following change of variable:

$$\Phi(z) = \frac{T_s - T(z)}{T_s - T_0}$$

chapter 2

Simultaneous Differential Equations

In this chapter the concepts presented in Chapter 1 are extended to include more complex processes whose complete description requires more than one ordinary differential equation. Although analytic solutions will be presented for the simple equations derived in this chapter, several of the solutions are obtained by means of a computer. As more realistic problems are considered, computer solutions become essential. For this reason, Chapters 5 and 6 of this text are devoted to analog-computer and numerical-solution techniques, respectively. For the present, the student should devote his attention to formulating the equations and their boundary conditions. Correct formulations and complete descriptions obviously are prerequisites to meaningful solutions.

2-1. NUMBER OF EQUATIONS REQUIRED AND THEIR BOUNDARY CONDITIONS

To obtain a unique solution for algebraic equations, the number of independent equations must equal the number of unknowns appearing in these equations. This same concept also applies to sets of equations that include differential equations. If two dependent variables appear in the formulation of a differential equation describing a system, a second independent equation must be obtained in order to have a unique solution. When this second equation is algebraic, it can often be substituted into the differential equation to eliminate one of the variables, thereby leaving only one dependent variable in the final equation. As in Chapter 1, a single ordinary differential equation is obtained.

Frequently the second independent equation is also a differential equation involving the two dependent variables. In such case it is often impossible to eliminate one of the dependent variables by substitution. The only alternative is to solve these two differential equations simultaneously. Analytic solutions are occasionally possible, but numerical or analog-computer approaches are required more frequently.

Boundary-condition requirements are essentially the same as for single ordinary differential equations. In most cases the number of boundary conditions on each dependent variable equals the order of the highest derivative to which the dependent variable appears in any of the equations.

2-2.   MATHEMATICAL DESCRIPTION OF A MULTIPASS HEAT EXCHANGER [1,2]

In the analysis of complex heat exchangers, more than one dependent variable typically appears. The 1–2 heat exchanger in Fig. 2-1 has two tube passes and one shell pass. The temperatures of the fluid at the shell-side inlet and outlet are T_{s_1} and T_{s_2}, respectively. Similarly, the temperatures of the fluid at the tube-side inlet and outlet are T_{t_1} and T_{t_2}, respectively. The flow rates of the shell-side fluid

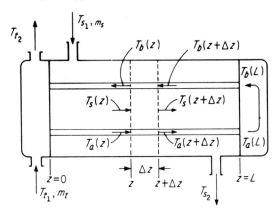

FIG. 2-1.   One-shell-pass, two-tube-pass exchanger.

and tube-side fluid are m_s and m_t, respectively. Likewise, the heat capacities are c_{ps} and c_{pt}. The flow rates are constant, no heat is lost to the surroundings, and variations in physical properties and the heat-transfer coefficient U with temperature are neglected. For simplicity, the heat-transfer area A per unit length per pass is assumed to be the same for each pass. The fluid temperatures in the shell and in each tube pass are assumed to be constant over a cross section of the exchanger at any distance from the inlet. Under this assumption, this is a one-dimensional problem. Entrance and exit effects are neglected.

Quantity Conserved: Thermal energy.

Control Volume: A cross section of length Δz of the exchanger in Fig. 2-1.

Independent Variable: Distance z from the inlet.

Dependent Variables: $T_s(z)$, the temperature of the shell-side fluid; $T_a(z)$, the temperature of the tube-side fluid in the first pass; and $T_b(z)$, the temperature of the tube-side fluid in the second pass.

Balance for the Shell-Side Fluid

To formulate the individual terms, assume the shell-side fluid is the hot fluid. The input term is the enthalpy of the fluid flowing into the control volume:

$$q_s(z) = m_s c_{ps} [T_s(z) - T_R]$$

where T_R is the reference temperature for computing the enthalpy. One of the output terms is the bulk flow of fluid leaving the control volume:

$$q_s(z + \Delta z) \cong q_s(z) + \frac{dq_s(z)}{dz} \Delta z$$

The other two output terms represent the rate of heat transfer to the fluid in each pass:

$$q_1(z) = UA[T_s(z) - T_a(z)] \Delta z$$

$$q_2(z) = UA[T_s(z) - T_b(z)] \Delta z$$

The energy balance is

$$\text{In} - \text{Out} = 0$$

$$q_s(z) - \left[q_s(z) + \frac{dq_s(z)}{dz} \Delta z \right] - q_1(z) - q_2(z) = 0$$

Simplifying and substituting the above relationships,

$$m_s c_{ps} \frac{dT_s(z)}{dz} + UA[T_s(z) - T_a(z)] + UA[T_s(z) - T_b(z)] = 0 \qquad (2\text{-}1)$$

Balance for Fluid in First Tube Pass

There are two input and one output terms in this balance. One input term is the enthalpy of the fluid flowing into the control volume:

$$q_a(z) = m_t c_{pt} [T_a(z) - T_R]$$

The output term for the bulk flow is

$$q_a(z + \Delta z) \cong q_a(z) + \frac{dq_a(z)}{dz} \Delta z$$

The second input term is $q_1(z)$, the heat received from the shell-side fluid. From these three terms, the energy balance for the fluid in the first tube pass can be made as follows:

$$\text{In} - \text{Out} = 0$$

$$q_a(z) + q_1(z) - \left[q_a(z) + \frac{dq_a(z)}{dz} \Delta z \right] = 0$$

Simplifying and substituting for $q_1(z)$ and $q_a(z)$,

$$m_t c_{pt} \frac{dT_a(z)}{dz} - UA[T_s(z) - T_a(z)] = 0 \qquad (2\text{-}2)$$

Balance for the Fluid in the Second Tube Pass

Although the energy balance for this pass is very similar to that for the first tube pass, one difference should be given special attention. In all previous problems the direction of flow has been in the positive direction of the independent variable. As in previous examples, the enthalpy of the fluid crossing the face at z is designated as

$$q_b(z) = m_t c_{pt}[T_b(z) - T_R]$$

But since the fluid is flowing out at this face, this is an output term. The enthalpy of the fluid crossing the face at $z + \Delta z$ is given by

$$q_b(z + \Delta z) \cong q_b(z) + \frac{dq_b(z)}{dz} \Delta z$$

and is an input term. The remaining term to complete the balance is $q_2(z)$, the heat received from the shell-side fluid. Making the balance,

$$\text{In} - \text{Out} = 0$$

$$q_2(z) + \left[q_b(z) + \frac{dq_b(z)}{dz} \Delta z \right] - q_b(z) = 0$$

Simplifying and substituting yields

$$m_t c_{pt} \frac{dT_b(z)}{dz} + UA[T_s(z) - T_b(z)] = 0 \qquad (2\text{-}3)$$

Boundary Conditions

First, consider the case in which the exchanger is in existence, and consequently A and L are known. Furthermore, assume that both inlet flows (m_t and m_s) and both inlet temperatures (T_{t_1} and T_{s_1}) are known. In this case it can be readily verified that the boundary conditions are

$$T_s(0) = T_{s_1}$$

$$T_a(0) = T_{t_1}$$

$$T_a(L) = T_b(L)$$

Thus there are three differential equations and three boundary conditions.

Next, consider the case in which the exchanger is to be designed. Assume that, as before, both inlet flows (m_t and m_s) and both inlet temperatures (T_{t_1} and T_{s_1}) are known. Furthermore, assume that the exchanger is to be designed to give a desired tube-side outlet temperature T_{t_2}, and let the length L also be specified. Now the following boundary conditions must be met:

$$T_s(0) = T_{s_1}$$
$$T_s(0) = T_{t_1}$$
$$T_b(0) = T_{t_2}$$
$$T_a(L) = T_b(L)$$

Four boundary conditions are available, apparently one more than necessary. However, as the exchanger is not yet designed, the value of A is unknown. Three boundary conditions are sufficient to solve the equations, and A must be selected so that the fourth is satisfied.

In Fig. 2-2 are shown the temperature profiles for a typical 1–2 exchanger.

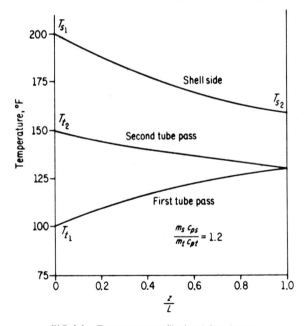

FIG. 2-2. Temperature profiles in a 1-2 exchanger.

The equations derived above can also be solved analytically [1] to give the F-charts to correct the logarithmic mean temperature difference used in heat-exchanger design.

2-3.   SIMULTANEOUS HEAT AND MASS TRANSFER
IN A PACKED TOWER [3,4]

In continuous contact humidification or dehumidification operations liquid is admitted at the top of a packed column and contacts a gas passing upward through the column.   A schematic drawing of such a column is shown in Fig. 2-3. The following nomenclature is used for this example:

$L(z)$ = Liquid rate, mole/hr.; a dependent variable

V_G = Flow rate of noncondensables, mole/hr

$Y(z)$ = Moles vapor per mole noncondensable gas; a dependent variable

$H_G(z)$ = Enthalpy of gas stream, Btu/mole noncondensable gas; a dependent variable.

$H_L(z)$ = Enthalpy of liquid stream, Btu/mole; a dependent variable

$T_G(z)$ = Temperature of gas stream, °F; a dependent variable

$T_L(z)$ = Temperature of liquid stream, °F; a dependent variable

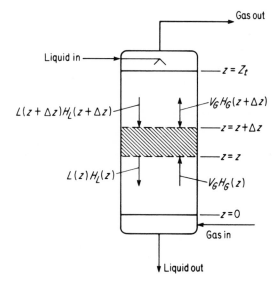

FIG. 2-3.   Continuous-contact humidification tower, showing terms for enthalpy balance.

T_R = Reference temperature, °F

a = Interfacial area, sq-ft/cu-ft of packing

A = Tower cross section, ft²

h_g, h_L = Heat-transfer coefficient, Btu/hr-ft²-°F

k_Y = Mass-transfer coefficient, mole/hr-ft²-mole ratio

C_h = Heat capacity of gas stream, Btu/°F-mole noncondensable

c_p = Heat capacity of liquid, Btu/mole-°F

λ_R = Heat of vaporization of liquid at T_R, Btu/mole

z = Distance from bottom of tower, ft; the independent variable

Z_t = Height of tower, ft

The objective of the ensuing discussion is to derive the necessary equations to describe the temperature and concentration profiles in the tower.

Before making any balances, a discussion of the assumed heat- and mass-transfer mechanism at the interface is in order. Both heat and mass are transferred between the liquid and gas phases. The following mechanism is assumed (refer to Fig. 2-4):

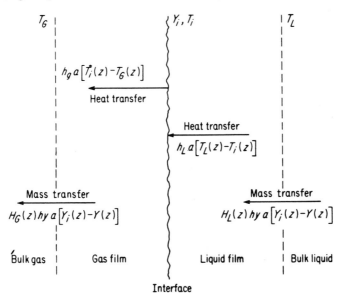

FIG. 2-4.   Heat- and mass-transfer mechanisms at the interface.

1. Heat is transferred by convection from the bulk liquid to the interface, and is represented by

$$q(z) = h_L a \lfloor T_L(z) - T_i(z) \rfloor$$

    where $q(z)$ is the rate of heat transfer per unit volume at point z and $T_i(z)$ is the interfacial temperature at z.

2. The mass-transfer per unit volume is given by

$$m(z) = k_Y a \lfloor Y_i(z) - Y(z) \rfloor$$

    where $Y_i(z)$ is the equilibrium mole ratio at the interface at point z and $m(z)$ is the rate of mass transfer per unit volume.

3. The heat transfer associated with the mass transfer is $H_L(z)m(z)$ for the liquid phase and $H_G(z)m(z)$ for the gas phase.

4. There is also heat transfer by convection from the interface to the bulk gas stream, and is given by

$$h_g a[\, T_i(z) - T_G(z)]$$

For many cases, a much simpler mechanism would suffice. However, this mechanism makes a good illustration.

Overall Enthalpy Balance

For the first equation an enthalpy balance is made over *both* phases entering and leaving the control volume of thickness Δz shown in Fig. 2-3. The enthalpy of the liquid entering the control volume is

$$L(z + \Delta z)\, H_L(z + \Delta z)$$

The enthalpy of the liquid leaving the control volume is simply $L(z)H_L(z)$. For the gas stream the enthalpy of gas entering the control volume is $V_G H_G(z)$ and the enthalpy of the gas leaving is

$$V_G H_G(z + \Delta z)$$

From these terms the equation is formulated as follows:

$$\text{In} - \text{Out} = 0$$

$$L(z + \Delta z)\, H_L(z + \Delta z) + V_G H_G(z) - L(z)H_L(z) - V_G H_G(z + \Delta z) = 0$$

or

$$\frac{d[L(z)H_L(z)]}{dz} = \frac{d[V_G H_G(z)]}{dz} \tag{2-4}$$

The following substitutions and simplifications can be made:

1. The rate of flow V_G of noncondensables is constant. Although this is pratically obvious, it can be derived by making a material balance on the noncondensables within the control volume.
2. The enthalpy per mole of liquid is

$$H_L(z) = c_p[\, T_L(z) - T_R] \tag{2-5}$$

3. The enthalpy of the gas stream is

$$H_G(z) = C_h[\, T_G(z) - T_R] + \lambda_R\, Y(z) \tag{2-6}$$

To be precise, c_p is a function of temperature, and C_h is a function of both mole ratio and temperature. However, the mole ratios encountered in such columns are usually rather low, and assuming both c_p and C_h constant simplifies the resulting equations considerably with little loss in accuracy. Substituting Eq. 2-5 and 2-6 into Eq. 2-4 and simplifying,

$$L(z)c_p \frac{dT_L(z)}{dz} + c_p[T_L(z) - T_R] \frac{dL(z)}{dz} = V_G C_h \frac{dT_G(z)}{dz}$$

$$+ V_G \lambda_R \frac{dY(z)}{dz} \qquad (2\text{-}7)$$

Enthalpy Balance Around Gas Phase

For this equation an enthalpy balance is made around the gas phase within the control volume in Fig. 2-3. The input and output terms for the bulk-gas flow are the same as for the previous balance. In addition, heat transfer occurs between the gas and liquid phases. Assuming the gas phase is warmer than the liquid phase, this is an output term and is given by

$$h_g a A [T_G(z) - T_i(z)] \Delta z$$

Furthermore, the liquid being transferred to the gas stream also contributes to the enthalpy of the gas stream. Assuming humidification is occurring, the rate of increase of the enthalpy in the bulk stream is

$$H_g(z) m(z)$$

The equation can now be formulated as follows:

$$\text{In} - \text{Out} = 0$$

$$V_G H_G(z) + H_G(z) m(z) - h_g a A [T_G(z) - T_i(z)] \Delta z - V_G H_G(z + \Delta z) = 0$$

or

$$\frac{V_G}{A} \frac{dH_G(z)}{dz} = h_g a [T_i(z) - T_G(z)] + H_G(z) m(z)$$

Substituting Eq. 2-6 for H_G in the derivative:

$$\frac{V_G}{A} C_h \frac{dT_G(z)}{dz} + \frac{V_G}{A} \lambda_R \frac{dY(z)}{dz} = h_g a [(T_i(z) - T_G(z)] + H_G(z) m(z)$$

$$(2\text{-}8)$$

Enthalpy Balance for the Liquid Phase

The input and output terms for the enthalpy of the bulk-liquid stream are the same as in the overall enthalpy balance. The output of sensible heat from bulk liquid to the interface is given by

$$g_L(z) = h_L a A [(T_L(z) - T_i(z)] \Delta z + A H_L(z) m(z)$$

This equation is formulated as follows:

$$\text{In} - \text{Out} = 0$$

$$L(z + \Delta z)H_L(z + \Delta z) - g_L(z)\Delta z - L(z)H_L(z) = 0$$

or

$$\frac{d[L(z)H_L(z)]}{dz} = h_L aA[T_L(z) - T_i(z)] - AH_L(z)m(z)$$

Substituting Eq. 2-5 for $H_L(z)$ in the derivative

$$L(z)c_p \frac{dT_L(z)}{dz} + c_p[T_L(z) - T_R]\frac{dL(z)}{dz} = h_L aA[T_L(z) - T_i(z)] - AH_L(z)m(z)$$

$$(2-9)$$

Actually, this equation could be replaced by an algebraic equation. Equating heat-flow terms for the mechanism in Fig. 2-4 yields

$$h_L a[T_L(z) - T_i(z)] + H_L(z)m(z) = h_g a[T_i(z) - T_G(z)] + H_G(z)m(z)$$

$$(2-10)$$

This equation can be used to derive either Eq. 2-7, 2-8, or 2-9 from the other two. As it is an algebraic equation, it will be used instead of Eq. 2-9.

Mass Balance on Distributed Component for Both Phases

For the liquid stream the mole/hr of liquid entering the control volume is $L(z + \Delta z)$, and the mole/hr of liquid leaving is $L(z)$. For the gas phase the mole/hr vapor entering is $V_G Y(z)$, and the mole/hr vapor leaving is $V_G Y(z + \Delta z)$. The equation is formulated as usual:

$$\text{In} - \text{Out} = 0$$

$$L(z + \Delta z) + V_G Y(z) - L(z) - V_G Y(z + \Delta z) = 0$$

$$\frac{dL(z)}{dz} = V_G \frac{dY(z)}{dz} \qquad (2-11)$$

Material Balance on Distributed Component in Gas Phase

The moles/hr of vapor in and out with the bulk streams is the same as for the previous balance. The rate of transfer of the distributed component from the liquid stream to the vapor stream is

$$k_Y aA|Y_i(z) - Y(z)|\Delta z$$

As before, the formulation of the equation is as follows:

$$\text{In} - \text{Out} = 0$$

$$V_G Y(z) + k_Y aA[Y_i(z) - Y(z)]\Delta z - V_G Y(z + \Delta z) = 0$$

$$V_G \frac{dY(z)}{dz} = k_Y aA[Y_i(z) - Y(z)] \qquad (2-12)$$

Problem Description

From the above procedure five equations, 2-7, 2-8, 2-9, 2-11, and 2-12, with six dependent variables, T_G, T_L, L, Y, T_i and Y_i, have been derived. Since z is the only independent variable, all five equations are ordinary differential equations. In addition to these equations, the relationship between T_i and Y_i must also be specified. Since the interface concentration is normally the concentration that would be in equilibrium at that temperature, this relationship is typically vapor pressure vs. temperature data. This effectively gives six equations and six unknowns.

The following boundary conditions are required:
1. One on T_L, namely, the temperature $T_L(Z_t)$ at the liquid inlet.
2. One on L, namely, the flow rate $L(Z_t)$ of liquid entering the column.
3. One on T_G, namely, the temperature $T_G(0)$ of the inlet gas.
4. One on Y, namely, the mole ratio $Y(0)$ of condensables to noncondensables in the inlet gas.

2-4. SERIES CHEMICAL REACTIONS [1]

An important class of chemical reactions is the following type, known as series reactions:

$$A \xrightarrow{k_1} B \xrightarrow{k_2} C \qquad (2\text{-}13)$$

One interesting case for which we will examine this type of reaction is when B, the intermediate product in the reaction sequence, is the valuable product, i.e., the product whose recovery is to be maximized. If the reaction begins in a vessel containing pure A, the first reaction proceeds to form B. Since no B is present initially, the second reaction is not occuring at time zero. Consequently, the concentration of B increases until the rate of consumption by the second reaction equals its rate of production by the first reaction. From this point on, the consumption of B exceeds its production, and the concentration of B in the reactor decreases. If the reacting material remains in the reactor for a long period of time, essentially pure C will be present.

If the above reaction is carried out in a batch reactor, there will be an optimum time for dumping the contents if the production of B is to be maximized. Similarly, in a tubular reactor of given size, there is a volumetric flow rate that optimizes the production of B. To better illustrate this point, an example will be presented in which the concentration profiles for components A, B, and C in the above reaction are determined for a tubular reactor.

The specific tubular reactor to be considered is shown in Fig. 2-5. The cross-sectional area of the reactor is A_c, and its feed is pure A at concentration C_{Af} and volumetric flow rate v_0. The reaction is carried out under steady state, isothermal conditions, and the volume change upon reaction is assumed to be negligible.

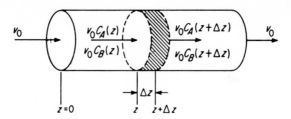

FIG. 2-5.   Tubular reactor in which the series reaction in Eq. 2-13
occurs.

For the reactions in Eq. 2-13 each reaction is assumed to be first-order with
respect to each of the reactants.

Since the concentrations of components A, B, and C vary continuously from
the inlet to the outlet of the reactor, the equations must be developed for the
infinitesimal element shown in Fig. 2-5.   For component A one of the input terms
is the input of component A with the bulk flow into the element, which is given
by $v_0 C_A(z)$.   Since component A is involved only in the first reaction, its rate of
appearance r_A is given by

$$r_A = -k_1 C_A(z)$$

The rate of input of component A by reaction in the element is given by $r_A A_c \Delta z$,
where $A_c \Delta z$ is the volume of the element.   The output term in the mass balance
for component A is

$$v_0 C_A(z + \Delta z)$$

Since the reactor is in steady-state operation, the accumulation term is zero.
Now formulating a material balance from these terms,

$$\text{In} - \text{Out} = 0$$

$$v_0 C_A(z) + [-k_1 C_A(z)] A_c \Delta z - v_0 C_A(z + \Delta z) = 0$$

Since the volume change upon reaction is negligible, v_0 is constant.   Thus the
above equation reduces to

$$-k_1 C_A(z) A_c = v_0 \frac{dC_A(z)}{dz}$$

Since the concentration of component A is C_{Af} at the inlet ($z = 0$) of the reactor,
the above equation may be separated and integrated to obtain

$$C_A = C_{Af} e^{-\frac{A_c k_1}{v_0} z} \tag{2-14}$$

A new independent variable τ, the time required for fluid to flow from entrance to
point z, is defined as

$$\tau = \frac{A_c z}{v_0} \tag{2-15}$$

Equation 2-14 now becomes

$$C_A = C_{Af} e^{-k_1 \tau}$$

At the end of this section, the concentration profile for A will be presented graphically and compared to the concentration profiles for component B.

For this reaction system it is not possible to express the relationship between the concentration $C_A(z)$ of component A and the concentration $C_B(z)$ of component B by an algebraic equation. Thus a mass balance must be made for component B for the differential element in Fig. 2-5. The following terms appear in this balance:

Input of component B with bulk-flow stream:

$$v_0 \, C_B(z)$$

Output of component B with bulk-flow stream:

$$v_0 \, C_B(z + \Delta z)$$

Rate of appearance of component B by reaction within the element:

$$r_B \, A_c \Delta z = [k_1 \, C_A(z) - k_2 C_B(z)] \, A_c \Delta z$$

Since the accumulation term is zero for steady-state operation, the differential equation for concentration $C_B(z)$ of component B can be formulated as follows:

$$\text{In} - \text{Out} = 0$$

$$v_0 C_B(z) + [k_1 C_A(z) - k_2 C_B(z)] \, A_c \Delta z - v_0 C_B(z + \Delta z) = 0$$

Since v_0 is constant, this equation reduces to

$$\frac{v_0}{k_2 A_c} \frac{dC_B(z)}{dz} + C_B(z) = \frac{k_1}{k_2} C_A(z)$$

Substituting the expression for $C_A(z)$ in Eq. 2-14 yields

$$\frac{v_0}{k_2 A_c} \frac{dC_B(z)}{dz} + C_B(z) = \frac{k_1 C_{Af}}{k_2} e^{-\frac{A_c k_1}{v_0} z}$$

The boundary condition is that $C_B(0) = 0$. This equation is a first-order nonhomogeneous differential equation which has the solution

$$C_B(z) = \frac{C_{Af} k_1}{k_2 - k_1} \left[e^{-\frac{A_c k_1}{v_0} z} - e^{-\frac{A_c k_2}{v_0} z} \right] \tag{2-16}$$

In terms of τ, this equation is

$$C_B = -\frac{C_{Af}k_1}{k_2 - k_1}\left[e^{-k_1\tau} - e^{-k_2\tau}\right] \tag{2-17}$$

which may be rearranged to

$$\frac{C_B}{C_{Af}} = \frac{1}{\left(\dfrac{k_2}{k_1} - 1\right)}\left[e^{-(k_1\tau)} - e^{-\left(\frac{k_2}{k_1}\right)(k_1\tau)}\right] \tag{2-18}$$

Using this form of the equation, we obtain the plot in Fig. 2-6. On this graph the concentration profile is plotted as (C_B/C_{Af}) vs. $(k_1\tau)$ for various values of (k_2/k_1). Also, the profile for (C_A/C_{Af}) as a function of $(k_1\tau)$ is plotted.

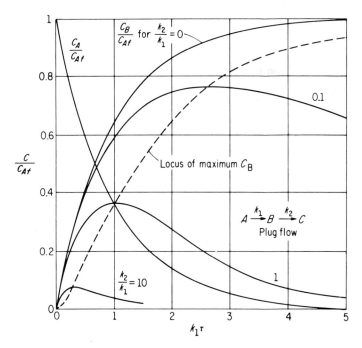

FIG. 2-6. Concentration profiles for a series reaction. (Reprinted with permission from Octave Levenspiel, *Chemical Reaction Engineering*, John Wiley & Sons, Inc., New York, 1962.)

If the reactor is to be operated such that the concentration of C_B is to be maximized when the contents are removed, the time from entrance to exit (the residence time) and the value of C_B can be determined from the line of maximum C_B in Fig. 2-6. Since C_B is a maximum at dump time, $dC_B/d\tau$ is zero. Consequently, an analytic expression for the line of maximum C_B can be determined by differentiating Eq. 2-17 with respect to τ and setting the result equal to zero. Thus

$$\frac{dC_B}{d\tau} = 0 = \frac{C_A k_1}{k_2 - k_1}\left[-k_1 e^{-k_1\tau} + k_2 e^{-k_2\tau}\right]$$

or

$$e^{(k_2 - k_1)\tau} = \frac{k_2}{k_1}$$

Taking logarithms and rearranging;

$$k_1\tau = \frac{ln\left(\dfrac{k_2}{k_1}\right)}{\dfrac{k_2}{k_1} - 1}$$

Thus, given a value of k_2/k_1, the value of $k_1\tau$ at which C_B is a maximum can be determined. Substituting this value into Eq. 2-18 gives the maximum value of C_B.

2-5. NONISOTHERMAL REACTORS

In operating many chemical reactors, the temperature is varied intentionally to obtain favorable yields or reaction times, while in other cases the temperature either rises or falls due to the nature of the reaction occurring. In many cases the primary factor in the design of the reaction vessel is the heat-transfer aspects. The selection of a temperature for operating the reactor depends upon such factors as the variation of the equilibrium yield with temperature, the reaction rate (the rate constant generally increases with temperature), and the importance of side reactions. However, the primary purpose of this section is not to discuss the design of a reaction system, but to consider the determination of the temperature and concentration profiles in an existing or proposed reaction system.

In reality, the selection of operating conditions for a reaction system is often a trial-and-error procedure. Usually an estimate of the optimum operating conditions is made, and the concentration and temperature profiles calculated. Next, this estimate is checked by varying the estimated operating conditions and repeating the calculations. This is normally continued until either a constraint, e.g., maximum cooling-water outlet temperature or flow rate, is encountered or a true maximum is determined. Sophisticated procedures for carrying out this optimization are discussed in later chapters of this book.

However, our purpose is to develop the equations describing a given reaction system. The principles developed previously can be used except for one difference—the variation of physical properties and the reaction-rate constant with temperature are often important, and considerable work has been done in the area of predicting the temperature dependence of the rate constant. Probably the most common expression for describing this dependence is the one originally suggested by Arrhenius:

$$k = k_0 e^{-\frac{E}{RT}}$$

where k_0 is called the *frequency factor,* E is called the *activation energy* of the reaction, R is the *gas law constant,* and T is the absolute temperature.

A variety of other expressions for the temperature dependence of the reaction rate constant are available. Using the concepts developed in collision theory for molecules, the rate constant would be predicted to behave as follows:

$$k = k_1 T^{1/2} e^{-\frac{E}{RT}}$$

where k_1 is a proportionality constant. On the other hand, transition-state theory predicts that the rate constant should vary with temperature according to

$$k = k_2 T e^{-\frac{E}{RT}}$$

where k_2 is also a proportionally constant. Another approach for complex reactions is to simply fit the experimental data to a polynomial in T, such as

$$k = a_0 + a_1 T + a_2 T^2 + \cdots + a_n T^n$$

where $a_0, a_1, ..., a_n$ are determined from the experimental data. No matter which expression is used, the constants are frequently evaluated from the experimental data, although progress is being made toward predicting them from basic concepts.

The general approach to describing nonisothermal reactors is to make a heat balance in addition to the component balances discussed previously. Most reactions involve either a release (exothermic reactions) or an absorption (endothermic reactions) of heat, and this must be included in the heat balance. This heat of reaction is usually denoted by ΔH, the enthalpy change for the reaction. For an exothermic reaction the numerical value of ΔH is negative; for an endothermic reaction it is positive. Therefore, in developing the equations the reaction is typically assumed to be endothermic and is consequently a rate of heat output from the system. If the reaction is exothermic, the signs will take care of themselves.

These concepts are well illustrated by considering the jacketed tubular reactor of length L in Fig. 2-7 in which the following first-order irreversible reaction is occurring:

$$A \rightarrow B + C$$

The temperature dependence of the rate constant for this reaction is assumed to obey the Arrhenius expression. The heat of reaction is ΔH, and the physical properties of both the reacting fluid and the fluid in the jacket are assumed to be constant. For many reactions the variation in ΔH with temperature is significant, but is ignored in this problem. The reacting fluid enters the reactor with volumetric flow rate v_1, temperature T_f, and concentration C_{Af} of component A. Components B and C are not present in the feed. The jacket fluid flows countercurrent to the reacting fluid and enters the jacket at volumetric flow rate v_2 and temperature t_f. The radius of the inside pipe is R_i, and the cross-sectional area for flow through the jacket is A_j. The overall heat transfer coefficient between the jacket fluid and the reacting fluid is U_i, based on the inside area of the pipe. Heat transfer between the jacket and surrounding air will be neglected. For this reactor expressions for the steady-state temperature and concentration profiles are to be developed.

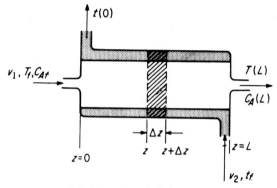

FIG. 2-7. Jacketed tubular reactor.

For this problem the independent variable is the distance z from the end at which the reacting fluid enters. The dependent variables are the reacting fluid temperature $T(z)$, the jacket fluid temperature $t(z)$, and the reacting fluid concentrations $C_A(z)$, $C_B(z)$, and $C_C(z)$. Since these dependent variables are continuous functions of z, the equations must be formulated by examining the infinitesimal element show in Fig. 2-7.

Material Balance on Component A

To formulate the equation for the concentration profile, a component mass balance must be made for component A. For the infinitesimal volume of the reactor in Fig. 2-7, the terms in this balance are

Input of A by bulk flow:

$$v_1 C_A(z)$$

Input of A by chemical reaction:

$$\pi R_i^2 r_A \Delta z = -\pi R_i^2 k_1 C_A(z) \Delta z$$

Output of A by bulk flow:

$$v_1 C_A(z) + \frac{d}{dz}[v_1 C_A(z)] \Delta z$$

Since the reactor is in steady-state operation, the accumulation term is zero. Combining these terms in a mass balance for component A yields

$$v_1 C_1(z) - \pi R_i^2 k_1 C_1(z) \Delta z - \{ v_1 C_1(z) + \frac{d}{dz} [v_1 C_A(z)] \Delta z \} = 0$$

This equation simplifies to

$$\frac{dC_A(z)}{dz} + \frac{\pi R_i^2 k_1}{v_1} C_1(z) = 0 \qquad (2\text{-}19)$$

The boundary condition is that the concentration of A in the feed is C_{Af}, or expressed mathematically,

$$C_A(0) = C_{Af} \qquad (2\text{-}20)$$

In the previous sections we have been able to integrate Eq. 2-19 with the boundary condition in Eq. 2-20. However, for this problem recall that k_1 is a function of the temperature T, which is a function of z. Consequently, k_1 is also a function of z, and the nature of this dependence must be known before Eq. 2-19 can be integrated. However, this dependence is not known until the temperature profile is determined, but the temperature profile cannot be determined without considering the concentration profile. The solution for this problem must be obtained by solving the equations for the temperature and concentration profiles simultaneously.

Before leaving the subject of the concentration profiles, the relationships between $C_A(z)$, $C_B(z)$, and $C_C(z)$ will be determined. For $C_B(z)$, reasoning similar to that given above for $C_A(z)$ gives the equation

$$\frac{dC_B(z)}{dz} - \frac{\pi R_i^2 k_1}{v_1} C_1(z) = 0$$

Although this equation could be used directly, it can be replaced by an algebraic equation in this case. To do this, first add this equation to Eq. 2-19 to obtain

$$\frac{dC_1(z)}{dz} = -\frac{dC_B(z)}{dz}$$

Now integrate both sides from 0 to z to obtain

$$C_A(z) - C_{Af} = C_B(z)$$

Similar reasoning for $C_C(z)$ gives

$$C_C(z) = C_B(z) = C_A(z) - C_{Af}$$

This procedure is equivalent to making a component mass balance over the section of the reactor from inlet to a distance z from the inlet.

Enthalpy Balance for Reacting Fluid

Now consider the temperature profile in the reacting fluid. To formulate this equation a heat balance must be made for the fluid in the element of reactor volume in Fig. 2-7. The significant terms in this equation are the heat input and output with the bulk flow, the heat absorbed or liberated upon reaction, and the heat transferred to or from the jacket fluid. The terms in the equation will be written as if the reaction is endothermic (the numerical value of ΔH is positive) and the temperature of the jacket is higher than the temperature of the reacting fluid. Of course, the equations developed will be valid even if these conditions are not met.

The following terms must be included in a heat balance for the reacting fluid in the element in Fig. 2-7:

(a) Sensible heat in by bulk flow:

$$v_1 \rho c_p [T(z) - T_R]$$

where T_R is a reference temperature.

(b) Sensible heat out by bulk flow:

$$v_1 \rho c_p [T(z) - T_R] + \frac{d}{dz} \{v_1 \rho c_p [T(z) - T_R]\} \Delta z$$

(c) Heat received from the jacket fluid:

$$U_i (2\pi R_i \Delta z) [t(z) - T(z)]$$

(d) Heat absorbed by chemical reaction:

$$k_1 C_A(z) (\pi R_i^2 \Delta z) \Delta H$$

The units on ΔH are typically Btu per mole of reactant consumed. The rate at which reactant A is consumed is r_A, or $k_1 C_A(z)$.

Combining these terms into a heat balance yields

$$v_1 \rho c_p [T(z) - T_R] + U_i (2\pi R_i \Delta z) [t(z) - T(z)]$$

$$- \{v_1 \rho c_p [T(z) - T_R] + \frac{d}{dz} [v_1 \rho c_p (T(z) - T_R)] \Delta z\}$$

$$- k_1 C_A(z)(\pi R_i^2 \Delta z)(\Delta H) = 0$$

This equation can be simplified to:

$$v_1 \rho c_p \frac{dT(z)}{dz} - 2\pi R_i U_i [t(z) - T(z)] + \pi R_i^2 (\Delta H) k_1 C_A(z) = 0 \qquad (2\text{-}21)$$

The boundary condition is

$$T(0) = T_f \qquad (2\text{-}22)$$

The solution of this equation must be obtained by solving simultaneously with Eq. 2-19 for the concentration profile and the equation for the jacket-temperature profile, which is developed next.

Enthalpy Balance on Jacket Fluid

Noting that the jacket fluid is flowing in the negative direction of the independent variable, the terms in the heat balance for the jacket fluid are

(a) Sensible heat out by bulk flow:

$$v_2 \rho' \, c_p' \, [t(z) - t_R]$$

where the primes denote the jacket fluid and t_R is the reference temperature for calculating enthalpy.

(b) Sensible heat in by bulk flow:

$$v_2 \rho' c_p' [t(z) - t_R] + \frac{d}{dz} \{v_2 \rho' c_p' [t(z) - t_R]\} \Delta z$$

(c) Heat transferred to the reacting fluid:

$$(2\pi R_i \Delta z) \, U_i [t(z) - T(z)]$$

Combining these terms in a steady-state heat balance yields

$$v_2 \rho' c_p' [t(z) - t_R] + \frac{d}{dz} \{v_2 \rho' c_p' [t(z) - t_R]\} \Delta z$$

$$- v_2 \rho' \, c_p' [t(z) - t_R] - (2\pi R_i \Delta z) \, U_i [t(z) - T(z)] = 0$$

This equation can be simplified to

$$v_2 \rho' c_p' \cdot \frac{dt(z)}{dz} - 2\pi R_i U_i [t(z) - T(z)] = 0 \tag{2-23}$$

The boundary condition is

$$t(L) = t_f \tag{2-24}$$

To summarize, the equations and their boundary conditions are as follows:

$$\frac{dC_A(z)}{dz} + \frac{\pi R_i^2 k_1}{v_1} \, C_A(z) = 0 \tag{2-19}$$

$$v_1 \rho' c_p' \, \frac{dT(z)}{dz} - 2\pi R_i U_i [t(z) - T(z)] + \pi R_i^2 (\Delta H) k_1 \, C_A(z) = 0 \tag{2-21}$$

$$v_2 \rho' c_p' \, \frac{dt(z)}{dz} - 2\pi R_i U_i [t(z) - T(z)] = 0 \tag{2-23}$$

$$k_1 = k_0 e^{-E/RT(z)} \quad \text{(Arrhenius expression).}$$

$$C_A(0) = C_{Af} \qquad\qquad\qquad [2\text{-}20]$$

$$T(0) = T_f \qquad\qquad\qquad [2\text{-}22]$$

$$t(L) = t_f \qquad\qquad\qquad [2\text{-}24]$$

These equations could be readily solved either numerically or with an analog computer except for one problem. Two of the boundary conditions are at the inlet and one is at the outlet. For most numerical methods and the analog computer, all conditions at one end must be specified to obtain a solution. Consequently, the approach to the solution is to assume the outlet temperature of the jacket water and solve the problem, thus obtaining a value for the inlet jacket-water temperature. This value is compared to the desired value given by the boundary condition, a new value of the inlet temperature is assumed, and the calculations are repeated. This procedure is continued until the outlet temperature is found that gives the desired inlet temperature. Optimization procedures, such as the Golden Section technique, are helpful in obtaining convergence.

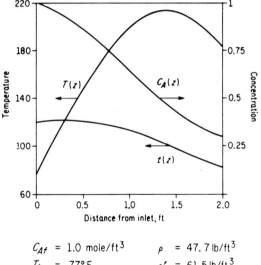

C_{Af} = 1.0 mole/ft^3	ρ = 47.7 lb/ft^3
T_f = 77°F	ρ' = 61.5 lb/ft^3
t_f = 82°F	c_p = 0.47 Btu/lb-°F
R_i = 0.565 ft	c_p' = 1.0 Btu/lb-°F
U_i = 100 Btu/hr-ft^2-°F	v_1 = 10 ft^3/hr
k_1 = 3000 exp(-4000/T)	v_2 = 20 ft^3/hr
per hr, T in °R	ΔH = -10,500 Btu/mole A

FIG. 2-8. Temperature and concentration profiles in a jacketed tubular reactor.

In Fig. 2-8 are shown the concentration and temperature profiles for the given set of boundary conditions and parameters. This solution was obtained using a digital simulation program.

2-6. A PROBLEM IN FREE CONVECTION [5]

Consider two vertical walls a distance $2b$ apart as shown in Fig. 2-9. The heated wall at $x = -b$ is maintained at temperature T_h, and the cooled wall at $x = +b$ is maintained at temperature T_c. A fluid with density ρ and viscosity μ is placed between the two walls. The viscosity of the fluid is assumed to be constant, but the density ρ depends upon the temperature as follows:

$$\rho = \rho_R[1 - \beta(T - T_R)] \tag{2-25}$$

where T_R is a reference temperature, ρ_R is the density at T_R, and β is the coeffi-

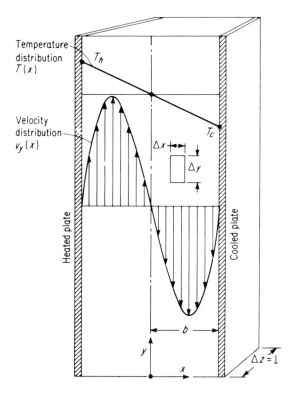

FIG. 2-9. Free-convection flow between two vertical plates at different temperatures. (Reprinted with permission from R.B. Bird, W.E. Stewart, and E.N. Lightfoot, *Transport Phenomena*, John Wiley & Sons, Inc., New York, 1960.)

cient of volumetric expansion.   At this point T_R need not be specified, but can be specified later in such a manner that the final solution is in its simplest form.

Because of the temperature difference between the plates, the fluid near the hot wall is warmer and consequently less dense than the fluid near the cold wall. Therefore the fluid near the hot wall rises, and the fluid near the cold wall descends.   Furthermore, it is assumed that the system is so constructed that the volumetric flow rate in the upward direction is the same as that in the downward direction.

Enthalpy Balance

If the walls are very long in the y-direction, at some distance from the ends the temperature of the fluid will not vary with height y; that is, temperature T is a function of x only.   To solve for the velocity profile it is necessary to first determine the temperature profile.   If a control volume of height Δy, as shown in Fig. 2-9, is examined, the input and output terms in a thermal energy balance are

(a) Input by conduction at x:

$$k\Delta y \frac{dT(x)}{dx}$$

(b) Input by bulk flow at y equals the output by bulk flow at $y + \Delta y$, so these two terms cancel.

(c) Output by conduction at $x + \Delta x$:

$$k\Delta y \frac{dT(x + \Delta x)}{dx} \cong k\Delta y \frac{dT(x)}{dx} + \frac{d}{dx}\left[k\Delta y \frac{dT(x)}{dx} \right] \Delta x$$

The equation is now formulated as follows:

$$k\Delta y \frac{dT(x)}{dx} - \left\{ k\Delta y \frac{dT(x)}{dx} + \frac{d}{dx}\left[k\Delta y \frac{dT(x)}{dx} \right] \Delta x \right\} = 0$$

This equation simplifies to

$$\frac{d^2 T(x)}{dx^2} = 0 \tag{2-26}$$

The boundary conditions are

$$T(-b) = T_h$$
$$T(b) = T_c$$

Integrating Eq. 2-26 twice yields

$$T(x) = C_1 x + C_2$$

Solving for C_1 and C_2 from the boundary conditions gives

$$T(x) = \frac{T_h + T_c}{2} - \left(\frac{T_h - T_c}{2}\right)\left(\frac{x}{b}\right)$$

Momentum Balance

The following terms appear in the momentum balance:

(a) Rate of momentum transfer in across the face at x:

$$\tau_{yx}(x)\Delta y$$

(b) Rate of momentum transfer out across the face at $x + \Delta x$:

$$\tau_{yx}(x)\Delta y + \frac{d}{dx}[\tau_{yx}(x)\Delta y]\Delta x$$

(c) Rate of momentum in by bulk flow:

$$\rho v_y(x)\Delta x$$

where $v_y(x)$ is taken as positive in the positive y-direction.

(d) Rate of momentum out by bulk flow:

$$\rho v_y(x)\Delta x$$

Since the temperature is not changing in the y-direction, the product $\rho v_y(x)$ does not change either.

(e) Force on face at y due to static pressure p:

$$p\Delta x$$

(f) Force (in negative y-direction) on face at $y + \Delta y$ due to static pressure:

$$-p\Delta x - \frac{d}{dy}(p\Delta x)\Delta y$$

(g) Gravitational force on fluid in control volume (in negative y-direction):

$$-\rho g\Delta x\Delta y$$

Substituting the above quantities into the steady-state momentum balance yields:

$$\tau_{yx}(x)\Delta y - \left\{\tau_{yx}(x)\Delta y + \frac{d}{dx}[\tau_{yx}(x)\Delta y]\Delta x\right\}$$

$$+ \rho v_y(x)\Delta x - \rho v_y(x)\Delta x + p\Delta x$$

$$- \left\{p\Delta x + \frac{d}{dy}[p\Delta x]\Delta y\right\} - \rho g\Delta x\Delta y = 0$$

This equation reduces to

$$-\frac{d\tau_{yx}(x)}{dx} = \frac{dp}{dy} + \rho g$$

Substituting Newton's law of viscosity for $\tau_{yvx}(x)$ and rearranging (g_c is omitted),

$$\mu \frac{d^2 v_y(x)}{dx^2} = \frac{dp}{dy} + \rho g$$

Although the above equation contains two independent variables, the use of partial derivatives instead of total derivatives is unnecessary because v_y is a function of x alone and p is a function of y alone.

Since the y-velocity is zero at both walls, the mathematical statements for the boundary conditions are

$$v_y(+ b) = 0$$
$$v_y(- b) = 0$$

The solution is begun by substituting the expression in Eq. 2-25 for ρ as a function of T, followed by substituting the solution to Eq. 2-26 for T as a function of x. The solution involves several substitutions and manipulations, the more significant being:

1. Making the restriction that the total volumetric flow in the upward direction equals the total volumetric flow in the downward direction, or mathematically,

$$\int_{-b}^{b} v_y(x)\, dx = 0$$

2. Noting that the form of the equation is simplified by choosing T_R in Eq. 2-25 equal to $(T_h + T_c)/2$.

The analytic solution is [5]

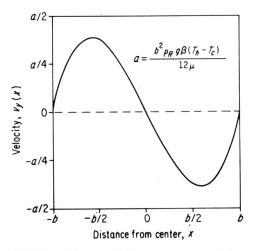

$$a = \frac{b^2 \rho_R\, g\beta (T_h - T_c)}{12\mu}$$

FIG. 2-10.   Velocity profile between the plates in Fig. 2-9.

$$v_1(x) = \frac{b^2 \rho_R g \beta (T_h - T_c)}{12\mu} \left[\left(\frac{x}{b}\right)^3 - \frac{x}{b} \right]$$

A graphical representation is shown in Fig. 2-10.

2-7. THE LAGRANGE POINT OF VIEW

In all previous examples the control volume has been of fixed size and fixed in space.   This is as would be seen by a stationary observer, and is called the *Euler point of view*.   This is not the only way to look at things, and indeed for many flow systems there is a better approach.   This is from the viewpoint of an observer moving with a differential *mass* (not volume) of the fluid, and is known as the *Lagrange point of view*.

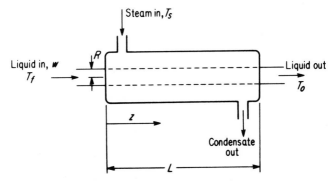

FIG. 2-11.   Steam-jacketed heat exchanger.

To illustrate the concepts, consider the steam-jacketed heat exchanger in Fig. 2-11, which is the same exchanger considered in Chapter 1.   However, in this case let the steam temperature be a function of time i.e., $T_s(t)$.   Now the temperature of the fluid in the pipe is a function of position and time, i.e., $T(z,t)$.   This function is defined for all time t and for z on the interval $|0,L|$.

Now consider the following three ways to view this problem.

1. *As an observer at point Z_0 in the exchanger*.   This observer sees only the temperature at point Z_0, but can monitor the variations of temperature with time at this point.   See the graphical representation in Fig. 2-12. Thus, he sees $T(Z_0, t)$, and the rate of change (derivative) that he observes

    is the rate of change with time, $\dfrac{\partial T(Z_0,t)}{\partial t}$

2. *From a "snapshot" taken at time t_1*.   From examination of the snapshot, the temperature at all points in the exchanger at time t_1 can be seen, as

is illustrated graphically in Fig. 2-12. Thus the observer sees $T(z,t_1)$, and the rate of change observed is $\dfrac{\partial T(z,t_1)}{\partial z}$.

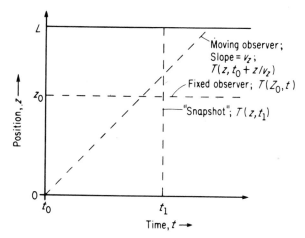

FIG. 2-12. Viewpoints of different observers.

3. As an observer riding on a small mass of fluid passing through the exchanger. If he enters at time t_0 and is moving with velocity v_z, the temperatures seen by the observer are those along the slanted line in Fig. 2-12, or $T(z,t_0 + z/v_z)$. The rate of change observed is the directional derivative along the path he takes, or more commonly called the *substantial derivative*. This manner of observing the system is that attributed to Lagrange. This derivative does have a physical meaning, and is commonly denoted as

$$\frac{D}{Dt}[T(z,t)]$$

It is equivalent to the directional derivative $\dfrac{\partial}{\partial s}[T(z,t)]$, where s is the direction of the line $\tau = t + \dfrac{z}{v_z}$. This relationship can be used to show that the substantial derivative is related to the partial derivatives as follows:

$$\frac{D}{Dt}[T(z,t)] = \frac{\partial T(z,t)}{\partial t} + v_z\frac{\partial T(z,t)}{\partial z} \qquad (2\text{-}27)$$

This definition is readily extended to three-dimensional systems, and in rectangular coordinates the substantial derivative of some quantity Q is defined as follows:

$$\frac{DQ}{Dt} = \frac{\partial Q}{\partial t} + v_x \frac{\partial Q}{\partial x} + v_y \frac{\partial Q}{\partial y} + v_z \frac{\partial Q}{\partial z}$$

In cylindrical coordinates:

$$\frac{DQ}{Dt} = \frac{\partial Q}{\partial t} + \frac{v_r}{r} \frac{\partial (rQ)}{\partial r} + \frac{v_\theta}{r} \frac{\partial Q}{\partial \theta} + v_z \frac{\partial Q}{\partial z}$$

In spherical coordinates:

$$\frac{DQ}{Dt} = \frac{\partial Q}{\partial t} + \frac{v_r}{r^2} \frac{\partial (r^2 Q)}{\partial r} + \frac{v_\theta}{r \sin \theta} \frac{\partial (Q \sin \theta)}{\partial \theta} + \frac{v_\phi}{r \sin \theta} \frac{\partial Q}{\partial \varphi}$$

As nothing has been said about the accumulation term up to this point, a few words are in order. The accumulation term is simply the rate of change of the quantity conserved within the system. Using the Euler point of view, this rate of change is the partial with respect to time; using the Lagrange point of view, it is the substantial derivative. If the quantity conserved is thermal energy or enthalpy, the accumulation term is simply the rate of change of the enthalpy of the material in the element. In all previous problems the accumulation term was zero due to the steady-state conditions. But when using the Lagrange point of view, the substantial derivative (D/Dt) must be used, which may not be zero at steady-state (the term $\partial/\partial z$ in Eq. 2-27 may be nonzero). Thus the accumulation term must be included in the following example.

Consider the single-pass steam heater considered in Sec. 1-7. Recall that when the Euler point of view was used, the system was defined as a section of the tube of length Δz at a distance z from the inlet. But when using the Lagrange point of view, the system is defined as an infinitesimal amount of fluid Δm. This can be related to Δz as follows:

$$\Delta m = \text{const} = \rho \cdot (\pi R^2) \cdot \Delta z$$

As for the Euler point of view, the differential need only be in the direction of change of the dependent variable. Note that Δm is constant, but Δz is not if either ρ or R vary.

Quantity Conserved: Heat.
Independent Variable: Distance z from inlet.
Dependent Variable: Temperature T.
Input Term: The only input of heat to the mass Δm is that transferred from the steam, which equals $U \cdot (2\pi R \Delta z) \cdot (T_s - T)$.
Output Term: There are no output terms.
Accumulation Term: The enthalpy of the mass Δm is given by

$$\Delta m \, c_p \, (T - T_R)$$

where T_R is the reference temperature. Thus the accumulation term is

$$\frac{D}{Dt}[\Delta m \cdot c_p (T - T_R)]$$

Formulation of the Equation:

$$\text{In} - \text{Out} = \text{Accumulation}$$

$$U \cdot (2\pi R \Delta z) \cdot (T_s - T) = \frac{D}{Dt}[\Delta m \cdot c_p (T - T_R)]$$

As the substantial derivative behaves like other derivatives as regards constants, this equation becomes

$$U \cdot (2\pi R \Delta z) \cdot (T_s - T) = \Delta m \cdot c_p \frac{DT}{Dt}$$

Substituting Eq. 2-23 for Δm and simplifying:

$$2U(T_s - T) = \rho c_p R \frac{DT}{Dt}$$

Substituting Eq. 2-27 for the substantial derivative and recalling that $\frac{\partial T}{\partial t}$ is zero due to steady-state conditions:

$$2U(T_s - T) = \rho c_p R v_z \frac{dT}{dz}$$

This is the same result as obtained in Sec. 1-7.

Next consider the laminar-flow problem discussed in Sec. 1-12 and illustrated in Fig. 1-17. Instead of developing equations for a fixed volume of fluid, the equations in this example are developed for a fixed mass of fluid Δm, which is related to Δr as follows:

$$\Delta m = \text{const} = 2\pi \rho r \Delta z \Delta r$$

where Δz is the length of the element in the axial direction.

Quantity Conserved: Momentum
Independent Variables: Radial distance r from center; distance z from inlet.
Dependent Variables: τ_{rz}, v_z
Viscous Terms: The viscous terms are analogous to those derived in Sec. 1-12 the input term being

$$2\pi \tau_{rz} \Delta z$$

For the output term:

$$2\pi r \tau_{rz} \Delta z + \frac{d}{dr}(2\pi r \tau_{rz} \Delta z)\Delta r$$

Bulk-Flow Terms: Since the equations are being developed for a fixed quantity of mass, there is no input or output of mass from the infinitesimal element of mass. Consequently there are no bulk-flow terms. Whereas these are present when using the Euler point of view, they are included in the substantial derivative where using the Lagrange point of view.

Pressure Forces: If the pressure at z is $P(z)$, the pressure force on the face at z is

$$(2\pi rg_c \Delta r)P(z)$$

Similarly, the pressure force on the face at $z = z + \Delta z$ is

$$-(2\pi rg_c \Delta r)P(z + \Delta z)$$

Gravitational Forces: The force exerted by gravity on this differential mass of fluid is

$$-2\pi r\rho g \Delta r \Delta z$$

Accumulation Term: When taking the Lagrange point of view, the rate of accumulation of momentum of the infinitesimal element of mass is:

$$\frac{D}{Dt}[v_z \Delta m]$$

Formulation of the Equation: Substituting the above terms into a momentum balance yields

$$2\pi r\tau_{rz}\Delta z - \left[2\pi r\tau_{rz}\Delta z + \frac{d}{dr}(2\pi r\tau_{rz}\Delta z)\Delta r \right]$$

$$+ 2\pi rg_c P(z)\Delta r - 2\pi rg_c P(z + \Delta z)\Delta r$$

$$- 2\pi r\rho g \Delta r \Delta z = \Delta m \frac{Dv_z}{Dt}$$

This equation reduces to

$$-\frac{1}{r}\frac{d}{dr}(r\tau_{rz}) - g_c \frac{d\,P(z)}{dz} - \rho g = \rho \frac{Dv_z}{Dt} \tag{2-28}$$

The substantial derivative term can be expanded as follows:

$$\frac{Dv_z}{Dt} = \frac{\partial v_z}{\partial t} + \frac{v_r}{r}\frac{\partial (rv_z)}{\partial r} + v_z \frac{\partial v_z}{\partial z}$$

Since the system is at steady-state, the term $\partial v_z/\partial t$ is zero. The second term is also zero because there is no velocity component in the r-direction ($v_r = 0$). In Sec. 1-12 it was concluded that v_z is a function of r only, and thus $\partial v_z/\partial z$ is also zero. Consequently, we conclude that Dv_z/Dt is zero, and Eq. 2-28 becomes

$$-\frac{1}{r}\frac{d}{dr}(r\tau_{rz}) - g_c\frac{dP(z)}{dz} - \rho g = 0$$

It can be readily shown that this equation is equivalent to the result derived in Sec. 1-12, using the Euler point of view.

2-10. SUMMARY

In this chapter the concepts used in Chapter 1 to formulate descriptive equations have been extended to include situations requiring simultaneous differential equations. Although nothing has been said about solution techniques, these will be examined more carefully in Chapters 5 and 6. Either the Lagrange or Euler points of view could be used in subsequent chapters, but the Euler point of view will be used primarily because it is generally simpler for engineers to visualize.

REFERENCES

1. D. Q. Kern, *Process Heat Transfer,* McGraw-Hill, New York, 1950.
2. F. Kreith, *Principles of Heat Transfer,* 2d ed., International Textbook, Scranton, Pa., 1965.
3. A. S. Foust, et al., *Principles of Unit Operations,* Wiley, New York, 1960.
4. W. L. McCabe and J. C. Smith, *Unit Operations of Chemical Engineering,* McGraw-Hill, New York, 1956.
5. R. B. Bird, W. E. Stewart, and E. N. Lightfoot, *Transport Phenomena,* Wiley, New York, 1960.
6. C. O. Bennett and J. E. Myers, *Momentum, Heat, and Mass Transfer,* McGraw-Hill, New York, 1962.
7. H. S. Mickley, T. K. Sherwood, and C. E. Reed, *Applied Mathematics in Chemical Engineering,* McGraw-Hill, New York, 1957.

PROBLEMS

2-1. Derive the differential equations and boundary conditions describing an exchanger similar to the one in Fig. 2-1 except that the shell-side nozzles are reversed. Assume the area A is fixed in this case.

2-2. Consider the wetted-wall column shown. Pure component B enters as a gas at the bottom of the column, and the liquid flowing down the sides is pure A. Component B is insoluble in A, but at the temperature of column operation the mole fraction vapor in equilibrium with liquid A is $x_A{}^*$ The thickness of the liquid film is small in comparison to the diameter D of the column. At the top of the column, the flow rate of liquid is l_0. If the mass transfer coefficient is k, determine the differential equations and boundary conditions that describe the total gas flow w, liquid flow l, and concentration x_A as a function of z. Assume isothermal operation. Neglect concentration gradients in the radial direction and diffusion in the direction of flow.

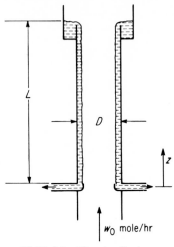

w_0 mole/hr

PROB. 2-2.   Wetted-wall column.

2-3. In the previous problem, assume the column does not operate isothermally. The inlet gas temperature is T_{g_0}, and the inlet liquid temperature is T_{l_0}. The specific heats of the liquid and gas do not vary with temperature or composition. At temperature T_R the latent heat of vaporization is λ_R. The convective heat-transfer coefficient between gas and liquid is h. Determine the equations and boundary conditions in addition to those derived for the previous problem required to describe the temperature profile in the column.

2-4. In direct-contact heat transfer, two immiscible liquids at different temperatures

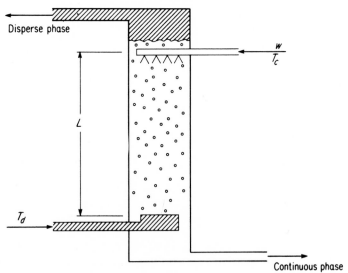

Disperse phase

w
T_c

L

T_d

Continuous phase

PROB. 2-4.   Direct-contact heat transfer.

are mixed directly in a countercurrent column such as shown. The number of bubbles per unit height of column is N, their upward velocity is v, and the average mass per bubble is m. The flow rate of the continuous phase is w. The inlet temperatures of the disperse and continuous phases are T_d and T_c, respectively. The heat-transfer coefficient between the phases is h, and all physical properties are constant. Neglect temperature variations over a cross section of the continuous phase, and assume each drop to be of uniform temperature. Determine the equations and boundary conditions necessary to describe the temperature profiles in each phase. Assume spherical bubbles.

2-5. Consider gas bubbles of component A rising through a liquid of constant temperature. The number of bubbles per foot of column height is N, and their upward velocity is v. When formed, an average bubble contains m_0 moles of gas at temperature T_i. The liquid (density $= \rho_L$) is pure component B at temperature T_f. Component A is insoluble in B, and the mole fraction of B in the vapor in equilibrium with liquid at T_L is x_{Bi}. The convective heat-transfer coefficient between the two phases is h, and the convective mass-transfer coefficient is k. The enthalpy of B vapor at T_L is H_v. Neglect temperature variations over a cross section of the column, and consider each bubble to be spherical and well-mixed. The gas in the bubble can be assumed to follow the ideal gas law. Determine the differential equations and boundary conditions that describe the temperature and concentration of a bubble as a function of z.

2-6. A gas probe for taking samples of high-temperature combustion products must often be cooled in order for the probe to have a reasonable usable lifetime. One arrangement for cooling the probe without contaminating the gas sample is to use an arrangement as shown. Let the flow rate of coolant be w, and the flow rate of gas be m. The gas temperature is T_G, and the inlet coolant temperature is T_i. The heat-transfer coefficient between coolant and outside gas is h_1, between coolant and coolant is h_2, and between coolant and the gas in the probe is h_3. Determine the equations and boundary conditions describing the temperature profiles in the probe. Neglect heat transfer at the end.

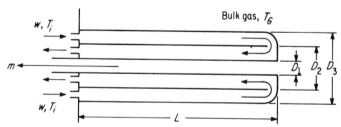

PROB. 2-6. Gas probe. (Reprinted with permission from Mickley, Sherwood, and Reed, *Applied Mathematics in Chemical Engineering.* McGraw-Hill, New York, 1957.)

2-7. The following reactions occur in a tubular reactor:

$$A + B \xrightarrow{k_1} C \qquad k_1, \text{ second-order} \qquad \Delta H_1$$

$$A + D \underset{k_2'}{\overset{k_2}{\rightleftarrows}} E \qquad \begin{array}{l} k_2, \text{ second-order} \\ k_2', \text{ first-order} \end{array} \qquad \Delta H_2$$

$$C \underset{k_3'}{\overset{k_3}{\rightleftarrows}} E \qquad \begin{array}{l} k_3, \text{ first-order} \\ k_3', \text{ first-order} \end{array} \qquad \Delta H_3$$

The tubular reactor of cross-sectional area A_c is insulated. The feed enters at velocity v, temperature T_0, and concentrations C_{A_0}, C_{B_0} and C_{D_0} (other components being absent). Determine the differential equations describing the concentration and temperature profiles in the reactor. The volume change on reaction can be neglected.

2-8. The reaction $2A + B \rightarrow C + D$ occurs in the gas phase, and is second-order with respect to A and first-order with respect to B. The heat of reaction is ΔH, and the heat capacity per mole of reactants and products is relatively constant. If an insulated tubular reactor is fed at a rate of n_{A_0} and n_{B_0} mole/hour of components A and B, respectively, at temperature T_0, determine the differential equations that describe the temperature and concentration profiles in the reactor. The reactor cross-sectional area is A_c. The reactants and products obey the ideal gas law, and the pressure in the reactor is constant at P.

2-9. Consider the liquid film adjacent to a liquid-gas interface as suggested by the film theory. The concentration of component A at the interface is x_{A_i}, and in the bulk liquid the concentration is x_{A_b}. The diffusivity of A through the liquid is D. As it diffuses through the film, component A is consumed by a first-order irreversible reaction $A \rightarrow B$ with rate constant k_1. Determine an expression for the concentration profile in the film, and the rate of absorption of component A from the gas.

2-10. Many decomposition reactions are carried out in tubes heated from the outside. Consider the general reaction

$$A \rightarrow B + C$$

An industrial example is the decomposition of ethylene to acetylene:

$$C_2H_4 \rightarrow C_2H_2 + H_2$$

Suppose the reactants enter a tube of radius R and length L at temperature T_0, pressure P_0, flow M_0 mole/hr, and mole fraction x_{A_i} of A, the remaining feed being inerts. Assume the reaction is first-order, irreversible, and obeys the Arrhenius expression for temperature. The heat of reaction is ΔH Btu/mole A, and the heat-transfer coefficient between the reacting mass and surrounding medium at T_R is U. The pressure gradient is given by the Fanning friction factor:

$$\frac{dP}{dz} = -\rho \frac{fV^2}{2R}$$

Determine the differential equations and boundary conditions describing the temperature, pressure, and concentration profiles down the tube.

2-11. Consider the gas-absorption column shown. A stream of inert gas and solute A enter the bottom of the column, and the solvent (which may contain some A) enters the top. A negligible amount of solvent leaves with the gas stream. The flow rate of inert gas through the column is G, and the flow rate of inert liquid is L. The convective mass-transfer coefficient based on the gas phase driving force $(y_A - y_A^*)$ is k_y, where y_A^* is the mole ratio of gas in equilibrium with liquid of mole ratio x_A. Let the interfacial area per unit volume be a, and let the tower cross-sectional area be A_c. Assuming isothermal operation, what equations are necessary to describe the variation of y_A with the height up the column z?

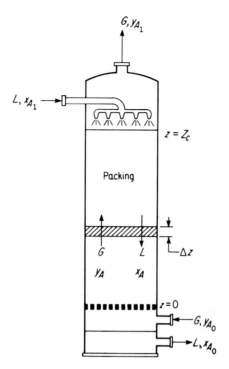

PROB. 2-11. Absorption tower. (Reprinted by permission from C.O. Bennett and J. E. Myers, *Momentum, Heat, and Mass Transfer,* McGraw-Hill, New York, 1963.)

2-12. For nonisothermal operation energy balances are also required to describe the absorption column in addition to the equations developed in the above problem. Assume the heat capacities c_{p_g} and c_{p_L} of the gas and liquid phases do not vary appreciably, and that the latent heat of vaporization of component A at temperature T_R is λ_R. If the gas and liquid streams enter at T_{G_0} and T_{L_0}, respectively, what equations and boundary conditions are required to describe the thermal gradients in the column? Let the convective heat-transfer coefficient between gas and liquid be h.

2-13. Consider an absorption column of cross-sectional area A_c in which the following reaction occurs in the liquid phase:

$$A + 2B \rightarrow C, \; r_A = -k_1 \, C_A^n \, C_B^m$$

where r_A = mole A/hr-ft² interfacial area. Component A enters with the gas stream, the concentration at the inlet being y_{A_0}. The inlet liquid is pure component B. Gas and liquid molar densities are independent of concentration. The mass-transfer rate is given by $k_y a(y_A - y_A^*)$ where y_A^* is the concentration of vapor in equilibrium with liquid of concentration x_A. The inlet liquid and gas molar flow rates are L_0 and G_0 respectively. Assuming isothermal operation, develop the necessary differential equations and boundary conditions to give the liquid and vapor concentration profiles in the column.

2-14. The flow in the kidney can be idealized to the flow between two horizontal flat plates made of a porous material. Fluid flows into the bulk flow on one side and out through the porous wall on the other side. In this case there is two-dimensional flow as shown. Selecting the y-axis as the direction of bulk flow, fluid is injected normal to the bulk flow with a velocity V at $x = -B$ and is removed at the same rate at $x = +B$. The pressure gradient in the y-direction is constant. Neglecting end effects, note that $v_y = v_y(z)$ only and $v_x = v_x(z)$ only. Derive an expression for the velocity profile between the plates.

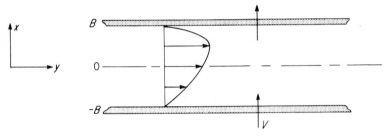

PROB. 2-14

2-15. The mean heat-transfer coefficient for condensation of a pure vapor on a vertical tube can be calculated from Nusselt's equation:

$$h_m = 0.943 \frac{k^3 \rho^2 g \lambda}{\mu L g_c \Delta T}$$

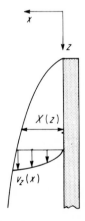

PROB. 2-15

where L is the tube length and ΔT is the temperature drop across the condensate film, which varies with height as shown. Nusselt's equation can be derived in the following manner:

(a) Make a momentum balance over a differential element in the x-direction at some distance from the top of the tube to relate the film thickness $X(z)$ to the flow rate of condensate $w(z)$ at that point. Neglect the curvature of the film, i.e., consider the condensation to occur on a flat plate of width πD_0. The answer should be

$$w(z) = \frac{\pi D_0 \rho^2 g X(z)^3}{3\mu g_c}$$

(b) Make a material balance over a differential element in the z-direction to obtain $X(z)$:

$$X(z) = \left[\frac{4k\mu g_c \Delta T}{\rho^2 g\lambda} z \right]^{1/4}$$

(c) Note that the heat-transfer coefficient $h(z)$ at point z equals $k/X(z)$. Calculate the average value of $h(z)$ over the tube to obtain Nusselt's equation.

2-16. As shown schematically in the figure, water flows from the infinitely large reservoir at pressure P_1 through a tube of cross-sectional area A filled with a material of porosity ϵ and permeability k. The pressure gradient is related to the superficial velocity by Darcy's law:

$$v_0 = -\frac{k}{\mu} \frac{dP}{dz}$$

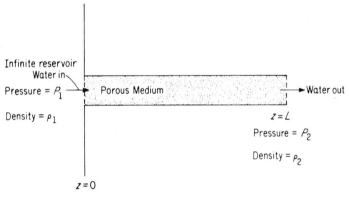

PROB. 2-16

The above situation is a simple representation to the flow of water through strata below the earth's surface. At the high pressures occurring in these strata, the variation of density with pressure is significant, being given by

$$\rho = \rho_0 \, e^{\kappa P}$$

where κ is the compressibility factor. Derive the following expression for the flow W (lb/hr) through the tube:

2-17. The term *transpiration cooling* applies to the case in which the outward flow of gas through a porous medium subjected to a temperature gradient is used to reduce the heat transfer. Typical applications range from storage vessels for cryogenic liquids to ablation materials for reentry vehicles. Consider the following case. The slab shown in

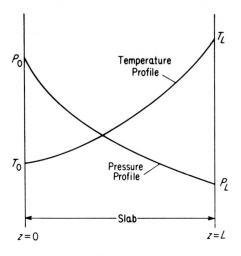

PROB. 2-17

the sketch is composed of a material of porosity ϵ and permeability k. The velocity v_z (based on actual flow area) through such a material is related to the pressure gradient dP/dz by Darcy's law:

$$v_z = -\frac{k}{\mu\epsilon}\frac{dP}{dz}$$

where w is based on actual area for flow. Assuming that the gas and solid are at the same temperature, the effective thermal conductivity k_c can be related to the thermal conductivities of the gas (k_g) and solid (k_s) as follows:

$$k_c = \epsilon k_g + (1-\epsilon)k_s$$

Assuming that the gas obeys the ideal gas law, develop the differential equations and boundary conditions that can be solved for the pressure and temperature profiles. Assume all physical properties are constant.

2-18. Suppose a compressible fluid flows down a pipe of length L which is heated on the exterior by steam at temperature T_s. The pressure gradient at point z is given by:

$$\frac{dP(z)}{dz} = K_1\frac{\rho[V(z)]^2}{D^s}$$

where K_1 = proportionality constant
   ρ = density
   $V(z)$ = velocity at point z
   D = diameter

Make the following assumptions:

1. Heat-transfer coefficient U between fluid and steam is a function, $f[V]$, of the fluid velocity.
2. The gas obeys the ideal gas law.
3. Let the molecular weight be M.
4. Let c_p be constant.

The pressure at inlet and outlet are P_1 and P_2, respectively. The gas enters at temperature T_1. Determine the equations and boundary conditions necessary to solve for the mass-flow rate w.

2-19. Osmotic pressure is caused by the tendency of a component to move from a region of high concentration to a region of low concentration across a semipermeable material. Suppose such a membrane separates water from a brine of concentration C lb salt/ft³. Suppose the osmotic pressure P_0 is some known function $f_0(C)$. The rate of transfer of water across the semipermeable material is given by:

$$w = K(P_0 - P_w)$$

where P_w = pressure differential across the wall
k = permeation coefficient, lb H_2O/hr-ft²-psi
w = rate of transfer of water into brine solution, lb H_2O/hr-ft²

Suppose the brine enters a horizontal tube of radius R at concentration C_i, total flow W_i, and pressure P_i above the pressure of the water outside the tube. The pressure drop in the pipe is given by the Fanning friction factor:

$$\frac{dP}{dx} = -\frac{\rho f V^2}{2R}$$

where f is a function of the Reynolds number. Neglecting variations of density ρ with concentration C, develop the equations that give the flow, concentration, and pressure profiles. Note that if the inlet pressure is sufficiently high, this process could be used to obtain water from brine.

2-20. Consider water flowing over a hot plate as shown. The width of the plate is W. The plate is maintained at temperature T_P, and the water enters at temperature T_0. The heat transfer coefficient between the plate and the water is h. Neglect sensible heat transfer between water and air. The mass transfer coefficient between the water and air is K_Y mole/hr-ft²-mole fraction. The mole fraction water vapor in the air is x_W. The total pressure is P, and the vapor pressure of the water is given by $P_. = f_1(T)$. The latent heat of vaporization is $\lambda = f_2(T)$. Develop expressions for the temperature T and flow rate w as functions of distance down the plate.

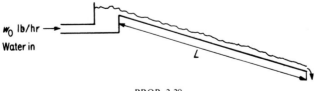

w_0 lb/hr → Water in

L

PROB. 2-20

2-21. Consider the condensation of a mixed vapor on a flat surface at temperature T_c as shown. The more volatile species will accumulate near the surface, thus forming a film over which both the concentration and temperature vary. Specifically, consider a binary vapor whose bulk temperature is T_b and mole fraction is x_{B_b}. The thermal conductivity of the gas mixture is independent of composition and temperature, and the mean heat capacities are the weighted average of the heat capacities of the individual components. The gas film thickness is δ, and the temperature variation across the liquid is small. The vapor pressure of B at T_c is P_B, and the total pressure is P_T. Determine differential equations, boundary conditions, and other necessary relationships which can be solved for the concentration profiles in the film, the rate of heat transfer across the liquid film, and the rate of condensation of each component.

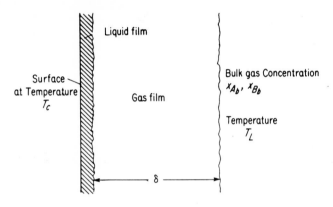

PROB. 2-21.   Condensation of mixed vapors.

2-22. Consider a multicomponent liquid being vaporized from a hot surface as illustrated.

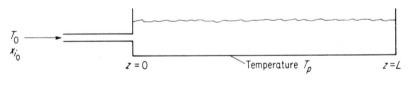

PROB. 2-22

The feed is maintained at such a rate that the level of liquid on the plate is constant. The heat-transfer coefficient between plate and liquid is h. The liquid enters at temperature T_0 (below bubble point) and mole fraction x_{i_0} of each of the n components. Assuming no backmixing in the liquid, determine the concentration and temperature profiles on the plate. The enthalpy H_v of the vapor can be calculated from the vapor temperature and composition. Neglect heat transfer by conduction in the liquid, and assume c_p is constant. The vapor composition is related to the liquid composition by the equilibrium constants, which are known functions of temperature for each component. For problems of this type, it is convenient (both mathematically and computationally) to assume that the rate of change of the rate of vaporization $m(z)$ per unit area is given by the following relationship:

$$\frac{dm(z)}{dz} = k|\Sigma y_i - 1|$$

If Σy_i exceeds 1 at some point, $m(z)$ is increased, which tends to deplete the more volatile component and cool the liquid, both of which tend to drive Σy_i toward 1, i.e., toward equilibrium. In order to effectively maintain equilibrium at all points, k should be large; however, exceedingly large values may cause numerical problems.

2-23. One type of industrial alkylation reactor is essentially a series of tanks arranged in one vessel as shown. Although a wide variety of reactions occur, the main reaction is that of isobutylene with isobutane in the presence of a sulfuric acid catalyst to form isooctane;

$$\underset{\text{isobutylene}}{CH_3 - \overset{\overset{\displaystyle CH_3}{|}}{CH} - CH_3} + \underset{\text{isobutylene}}{H_2C = \overset{\overset{\displaystyle CH_3}{|}}{C} - CH_2 - CH_2} \xrightarrow[H_2SO_4]{} \underset{\text{isooctane}}{H_3C - \overset{\overset{\displaystyle CH_3}{|}}{\underset{\underset{\displaystyle CH_3}{|}}{C}} - \overset{\displaystyle \cdot}{CH_2} - \overset{\overset{\displaystyle CH_3}{|}}{CH} - CH_3}$$

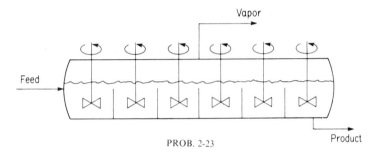

PROB. 2-23

This reaction is exothermic, the heat of reaction being removed by flashing propane. (The reaction temperature is in the neighborhood of 43°F.) As will be shown in Chapter 3, a sequence of mixed tanks in series approximates the plug-flow condition, and vice versa. Consider the model shown.

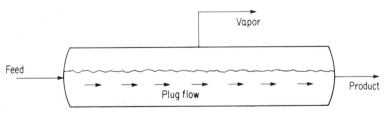

PROB. 2-23

Consider the following simplified case.

Feed:

M lb/hr acid of density ρ_A and heat capacity c_{p_1}

F mole/hr feed of molar density ρ, heat capacity c_p, and composition as follows:

No.	Component	Mole Fraction
1	Propane	x_{1f}
2	Isobutane	x_{2f}
3	Isobutylene	x_{3f}
4	Isooctane	0

Heat of reaction: ΔH Btu/mole

Length of reactor: L

Cross-sectional area for flow: A_c

Width of reactor at liquid surface: W

Equilibrium constants: $K_i(T)$

Feed temperature: T_f (the bubble point). If T_f is above bubble point, an equilibrium flash would be calculated at the entrance. If below bubble point, no vaporization occurs until the reacting mass reaches the bubble point.

Rate of formation of isooctane: $r_4 = f(\rho x_2, \rho x_3, \rho x_4, T)$

Vapor enthalpy: $H_V = f(x_1, x_2, x_3, x_4, T)$

Pressure in vapor space: P

Acid and hydrocarbon are at the same temperature

In reactors of this type, vapor-liquid equilibria is assumed. That is, the liquid temperature is such that the liquid is in equilibrium with the vapor leaving, or mathematically, $\Sigma y_i = 1$. This is not, however, convenient to handle mathematically in a direct fashion. Instead, the following mechanism is convenient mathematically and gives the same results:

1. Assume that $V(z)$, the rate of vaporization per square foot surface, is given by

$$\frac{dV(z)}{dz} = k_v [\Sigma y_i - 1]$$

That is, the sum of mole fractions is the driving force for increasing the rate of vaporization.

2. This equation inherently drives the system to equilibrium, as increasing the rate of vaporization $V(z)$ cools the liquid and depletes the more volatile components, thus driving Σy_i toward 1.0.

3. By making k_v very large, the system is always at equilibrium for all practical purposes.

4. Note that $V(z)$ has a real physical meaning—the rate of vaporization. It is the constant k_v that does not.

A similar approach to equilibrium problems is the subject of Prob. 3-16 at the end of Chapter 3.

Determine the equations and boundary conditions that can be solved for the temperature and concentration profiles in the reactor.

Lumped-Parameter Systems

As discussed in Chapter 1, systems can be classified into two broad categories: distributed-parameter systems (one-dimensional or higher) and lumped-parameter systems (zero-dimensional). The previous two chapters considered one-dimensional steady-state problems described by ordinary differential equations in which a space parameter is the independent variable. This chapter considers zero-dimensional or lumped-parameter systems which are described by algebraic equations if at steady-state or by ordinary differential equations (independent variable being time) if at unsteady state.

This chapter starts by discussing the accumulation term and system definition for this type of problem, and then three examples of true lumped-parameter systems are discussed. Since one important application of lumped-parameter systems is as approximations to distributed-parameter systems, the last half of this chapter considers this concept. The initial portion is devoted to predicting stage efficiencies, followed by a similar application for stage dynamics. Finally the classical relaxation method in heat transfer is introduced as an application of lumped-parameter models.

3-1 THE DEVELOPMENT OF THE ACCUMULATION TERM [1,2,3]

In the previous problems the accumulation term was set equal to zero in all cases except when using the Lagrange point of view. In this section we will take a closer look at the accumulation term.

As discussed in Chapter 1, the general balance for unsteady-state problems is

$$\begin{bmatrix} \text{Rate of input} \\ \text{of quantity conserved} \end{bmatrix} - \begin{bmatrix} \text{Rate of output} \\ \text{of quantity conserved} \end{bmatrix} = \begin{bmatrix} \text{Rate of accumulation} \\ \text{of quantity conserved} \end{bmatrix}$$

The first two terms in this expression are derived for lumped-parameter systems in much the same manner as for the distributed-parameter systems discussed in the previous chapters. However, the accumulation term deserves additional explanation.

The following simple rules for developing the accumulation term apply to both lumped-and distributed-parameter systems:

1. Derive the mathematical expression for the quantity conserved in the system as a function of the appropriate variables. For example, the quantity of fluid in a tank of cross-sectional area A is $\rho A L$, where ρ is the density and L is the liquid level. Any or all of these quantities could be dependent variables. For problems in which heat is the quantity conserved, the expression for the enthalpy of the material within the system or control volume is used in developing the accumulation term.

2. Take the derivative with respect to time of the expression for the quantity conserved in the system. This gives the accumulation term.

In general, formulation of unsteady-state problems is no more difficult than the formulation of steady-state problems, primarily because the expression for the accumulation term is developed so easily.

3-2. DEFINITION OF THE SYSTEM FOR A LUMPED-PARAMETER MODEL

For one-dimensional problems the control volume is defined as an element of infinitesimal thickness in the direction of the independent space variable. The reason for this is in order that the values of the dependent variables can be considered constant throughout the entity for which the equations are written. Since the dependent variables are functions of the independent variable, the only entity for which the above condition can hold is for an element of infinitesimal thickness in the direction of the independent variable.

For zero-dimensional problems (lumped-parameter systems) the dependent variables do not vary with position within the object under investigation. Therefore, the control volume and the subject of the investigation can be synonymous for zero-dimensional problems.

In this chapter these concepts are illustrated by applying them to several examples.

3-3. PERFECTLY MIXED TANK—THE CLASSICAL LUMPED-PARAMETER PROBLEM

Perhaps the classic unsteady-state problem is the change in concentration of a component in a perfectly mixed tank due to a change in the concentration of the feed stream. Consider the tank in Fig. 3-1 in which the contents can be considered perfectly mixed. In this case the concentration of the outlet stream is the same as the concentration of the fluid in the tank. Maintaining the concentration of the inlet stream constant at C_0 for a long period of time insures that the concentration of the fluid within the tank and the concentration of the fluid leaving will both be C_0. If for $t > 0$, the inlet concentration is some function $C_f(t)$ of time, the concentration $C_s(t)$ in the tank will also vary with time. Neglecting the variation of the density with concentration, the differential equation describing the variation of the concentration of the component in the tank with time will be developed. In order to visualize the problem, assume that the component in the tank is salt.

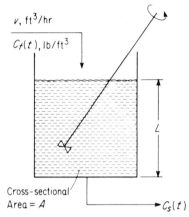

v, ft^3/hr

$C_f(t)$, lb/ft^3

L

Cross-sectional
Area $= A$

$C_s(t)$

FIG. 3-1.   Perfectly mixed tank.

Quantity Conserved: The units on the concentrations are assumed to be mass of component per unit volume, or specifically, lb/ft^3. Therefore, the pounds of salt can be conveniently used as the quantity conserved.

System: For a well-mixed tank, the concentration is uniform throughout the tank. Therefore, the equations can be written for the entire contents of the tank.

Independent Variable: Since this is an unsteady-state problem, the independent variable is the time t.

Dependent Variable: For time $t > 0$, the only variable is the concentration $C_s(t)$ of salt in the tank.

Input Term: For time $t > 0$, the rate of input of salt into the tank is $v C_f(t)$, where v is the volumetric flow rate (ft^3/hr) of the inlet stream.

Output Term: Since the concentration of the outlet stream is the same as the concentration of the fluid within the tank, the rate of output of salt is $v C_s(t)$.

Accumulation Term: At time t the amount of salt within the tank is

$$A L C_s(t)$$

where A is the cross-sectional area of the tank. The accumulation term is the time derivative of this expression, or

$$d[A L C_s(t)]/dt$$

Formulation of the Equation: Substituting these terms into a material balance yields

$$vC_f(t) - vC_s(t) = d[ALC_s(t)]/dt \qquad (3\text{-}1)$$

Since A and L are constant, this equation can be rearranged to

$$(AL/v)\,dC_s(t)/dt + C_s(t) = C_f(t) \qquad (3\text{-}2)$$

The quantity AL/v is often called the *time constant* (τ) of the system, and has the units of time. Note that the time constant is the amount of time required to fill the tank from empty to the level L if there is no outlet flow. Also, the function $C_f(t)$ is called the *forcing function,* because it is this function that "forces" $C_s(t)$ to vary with time.

Some textbooks and articles develop equations such as these in a slightly different fashion, and it seems appropriate to present this alternative approach. Consider the system to be defined as the empty tank. If it is filled at time t to level L with concentration $C_s(t)$, the input term is

$$ALC_s(t)$$

Now let the input and output streams flow as usual for time Δt. This gives the input term (salt entering over time Δt) as

$$vC_f(t)\Delta t$$

and the ouput term

$$vC_s(t)\Delta t$$

Now if the contents are removed to leave the empty tank, the output term is

$$ALC_s(t + \Delta t)$$

As the system is the empty tank, there can be no accumulation term.

Collecting the above terms into a material balance,

$$ALC_s(t) + vC_f(t)\Delta t - vC_s(t)\Delta t - ALC_s(t + \Delta t) = 0$$

Collecting terms and taking the limit as $\Delta t \to 0$ gives

$$\frac{AL}{v}\frac{dC_s(t)}{dt} + C_s(t) = C_f(t)$$

This is the same result as obtained earlier.

Initial Condition: Since Eq. 3-2 is a first-order differential equation, only one initial condition is required. The concentration of salt in the tank at time $t = 0$ is C_0, or expressed mathematically as

$$C_s(0) = C_0 \qquad (3\text{-}3)$$

Solution: Before a solution is attempted, the forcing function $C_f(t)$ must be specified. For this case, let $C_f(t)$ be a *step function* defined as follows:

$$C_f(t) = \begin{cases} C_0 & t \leq 0 \\ C_1 & t > 0 \end{cases} \tag{3-4}$$

The solution with the above boundary condition is

$$\frac{C_s(t) - C_0}{C_1 - C_0} = 1 - e^{-tv/AL} \tag{3-5}$$

and is plotted in Fig. 3-2. This response is characteristic of systems of this type, frequently called *first-order lags*. Defining a dimensionless time variable θ equal to the time divided by the time constant, or

$$\theta = t/\tau = tv/AL$$

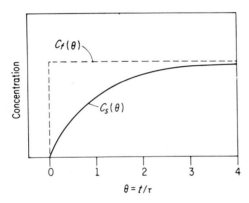

FIG. 3-2. Response of the tank in Fig. 3-1 to a step change in inlet salt concentration from C_0 to C_1.

gives the solution

$$\frac{C_s(\theta) - C_0}{C_1 - C_0} = 1 - e^{-\theta}$$

Note that the numerical value of the time constant does not affect the shape of the response, and only acts as a scale factor for the independent variable (time).

3-4. THE BATCH CHEMICAL REACTOR

In the batch reactor shown in Fig. 3-3 the reaction vessel is initially charged with the reactants, which are then agitated for a certain period of time during which they react. Upon completion of the reaction, the components are discharged. Since the composition within the reactor is varying with time as the

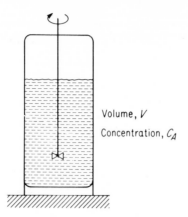

Volume, V

Concentration, C_A

FIG. 3-3.   Batch reactor.

reaction proceeds, this is an unsteady-state operation.   However, at a given instant of time the composition is uniform throughout the reactor.   Although this is an idealized example of a commercial batch reactor, many actual reactors approach these conditions very closely.

To illustrate the techniques for a batch reactor, assume that the reactor is initially filled to volume V with material of concentration C_{A_0} of component A. This component reacts to form component B via the following reaction:

$$A \rightarrow B \qquad (3\text{-}6)$$

The reaction-rate constant is k, and the reaction is nth order with respect to A. There is no volume change on reaction, and it is also assumed that the reactor is operated isothermally.

In developing the equation that describes the variation of C_A with time, the equation will be written in terms of r_A, and the rate expression for r_A substituted subsequently.   Since the reactor is assumed to be homogeneous, the system is defined as the volume V of reacting mass, and a molar balance is made for component A.   The rate of input of component A is solely due to the rate of appearance of A by reaction, which is $r_A V$.   Recall that the fact that component A is actually disappearing by reaction will be accounted for by the negative sign in the expression for r_A.   Since there are no output terms, the only remaining term is the accumulation term, which is

$$\frac{d}{dt}\left[VC_A(t) \right]$$

Substituting these terms into a material balance yields

$$r_A V = \frac{d}{dt}\left[VC_A(t) \right] \qquad (3\text{-}7)$$

Since V is constant, this equation reduces to

$$r_A = \frac{dC_A(t)}{dt} \tag{3-8}$$

Since A is a reactant, the expression for the rate of reaction is

$$r_A = -kC_A(t)^n \tag{3-9}$$

Substituting into Eq. 3-9 gives the first-order differential equation

$$-kC_A(t)^n = \frac{dC_A(t)}{dt}$$

Using the initial condition that the concentration at time zero is C_{A_0}, this equation can be separated and integrated to yield

$$\frac{C_A(t)}{C_{A_0}} = \left[(n-1)C_{A_0}^{\,n-1} kt + 1 \right]^{-\frac{1}{n-1}}, \quad n \neq 1 \tag{3-10}$$

or

$$\frac{C_A(t)}{C_{A_0}} = \exp(-kt), \quad n = 1 \tag{3-11}$$

These solutions are equivalent to the solutions obtained for the tubular reactor in Chapter 1.

In the study of chemical reactor design, the three classical types of reactors studied are:

1. *Tubular-flow reactor*, a continuous-flow reactor typically operated under steady-state conditions.   This reactor was studied in Sec. 1-8.
2. *Batch reactor*, typically operated under cyclic, unsteady-state conditions. This reactor was covered previously in this section.
3. *Backmix reactor*, a continuous-flow reactor typically operated under steady-state conditions.

A schematic drawing of the backmix reactor is shown in Fig. 3-4.   The contents

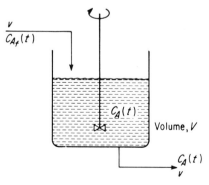

FIG. 3-4.   Backmix reactor.

of the reàctor are well-mixed, which is similar to the batch reactor. However, reactañts are added and products withdrawn on a continuous basis. Since the development of equations for this reactor is analogous to developing equations for the batch reactor, it will not be treated in detail at this point. The exercices at the end of the chapter contain several examples of such a reactor.

3-5. A LUMPED-PARAMETER SYSTEM REQUIRING SIMULTANEOUS DIFFERENTIAL EQUATIONS

In Fig. 3-5 is shown a well-agitated tank of diameter D surrounded by steam coils. The tank is open to the atmosphere, and water enters at the top and discharges at atmospheric pressure through a short line with a valve. The valve may

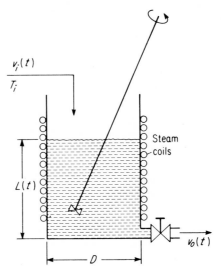

FIG. 3-5.   Variable-volume water heater.

be approximately represented by the orifice equation, and the volumetric flow rate $v_0(t)$ ft³/sec through the valve is related to the pressure drop across it by the following equation:

$$v_0(t) = C_v A_v \sqrt{\frac{2g_c[-\Delta p(t)]}{\rho}}$$

where C_v = valve coefficient

A_v = valve opening, ft²

$-\Delta p(t)$ = pressure drop across valve, lb$_f$/ft²

g_c = gravitational constant, ft-lb$_m$/lb$_f$-sec²

ρ = fluid density, lb$_m$/ft³

The water enters the top of the tank at temperature T_i, and for $t > 0$ its volumetric flow rate is represented by $v_i(t)$, a function of time. At $t = 0$, the temperature of the liquid in the tank is T_0 and the liquid level in the tank is L_0.

As the liquid level in the tank changes, the area for heat transfer also changes, but the temperature T_s of the steam and the heat-transfer coefficient h between the tank wall and the water are constant. Furthermore, the variation in the physical properties of the water in the tank will be neglected.

For this problem both an energy and a mass balance must be made. The mass balance will be made first, followed by the heat balance. In both cases the system will be defined as the contents of the tank. The independent variable is time t, and the dependent variables are the liquid level $L(t)$ and temperature $T(t)$ of the water in the tank.

Since the tank and its exit stream are both at atmospheric pressure, the pressure drop across the valve equals the pressure exerted on the bottom of the tank by the weight of the water. If the local acceleration due to gravity is g, this pressure is

$$- \Delta p(t) = \frac{g \rho L(t)}{g_c}$$

Substituting into Eq. 3-12 for the flow rate through the valve, the outlet flow $v_0(t)$ is related to the liquid level in the tank as follows:

$$v_0(t) = C_v A_v \sqrt{2gL(t)} \tag{3-13}$$

At time t the mass of fluid in the tank is $\pi D^2 \rho L(t)/4$. Since the accumulation of fluid in the tank equals the rate of change of the mass of fluid in the tank with time, the accumulation term for the balance is

$$\frac{d}{dt} \left[\frac{\pi D^2 \rho L(t)}{4} \right]$$

Making the material balance,

$$\rho v_i(t) - \rho C_v A_v \sqrt{2gL(t)} = \frac{d}{dt} \left[\frac{\pi D^2 \rho L(t)}{4} \right] \tag{3-14}$$

Since ρ and D are constant,

$$\frac{\pi D^2}{4 C_v A_v \sqrt{2g}} \frac{dL(t)}{dt} + \sqrt{L(t)} = \frac{v_i(t)}{C_v A_v \sqrt{2g}} \tag{3-15}$$

Note the nonlinear nature of this equation introduced by the $\sqrt{L(t)}$ term. In numerical or analog approaches to the solution, this usually causes little difficulty. On the other hand, it practically dispels any possibility of an analytic solution. If some error is acceptable, the mathematical convenience of a linear

equation can be obtained by linearizing the $\sqrt{L(t)}$ term with a Taylor series truncated after two terms:

$$f(x) = f(x_b) + \frac{df(x_b)}{dx} (x - x_b)$$

Letting x be L and $f(x)$ be $\sqrt{L}$, this becomes

$$\sqrt{L} = \sqrt{L_b} + \frac{1}{2\sqrt{L_b}} (L - L_b)$$

$$= \frac{L + L_b}{2\sqrt{L_b}}$$

where L_b is the liquid level about which the equation is linearized.

Substituting into Eq. 3-15 yields the linear equation

$$\frac{\pi D^2}{2C_v A_v \sqrt{2g/L_b}} \frac{dL(t)}{dt} + L(t) = \frac{2v_i(t)}{C_v A_v \sqrt{2g/L_b}} - L_b$$

Whereas the "time constant" for nonlinear equations such as Eq. 3-15 has little
meaning, it does for their linear counterparts.   For example, the time constant of
the above equation is $\dfrac{\pi D^2}{2C_v A_v \sqrt{2g/L_b}}$, and the solution for a step change in $v_i(t)$
would be a first-order lag as in Fig. 3-2.

The boundary condition for either of these equations is

$$L(0) = L_0 \tag{3-16}$$

Figure 3-6 shows the variation in $L(t)$ for a typical forcing function $v(t)$ as
computed from Eq. 3-15.

Since the variation of the liquid level in the tank also causes the area for heat
transfer to vary, the temperature fluctuates due to changes in the inlet flow $v_i(t)$.
In this case the area for heat transfer is related to the liquid level and tank diameter as follows:

$$A_t = \pi D L(t) \tag{3-17}$$

If T_R is defined as the reference temperature, the inlet and outlet terms in the heat
balance are as follows:

Enthalpy of Inlet Stream:

$$\rho c_p (T_i - T_R) v_i(t)$$

Enthalpy of Outlet Stream:

$$\rho c_p [T(t) - T_R] v_o(t) = \rho c_p C_v A_v [T(t) - T_R] \sqrt{2gL(t)}$$

Rate of Heat Input from the Coils:

$$h[\pi D L(t)] [T_s - T(t)]$$

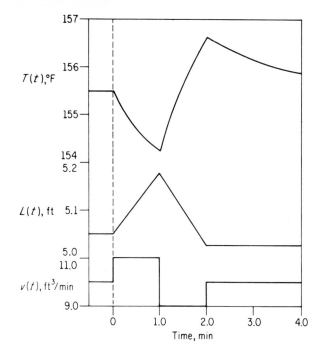

FIG. 3-6.   Response of hot-water tank to changes in input flow.

Accumulation Term:

$$\frac{d}{dt}\{\rho c_p [\pi D^2 L(t)/4][T(t) - T_R]\}$$

From these terms, the equation is formulated as follows:

$$\rho c_p(T_i - T_R)v_i(t) - \rho c_p C_v A_v [T(t) - T_R]\sqrt{2gL(t)}$$

$$+ h[\pi DL(t)][T_s - T(t)] = \frac{d}{dt}\{\rho c_p [\pi D^2 L(t)/4][T(t) - T_R]\} \qquad (3\text{-}18)$$

Multiplying Eq. 3-14 by $[T(t) - T_R]$ and subtracting from Eq. 3-18 yields

$$\frac{dT(t)}{dt} = \frac{4[T_i - T(t)]v_i(t)}{\pi D^2 L(t)} + \frac{4h}{D\rho c_p}[T_s - T(t)] \qquad (3\text{-}19)$$

Note that this equation is also nonlinear.   In order to linearize, the term $\dfrac{[T_i - T(t)]v_i(t)}{L(t)}$ must be linearized.   Again using the Taylor series and truncating after the linear terms;

$$f(x,y,z) = f_b + \frac{\partial f_b}{\partial x}(x - x_b) + \frac{\partial f_b}{\partial y}(y - y_b) + \frac{\partial f_b}{\partial z}(z - z_b)$$

where f_b is $f(x_b, y_b, z_b)$.   Applying to $\dfrac{[T_i - T(t)]v_i(t)}{L(t)}$ gives

$$\frac{[T_i - T(t)]v_i(t)}{L(t)} = \frac{(T_i - T_b)v_b}{L_b} - \frac{(T_i - T_b)v_b}{L_b^2}[L(t) - L_b]$$

$$- \frac{v_b}{L_b}[T(t) - T_b] + \frac{T_i - T_b}{L_b}[v_i(t) - v_b]$$

$$= \frac{(T_i - T_b)v_b}{L_b}\left[\frac{v_i(t)}{v_b} - \frac{L(t)}{L_b} - \frac{T(t) - T_b}{T_i - T_b}\right]$$

Substituting into Eq. 3-19 now yields a linear differential equation.
    The boundary condition for either equation is

$$T(0) = T_0 \tag{3-20}$$

Solving Eq. 3-19 gives the solution for $T(t)$ in Fig. 3-6 along with $L(t)$ and $v_i(t)$.

3-6.   LUMPED-PARAMETER MODEL OF A COLUMN TRAY

    In the previous two chapters one-dimensional steady-state problems were examined in which the variables were continuous functions of the position in the system.   For such systems the differential equations were developed by examining an element of infinitesimal thickness in the direction of change.   Although the representation so obtained is exact, an approximate representation can be obtained by modifying this procedure slightly.   Instead of examining an element of infinitesimal size, the system is divided into a series of small but finite elements. Each element is then examined individually, and the descriptive equations are written assuming that the values of the dependent variables are constant over the element.   In essence the result is to consider the parameters in the element as lumped at the center, in contrast to the more exact approach of considering the parameters distributed throughout the element.   Hence, by dividing the system into a finite number of elements, a lumped-parameter model of a distributed parameter system is obtained.
    By this procedure, an approximation to the distributed-parameter system is made as accurate as desired by choosing the size of the elements to be very small. One of the advantages of the lumped-parameter model is that the equations describing it are often easier to solve than the equations for the distributed-parameter system, especially if the system is of a practical nature.   In reality, this same result is obtained when finite differences are used to numerically solve differential equations.
    The major usefulness of lumped-parameter models is for systems that are so complex that the equations for the distributed-parameter model cannot be

accurately derived. The example to be considered shortly is that of a tray in a distillation column. In industrial columns the heat, mass, and momentum-transfer
aspects are so involved that a rigorous mathematical description is impossible at
the present state of the art. The liquid on the tray is in many respects really a
froth, and the level of this froth rises and falls with violent fluctuations. Above
the froth is a mist, some of which enters the tray above. The complexity of the
problem makes the prediction of such properties as the tray efficiency very
difficult, since the degree of mixing on the tray would have a large influence on
its value.

In attacking such a problem, the usual approach is to first examine the
extremes of operation. In Chapter 1 we examined the case of plug flow across
such a plate, and we concluded that the overall tray efficiency E_{MV} is related to
the point efficiency E_{OG} as follows:

$$E_{MV} = \frac{1}{\lambda} \left[e^{\lambda E_{OG}} - 1 \right]$$

Another extreme is that of perfect mixing on the tray, in which case the composition of the liquid would be the same at every point on the tray. For this case,
the overall tray efficiency and the point efficiency are synonymous, i.e.,

$$E_{MV} = E_{OG} \tag{3-21}$$

However, the true situation is actually intermediate between these two extremes.
In the A.I.Ch.E. method for predicting the tray efficiency, the degree of mixing
is taken into account by a parameter called the Peclet number[5]. A similar
result could have been obtained by using a lumped-parameter model.

Actually, the efficiency of one lumped-parameter model—that in which the
tray was taken to be one well-mixed pool—was discussed above. Next, the
efficiency will be determined for a lumped-parameter model in which the tray
is considered as two well-mixed pools in series as shown in Fig. 3-7. The
nomenclature for this example is very similar to that used in Sec. 1-7. The liquid
in the first pool is of concentration x_1, and the vapor leaving this pool is of
concentration y_1. Similarly, subscript 2, i.e., x_2 and y_2, is used for the second
pool. Since the composition of the liquid leaving the tray is the same as that of
the liquid in the pool, x_n equals x_2. However, the average composition y_n of the
vapor leaving tray n is the average of y_1 and y_2, assuming the two pools to be of
equal volume.

As in Sec. 1-7, the liquid flow rate down the column is L, and the vapor rate
up the column is G, both in mole/hr. The composition of liquid entering the tray
is x_{n-1}, and the composition of the vapor entering the tray is y_{n+1}. The equilibrium relationship used is also the same as in Sec. 1-7, namely

$$y^* = mx + b$$

The reader is also referred to Chapter 1 for a brief discussion of point efficiency

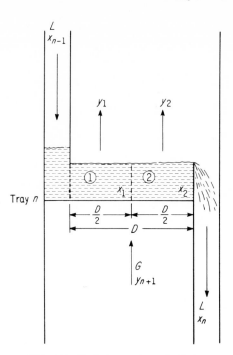

FIG. 3-7. Typical tray in a distillation
column.

and overall tray efficiency. Basically the same assumptions are made as in
Sec. 1-7, namely, equimolal overflow, isothermal operation, no concentration
gradients in the vertical direction, etc.

For the first pool, a steady-state material balance for the light component
yields

$$Lx_{n-1} + \frac{G}{2}y_{n+1} - Lx_1 - \frac{G}{2}y_1 = 0 \tag{3-22}$$

For the second pool,

$$Lx_1 + \frac{G}{2}y_{n+1} - Lx_2 - \frac{G}{2}y_2 = 0 \tag{3-23}$$

Adding these two equations yields (replacing x_2 by x_n)

$$L(x_{n-1} - x_n) = \frac{G}{2}\left[(y_1 - y_{n-1}) + (y_2 - y_{n+1})\right] \tag{3-24}$$

Recalling that y_n is the average of y_1 and y_2, this equation becomes

$$L(x_{n-1} - x_n) = G(y_n - y_{n+1}) \tag{3-25}$$

The same result could be obtained from a material balance over the entire tray.

Substituting the definition of E_{OG} and the equilibrium relationship into Eq. 3-24 yields

$$y_n - y_{n+1} = \frac{E_{OG}}{2} [m(x_1 + x_n) + 2b - 2y_{n+1}] \qquad (3-26)$$

This equation relates the change in vapor composition across the tray to the composition of the entering vapor, the composition of the liquid in the pools, the point efficiency, and the equilibrium relationship. Eliminating x_1 with Eq. 3-23, inserting $\lambda = mG/L$, and using the definition of E_{MV} yields

$$E_{MV} = E_{OG} + \frac{\lambda E^2_{OG}}{4} \qquad (3-27)$$

Recall that $E_{MV} = E_{OG}$ for a well-mixed tray. Consequently, the efficiency of a tray containing two well-mixed pools is higher than for a tray on which the entire liquid is well mixed. The other extreme condition is a tray on which the liquid flows across without any backmixing. The efficiency of this tray was derived in Sec. 1-7 to be

$$E_{MV} = \frac{1}{\lambda} \left[e^{\lambda E_{OG}} - 1 \right]$$

Expanding $e^{\lambda E_{OG}}$ in a power series yields

$$E_{MV} = E_{OG} + \frac{\lambda E^2_{OG}}{2} + \frac{\lambda^2 E^3_{OG}}{6} + \cdots \qquad (3-28)$$

Note that this expression gives a higher value for E_{MV} than the expression in Eq. 3-27 for a tray with two well-mixed pools.

For a tray with three-mixed pools the expression for the efficiency is

$$E_{MV} = E_{OG} + \frac{\lambda E^2_{OG}}{3} + \frac{\lambda^2 E^3_{OG}}{9} \qquad (3-29)$$

Note that the efficiency of this tray is higher than that for the tray with only two pools, but is still less than the efficiency for the plug-flow tray.

TABLE 3-1

Efficiencies of Various Lumped-Parameter Tray Models

$E_{OG} = 0.6, \quad \lambda = 1.44$

Number of Pools	Tray Efficiency	Number of Pools	Tray Efficiency
1	0.60000	10	0.89606
2	0.72960	15	0.91420
3	0.78939	20	0.92360
4	0.82391	30	0.93323
5	0.84639	40	0.93813
6	0.86222	50	0.94110
7	0.87395	75	0.94510
8	0.88301	100	0.94712
9	0.89020		

In Table 3-1 the efficiencies are given for lumped-parameter models with various numbers of well-mixed pools in series to approximate the flow patterns on the tray, and the concentration profiles over the tray are shown in Fig. 3-8 for

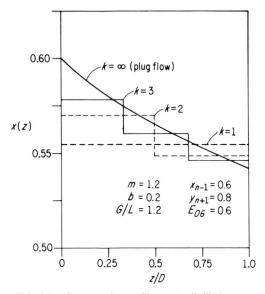

FIG. 3-8. Concentration profiles across distillation tray for various models.

three of these lumped-parameter models. Note that as the number of pools increases the concentration profile more closely approximates that for the distributed-parameter model.

3-7.   UNSTEADY-STATE OPERATION OF THE TRAY

Although the previous discussion was directed only to steady-state operation, an analogous approach can be used for unsteady-state operation [6]. In this section the lumped-parameter model will be developed for the unsteady-state operation of the distillation tray considered in the previous section. To simplify the discussion, assume that the vapor flowing through the plate is a noncondensable gas, as is the case in many laboratory experiments for investigating tray hydraulics. If a step change is made in x_{n-1} at time zero, the subsequent variation in x_n depends upon the tray hydraulics. Although this is also a distributed-parameter system, it is too complex to be adequately analyzed directly. Hence, a lumped-parameter model is again considered for the system.

As in the previous section, two limiting cases can be defined. First, the liquid may flow across the tray with little or no backmixing (plug flow), in which case the variation of $x_n(t)$ is the delayed step change shown in Fig. 3-9.

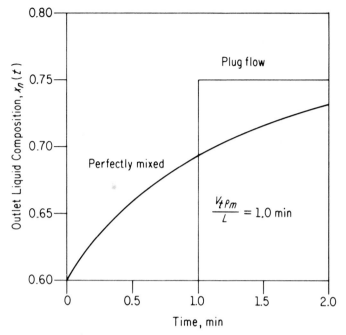

FIG. 3-9. Comparison of the responses of the plug-flow and the perfectly fixed models to a step change in inlet concentration from 0.60 to 0.75.

On the other hand, the liquid on the tray may be considered to be perfectly mixed. For this case, an unsteady-state material balance can be made for the liquid on the tray. Making a balance for the light component in the liquid on the tray, the equation for unsteady-state operation is derived as follows:

$$\text{In} \quad - \quad \text{Out} \quad = \text{Accumulation}$$

$$Lx_{n-1}(t) - Lx_n(t) = \frac{d}{dt}\left[V_t \rho_m x_n(t) \right] \tag{3-30}$$

where V_t is the volume of liquid on the tray and ρ_m is its molar density. As no condensable components enter or leave with the gas stream, G does not appear. Furthermore, the concentration of the liquid leaving the tray is the same as that of the liquid on the tray, since it is assumed to be well mixed.

If it is assumed that the volume of liquid V_t on the tray and its molar density ρ_m are constant, Eq. 3-30 reduces to

$$\frac{V_t \rho_m}{L} \frac{dx_n(t)}{dt} + x_n(t) = x_{n-1}(t) \tag{3-31}$$

This is a first-order differential equation whose time constant τ is $\dfrac{V_t \rho_m}{L}$ and in which $x_{n-1}(t)$ is the forcing function.

The solution to this equation for the step change in x_{n-1} is also shown in Fig. 3-9. Of course, the shape of the response is that of a first-order lag.

To account for intermediate cases, the liquid on the tray may be divided into two or more pools, as was done in the previous section. Assuming that the total volume of liquid on the tray remains constant, the volume of liquid per pool decreases as the total number of pools increases. In the examples in this section the tray is divided into pools of equal volume, although this is not at all necessary.

If the tray is divided into two pools as in Fig. 3-10, the differential equations describing each pool are as follows:

Pool 1
$$L x_{n-1}(t) - L x_1(t) = \frac{d}{dt}\left[\frac{V_t}{2}\rho_m x_1(t)\right]$$

or

$$\frac{V_t \rho_m}{2L}\frac{dx_1(t)}{dt} + x_1(t) = x_{n-1}(t) \tag{3-32}$$

Pool 2
$$L x_1(t) - L x_n(t) = \frac{d}{dt}\left[\frac{V_t}{2}\rho_m x_n(t)\right]$$

or

$$\frac{V_t \rho_m}{2L}\frac{dx_n(t)}{dt} + x_n(t) = x_1(t) \tag{3-33}$$

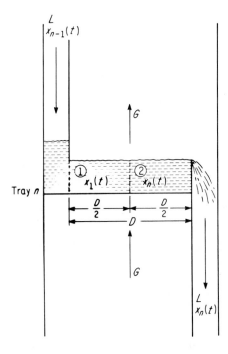

FIG. 3-10. Tray with two "pools" to approximate transient behavior.

The model is composed of two first-order lags in series, each with a time constant equal $\dfrac{V_t \rho_m}{2L}$

Equations 3-32 and 3-33 can be combined by eliminating x_t to yield

$$\left(\frac{V_t \rho_m}{2L}\right)^2 \frac{d^2 x_n(t)}{dt^2} + \left(\frac{V_t \rho_m}{L}\right)\frac{dx_n(t)}{dt} + x_n(t) = x_{n-1}(t) \qquad (3\text{-}34)$$

Note that this equation is second-order, whereas Eq. 3-31 for a tray with one pool is first-order. Since the system is at steady state at time zero, the boundary conditions are

$$\frac{dx_n(0)}{dt} = 0 \qquad (3\text{-}35)$$

$$x_n(0) = x_{n-1}(0) \qquad (3\text{-}36)$$

The last boundary condition is obtained by noting that the inlet liquid composition is the same as that of the outlet liquid at steady-state operation. The solution for this equation is given in Fig. 3-11 (curve corresponding to $k = 2$). At time zero, the derivative of the response is zero, which is typical of step responses of second- and higher-order lags.

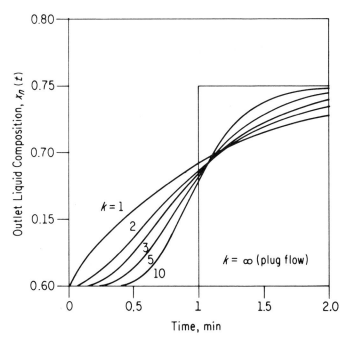

FIG. 3-11. Effect of the number of assumed pools on the calculated transient response to a step change from 0.60 to 0.75 ($V_t \rho_m / L = 1.0$ min).

In general, the tray can be divided into as many pools as desired, and in this section the general equation is derived for the ith pool on a plate divided into a total of k pools. The nomenclature and a sketch of the system are shown in Fig. 3-12. In this case the volume of liquid in the pool is V_t/k. The composition

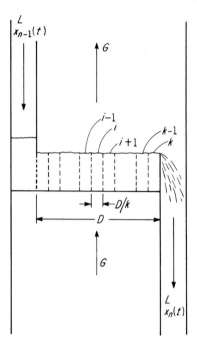

FIG. 3-12.   Tray divided into k pools.

of liquid entering the ith pool is $x_{i-1}(t)$, and the composition of the liquid in the pool is $x_i(t)$. Using this nomenclature, the equation for the ith pool is

$$Lx_{i-1}(t) - Lx_i(t) = \frac{d}{dt}\left[\frac{V_t\rho_m}{k} x_i(t)\right]$$

or

$$\frac{V_t\rho_m}{kL}\frac{dx_i(t)}{dt} + x_i(t) = x_{i-1}(t) \tag{3-37}$$

In this case there will be a total of k such differential equations, i.e., one for each of the k pools on the tray. If these k equations are combined into a single differential equation, its order will be k. Therefore, the solution can be obtained by solving one differential equation of order k, or by solving k first-order differential equations simultaneously. The boundary conditions for the kth order differential equation are analogous to those in Eqs. 3-35 and 3-36 for the second-order differential equation. For the k first-order equations, the boundary condition for the ith equation is

$$x_i(0) = x_{i-1}(0) \qquad (3\text{-}38)$$

The solutions for $x_n(t)$ for several values of k are shown in Fig. 3-11. Note that as k increases the solution approaches the solution for the plug-flow case. In formulating a model for a tray, the variation of $x_n(t)$ could be experimentally measured for a prescribed or recorded change in the forcing function $x_{n-1}(t)$. Next, the response of $x_n(t)$ to this same forcing is calculated for the model with various values of k. Such responses are shown in Fig. 3-11 for a step change in $x_{n-1}(t)$. The best value of k to be used in the model is determined by comparing the experimentally determined response to the model responses calculated for various values of k.

3-8. THE RELAXATION METHOD APPLIED TO HEAT TRANSFER [4]

The differential equation for the temperature profile for heat transfer by conduction in a one-dimensional solid is Laplace's equation,

$$\frac{d}{dx}\left[k \frac{dT(x)}{dx} \right] = 0$$

where k is the thermal conductivity. If k is not a function of x or T, this equation can be simplified further. The extension of Laplace's equation to higher dimensions is discussed in Chapter 7.

Although this equation seems rather simple, two difficulties arise. First, the boundary conditions may be on irregular boundaries, may be functions of temperature, or may involve other complications. Second, when k is a function of position or temperature, analytic solutions exist for only a few cases. The ultimate difficulty appears when k is a tabulated function, thus normally ruling out even the possibility of an analytic solution. These complications exist for many one-dimensional problems and are even more severe for higher-dimensional problems.

Solutions to such problems may be approached in a number of ways. For example, one approach is to write the analytic equations and solve them with the analog computer (Chapter 5) or by numerical procedures, such as finite differences (Chapter 6). Basically the same result is obtained if the system is divided into several small but finite elements, for which equations are developed by considering each individual element to be a lumped-parameter system. This is the basic concept of the relaxation method for solving heat-transfer problems. In this chapter it is not treated as such, but is used as another example of a lumped-parameter model for a distributed-parameter system. The relaxation method was developed for heat-transfer problems, but can be extended to other types of problems, as illustrated in the exercises at the end of the chapter.

This approach is applicable to both steady-and unsteady-state problems. A common technique for obtaining the steady-state solution to a problem is to obtain the unsteady-state solution for very large values of time. Therefore, since the steady-state problem is a special case of the unsteady-state problem, the discussion in this section will be devoted to the more general unsteady-state derivation.

The general approach of the relaxation method to heat-transfer problems can be outlined as follows:

1. Divide the system into a set of small but finite elements.
2. The temperature over the entire element is assumed to be the same as the temperature at the center.
3. A heat balance is made for each element.
4. The resulting set of equations is then solved to obtain the desired solution.

The technique is best discussed by illustration of a simple example, such as determining the temperature profile in a turbine blade (Fig. 3-13a) of length L, area A, and perimeter P. The base of the blade is at constant temperature T_b, and the temperature of the fluid surrounding the blade is T_f. The thermal conductivity of the metal in the blade is k, its heat capacity is c_p, and its density is ρ. The convective heat transfer coefficient between the blade and the fluid is h.

If the blade is relatively thin the temperature over a cross section can be assumed constant. Consequently, the dependence of temperature on position requires only the independent variable x, the distance from the base of the blade. Since this is a one-dimensional problem, the grid of elements needs to be established only in the x-direction. The number of elements to be used in the grid depends upon the desired accuracy of the final answer. The more elements in the grid, the more accurate the answer, but more effort is required to obtain the answer.

Suppose it has been decided to use ten elements in the lumped-parameter model. Although it is not necessary that the elements be uniform or regular in any manner, it is certainly convenient whenever possible. Furthermore, there is no one correct grid arrangement. For example, either of the arrangements in Fig. 3-13b or 3-13c could be used. They do not give exactly the same answer, but recall that we want *approximations* to the true temperature profile. Their approximations are about equivalent.

As for any lumped-parameter system, the element is of uniform temperature. For notational purposes, the temperature for element (node) i is T_i. As the original problem is a one-dimensional, unsteady-state problem, the temperature is a function of x and t, that is, $T(x,t)$. Thus the need for the partial derivatives in the next paragraph. However, the temperature T_i of node i is a function of t alone, that is, $T_i(t)$, and ordinary derivatives suffice.

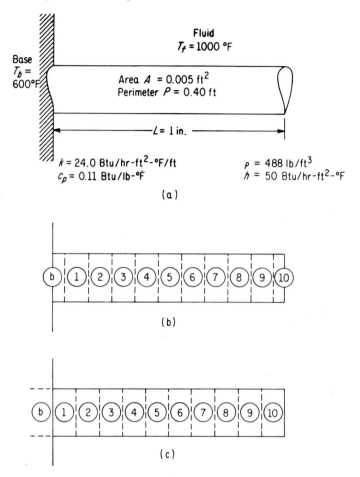

FIG. 3-13. Lumped-parameter model of turbine blade. (a) Turbine blade. (b) One possible grid arrangement. (c) Another possible grid arrangement.

For the elements in Fig. 3-13, heat is transferred by convection from the fluid to the blade, and it is conducted through the blade to the base. The equation for heat transfer by conduction is Fourier's law:

$$q = -kA \frac{\partial T(x,t)}{\partial x}$$

where q = rate of heat transfer by conduction, Btu/hr
    h = thermal conductivity, Btu/hr-ft²-°F/ft

A = area for conductive heat transfer, ft²
$\partial T(x,t)/\partial x$ = rate of change of temperature in the x-direction, °F/ft.
In terms of finite differences, this equation becomes

$$q = -kA \frac{\Delta T}{\Delta x}$$

where ΔT = temperature difference between nodes
     Δx = distance between centers of the nodes

This equation is used to describe the rate of heat transfer by conduction from one element to the next.

For sake of brevity, the model will be developed only for the grid arrangement in Fig. 3-13b. For element 1 the equation will be developed by assuming that the temperature of the element is less than the fluid temperature and the temperature of the adjoining elements. Although this assumption is known to be incorrect for the specific numerical values in Fig. 3-13, the resulting equation is independent of the directions assumed for heat transfer. Consequently, the rate of heat conduction into the element from the wall is

$$kA[T_b - T_1(t)]/\Delta x$$

The rate of heat conduction *into* element 1 from element 2 is

$$kA[T_2(t) - T_1(t)]/\Delta x$$

The rate of heat *in* by convection is

$$hP[T_f - T_1(t)]\Delta x$$

The accumulation of heat within the element is given by the rate of change of the enthalpy of the material within the element, which is

$$\frac{d}{dt}\left[\int_{T_R}^{T_1(t)} A(\Delta x)\rho c_p \, dT \right]$$

where T_R is the reference temperature for computing the enthalpy. Since $\rho, c_p, A, \Delta x$, and T_R are constant, this term simplifies to

$$A(\Delta x)\rho c_p \frac{dT_1(t)}{dt}$$

thus eliminating T_R.
Substituting these terms into a heat balance yields

$$\frac{kA[T_b - T_1(t)]}{\Delta x} + \frac{kA[T_2(t) - T_1(t)]}{\Delta x} + hP(\Delta x)[T_f - T_1(t)]$$

$$= A(\Delta x)\rho c_p \frac{dT_1(t)}{dt}$$

This equation can be rearranged to

$$\frac{dT_1(t)}{dt} = -\left[\frac{2k}{(\Delta x)^2\rho c_p} + \frac{hP}{\rho c_p A}\right] T_1(t) + \frac{k}{(\Delta x)^2\rho c_p} T_b$$

$$+ \frac{k}{(\Delta x)^2\rho c_p} T_2(t) + \frac{hP}{\rho c_p A} T_f$$

Define:

$$a = \frac{k}{(\Delta x)^2\rho c_p} \tag{3-39}$$

$$b = \frac{hP}{\rho c_p A} \tag{3-40}$$

Using the notation $\dot{T}_1(t)$ for the time derivative of $T_1(t)$, the differential equation becomes

$$\dot{T}_1(t) = -(2a + b)T_1(t) + aT_b + aT_2(t) + bT_f \tag{3-41}$$

By an analogous procedure, the equations describing the remaining nine elements can be derived. The ten equations—one for each node– for this problem can be represented as follows:

$$\dot{T}_i(t) = aT_{i-1}(t) - (2a + b) T_i(t) + aT_{i+1}(t) + bT_f$$

$$i = 1,2\cdots, 10$$

$$T_0 = T_b;\ T_{11} = T_9 \tag{3-42}$$

In deriving the equation for node 10, the rate of heat transfer by convection at the end of the blade is neglected. In Fig. 3-14 is shown the steady-state temperature profiles at values for T_f of 1000° F and 1200° F. The temperatures at the centers of the elements are given by the circles, and the continuous line drawn between them is an approximation to the actual temperature profile. These solutions can be obtained by setting the initial conditions in Eq. 3-42 to any value and solving until steady-state is reached. Alternatively, the derivative $\dot{T}_i(t)$ can be set equal to zero, leaving a set of ten algebraic equations that can be solved by Gauss reduction, matrix inversion, or other means.

As an example of an unsteady-state problem, consider the transient conditions during the transition from steady-state conditions at $T_f = 1000°$ F to steady-state conditions at $T_f = 1200°$ F. To determine the transient temperature profiles, the boundary conditions for Eq. 3-42 are the temperatures of the elements for $T_f = 1000°$ F, which are given in Fig. 3-14. With these boundary conditions these equations may be solved by several techniques, such as numerical methods, analog computer techniques, R-C electrical analogs, or matrix techniques. The solution for the transient profiles is given in Fig. 3-15.

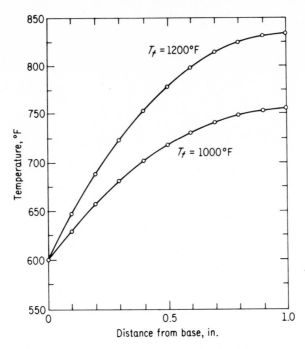

FIG. 3-14. Steady-state temperature profile for blade in Fig. 3-13 for fluid temperatures of 1000° F and 1200° F.

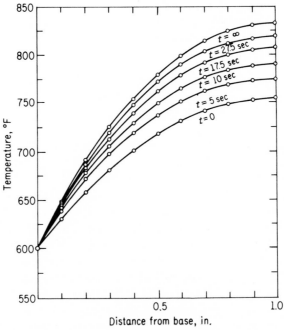

FIG. 3-15. Transient temperature profile in turbine blade for a change in fluid temperature from 1000° F to 1200° F at $t = 0$.

3-10.   SUMMARY

In this chapter the techniques for analyzing lumped-parameter systems and models have been presented.   These methods find extensive use in almost all fields of engineering.   Further discussion of using the computer as an aid to developing such models is given in Chapter 5.

REFERENCES

1. H. S. Mickley, T. K. Sherwood, and C. E. Reed, *Applied Mathematics in Chemical Engineering,* McGraw-Hill, New York, 1957.
2. B. G. Jenson, and G. V. Jeffreys, *Mathematical Methods in Chemical Engineering,* Academic Press, New York, 1963.
3. R. B. Bird, W. E. Stewart, and E. N. Lightfoot, *Transport Phenomena,* Wiley, New York, 1960.
4. F. Kreith, *Principles of Heat Transfer,* 2d ed., International Textbook, Scranton, Pa., 1965.
5. *Bubble Tray Design Manual,* A.I.Ch.E., New York, 1958.
6. J. A. Albritton, and P. W. Murrill, "Digital Computer Simulation of Distillation Column Responses," presented at 55th National A.I.Ch.E. Meeting, Houston, February 1965.

PROBLEMS

3-1.   At time $t = 0$ a billet of mass M, surface area A, and temperature T_o is dropped into a tank of water at temperature T_w.   Assume that the heat-transfer coefficient between the billet and water is h.   Furthermore, assume that the thermal conductivity of the metal in the billet is sufficiently high that the billet is of uniform temperature.   If the mass of water is large enough that its temperature is virtually constant, determine the equation describing the variation of temperature of the billet with time.

3-2.   Consider the direct contact hot water heater shown.   The water enters at temperature T_l and flow rate $L(t)$, and steam of enthalpy H_s is admitted at flow rate $S(t)$.

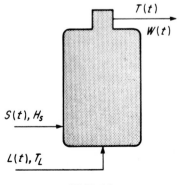

PROB. 3-2

The mass of water in the tank is M, initially at temperature T_i. The demand for the hot water varies, thus varying the water flow rate $L(t)$. A temperature control system is proposed to manipulate $S(t)$. Determine the differential equation relating the outlet water temperature to $S(t)$ and $L(t)$.

    3-3. Consider a tank of volume V placed in a process line carrying a gas stream as shown. The gas flow rate $m(t)$ into the tank varies with time, typically in a rather random fashion. The presence of the tank will dampen or attenuate the effects of these fluctuations on the flow out of the tank. If the gas in the tank is at constant temperature T and obeys the ideal gas law, develop an expression relating the output flow rate to the input flow rate (P_d is constant). The valve is described by the orifice equation as in Sec. 3-5.

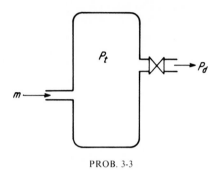

PROB. 3-3

    3-4. Consider the thermal bulb shown, immersed in a fluid of temperature T_i. The surface area of the bulb is A, its heat capacity is c_p, its mass is M, and the convective heat-transfer coefficient between the bulb and the fluid is h. Heat conduction from the bulb to the wall is negligible. After the bulb and fluid have come to equilibrium at temperature T_i, the fluid temperature changes suddenly to T_f. How does the temperature in the bulb vary?

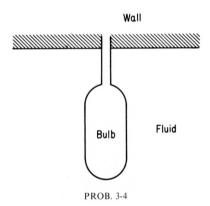

PROB. 3-4

    3-5. At the end of many industrial operations, one of the reactants or products must be removed from a solid product by leaching. One practical example is the leaching of

HCl from chlorinated polyethylene. Suppose the experiment were set up in the laboratory as shown. The mass of solids is M_s, and the mass of water is M_L. For each pound of solid, r pounds of water is entrapped in the solid. The HCl/H_2O ratio in this entrapped liquid is x_p. The rate equation for the transfer of HCl from the particles is

$$w = K(x_p - x_w)$$

where x_w is the HCl/H_2O ratio in the bulk liquid and w is the HCl removed per hour per pound of solid particles. Let the initial HCl/water ratio in the particles be x_{p_0}, and let the bulk liquid initially be pure water.

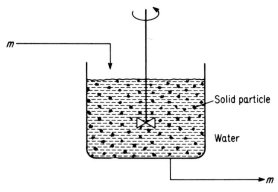

PROB. 3-5

Derive the equations describing the following three cases:
(a) The inlet and outlet flows (see accompanying figure) are zero.
(b) The inlet and outlet flows are m(lb H_2O/hr).
(c) The inlet and outlet flow are zero, but sufficient NaOH is added initially to neutralize all the HCl. In effect, this means $x_w = 0$ for all t. Neglect water created by reaction.

3-6. Problem 1-15 treated heat transfer to bubbles rising through a fluid at constant temperature. Taking $t = 0$ to be the time that the bubble is formed, derive the differential equation and boundary conditions describing the temperature of the bubble as a function of time.

3-7. Problem 1-26 treated the sheet formation on a fourdrinier paper machine. Define $t = 0$ as the time when a given point on the wire passes under the slice. Determine the basis weight of the sheet and the mass of liquid above the sheet as a function of time.

3-8. Problem 1-29 treated the case in which a reaction with volume change occurred in a tubular reactor. Rework this problem for a batch reactor (initial volume $= V_0$).

3-9. In the series of tanks shown in the sketch, the first tank is filled to a level h_i (the second tank being empty) before the valve is opened. The resistance in the line connecting the two tanks is negligible compared to that of the valve. The discharge line is of length L and inside diameter d, and the flow through this line may be assumed to be turbulent ($\Delta P = \dfrac{\rho f L V^2}{d}$). The physical properties of the fluid are constant. Develop the differential equations and boundary conditions which may be solved for the levels in the two tanks and flow rate in and out of the second tank as functions of time.

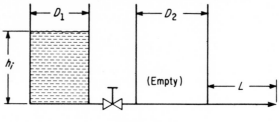

PROB. 3-9

3-10. Suppose that a backmix reactor of volume V were used for Prob. 2-7 instead of the tubular reactor. If the reactor were filled with the feed before "start-up," determine the equations and boundary conditions describing this process.

3-11. Suppose a backmix reactor of volume V is used in Prob. 2-8 instead of the tubular reactor. Let the reactor be initially filled with A and B in the same ratio as in the feed. The temperature is also the feed temperature. Determine the differential equations and boundary conditions that describe the variations of the temperature and concentrations with time.

3-12. Nitrogen pentoxide decomposes as follows:

$$N_2O_5 \xrightarrow{k_1} \frac{1}{2} O_2 + N_2O_4 \qquad \text{(first-order)}$$

$$N_2O_4 \rightleftharpoons 2NO_2$$

The last reaction is very fast, its equilibrium constant being K_e. If n_0 moles of pure N_2O_5 are charged to a cylinder of volume V, determine expressions for concentrations of N_2O_5, O_2, N_2O_4, and NO_2 as functions of time.

3-13. Consider the stirred tank reactor shown. The liquid feed is of temperature $T_0(°F)$, flow rate $w(t)$ (1b/min), and concentration C_{A_0} (1b-mole/cu ft) of reactant A. The outlet stream is of concentration $C_A(t)$, flow rate $w(t)$, and temperature $T_k(t)$. Since the contents of the tank are well mixed, the outlet concentration of A is approximately equal to the concentration of A in the tank. In the reactor component A reacts to yield B according to the following irreversible reaction:

$$2A \rightarrow B$$

This reaction is exothermic, the heat of reaction being ΔH Btu/1b mole A consumed. This reaction is second-order, and r_A, the rate of appearance of A, is given by

$$r_A = -kC_A{}^2$$

Although k is independent of C_A, its dependence upon temperature is given by the equation:

$$k = k_c \exp(-k_a/T_k)$$

There is no volume change upon reaction. To remove heat from the reacting mass, cooling water is admitted to the jacket at temperature T_{in} and flow rate $m(t)$ (1b/min). The

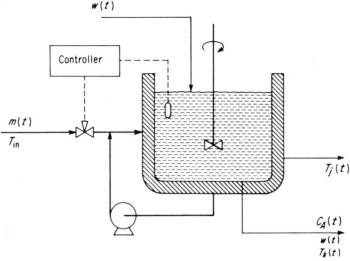

PROB. 3-13

contents of the jacket are circulated sufficiently rapidly to insure that the jacket temperature is uniform. The initial concentration of component A in the reactor is C_{A_i} and the initial temperatures of the reacting and jacket fluid are T_{j_i} and T_{k_i} respectively. Develop the differential equations and boundary conditions that describe the temperature and concentration in the reactor and the jacket due to changes in flow rates or inlet conditions.

3-14. Consider computation of the bubble point T of a multicomponent mixture under total pressure P_T. At equilibrium the following relationship must hold for each component:

$$y_i = K_i(T)x_i$$

where $K_i(T)$ is the equilibrium constant, which is a function of temperature. These equations constitute a set of nonlinear algebraic equations that must be solved by trial and error. Suppose this problem is viewed differently. The objective is to formulate a set of unsteady-state equations that can be solved on the analog computer. Consider the following point of view:

1. Suppose that we start with some mass M of liquid and vapor at temperature T_0. Now if T_0 is above the bubble point, liquid would vaporize, which would in turn cool the remaining liquid and lower T_0. Thus the temperature approaches the bubble point T. If T_0 is below T, the effect is similar.
2. In reality, the composition of the liquid would also change. However, as we know the desired liquid concentration, suppose we hold the liquid composition constant.
3. If $T_0 > T$, then $\Sigma y_i > 1$. Let the rate of vaporization be proportional to $[(\Sigma y_i) - 1]$. Let the latent heat of vaporization be λ.

Make an unsteady-state heat balance over the liquid to derive the equation

$$\frac{dT}{dt} = -G(\Sigma y_i - 1)$$

In practice, G (an arbitrary constant) is set equal to a large number, and the equation patched on the analog computer. In effect, the speed of the analog has been used to replace the trial-and-error procedure.

3-15. Consider the following equilibrium flash calculation. The feed to a flash chamber consists of M mole/hr of liquid at temperature T_f containing three components at mole fractions x_{1_f}, x_{2_f}, x_{3_f}. The desired rate of vapor leaving the flash chamber is V, and the vapor and liquid leaving are in equilibrium. For a given temperature and composition, the heat capacity of the liquid c_{p_L}, the heat capacity of the vapor c_{p_V}, and the latent heat of vaporization λ_R at temperature T_R can be calculated. The equilibrium constants and vapor pressures of each component are also available. From this information the composition and temperatures of each stream leaving and the pressure in the flash chamber are to be calculated. The steady-state equations describing the above problem can be written quite easily, but their solution must be by a trial-and-error procedure. Consequently, the solution may be difficult to obtain. However, the formulation of the problem may be changed slightly to eliminate the trial-and-error solution procedure. Formulate the unsteady-state equations for the problem, and then calculate the transient solution until it approaches the steady-state solution, which is the solution to the flash problem. To obtain this formulation, the holdup of liquid in the flash chamber is H_L, and the vapor holdup is neglected. This is sufficient to specify all but the boundary conditions, which can be specified as any value desired since only the steady-state solution (not the transient solution) is desired.

(a) Formulate the unsteady-state differential equations for the above problem. These should yield the composition, temperature, and pressure as functions of time.

(b) A slightly different flash problem occurs when the pressure instead of the split is specified. A proposed scheme is to use the same formulation as in part (a) with the following additional equation:

$$\frac{dV}{dt} = C_1(P_T - P_{T_s})$$

where $V(0)$ = arbitrarily specified
$\quad\quad P_T$ = total pressure calculated
$\quad\quad P_{T_s}$ = total pressure desired

Should C_1 be positive or negative? What determines the magnitude of C_1? Will the above scheme work?

3-16. Consider the reboiler shown, commonly used for a distillation column. The liquid holdup, i.e., the amount of liquid contained in the reboiler, is H_b moles, which is maintained constant by removing the bottom product. Initially the mole fraction of component A in the reboiler is x_{A_i} at temperature T_i. The steam temperature is T_s, and the product of the heat-transfer coefficient and area is UA. The heat capacity c_{p_L} and density ρ_L of the liquid are constant. The vapor leaving is in equilibrium with the liquid, the equilibrium constant K being a function of temperature and composition. The liquid in the reboiler can be assumed well-mixed, and the vapor holdup is negligible. The primary disturbances are changes in incoming liquid concentration x_{A_n} and temperature T_n. Make a component material balance and an energy balance around the reboiler to obtain equations relating x_A and T to changes in x_{A_n} and T_n. The latent heat of vaporization is λ, and may be considered constant over the range of interest. The temperature may be assumed to be the temperature at which the liquid and vapor are in equilibrium at the given composition. To incorporate this condition into the formulation, let the rate of vaporization V be incorporated by the equation

$$\frac{dV}{dt} = K(\Sigma y_i - 1)$$

where K is an arbitrary constant.

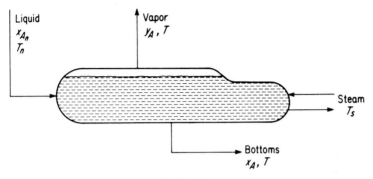

PROB. 3-16

3-17. In a batch-distillation process an initial volume V_L of liquid of composition x_{A_f} is charged to a still and the heat applied. The time at which the temperature of the liquid attains the bubble point will be called time zero. The vapor leaving the still is in equilibrium with the liquid, and the equilibrium constant is K. The latent heat of vaporization at temperature T_R (below bubble point) is λ, independent of composition. The heat capacities of liquid and vapor are c_{p_L} and c_{p_V}, and are essentially independent of composition. The heat is supplied by steam at temperature T_s, and the product of the heat transfer coefficient and area of the coils is UA. The liquid in the still is well-mixed, and vapor holdup is negligible. Write the necessary equations and boundary conditions to solve for the amount, composition, and temperature of the liquid in the still as time progresses. Incorporate equilibrium relationships as per Prob. 3-16.

3-18. Consider a plate in a distillation column upon which equimolal overflow cannot

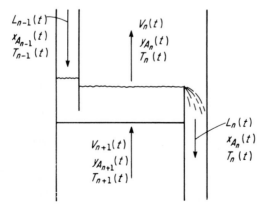

PROB. 3-18

be assumed. Suppose that the heat capacities of the liquid and vapor are substantially constant, but that the latent heats of components A and B are λ_A and λ_B at temperature T_R (below bubble point). The nomenclature is given in the sketch. Assume the pool is perfectly mixed and that the holdup is constant. Derive the unsteady-state differential equations describing this problem. Incorporate equilibrium relationships as per Prob. 3-16.

3-19. Consider the simplified cycle shown in the sketch, which represents the cycle for the cardboard cartons used for carriers of soft drinks. To test if a proposed change in the construction of the carrier would be justified, a certain number $I(t)$ of improved carriers are placed on the supermarket shelves over a period of time. The total number of all carriers on the shelf H_s is substantially constant. The total purchases $P(t)$ of all carriers can be readily measured, and the fraction of improved carriers in $P(t)$ is the same as the fraction $x(t)$ on the shelves. Some fraction k_L of the improved carriers in $P(t)$ disappear permanently, which comprises the loss $L(t)$ of good carriers, i.e.,

$$L(t) = k_L x(t) P(t)$$

The return $R(t)$ of good carriers can be measured at the bottling line, which is approximately described by a time delay of θ days. Determine the differential equation describing this situation. Rearrange such that only $R(t)$, $P(t)$, $I(t)$, k_L, θ, and H_s are involved. How can k_L be determined?

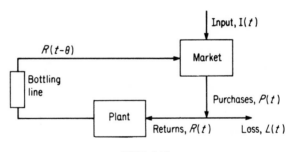

PROB. 3-19

3-20. The alkylation reactor described in Prob. 2-23 is basically a set of well-mixed tanks in series. Develop the unsteady-state equations describing one of the tanks. Incorporate equilibrium relationships as per Prob. 3-16.

3-21. Develop a lumped-parameter model for the tubular reactor considered in Prob. 1-28.

3-22. An alternative to solving the two-point boundary-value problem in Sec. 2-5 is to formulate the description in terms of a lumped-parameter model. If the reactor is divided into n equal sections, determine the unsteady-state equations for the ith general section $(1 < i < n)$ and the sections at each end.

3-23. For the heat exchanger considered in Sec. 2-2, divide the exchanger into twenty sections to develop a lumped-parameter model. Let the cross-sectional areas in the shell and tube passes be A_s and A_t, respectively. Note that the two-point boundary conditions are readily incorporated.

3-24. For the humidification column considered in Sec. 2-3, develop a lumped-parameter model.

3-25. Problem 1-30 illustrated the difficulties encountered in determining the analytic boundary conditions for a tubular reactor with longitudinal dispersion. Divide the reactor into 20 elements and derive the unsteady-state equations for each terminal node and a general interior node. Incorporate the restrictions that there is no diffusion into or out of the ends, and that the feed concentration is C_{A_0}.

3-26. Problem 1-31 illustrated the difficulties encountered in determining the analytic boundary conditions for an exchanger with appreciable heat transfer by conduction in the axial direction. Divide the exchanger into 20 elements and derive the unsteady-state equations for each terminal node and a general interior node. Incorporate the restrictions that no heat is conducted into or out of the ends of the exchanger and that the inlet temperature is T_0.

chapter 4

Simple Control-System Models

The objective of this chapter is to introduce the concept of control systems, to formulate mathematical models of these systems, and to illustrate the interaction that occurs between components within a control system. The general approach will be to accomplish these objectives through the development of conventional "block diagrams" for feedback-control systems. By doing this, it is possible to introduce a number of techniques that are very useful in many aspects of mathematical model formulation beyond modeling simple control systems.

4-1. CONCEPT OF FEEDBACK CONTROL

Feedback control is the achievement and maintenance of a desired condition by using an actual value of this condition and *comparing* it to a reference value and then using the difference between these to eliminate any difference between them. It this is done without any human intervention, then the control may be termed automatic feedback control.

In order to introduce the concept of control and some of the associated terminology, consider the home-heating system shown schematically in Fig. 4-1. The overall objective for this system is to maintain the room temperature at some desired value–typically this is 72ºF. This desired value of room temperature is referred to as the *set-point,* the *reference condition,* or the *desired value.*

The actual temperature in the room is referred to as the *controlled variable.* This controlled variable is measured by the thermostat and compared to a set-point (often introduced by rotation of the thermostat casing), and the difference between the set-point and the controlled variable is used to actuate the fuel solenoid valve. This difference is referred to as the *error signal* or the *actuating signal.* The fuel supply is adjusted to eliminate the error signal, and therefore, the fuel supply is referred to as the *manipulated variable.* Since the thermostat is used to compare the controlled variable and the set-point, it is referred to as the *error detector* or *comparator.* In terms of the above definitions, the process operates as shown in Fig. 4-2 for a change in the temperature set-point. Note that Fig. 4-2 emphasizes changes, i.e., the dynamic behavior of the variables.

There are, of course, many situations giving rise to a change in room temper-

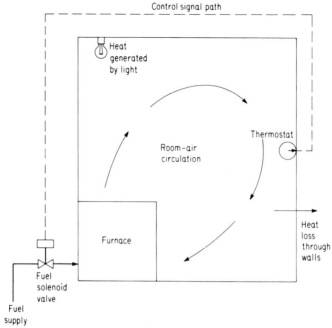

FIG. 4-1. Home-heating system.

ature other than the change in set-point noted in Fig. 4-2.   Examples of this might be a change in ambient temperature, an opening of a window, the cooking of a meal, and the like.   These other sources of potential change in room temperature are generally referred to as *disturbances*, and their effects definitely will affect the nature of the control actually achieved by the control system.   These effects are also shown in Fig. 4-2.

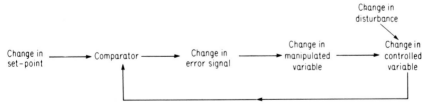

FIG. 4-2. Operation of home-heating system due to a change in set-point.

4-2. CONCEPT OF RESPONSE

Many characteristics of control systems and their study can be illustrated in terms of a familiar problem.   The system for study is the combination of a mass, a spring, and a dashpot as shown in Fig. 4-3a.   The basic problem is to character-

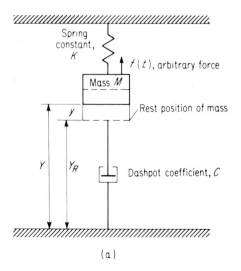

Spring constant, K

$f(t)$, arbitrary force

Mass M

Rest position of mass

y

Y Y_R Dashpot coefficient, C

(a)

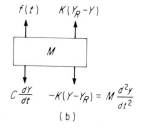

$f(t)$ $K(Y_R - Y)$

M

$C \frac{dY}{dt}$ $-K(Y - Y_R) = M \frac{d^2y}{dt^2}$

(b)

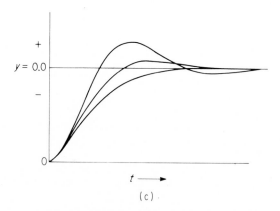

$+$

$y = 0.0$

$-$

0

$t \longrightarrow$

(c)

FIG. 4-3. Mass-spring-dashpot system. (a) Schematic diagram. (b) Free-body diagram. (c) Displacement of the mass in time.

ize the displacement of the mass with respect to time. In order to do this, Newton's second law is applied and all of the forces associated with the system are summed.

$$M\frac{d^2Y}{dt^2} + C\frac{dY}{dt} + K(Y - Y_R) = f(t) \qquad (4\text{-}1)$$

Since the rest position of the mass Y_R is a constant, Eq. 4-1 will not be changed if it is expressed in the form

$$M\frac{d^2(Y - Y_R)}{dt^2} + C\frac{d(Y - Y_R)}{dt} + K(Y - Y_R) = f(t) \qquad (4\text{-}2)$$

because the derivatives of Y_R will be zero. This is equivalent to saying that the velocity and acceleration of the mass are not dependent on the point from which the displacement of the mass is measured. It is possible, therefore, to express all displacements in terms of variations from a rest position. If y is defined as

$$y = Y - Y_R$$

then

$$M\frac{d^2y}{dt^2} + C\frac{dy}{dt} + Ky = f(t) \qquad (4\text{-}3)$$

This technique of measuring variables from an equilibrium, steady-state, or rest position is a very important and common technique in the analysis of control systems, and its use will be extensively employed. The primary reason for using this approach is found in the desirability for developing linear approximations to nonlinear functions, and this will soon become more apparent. As above, the nomenclature used for variables is to consider variations from an equilibrium, steady-state, or rest condition to be expressed in small (lowercase) letters, and variations from more classically defined bases to be expressed in capital letters.

One point that should be noted about the approach used in the analysis of Fig. 4-3 was the use of lumped parameters. This means that all of the mass of the system was considered to be lumped at one point, all of the friction of the system was lumped into the dashpot, and all of the inherent self-regulation of the system was lumped into the spring. This use of lumped parameters instead of distributed parameters simplified the mathematics of the model by producing ordinary differential equations insteady of partial differential equations. Because of this major simplification, the lumped-parameter approach is quite common in the analysis of control systems.

The general solution to Eq. 4-3 gives the displacement of the mass as a function of time, and in order to. get the specific solution applicable to a specific situation, it is necessary to define the initial conditions of the problem. For the sake of discussion, define the initial conditions to be

$$y(0) = \text{const.} \tag{4-4a}$$

and

$$\frac{d[y(0)]}{dt} = 0 \tag{4-4b}$$

and for the sake of defining a simple case, assume

$$f(t) = 0$$

The general solution to Eq. 4-3 is

$$y = C_1 e^{r_1 t} + C_2 e^{r_2 t} \tag{4-5}$$

where

$$r_{1,2} = \frac{-C \pm \sqrt{C^2 - 4KM}}{2M}$$

$$C_{1,2} = \text{const.}$$

$C_{1,2}$ are evaluated by use of Eqs. 4-4a and 4-4b and the resulting plot of y vs. t will have one of the general shapes shown in Fig. 4-3c. The particular shape or path which the mass will follow will depend on the relative magnitude and nature, i.e., real or complex, of r_1 and r_2. These are, of course, set by C, K, and M, the descriptive parameters of the system. The very important point to be noted is that the descriptive parameters of the system govern the particular response or behavior of the system.

4-3. HEAVISIDE NOTATION

Before proceeding further it is advisable to introduce a mathematical notation which will be used extensively. The notation to be employed will be based on the following differential-operator substitution:

$$p^m = \frac{d^m}{dt^m} \tag{4-6}$$

From Eq. 4-6 it can be shown that

$$\frac{1}{p^n} = \int \cdots \int \cdots (dt)^n \tag{4-7}$$

In Eqs. 4-6 and 4-7 both n and m must be integers. The differential operator follows the distributive law for multiplication:

$$p(x + y) = px + py$$

The operator follows the commutative law for multiplication:

$$(p + a)\,(p + b) = (p + b)\,(p + a)$$

where a, b are constants.   It also follows the law of exponents:

$$p^n p^m y = p^{n + m} y$$

where n, m are positive real numbers.

The operator does, in general, follow all algebraic laws except it does *not*, in general, follow cancellation since:

$$\frac{1}{p} \cdot p \cdot f(t) = \int pf(t)\,dt = f(t) + C_1$$

Simple cancellation would have given $f(t)$ which is *only* true if the initial conditions had yielded $C_1 = 0$.   This particular problem concerning cancellation is not too serious, however, because the description of physical problems seldom brings about this situation.   This restriction should be recalled however, whenever algebraic manipulations of p are attempted.

Using this notation, Eq. 4-3 would be stated:

$$(Mp^2 + Cp + K)\,y = f(t)$$

and the initial conditions would be

$$py(0) = 0$$
$$y(0) = \text{const.}$$

4-4.   BLOCK DIAGRAMS

The use of mathematical equations in their conventional form is not generally the most convenient technique to employ in control studies.   The so-called block-diagram representation is more desirable because:

1. It provides a type of road map to the determination of needed relationships.
2. It facilitates the simultaneous solution of the equations which describe the system.
3. It is not as vague as pure mathematical representation, and it indicates more closely what is actually taking place in the system.

There are two symbols used in the construction of block-diagram schematics. The first symbol is a circle which indicates an algebraic summation.

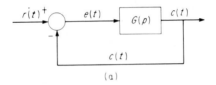

or $r - c = e$

(4-8)

The second symbol is a square which indicates multiplication:

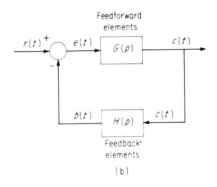

or $c = G(p)e$

(4-9)

These two symbols can be combined to give feedback-control loops as shown in Fig. 4-4. In general, any feedback-control scheme can be represented as shown

(a)

Feedforward elements

Feedback elements

(b)

FIG. 4-4. Feedback control systems. (a) Simple unity feedback system. (b) More general feedback system.

in Fig. 4-5. Note that the multiplication quantities within the blocks are, in general, functions of the differential operator p, and consequently, they indicate the possible integration of differentiation of the input with respect to time to yield the output.

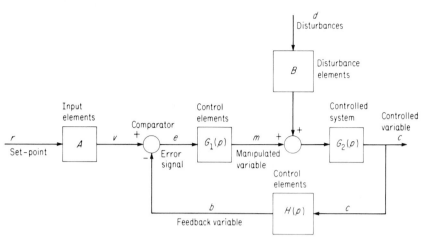

FIG. 4-5. General block diagram for a control system.

Equation 4-3 can be written in block-diagram form in one of two ways (the choice is a matter of convenience, but the lower form more closely represents the real physical system.)

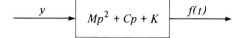

$$y \longrightarrow \boxed{Mp^2 + Cp + K} \xrightarrow{f(t)}$$

or

$$f(t) \longrightarrow \boxed{\dfrac{1}{Mp^2 + Cp + K}} \xrightarrow{y}$$

In order to make maximum usage of block diagrams it is imperative that the user be able to convert from one form of the diagram to an equivalent form without destroying the validity of the representation. Manipulations of this sort are referred to under the general heading of "block-diagram algebra" and are restricted to systems which can be described by linear algebraic equations and/or by linear differential equations with constant coefficients. Some examples of these manipulations are given in Fig. 4-6.

Based on the type of illustration given in Fig. 4-6, it is possible to see that Fig. 4-5 is a block-diagram representation of the mathematical equation:

$$[[rA - cH(p)] \, [G_1(p)] + Bd] \, [G_2(p)] = c \qquad (4\text{-}10)$$

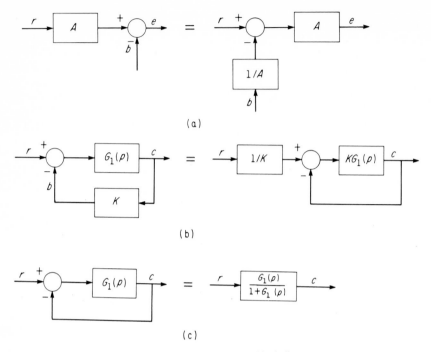

FIG. 4-6.   Examples of equivalent block diagrams.

This equation gives c as an implicit function of r and d, and it can be rearranged to the explicit form

$$c = \frac{A\,G_1(p)\,G_2(p)}{1 + G_1(p)\,G_2(p)\,H(p)}\,r + \frac{B\,G_2(p)}{1 + G_1(p)\,G_2(p)\,H(p)}\,d \qquad (4\text{-}11)$$

It will be noted that

$$c = \left(\frac{\text{Product of feedforward blocks}}{1 + \text{product of closed-loop blocks}}\right)\left(\text{input}\right) \qquad (4\text{-}12)$$

$$+ \left(\frac{\text{Product of feedforward blocks}}{1 + \text{product of closed-loop blocks}}\right)\left(\text{disturbance}\right)$$

In general, any block diagram fits Eq. 4-12 and its usage can save much time.

It should be noted that when an equation such as Eq. 4-12 is used to establish the controlled variable as a function of the command and disturbance signals, the real accomplishment is a simultaneous solution of all the equations represented by all of the elements in the block diagram. In the case of even the simpler diagrams, this can be a major accomplishment.

4-5. SIMPLE LINEAR APPROXIMATIONS

The great majority of the techniques which can be brought to bear on automatic control problems are dependent on the existence of the mathematical equation in a linear form. By their very nature, most systems are nonlinear. To reconcile this dilemma it is often necessary to use linear approximations of nonlinear functions. A simple case of this is illustrated in Fig. 4-7.

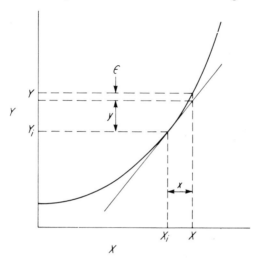

FIG. 4-7. Simple linear approximation.

Consider the initial value of X to be X_i. If X is then increased or allowed to vary by an amount of x, then

$$X = X_i + x$$

There was a value of Y corresponding to X_i which can be defined as Y_i. When X is increased, the true value of Y is

$$Y = Y_i + y + \epsilon$$

The assumption is made that

$$Y \approx Y_i + y = Y_i + \frac{dY}{dX}\bigg|_i x \qquad (4\text{-}13)$$

where $\dfrac{dY}{dX}\bigg|_i$ implies the evaluation of $\dfrac{dY}{dX}$ at an initial, steady-state, or equilibrium point. This involves an error ϵ, and the further away from the point Y_i, the worse the error becomes. It is fortunate that in most control work the range of values of variables is restricted to a small band about an operating point, and in many situations the error involved in making these linear approximations is acceptable.

Example 4-1.

An an example of the error involved in simple linearization, consider the nonlinear expression

$$Y = 2X^2 + 3$$

If $X_i = 10$, then $Y_i = 203$. The linear approximation to this expression is

$$Y \approx Y_i + y = Y_i + \left(\frac{dY}{dX}\right)\bigg|_i x$$

$$= Y_i + 4X_i x$$

$$= 203 + (4)(10)(x)$$

If $X = 12$, then $Y = 291$ by the actual expression and the linear approximation predicts

$$y \approx 203 + (40)(2) = 283$$

The error involved in the linear approximation is

$$\% \text{ error} = \left(\frac{\text{Predicted value} - \text{actual value}}{\text{actual value}}\right) 100$$

$$= \left(\frac{283 - 291}{291}\right) 100 = \frac{-800}{291}$$

$$= -2.75\%$$

In general, this linearization technique is not sufficient for most problems, however, because of the existence of nonlinear functions of several independent variables. The technique can be extended to these situations, however, and can best be illustrated by consideration of the following situation:

$$Y = \varphi(X_1, X_2, X_3, \ldots) \tag{4-14}$$

where Y = dependent variable
$X_1, X_2, X_3, \ldots$ = independent variables
φ = a function

The total differential of Y is

$$dY = \left(\frac{\partial Y}{\partial X_1}\right) dX_1 + \left(\frac{\partial Y}{\partial X_2}\right) dX_2 + \left(\frac{\partial Y}{\partial X_3}\right) dX_3 + \cdots \qquad (4\text{-}15)$$

These differentials dY, dX_1, dX_2, dX_3,... are really infinitesimal changes in these variables, and thus written in terms of variations from an equilibrium, steady-state, initial, or rest point, they could be written

$$dY \approx y \qquad (4\text{-}16)$$
$$dX_1 \approx x_1$$
$$dX_2 \approx x_2$$
$$dX_3 \approx x_3$$

Substituting these finite variations into Eq. 4-15 yields

$$y = \frac{\partial Y}{\partial X_1}\bigg|_i x_1 + \frac{\partial Y}{\partial X_2}\bigg|_i x_2 + \frac{\partial Y}{\partial X_3}\bigg|_i x_3 + \cdots \qquad (4\text{-}17)$$

where the partial derivatives must all be evaluated at the initial point about which linearization is being made.

Example 4-2

The application of this and the error involved in this type approximation can be seen very lucidly in a consideration of the area A of a rectangle:

$$A = LW$$

where L = length
$\quad\quad\;\; W$ = width

It can be seen that the area is a function of both length and width, and a linear approximation expression for any variation in area would be

$$a = \frac{\partial A}{\partial L}\bigg|_i l + \frac{\partial A}{\partial W}\bigg|_i w$$

$$= W_i l + L_i w$$

Therefore

$$A_{\text{approx.}} = A_i + a$$
$$= L_i W_i + W_i l + L_i w$$

The various quantities involved and the error experienced can be seen best in Fig. 4-8. If $L_i = 20$, $W_i = 15$, $L = 22$, and $W = 17$,

$$A_{\text{actual}} = (17)(22) = 374$$

$$A_{\text{approx.}} = (20)(15) + (15)(2) + (20)(2)$$
$$= 300 + 30 + 40$$
$$= 370$$

$$\text{Percent error} = \left(\frac{370 - 374}{374}\right) 100 = -1.07\%$$

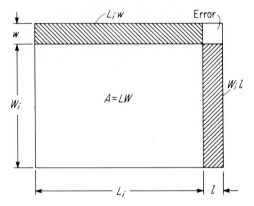

FIG. 4-8.   Linear approximations to area of a rectangle.

It is obvious that the further away from the initial point the approximation is employed, the greater will be the error involved.   This is the primary reason for defining all variables in terms of variations from an equilibrium, initial, steady-state, or operating point.   It should also be noted that the more nonlinear the original function might be, the poorer the approximation.

The use of this type of linearization in the analysis of control systems will be illustrated throughout the remainder of this chapter and in several additional chapters.   It should be emphasized that linearization itself is a necessary step in the application of virtually all of the analytical tools presented.

4-6.   LEVEL-CONTROL PROBLEM

The analysis of control systems and the development of their models can best be shown by example.   Consider the level-control system shown in Fig. 4-9.   The input to the first tank is a controlled flow rate Q_{in} of liquid.   The output of this tank Q_1 flows into a second tank along with an uncontrolled stream Q_d.   The object of the control system is to maintain the level H_2 in the second tank constant which will insure a reasonable steady flow Q_2 from this tank.   For sake of defining an example, assume both tanks are open to the atmosphere, and both discharge at atmospheric pressure.   The sizes of the various tanks and the fluid properties are known.

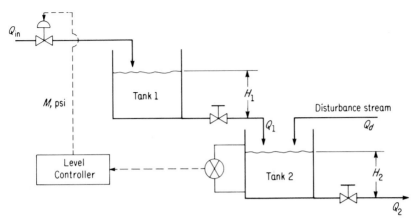

FIG. 4-9. Liquid-level control system.

The object of the investigation is to determine the variation in time of the level in the second tank when the uncontrolled flow into this tank Q_d changes in some manner.

The next step is to obtain the mathematical equations describing this system. Since the investigation is primarily concerned with control aspects, an expression is needed that relates the controlled variable (the liquid level H_2) to the manipulated variable (the controller output M psi). The flow rate Q_{in} into the first tank can be related to the are A_v of the valve opening by the orifice equation:

$$Q_{in} = C_d A_v \sqrt{\frac{2g_c(-\Delta P)}{\rho}}$$
(4-18)

where g_c = gravitational constant
C_d = orifice coefficient
ρ = liquid density
$-\Delta P$ = pressure drop

Since the problem has been defined so that all of these are constant except A_v, Eq. 4-18 reduces to

$$Q_{in} = K_1 A_v$$
(4-19)

where

$$K_1 = C_d \sqrt{\frac{2g_c(-\Delta P)}{\rho}} = \text{const.}$$

Depending on the specific characteristics of the particular control value used, the area A_v may vary with signal M in a variety of ways, many of which are nonlinear. In any case, a linear approximation of the type discussed in the previous section may be used. This yields

$$A_v = K_2 M + K_3 \tag{4-20}$$

where K_2 and K_3 are constants.

Combining Eqs. 4-19 and 4-20:

$$Q_{in} = K_1 K_2 M + K_1 K_3 = K_4 M + K_5 \tag{4-21}$$

Equation 4-21 relates the actual magnitude of Q_{in} to M. It is more convenient to use such relationships when they are in terms of deviations from a steady-state value. If m is defined as the deviation of the controller output M from an initial value of M_i, then

$$M = M_i + m \tag{4-22}$$

If the value of Q_{in} corresponding to M_i is $(Q_{in})_i$ and q_{in} is the deviation of Q_{in} from $(Q_{in})_i$, then

$$Q_{in} = (Q_{in})_i + q_{in} \tag{4-23}$$

By Eq. 4-21, the value of $(Q_{in})_i$ corresponding to M_i is

$$(Q_{in})_i = K_4 M_i + K_5$$

Substituting Eqs. 4-22 and 4-23 into Eq. 4-21 yields

$$(Q_{in})_i + q_{in} = K_4 (M_i + m) + K_5 \tag{4-24}$$

Combining Eqs. 4-23 and 4-24 indicates

$$q_{in} = K_4 m \tag{4-25}$$

This equation relates *changes* in the flow rate to *changes* in the controller signal. Since most control systems operate about some definite set of values, this form of the equation is often more useful than the form of Eq. 4-21. This can be expressed in block diagram form as

$$\tag{4-26}$$

To develop the differential equation relating changes in the liquid level in the tank to the liquid flow into and out of the tank, the unsteady-state material balance may be written:

$$\begin{bmatrix} \text{Rate of} \\ \text{liquid in} \end{bmatrix} - \begin{bmatrix} \text{Rate of} \\ \text{liquid out} \end{bmatrix} = \begin{bmatrix} \text{Rate of accumulation} \\ \text{of liquid in the tank} \end{bmatrix}$$

or,

$$Q_{in} - Q_1 = A_{t_1} \frac{dH_1}{dt} \tag{4-27}$$

where H_1 = liquid level in tank 1

A_{t_1} = cross-sectional area of tank 1

This equation also may be expressed in terms of changes from steady-state values. At steady-state,

$$\frac{dH_1}{dt} = 0 \tag{4-28}$$

Therefore, by Eq. 4-27,

$$(Q_{in})_i - (Q_1)_i = 0 \tag{4-29}$$

Defining q_1 and h_1 as deviations from the steady-state values of Q_1 and H_1, respectively, yields the following two relationships:

$$Q_1 = (Q_1)_i + q_1 \tag{4-30}$$
$$H_1 = (H_1)_i + h_1 \tag{4-31}$$

Substituting Eqs. 4-23, 4-30, and 4-31 into Eq. 4-27 yields

$$(Q_{in})_i + q_{in} - (Q_1)_i + q_1$$
$$= A_{t_1} \frac{d}{dt}\left[(H_1)_i + h_1\right] \tag{4-32}$$

Recalling that $(H_1)_i$ is a constant and combining Eq. 4-32 with 4-29:

$$q_{in} - q_1 = A_{t_1} \frac{dh_1}{dt} = A_{t_1} ph_1 \tag{4-33}$$

In block-diagram form Eq. 4-33 is

$$\tag{4-34}$$

As done in developing the equation relating controller output and the input flow rate, the outlet flow rate Q_1 may be related to liquid level in the tank by using the orifice equation

$$Q_1 = C_d A_1 \sqrt{\frac{2g_c(-\Delta P)}{\rho}}$$

In this case

$$-\Delta P = \rho H_1 \frac{g}{g_c}$$

Therefore

$$Q_1 = C_d A_1 \sqrt{2gH_1} = K_6\sqrt{H_1} \tag{4-35}$$

Again, it is preferable to linearize this equation about the initial operating point:

$$Q_1 = (Q_1)_i + \frac{dQ_1}{dH_1}\bigg|_i h_1 \tag{4-36}$$

According to Eq. 4-30

$$Q_1 - (Q_1)_i = q_1 \tag{4-37}$$

Therefore

$$q_1 = \frac{dQ_1}{dH_1}\bigg|_i h_1 = K_7 h_1 \tag{4-38}$$

where

$$K_7 = \frac{dQ_1}{dH_1}\bigg|_i = \frac{K_6}{2\sqrt{(H_1)}}$$

In terms of block diagrams this is

$$\tag{4-39}$$

For the second tank the derivation of the descriptive equations is analogous to that for the first tank.

The material balance on the second tank yields

$$Q_1 + Q_d - Q_2 = A_{t_2}\frac{dH_2}{dt} \tag{4-40}$$

The variables in this equation may be expressed in terms of deviations from their initial values:

$$q_1 + q_d - q_2 = A_{t_2}\frac{dh_2}{dt} \tag{4-41}$$

The reasoning used to develop this equation is analogous to that used to derive Eq. 4-33 from Eq. 4-27.   It can be shown in block-diagram form:

$$\tag{4-42}$$

Similarly, the expression for q_2 in terms of h_2 is

$$q_2 = K_8 h_2 \tag{4-43}$$

which is analogous to Eq. 4-38.   It can be expressed in block-diagram form as

$$\xrightarrow{\quad h_2 \quad} \boxed{K_8} \xrightarrow{\quad q_2 \quad} \qquad (4\text{-}44)$$

It would be wise to stop at this point and summarize results. Basically, all that has been done to this point is to formulate the unsteady material balance on each of the two tanks in the system. These results can be combined into a single differential equation giving the effect in h_2 (the controlled variable) for a change in m and/or q_d (the manipulated variable and/or the disturbance). Equations 4-25, 4-33, 4-38, 4-41, and 4-43 may be combined into this single differential equation

$$A_{t_2} A_{t_1} \frac{d^2 h_2}{dt^2} + \left[A_{t_2} K_7 + A_{t_1} K_8 \right] \frac{dh_2}{dt} + K_7 K_8 h_2$$

$$= \frac{A_{t_1}}{K_7} \frac{dq_d}{dt} + q_d + K_4 m \qquad (4\text{-}45)$$

or using Heaviside notation,

$$\left\{ A_{t_2} A_{t_1} p^2 + \left[A_{t_2} K_7 + A_{t_1} K_8 \right] p + K_7 K_8 \right\} h_2$$

$$= \left| \frac{A_{t_1}}{K_7} p + 1 \right| q_d + K_4 m \qquad (4\text{-}46)$$

This equation can be solved for h_2:

$$h_2 = \frac{1/K_8}{\left(1 + \dfrac{A_{t_2}}{K_8} p\right)} q_d + \frac{K_4}{\left(1 + \dfrac{A_{t_1}}{K_7} p\right)} \cdot \frac{1/K_8}{\left(1 + \dfrac{A_{t_2}}{K_8} p\right)} m \qquad (4\text{-}47)$$

Defining two functions,

$$G_1(p) = \frac{K_4}{1 + \dfrac{A_{t_1}}{K_7} p} \qquad (4\text{-}48)$$

where A_{t_1}/K_7 is the time constant of tank 1.

$$G_2(p) = \frac{1/K_8}{1 + \dfrac{A_{t_2}}{K_8} p} \qquad (4\text{-}49)$$

where A_{t_2}/K_8 is the time constant of tank 2.

Equation 4-47 becomes

$$h_2 = G_2(p)q_d + G_1(p)G_2(p)m \qquad (4\text{-}50)$$

Equation 4-50 could be achieved through block-diagram techniques just as readily. If Eqs. 4-26, 4-34, 4-39, and 4-44 are combined, the result would appear as shown in Fig. 4-10.

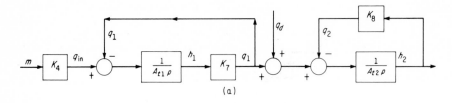

(a)

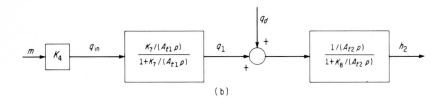

(b)

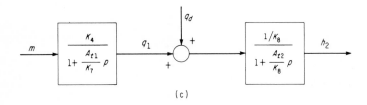

(c)

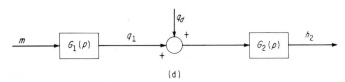

(d)

FIG. 4-10. Block diagram of two tanks. (a) Direct combination of equations. (b) Elimination of loops in (a) by Eq. 4-12. (c) Combining blocks in (b). (d) Using Eqs. 4-48 and 4-49.

It should be noted that these system equations were intentionally arranged in the general form

$$c(t) = \frac{A(p)}{B(p)} r(t) + \frac{A(p)}{B(p)} d(t)$$

where $c(t)$ = the output, response, or controlled variable
$r(t)$ = the reference input; a forcing function
$d(t)$ = the disturbance input; also a forcing function
$A(p), A(p),$ and $B(p)$ = polynomials in p

Also, the terms in the $B(p)$ polynomials were arranged in terms of the form $(1 + \tau p)$ where τ was the time constant. The significance of the time constant can best be appreciated in terms of a first-order component or system where

$$c(t) = \frac{K}{1 + \tau p} r(t)$$

If all initial conditions are zero and $r(t)$ is a step input of height R, the solution to this equation is

$$c(t) = KR(1 - e^{-t/\tau})$$

when $t = \tau$,

$$c(t) = KR(1 - e^{-t/\tau})$$

$$= KR(1 - e^{-1}) = 0.632KR$$

Therefore, 63.2% of the change that will occur in $c(t)$ will have occurred in $t = \tau$. During the next period of time equal to τ there will occur 63.2% of the remaining 36.8% change. The continuation of this reasoning is illustrated in Table 4-1, and it can also be seen in Fig. 4-11. The important point to note is that this fractional-response curve is unchanged regardless of the height of input unless the input is so great that it changes the value of one of the parameters which constitute the time constant. The generalization of these ideas to other first-order systems follows immediately.

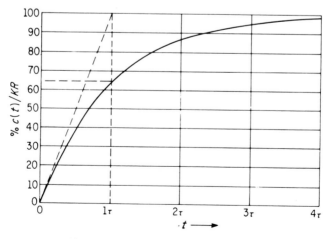

FIG. 4-11. First-order response to a step change.

TABLE 4-1

Response of a First-Order Component to a Step Change

Elapsed Time	Percent of Total Response	Response Remaining	63.2 Percent of Response Remaining
1τ	63.2	36.8	23.2
2τ	86.4	13.6	8.6
3τ	95.0	5.0	3.16
4τ	98.16	1.84	1.16
5τ	99.32	0.68	0.429

The derivative of the controlled variable with respect to time is almost as important as the value of the controlled variable itself. This is true because of the basic interest in the dynamics of $c(t)$. This derivative is

$$\frac{dc}{dt} = \frac{d}{dt}\left[KR(1 - e^{-t/\tau}) \right]$$

$$= KR\, e^{-t/\tau}\left(\frac{1}{\tau}\right)$$

$$= \frac{KR}{\tau}\, e^{-t/\tau}$$

At time equal to zero this is a maximum and is equal to

$$\left.\frac{dc}{dt}\right|_{t=0} = \frac{KR}{\tau}$$

This maximum rate of change at time equal to zero can be seen in Fig. 4-11. This is an important characteristic of first-order systems, and it is a very useful characteristic in terms of quick detection of variations. It might also be noted that if the rate of change of $c(t)$ continued at this maximum rate, the total change in $c(t)$ would be accomplished in an amount of time equal to the time constant. This can also be seen in Fig. 4-11.

4-7. ADDING CONTROL COMPONENTS

The previous section developed the model to describe the tanks and their behavior. Now the control components must be added.

The error detector compares the feedback value of h_2 to its desired value r which is the reference input or set-point and this generates an error signal e. This part of the industrial controller is usually denoted as the comparator and is described by the equation

$$e = r - h_2 \qquad\qquad (4\text{-}51)$$

The controller is that portion of the control elements which generates the manipulated output m from the error signal e. The equation for this part of the industrial controller depends on the various modes of control employed. At this point the equation relating e and m is denoted simply as

$$m = G_c(p)\,e \tag{4-52}$$

For the most general three-mode industrial controller containing "proportional" plus "integral" (commonly called "reset") plus "derivative" (commonly called "rate") action:

$$G_c(p) = K_c\left(1 + \frac{1}{T_i p} + T_d p\right) \tag{4-53}$$

where K_c = proportional gain
T_i = reset time
T_d = derivative time

In many instances the controller will have only one mode of action, e.g., proportional only or reset only (commonly called "floating"). The rate mode is never found by itself because the controller would not have any output for a constant value of e, the error signal. One common combination of controller modes is proportional plus reset. The action of such a controller is shown in Fig. 4-12. Note that in the time t equal to the integral time T_i, the reset portion

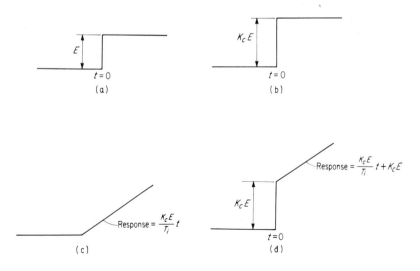

FIG. 4-12. Response of proportional plus reset controller. (a) Input e, a step increase. (b) Response of proportional term. (c) Response of integral term. (d) Total output m.

of the output equals the proportional portion of the output.   Another common combination of modes is proportional plus derivative.   The response of such a controller is shown in Fig. 4-13.   Here the combination tends to anticipate

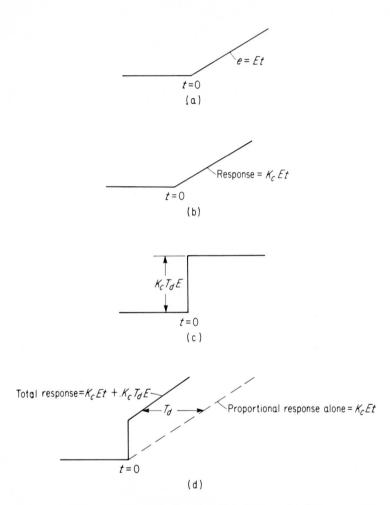

FIG. 4-13. Proportional-plus-derivative controller response. (a) Input e, a ramp increase. (b) Proportional-mode response. (c) Derivative-mode response. (d) Proportional-plus-derivative response.

"proportional only" action by an amount of time t equal to the derivative time T_d. The most general controller is the three-term controller, and its output for a typical input is shown in Fig. 4-14.

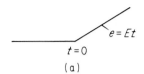

$e = Et$

$t = 0$

(a)

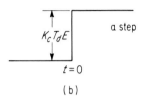

$K_c T_d E$

a step

$t = 0$

(b)

Response = $K_c Et$

= a ramp

$t = 0$

(c)

Response = $\dfrac{K_c Et^2}{T_i}$

= a parabola

$t = 0$

(d)

$K_c T_d E + \dfrac{K_c Et^2}{T_i} + K_c Et$

= a step + a ramp + a parabola

(e)

FIG. 4-14. Response of a controller containing three modes. (a) The error input e, a ramp increase. (b) Derivative-mode response. (c) Proportional-mode response. (d) Integral-mode response. (e) Proportional-plus-reset-plus-rate-mode response.

The complete block diagram for the control system can now be constructed and is shown in Fig. 4-15. The first summer corresponds to the comparator in the controller, which is described by Eq. 4-51. The block containing $G_c(p)$ corresponds to the modes in the controller. The block containing K_4 corresponds to the valve, the block relating q_1 to q_{in} expresses the dynamics of the first tank, the second summer corresponds to the entry point of q_d to the system, and the remaining block corresponds to the second tank. The output of the last block is h_2 which is fed back to the comparator. Note the similarity between this block diagram and the general block diagram of Fig. 4-5. (Implicit in Fig. 4-5 are feedback elements in which there is a proportionality constant of 1.)

From this block diagram it can be verified that the equation relating r, q_d, and h_2 is

$$h_2 = \cfrac{\cfrac{K_4 G_c(p)/K_8}{\left(\dfrac{A_{t_1}}{K_7}p+1\right)\left(\dfrac{A_{t_2}}{K_8}p+1\right)}}{1+\cfrac{K_4 G_c(p)/K_8}{\left(\dfrac{A_{t_1}}{K_7}p+1\right)\left(\dfrac{A_{t_2}}{K_8}p+1\right)}}\, r$$

$$+\cfrac{\cfrac{1/K_8}{\left(\dfrac{A_{t_2}}{K_8}p+1\right)}}{1+\cfrac{K_4 G_c(p)/K_8}{\left(\dfrac{A_{t_1}}{K_7}p+1\right)\left(\dfrac{A_{t_2}}{K_8}p+1\right)}}\, q_d$$

This equation can be simplified to

$$h_2 = \cfrac{K_4 G_c(p)/K_8}{\left(\dfrac{A_{t_1}}{K_7}p+1\right)\left(\dfrac{A_{t_2}}{K_8}p+1\right)+K_4 G_c(p)/K_8}\, r$$

$$+\cfrac{\left(\dfrac{A_{t_1}}{K_7}p+1\right)\Big/K_8}{\left(\dfrac{A_{t_1}}{K_7}p+1\right)\left(\dfrac{A_{t_2}}{K_8}p+1\right)+K_4 G_c(p)/K_8}\, q_d \qquad (4\text{-}54)$$

Equation 4-54 represents the mathematical model of the liquid-level control process. For various values of r and q_d the solution of Eq. 4-54 will give the variation of h_2 in time, i.e., will yield the process dynamics.

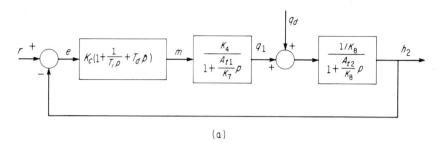

(a)

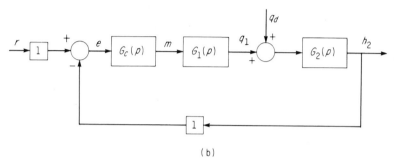

(b)

FIG. 4-15. Block diagram of liquid-level control system. (a) Actual equations. (b) Equivalent functions.

4-8. FURTHER LINEARIZATIONS

In the system of the previous section the product of two variables did not appear due to the nature of the assumptions made. This could have arisen if the area A_1 of the valve opening on the outlet of the first tank were varied. From Eq. 4-35,

$$Q_1 = C_d A_1 \sqrt{2gH_1} \qquad [4\text{-}35]$$

Assuming A_1 is constant, it was shown that:

$$q_1 = K_7 h_1 \qquad [4\text{-}36]$$

If A_1 had been a variable, Eq. 4-36 is not correct, but a linearization could have been made:

$$q_1 = K_9 a_1 + K_{10} h_1$$

where

$$K_9 = C_d \sqrt{2g(H_1)_i} = \frac{\partial Q_1}{\partial A_1}\bigg|_i$$

$$K_{10} = \frac{C_d(A_1)_i}{2}\sqrt{\frac{2g}{(H_1)_i}} = \frac{\partial Q_1}{\partial H_1}\bigg|_i$$

These same concepts may be extended to functions of more than two variables.

4-9. SUMMARY

This chapter has presented the basic tools and techniques necessary to develop a mathematical model of a control system. In addition, the concept of feedback control itself and its associated jargon were illustrated and used. As has been illustrated, this involves the determination of the mathematical models of all of the components of the control system. To facilitate this process, the use of block-diagram techniques has been presented along with the use of Heaviside notation to allow the block-diagram representation of differential equations. Since block-diagram techniques are restricted to linear equations, a technique for the linearization of nonlinear functions also was discussed and presented in terms of the convenient consideration of variables being expressed as variations from a steady-state value. The conventional feedback controller also was discussed from a mathematical basis and its typical dynamics were illustrated.

REFERENCES

1. D. P. Eckman, *Automatic Process Control,* Wiley, New York, 1958.
2. H. L. Harrison and J. G. Bollinger, *Introduction To Automatic Controls,* 2d ed., International Textbook, Scranton, Pa., 1969.
3. P. W. Murrill, *Automatic Control of Processes,* International Textbook, Scranton, Pa., 1967.
4. C. L. Smith and P. W. Murrill, "A Way to Handle Process Dynamics," *Hydrocarbon Processing & Petroleum Refiner,* February, 1967, Vol. 46, No. 2, pp. 105-122.

PROBLEMS

4-1. Given the function $r(t) = t^2 + 5$ for all positive time with $c(0) = 3$ and $pc(0) = 1$. Find the following:

(a) $c(t) = pr(t)$

(b) $c(t) = p^2 r(t)$

(c) $c(t) = \dfrac{1}{p} r(t)$

(d) $c(t) = \dfrac{1}{p^2} r(t)$

(e) $c(t) = \dfrac{1}{p} pr(t)$

(f) $c(t) = \dfrac{1}{p^2} p^2 r(t)$

(g) $c(t) = \dfrac{1}{p} p^2 r(t)$

(h) $c(t) = \dfrac{1}{p^2} pr(t)$

4-2. Figure 4-6 shows three pairs of equivalent block diagrams. Show that each member of these pairs is equivalent to the other by comparing the differential equations represented by the block diagrams.

4-3. By block-diagram algebra, rearrange each of the diagrams below into equivalent diagrams which are unity feedback diagrams, i.e., where the feedback elements $H(p)$ are equal unity.

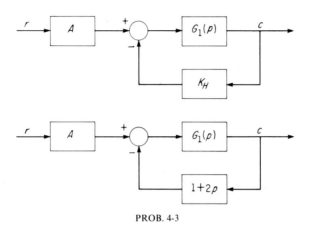

PROB. 4-3

4-4. Many control engineers prefer to develop their block diagrams into the general form shown below:

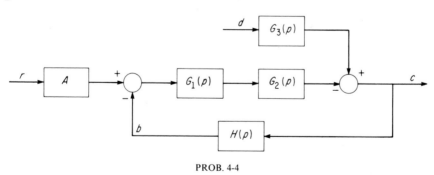

PROB. 4-4

What is $G_3(p)$ in terms of the disturbance elements B as shown in Fig. 4-5?

4-5. The volume V of a sphere is a nonlinear function of its radius R. Derive a simple linear approximation for the volume as a function of the radius. If the initial radius of the sphere is 10, what percent error will the simple approximation give when the radius is 11? When 15?

4-6. The flow of a liquid through an orifice is given by

$$Q = C_d A \sqrt{\frac{2g_c}{\rho} (P_2 - P_1)}$$

Derive a linear approximation for the flow if the pressure P_1 on the downstream side of the orifice tends to vary. What is the error in this approximation when P_1 is 22 psi if the normal value of P_1 is 20 psi?

Given C_d = 0.61

$\quad\quad \rho$ = 62.4 lb_m/ft^3

$\quad\quad gc$ = 32.2 (ft-lb mass)/(lb_f-sec^2)

$\quad\quad P_2$ = 40 psi

$\quad\quad A$ = 0.01 ft^2

4-7. The volume V of a cylinder is a function of its diameter D and its height H. Derive a linear approximation for the volume. What is the error when $D = 6$ and $H = 7$ if the initial value of these variables had been $D = 5$ and $H = 8$?

4-8. The flow of a liquid through an orifice is given by

$$Q = C_d A \sqrt{\frac{2g_c}{\rho}(P_2 - P_1)}$$

Derive a linear approximation for Q if P_1 and A both tend to vary.

4-9. Given the catalyst concentration control scheme shown. The tank is well mixed and an insignificant flow Q of constant catalyst concentration C_m parts per cubic foot (ppcf) is added to control catalyst concentration at its desired value C_1. The controller is a "proportional" device. F and H are constant. The input catalyst concentration C_i varies. Derive the block diagram for this system.

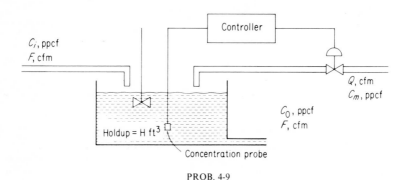

PROB. 4-9

4-10. Refer to the control system of Prob. 4-9.

(a) Find the tank time constant and the overall system time constant.

(b) Check the units of each.

(c) Prove which is larger.

4-11. Given the level control system for a distillation column's reboiler as shown. The liquid rate L varies but the vapor rate V is constant. The bottoms flow rate B is manipulated to control the level H. The diameter D of the column is constant. A proportional plus reset controller is used. Derive the block diagram for this system.

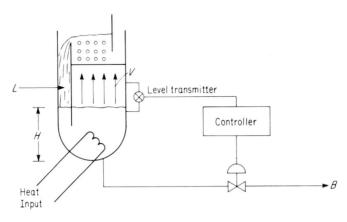

PROB. 4-11

4-12. Does the reboiler have a time constant in the level control system defined in Prob. 4-11? Does the overall control system have a time constant? What are the units of the coefficients of p in each case? How is this different?

4-13. Given the temperature control system as shown. An endothermic chemical reaction is taking place in the vessel, and a controlled amount of steam is admitted to the heat exchanger tubes to keep the reaction mass M at a controlled temperature C. The reaction goes to completion very rapidly. The latent heat of condensation of the steam ΔH_c is constant, the rate of addition of reactants R is constant, the heat of reaction ΔH_R is constant, and the heat capacities of the reactants and products are equal and are given as K Btu/lb-°F.

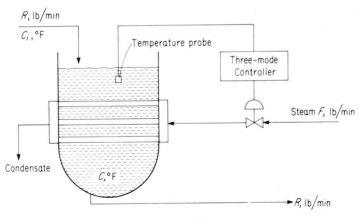

PROB. 4-13

The inlet temperature of the reactants C_i varies. The controller is a proportional plus reset plus rate device. The set-point on the controller is C_R.

Derive the block diagram to describe this control system.

4-14. For the system of Prob. 4-13 the variation in steam flow f is used in the block diagram. Explain the relationship between f and the steam flow that must be supplied for the heat of reaction.

4-15. Redo Prob. 4-9 if C_m is a variable.

4-16. Redo Prob. 4-11 if V is a variable.

4-17. Redo Prob. 4-13 if R and ΔH_c are variables.

chapter 5

Analog-Computer Solution Techniques

In studying analog computers and their use it is very helpful to keep the word "analog" firmly in mind. An analog computer is a combination of components whose descriptive equations are the same as those of the model to be solved and studied. A mathematical analog is drawn between the system whose model is to be studied and the components comprising the computer. There are many types of "special-purpose" analog computers such as slide rules, wind tunnels, and others, but this chapter will be concerned only with "general-purpose" analog computers which are formally referred to as electronic differential analyzers. In this type of analog computer the voltages at various points in the electrical network of the computer are used to represent the actual variables of the model being studied. The history of these computers only dates back to World War II, but they are a widely accepted tool today.

The analog computer (following common usage, the electronic differential analyzer will hereafter be referred to as the analog computer) provides an excellent tool for the solution of differential equations, and in certain applications, it is unsurpassed. This chapter will give some general fundamentals of how the computer works and how it can be applied to problems concerning model solution, analysis, and study.

5-1. MULTIPLICATION BY A CONSTANT[5,9]

The central component of the analog computer is a high-gain operational amplifier. This type of amplifier is shown schematically in Fig. 5-1a and its equation of operation is

$$E_o = -AE_i \tag{5-1}$$

where E_i = input voltage
E_o = output voltage
A = amplification factor, typically 10^8

There is a minus sign associated with the amplification factor in Eq. 5-1 because the amplifier imparts a 180° phase shift of the output voltage with respect to the input voltage.

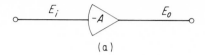

(a)

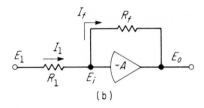

(b)

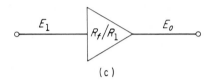

(c)

FIG. 5-1.   Multiplication by a constant. (a)
Operational amplifier. (b) Amplifier with
input and feedback resistors. (c) Programming symbol for circuit in Fig. 5-1b.

For computer components with a $\pm$ 100-volt range, an input voltage of 10^{-6} volts would saturate the circuit at the output of the amplifier. Since physical variables are represented by voltages, 10^{-6} volts are certainly negligible as compared to 100 volts. Furthermore, 10^{-6} volts are too small to measure with most voltmeters. For these reasons the input to the amplifier is zero *for all practical purposes.*

Operational amplifiers are commonly used in conjunction with input and feedback resistances as shown in Fig. 5-1b. To understand the operation of this amplifier, it must be understood that the input grid of the amplifier will draw little or no current. In most practical amplifiers this grid current is of the order of 10^{-11} ampere. Therefore

$$I_1 = I_f$$

or

$$\frac{E_i - E_1}{R_1} = \frac{E_o - E_i}{R_f}$$

But as was pointed out earlier, $E_i \approx 0$.   Therefore

$$E_o = -\frac{R_f}{R_1} E_1 \qquad\qquad (5\text{-}2)$$

Thus the multiplication of the input voltage by the factor $-R_f/R_1$ is accomplished. If $R_f = R_1$, the constant is -1. It has become common practice in many instances not to show the input and feedback resistors of Fig. 5-1b but to replace the entire network with one equivalent symbol as shown in Fig. 5-1c.

The operational amplifier also can be used to sum two or more voltages. This is shown in Fig. 5-2a where, as before

$$I_1 + I_2 + I_3 = I_x$$

or

$$\frac{E_i - E_1}{R_1} + \frac{E_i - E_2}{R_2} + \frac{E_i - E_3}{R_3} = \frac{E_o - E_i}{R_f}$$

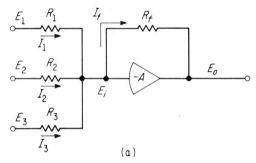

(a)

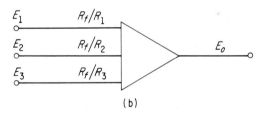

(b)

FIG. 5-2. Summing several voltages. (a) Summation of three voltages. (b) Programming symbol for Fig. 5-2a.

Taking $E_i = 0$ and solving for E_o,

$$E_o = -\left(\frac{R_f}{R_1}E_1 + \frac{R_f}{R_2}E_2 + \frac{R_f}{R_3}E_3\right) \tag{5-3}$$

By inspection of Eq. 5-3 it is seen that the amplifier not only sums the voltages E_1, E_2, and E_3 but also multiplies each of these input voltages by a constant. By adjusting the relative values of the resistors, these multiplication factors are adjusted.

With a little reflection it is quickly seen that it would be impossible to provide a complete range of resistors capable of giving any multiplication factor whatsoever. This flexibility is not necessary, since multiplication by a factor less than one can be achieved in potentiometers, and therefore only a few standard values of resistors are provided in conventional analog equipment. As an example, if

$$R_f = 1 \text{ megohm}$$
$$\text{Input } R\text{'s} = 1, 0.25, \text{ and } 0.10 \text{ megohm}$$

then multiplication factors or gains of 1, 4, and 10 are available.

Two different philosophies exist as to the best way to wire analog computers. Some manufacturers furnish a range of external resistors to be wired to the amplifier; others permanently wire the resistors into place. In this latter case there are no external wires and the use of the amplifier is convenient and simple, but obviously a great deal of flexibility is sacrificed. It seems that the optimum arrangement is some combination of the above extremes.

It might also be noted that some manufacturers differentiate between amplifiers used for general summing of voltages and those where gains are limited to 1. If the gain is limited to 1, the amplifier is often referred to as an "inverter."

5-2.   AN INTEGRATION NETWORK[6,9]

A modification of the network associated with the operational amplifier can be achieved by employing a capacitor in the feedback path instead of the resistor considered earlier. Refer to Fig. 5-3a. Employing the method of analysis used earlier,

$$I_1 = I_f$$

or

$$\frac{E_i - E_1}{R_1} = C_f \frac{d(E_o - E_i)}{dt}$$

and since E_i can be taken as zero,

$$\frac{E_1}{R_1} = - C_f \frac{dE_o}{dt}$$

Rearranging and solving this equation,

$$\int_{E_o(0)}^{E_o} dE_o = - \frac{1}{R_1 C_f} \int_o^t E_1 \, dt$$

$$E_o = - \frac{1}{R_1 C_f} \int_o^t E_1 \, dt + E_o(0) \tag{5-4}$$

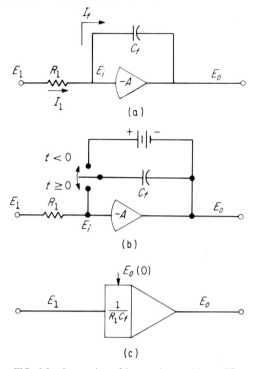

FIG. 5-3. Integration of input voltage. (a) Amplifier with a feedback capacitor. (b) Addition of initial condition circuit. (c) Programming symbol for Fig. 5-3b.

Equation 5-4 shows that the network will integrate the input voltage E_1 with respect to time and multiply it by a constant equal to $-1/R_1 C_f$.

After the resistors and capacitors to be used in the network in Fig. 5-3a have been selected, one parameter in Eq. 5-4 remains to be determined, i.e., $-E_o(0)$. In the physical system $E_o(0)$ represents an initial condition for the differential equation to, be solved, and it must be known before the problem can be solved. The value of $E_o(0)$ is introduced into the analog computer by charging the capacitor as illustrated in Fig. 5-3b. When the machine is idle, the switch is closed, and it is opened at the start of the solution. If $E_o(0)$ is to be zero, a short is substituted for the battery to insure that the capacitor is discharged before starting the solution.

Figure 5-3c shows the programming symbol for the integrating network and its initial condition circuit. This is the same type of programming symbol shown in Fig. 5-1c. These symbols are helpful during the early stages of the development of an analog-computer program. When it comes to actually wiring a program, however, these symbols are too vague for anyone but the most accomplished practitioner, and it is desirable to draw schematically both the input resistor and the feedback impedance.

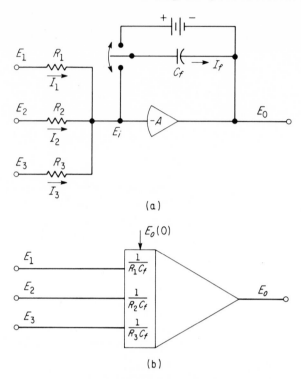

FIG. 5-4.   Integrating several inputs.   (a) Network used.   (b) Equivalent programming symbol.

An integrator network also can be used to integrate several input voltages
as shown in Fig. 5-4a.   As before.

$$I_1 + I_2 + I_3 = I_f$$

$$\frac{E_i - E_1}{R_1} + \frac{E_i - E_2}{R_2} + \frac{E_i - E_3}{R_3} = C_f \frac{d(E_o - E_i)}{dt}$$

and taking E_i as zero, rearranging, and solving,

$$E_o = -\left[\frac{1}{R_1 C_f}\int_0^t E_1\,dt + \frac{1}{R_2 C_f}\int_0^t E_2\,dt + \frac{1}{R_3 C_f}\int_0^t E_3\,dt\right] + E_o(0) \qquad (5\text{-}5)$$

The equivalent programming symbol is given in Fig. 5-4b.

5-3.   POTENTIOMETERS [7,9]

Potentiometers are commonly used in analog computers as adjustable
voltage dividers to accomplish multiplication by a constant less than one.

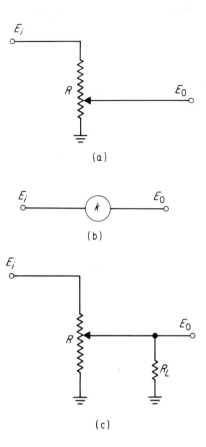

(a)

(b)

(c)

Refer to Fig. 5-5a. If the dial setting on the potentiometer is k, then the output voltage of Fig. 5-5a will be measured across a resistance of kR ohms. The dial setting k is

$$k = \frac{\text{Resistance over which } E_o \text{ is measured}}{\text{Total resistance of the potentiometer}} \qquad (5\text{-}6)$$

The equivalent programming symbol for a potentiometer is given in Fig. 5-5b.

A potentiometer dial scale is linear unless the potentiometer has a large load from some other element connected to its output terminal as shown in Fig. 5-5c. In this case a summation of the currents at the output terminal gives

$$\frac{E_o - E_i}{(1-k)R} + \frac{E_o}{kR} + \frac{E_o}{R_L} = 0$$

Rearranging,

$$\frac{E_o}{E_i} = \frac{k}{1 + k(1-k)R/R_L} \tag{5-7}$$

The difference between Eq. 5-7 and k is an error of the dial setting on the potentiometer.

$$k - \frac{E_o}{E_i} = k - \frac{k}{1 + k(1-k)R/R_L} \tag{5-8}$$

A plot of this error is shown in Fig. 5-6.

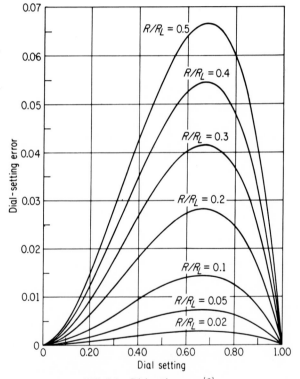

FIG. 5-6.   Dial-setting error [9].

It is easily possible to have dial errors of 100% or more, and thus, if accuracy is desired, it is necessary to make an electrical check of the actual attenuation of a potentiometer. This can be done, for example, by nulling the output voltage against the output voltage of a precise, unloaded potentiometer.

5-4. NONLINEAR COMPONENTS [2,7,9]

To this point in the study of analog computers all multiplication has been by a constant and either accomplished by an amplifier or by a potentiometer. It is also necessary to be able to multiply (or divide) one variable by another variable. To perform this nonlinear operation, two general types of analog components are available: servo-driven mechanical multipliers and electronic multipliers. The servo-multipliers are, of course, much slower than their electronic counterparts, and they experience definite frequency response limitations (rarely exceeding 20 cps). This has generated much greater interest in the electronic multipliers. A programming symbol for an electronic multiplier is given in Fig. 5-7a. One of the most popular methods of accomplishing electronic multiplication is by the quarter-square method which is based on the relation

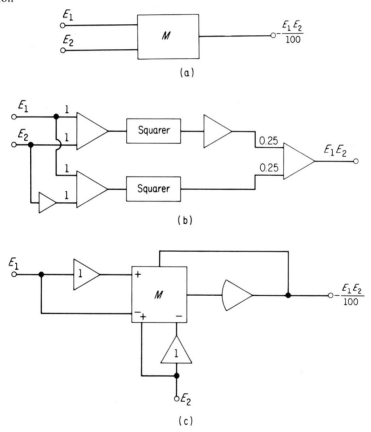

FIG. 5-7. Multiplication symbols and circuits. (a) Electronic multiplier symbol. (b) Quarter-square multiplication circuit. (c) Commercial quarter-square multiplier used with amplifier.

$$E_o = 1/4[(E_1 + E_2)^2 - (E_1 - E_2)^2]$$
$$= 1/4[(E_1)^2 + 2E_1E_2 + (E_2)^2 - (E_1)^2 + 2E_1E_2 - (E_2)^2]$$
$$= E_1E_2 \tag{5-9}$$

This is shown schematically in Fig. 5-7b and a commercial variety used in conjunction with a high-gain amplifier is shown in Fig. 5-7c.

The operation of division is normally carried out through the use of a multiplier in the feedback loop of a high-gain amplifier [2]. This is shown in Fig. 5-8 where it is desired to divide the positive or negative voltage E_1 by the *negative* voltage E_2. That this yields division can be shown, since.

$$I_1 + I_2 = \frac{E_1 - E_i}{R_1} + \frac{\dfrac{-E_2E_o}{100} - E_i}{R_2} = 0$$

or

$$\frac{E_1}{R_1} - \frac{E_2E_o}{100R_2} = 0$$

Rearranging,

$$E_o = \frac{100 \; R_2E_1}{R_1E_2} \qquad (E_2 \neq 0) \tag{5-10}$$

Another nonlinear operation which is commonly encountered is the development of a voltage which is a nonlinear function of another voltage. The

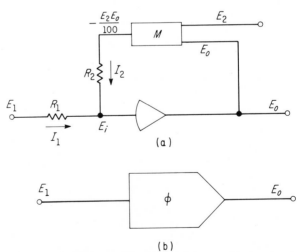

FIG. 5-8. Additional nonlinear components. (a) Division circuit. (b) Function generator symbol, $E_o = \varphi\,(E_1)$.

analog component provided for such a task is commonly referred to as a *function generator* and may be of several different types. The basic function generation problem may be either an analytic relationship between input and output such as $E_o = \sqrt{E_i}$ or it may be a table of values of E_o corresponding to values of E_i. Most methods of generating functions are either curve-following devices or line-segment approximations to the arbitrary function. The curve-following devices are often difficult to set up, but they are capable of reproducing a curve to within roughly $\pm 1\%$. The generators which use straight-line approximations to a curve employ some type of switching device such as a relay or diode. In the case of diodes there are no moving parts and, of course, the frequency response of the generator is a good as that of the operational amplifiers. There are, in addition to the general-purpose generators described above, a number of special-purpose function generators for commonly encountered functions such as the sine, the square, or the square root of a variable. The symbol for a function generator is shown in Fig. 5-8b.

5-5. OUTPUT DEVICES

To this point the emphasis has been on the components of the computer which perform the actual solution of the problem being studied on the analog computer. In addition to these devices, it is necessary to describe briefly the output components which are commonly encountered in analog-computer work. Since all of the variables of the problem will be represented as voltages, any output device must then be an indicating or recording voltmeter. These voltmeters are used to plot or indicate one variable vs. time or another variable, These devices may be servo-driven plotting boards, channel recorders, or the common cathode-ray oscilloscope. The use of an oscilloscope as an output device is particularly desirable when the computer is used in the mode of operation referred to as repetitive operation. In this mode of operation the computer is made to solve the problem and present the solution on the oscilloscope at a very high rate of speed. Through appropriate switching arrangements the solution of the problem is repeated over and over and the oscilloscope will present these solutions as a continuous curve. This mode of operation allows very quick and simple investigation of the variation in the solution of a problem as some parameter of the mathematical model is varied through the adjustment of a potentiometer.

To the novice in analog computation it might appear that the interconnection of all of the necessary components for problem solving will, in itself, make the programming of the computer impractical. All of the necessary wiring is normally accomplished through the use of removable patchboards. These patchboards contain the input and output terminals for the amplifiers, potentiometers, and other components of the computer. These are interconnected by external leads. By the use of serveral patchboards, one problem may be on the machine while several others are being programmed.

5-6. PROGRAMMING [2,3,7,9]

To understand the use of an analog computer, the development of a computer program is most helpful. For the system to be analyzed, consider the general second-order differential equation used to describe a damped harmonic oscillator—usually derived in terms of a mass-spring-dashpot arrangement. The equation is

$$M\frac{d^2y}{dt^2} + C\frac{dy}{dt} + Ky = f(t)$$

where y = displacement
     t = time
    M = mass
    K = spring constant
    C = coefficient of friction
  $f(t)$ = applied force

For the sake of example, consider the force applied to the system to be sinusoidal in nature. Assume $f(t) = f_o \sin \omega t$. The task at hand is to program the analog computer to solve the equation

$$M\frac{d^2y}{dt^2} + C\frac{dy}{dt} + Ky = f_o \sin \omega t$$

The following general steps are employed.

1. Solve the given equation for its highest derivative in terms of all other derivatives. For the present example, this is

$$\frac{d^2y}{dt^2} = -\frac{C}{M}\frac{dy}{dt} - \frac{K}{M}y + \frac{f_o}{M}\sin \omega t \tag{5-11}$$

2. Assume that this highest derivative exists and then integrate it a number of times that will be sufficient to produce the dependent variable. This is shown in Fig. 5-9a.

3. Use the dependent variable and its derivatives to produce the terms on the right-hand side of the equation for the highest derivative—Eq. 5-11. This is shown in Fig. 5-9b.

4. Use the terms generated in step 3 to obtain the highest derivative itself. The arbitrary forcing function $f_o \sin \omega t$ can be assumed to exist (for the present) and introduced into the summing amplifier. This is shown in Fig. 5-9c.

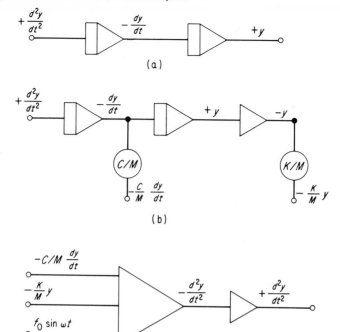

FIG. 5-9. Steps in preparing the program. (a) Obtaining the dependent variable. (b) Obtaining terms of the highest derivative. (c) Generating the highest derivative.

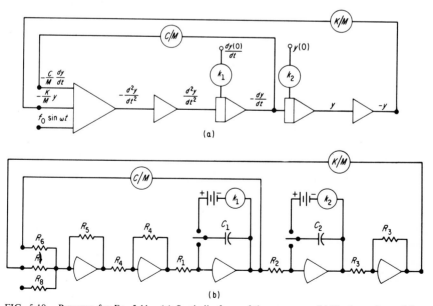

FIG. 5-10. Program for Eq. 5-11. (a) Symbolic form of the program. (b) Hardware form of the program.

5. Provide the initial conditions for the integrating amplifiers. Often it is helpful to use potentiometers on the initial condition inputs so that the initial conditions on the problem can be varied easily.

The complete diagram for the system is shown in Fig. 5-10. (It should be noted that the actual number of amplifiers needed to solve this problem for the dependent variable can be reduced to 3.)

5-7. SCALING VOLTAGES[4,7]

Now the problem of magnitude scaling of voltages must be considered. The variables of the system being studied may have any units imaginable, but when they are used on the computer they must be expressed in terms of volts. It is necessary, therefore, to have some scale factor to translate each variable from its natural units to volts and back again. One common range of voltages available on computers is $\pm$ 100 volts (some solid state units use $\pm$ 10 volts), and scale factors must be selected so that all the system variables are within this range. It is necessary, therefore, to estimate the maximum value which the system variable will attain. A general stepwise procedure for obtaining these scale factors might be as follows.

1. Estimate the maximum value which will be attained by each variable of the problem. To do this it is necessary to consider specific values for the parameters of the system. The application of this step to the mass-spring-dashpot system is given in Table 5-1. Note that these estimates do not need to be precise, and they should be rounded off in a conservative, i.e., upward, manner.

2. Obtain the scale factors for the variables by the equation:

$$\frac{\text{Maximum computer voltage}}{\text{Maximum value of variable from Step 1}} \qquad (5\text{-}12)$$

This is also shown for the example system in Table 5-1.

3. List the scaled variables to be produced on the computer. This is also done for the mass-spring-dashpot system in Table 5-1 where scaled variables are shown in brackets.

Now to determine the actual values of the resistors and capacitors used throughout the computer program, the system equation must be converted to a scaled-voltage equation. This can be done in a stepwise manner.

1. Take the system's descriptive equation in the form which is explicit for the highest derivative, e.g., Eq. 5-11 and multiply the equation by the scale factor for the highest derivative which explicitly appears in the computer program. For the example system this implies

$$\left[0.5\frac{d^2y}{dt^2}\right] = -\frac{dy}{dt} - 50y + 60\sin\omega t \qquad (5\text{-}13)$$

TABLE 5-1

Obtaining Scale Factors

Assumed parameters . . . . $\begin{cases} M = \dfrac{5 \text{ lb}}{\text{in.}/\text{sec}^2} \\ K = \dfrac{500 \text{ lb}}{\text{in.}} \\ C = \dfrac{10 \text{ lb}}{\text{in.}/\text{sec}} \end{cases}$

Assumed initial conditions . . . . $\begin{cases} y(0) = 0 \\ \dfrac{dy(0)}{dt} = 0 \end{cases}$

Assumed forcing function . . . . $f(t) = 600 \sin \omega t$

$C/M = 2.0/\text{sec}$

$f_0/M = 120 \ \dfrac{\text{in}}{\text{sec}^2}$

$K/M = 100/\text{sec}^2$

Eq. 5-11: $\dfrac{d^2 y}{dt^2} = -2 \dfrac{dy}{dt} - 100y + 120 \sin \omega t$

Variable	Range	Scale Factors	Scaled Variable
y	0 to 2.0 in.	$\dfrac{100}{2} = 50 \ \dfrac{\text{volts}}{\text{in.}}$	$[50y]$
$\dfrac{dy}{dt}$	$\pm$ 20 in./sec	$\dfrac{100}{20} = 5 \ \dfrac{\text{volts}}{\text{in.}/\text{sec}}$	$\left[5 \dfrac{dy}{dt} \right]$
$\dfrac{d^2 y}{dt^2}$	$\pm$ 200 in./sec²	$\dfrac{100}{200} = 0.5 \ \dfrac{\text{volts}}{\text{in.}/\text{sec}^2}$	$\left[0.5 \dfrac{d^2 y}{dt^2} \right]$
$f(t)$	$\pm$ 600 lb	$\dfrac{100}{600} = 0.167 \ \dfrac{\text{volts}}{\text{lb}}$	$[0.167 f(t)]$ $= [100 \sin \omega t]$

Now each of the terms of the equation has been converted from the units of the variable involved to volts.

2. Replace the problem variables on the right-hand side of the resultant equation with the scaled variables (in brackets) and multiply each term by the reciprocal of the scale factor.

$$\left[0.5 \frac{d^2 y}{dt^2} \right] = -\frac{1}{5} \left[5 \frac{dy}{dt} \right] - [50y] + \frac{[100 \sin \omega t]}{(0.167)(10)} \tag{5-14}$$

Each term is now a coefficient and a scaled variable.

3. Factor the coefficient of each scaled variable into an amplifier gain and a potentiometer setting. For the example system this is shown in Fig. 5-11

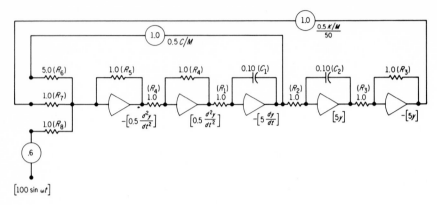

FIG. 5-11.   Voltage-scaled program.

where all the capacitors are measured in microfarads and all the resistors are measured in megohms so that it is not necessary to use the 10^{-6} and 10^6 factors. Figure 5-11 is a direct interpretation of Eq. 5-14, and it will be noted that still further modification of R_6 and R_7 must be carried out if a study of a variation of parameters is considered in which K/M and C/M exceed 100 and 2 respectively.

Example 5-1

As an exercise in programming and magnitude scaling, consider a pressure-control system whose descriptive differential equation is

$$c(t) = \frac{1.2}{p^3 + 4p^2 + 1.6p + 2.1}\, r(t)$$

and where all initial conditions are zero and $r(t)$ is a step change of 10.0 psi.

Derive and magnitude-scale the schematic diagram for determining the response $c(t)$ of this system on a $\pm$ 100-volt analog computer.

Solving for $\dfrac{d^3 c(t)}{dt^3}$,

$$\frac{d^3 c(t)}{dt^3} = -4\frac{d^2 c(t)}{dt^2} - 1.6\frac{dc(t)}{dt} - 2.1c(t) + 1.2r(t)$$

The unscaled computer program to solve this is shown in Fig. 5-12a.

In order to magnitude-scale the program, prepare the following table where the range of $\dfrac{dc(t)}{dt}$ and $\dfrac{d^2 c(t)}{dt^2}$ are assumed based on previous experience.

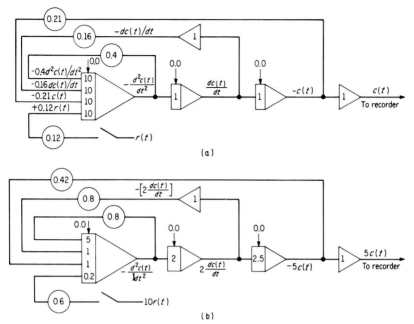

FIG. 5-12. Programs for Example 5-1. (a) Unscaled program. (b) Magnitude-scaled program.

Variable	Range	Scale Factor	Scaled Variable
$r(t)$	0-10 psi	$\dfrac{100}{10} = 10 \dfrac{\text{volts}}{\text{psi}}$	$[10\,r(t)]$
$c(t)$	0-20 psi	$\dfrac{100}{20} = 5 \dfrac{\text{volts}}{\text{psi}}$	$[5\,c(t)]$
$\dfrac{dc(t)}{dt}$	0-50 psi/min	$\dfrac{100}{50} = 2 \dfrac{\text{volts}}{\text{psi/min}}$	$\left[2\,\dfrac{dc(t)}{dt}\right]$
$\dfrac{d^2c(t)}{dt^2}$	0-100 psi/min²	$\dfrac{100}{100} = 1 \dfrac{\text{volt}}{\text{psi/min}^2}$	$\left[\dfrac{d^2c(t)}{dt^2}\right]$

Rewriting the equation to be solved in equivalent terms of the scaled variables,

$$\frac{d^3c(t)}{dt^3} = -4\frac{d^2c(t)}{dt^2} - 1.6\frac{dc(t)}{dt} - 2.1c(t) + 1.2r(t)$$

$$= -4\left[\frac{d^2c(t)}{dt^2}\right] - 0.8\left[2\frac{dc(t)}{dt}\right] - 0.42[5\,c(t)] + 0.12[10\,r(t)]$$

This equivalent equation is shown programmed in terms of scaled variables in Fig. 5-12b. One additional step has been included in preparing Fig. 5-12b. All the potentiometer settings have been increased with a corresponding decrease in integrator gains in order for the potentiometer outputs to be a relatively large voltage. Further modification of Fig. 5-12b may be neccessary because of errors in judgment in assuming the range of $\dfrac{dc(t)}{dt}$ and $\dfrac{d^2c(t)}{dt^2}$ or because of the need for time scaling.

5-8. TIME SCALING [7,9]

Time scaling is often necessary in analog-computer programs because of one of two reasons: one, the dynamic response of computing or recording components limits the ability of the components to solve the problem; or two, the regular time of solution of the problem is too long to be practical. When either of these occurs it is necessary to scale the time of problem just as it was necessary in the previous section to scale the amplitudes of the various voltages generated within the machine.

Time scaling is really a matter of changing integrator gains. If the *rate* at which a voltage change occurs is either increased or decreased, there will be a resultant variation in the *amount* of time required for a given change to occur. Since the inputs to all integrating amplifiers are rates, then a change in these amplifier gains will alter the solution time of the problem. It is necessary to define:

$$t = \text{time in the physical system being studied}$$
$$\theta = \text{machine time}$$

and these can be related as

$$\theta = \beta t \tag{5-15}$$

where β is the time scale factor. Its effect is such that

$$\frac{1}{\beta} < 1 \text{ for slowing down a problem} \tag{5-16a}$$

$$\frac{1}{\beta} > 1 \text{ for speeding up a problem} \tag{5-16b}$$

For a change in time scale, the input to every integrating amplifier must be multiplied by $1/\beta$. In general it is desirable to have β such that the solution time of the problem is as short as possible without introducing limitations caused by the dynamic characteristics of the computing or recording components.

During the voltage scaling of a problem the gains of amplifiers and potentiometers often give a hint as to whether time scaling is really necessary. If very large integrating-amplifier gains are encountered, then a large value of β

would be desirable to slow the problem down. Is small potentiometer settings are encountered (less than 0.01, for example) then a small β (less than one) can be used to increase the speed of solution.

When the computer solution is obtained for a problem that has been time-scaled, it is necessary to use β to rescale the solution back to the original independent variable t.

For a more fundamental approach to the consideration of β in time scaling, it is sufficient to note that

$$\frac{d}{dt} = \frac{d}{d(\theta/\beta)} = \beta \frac{d}{d\theta}$$

and,

$$\frac{d^2}{dt^2} = \frac{d}{dt} \beta \frac{d}{d\theta} = \beta^2 \frac{d^2}{d\theta^2}$$

and in general

$$\frac{d^n}{dt^n} = \beta^n \frac{d^n}{d\theta^n} \tag{5-17}$$

Equation 5-17 can be used to rescale the actual equation being solved.

5-9. SOLUTION OF TWO-POINT BOUNDARY-VALUE PROBLEMS

On the analog computer the actual independent variable is time. This does not, however, limit the computer to such problems. One class of problems to which the computer is well suited is two-point boundary-value problems. Such problems have position (not time) as the independent variable, and one or more of the boundary conditions is at one point while another boundary condition(s) is at a different point. As will be shown below, these problems generally require an interactive procedure to solve, and the high-speed solution capability of the analog makes it an attractive tool to use in obtaining solutions.

As an examples of such a problem, consider the 1-2 heat exchanger treated in Sec. 2-2 (Fig. 2-1). The equations were

$$\frac{dT_s(z)}{dz} + C_1[T_s(z) - T_a(z)] + C_1[T_s(z) - T_b(z)] = 0 \tag{2-1}$$

$$\frac{dT_a(z)}{dz} - C_2[T_s(z) - T_a(z)] = 0 \tag{2-2}$$

$$\frac{dT_b(z)}{dz} + C_2[T_s(z) - T_b(z)] = 0 \tag{2-3}$$

where

$$C_1 = \frac{UA}{m_s C_{p_s}}$$

$$C_2 = \frac{UA}{m_t C_{p_t}}$$

The boundary conditions were

$$T_s(0) = T_{s_1}$$
$$T_a(0) = T_{t_1}$$
$$T_a(L) = T_b(L)$$

Note that two of these boundary conditions are at the shell-side inlet end ($z = 0$), while one is at the discharge end ($z = L$). Thus we have a two-point boundary-value problem.

It is also convenient to made a change of variable as described in Sec. 1-4. Thus, define dimensionless, normalized temperatures as follows:

$$\varphi_a(z) = \frac{T_a(z) - T_{t_1}}{T_{s_1} - T_{t_1}}$$

$$\varphi_b(z) = \frac{T_b(z) - T_{t_1}}{T_{s_1} - T_{t_1}}$$

$$\varphi_s(z) = \frac{T_s(z) - T_{t_1}}{T_{s_1} - T_{t_1}}$$

Note that the values of all of these new variables are restricted between zero and one, a fact that will greatly facilitate magnitude scaling on the analog.

Substituting these definitions into the original equations and boundary conditions yields

$$\frac{d\varphi_s(z)}{dz} + C_1 [\varphi_s(z) - \varphi_a(z)] + C_1 [\varphi_s(z) - \varphi_b(z)] = 0$$

$$\frac{d\varphi_a(z)}{dz} - C_2 [\varphi_s(z) - \varphi_a(z)] = 0$$

$$\frac{d\varphi_b(z)}{dz} + C_2 [\varphi_s(z) - \varphi_b(z)] = 0$$

$$\varphi_s(0) = 1$$
$$\varphi_a(0) = 0$$
$$\varphi_a(L) = \varphi_b(L)$$

It is these equations that will be "patched" on the analog.

The first step is to perform magnitude scaling. Assume we have a $\pm$ 10-volt analog computer. To take full advantage of the resolution, the dependent variables should span this range as nearly as possible. Preparing a table for magnitude scaling as in previous examples:

Variable	Maximum Value	Scaled Variable
$\varphi_a(z)$	1.0	$[10\varphi_a(z)]$
$\varphi_b(z)$	1.0	$[10\varphi_b(z)]$
$\varphi_s(z)$	1.0	$[10\varphi_s(z)]$

Inserting into the previous equations and solving for the highest derivative gives

$$\frac{d\varphi_s(z)}{dz} = -\frac{C_1}{10}\{[10\varphi_s(z)] - [10\varphi_a(z)]\} - \frac{C_1}{10}\{[10\varphi_s(z)] - [10\varphi_b(z)]\}$$

$$\frac{d\varphi_a(z)}{dz} = -\frac{C_2}{10}\{[10\varphi_s(z)] - [10\varphi_a(z)]\}$$

$$\frac{d\varphi_b(z)}{dz} = -\frac{C_2}{10}\{[10\varphi_s(z)] - [10\varphi_b(z)]\}$$

Patching these equations onto the analog computer produces the schematic shown in Fig. 5-13.

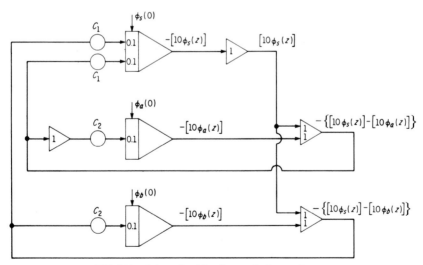

FIG. 5-13. Schematic wiring diagram for heat exchanger.

To this point nothing has been said about the actual independent variable on the computer being time, whereas the independent variable on the problem is distance. Suppose z is in feet, and the length of the exchanger is 10 ft. As the independent variable in the computer is time in seconds, then we are interested in only the first ten seconds of the solution.

Another examination of Fig. 5-13 shows that three initial conditions, namely, $\varphi_s(0)$, $\varphi_a(0)$, and $\varphi_b(0)$, are required. However, we do not have $\varphi_b(0)$. Instead, the third boundary condition is

$$\varphi_a(L) = \varphi_b(L)$$

However, this condition cannot be used directly in the analog computer.

One way to look at this situation is as follows: We have one unknown, $\varphi_b(0)$, and one equation, $\varphi_a(L) = \varphi_b(L)$, and the necessary relationships (a computer circuit) between the two. As the number of equations equals the number of unknowns, there should be a solution. We only have to find the value of $\varphi_b(0)$ such that when used to initiate the solution of the circuit in Fig. 5-13, the

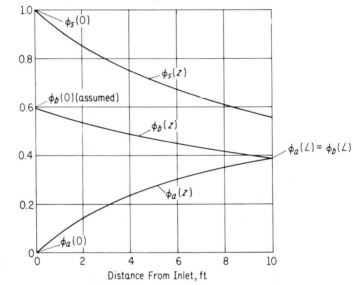

FIG. 5-14.   Temperature profiles in exchanger.

condition $\varphi_a(L) = \varphi_b(L)$ is met. This usually entails trial-and-error solutions. Figure 5-14 shows a typical solution and the significance of the various boundary conditions. It should be fairly obvious that any other assumption of $\varphi_b(0)$ would not satisfy the condition $\varphi_a(L) = \varphi_b(L)$.

The initial condition $\varphi_b(0)$ usually becomes a potential or "pot" setting on the computer. For computers on which the reptitive operation feature is available, the solution can be readily obtained by "knob twiddling." For more complex situations this can often be automated using one of the search techniques discussed in Chapter 16.

5-10.   SIMULATION OF CONTROL SYSTEMS[1,2,9]

One approach to the simulation of control systems on an analog computer is to determine the descriptive differential equation relating input to output for the system and then solve this equation on the computer.   The difficulty with this approach is that it becomes very difficult to study the effect of a change of a particular parameter of the system.

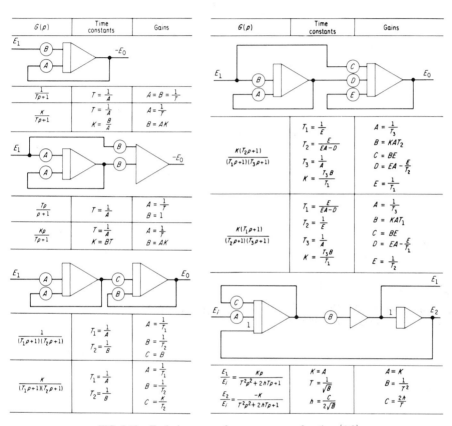

FIG. 5-15.   Typical programs for some common functions [1,9].

A much better approach is to program each block of the block diagram separately and then connect these small sections together to form the complete system simulation.   By so doing, there is a direct relationship between the computer program and the system's block diagram, and the effect of a variation in one of the system's components can be studied quite readily.   Since there are really only a few basic types of functions commonly encountered in block-diagram work, tables of analog-computer programs for these common functions have been prepared.   Representative functions are shown in Fig. 5-15.   In using these functions, the gains are calculated directly from the function time constants, gains, and so on.

5-11.   PHASE-SPACE CONCEPTS

In the material that has been presented thus far, it has been implied that the dependent variable would be the *one* voltage of the analog circuit that would be recorded.   This will not usually be the case.   It is of course desirable to record the dependent variable plotted vs. the independent variable to give a direct plot of the differential equation's solution.   As pointed out earlier, however, one of the prime advantages of the analog computer is its ability to handle nonlinear dynamic systems.   The analysis of nonlinear systems is conceptually much more difficult than the analysis of linear systems.   Many of the techniques for the analysis of nonlinear systems are based upon a simplification that involves changing to a coordinate system known as the *phase-space*.   In the phase-space type of coordinate system dependent variables are no longer explicitly plotted vs. the independent variable, but instead are plotted vs. some other parameter.   As an example, in the typical second-order equation describing the motion of the mass-spring-dashpot arrangement programmed for the analog computer in this chapter, the conventional output plot would give a measure of displacement vs. time.   In the phase-space analysis of this system, a plot might be made of velocity vs. distance with time as a coordinate along the resulting path, i.e., time could be replaced as a coordinate by velocity and time would then appear as a parameter along the new plot.   In actual application, the concept of phase analysis is applied to nonlinear systems and not to linear ones such as the mass-spring-dashpot system just mentioned. Phase-space analysis is an extremely important tool for the study of nonlinear behavior, and its utilization in connection with an analog computer is particularly important.   Its actual application cannot be further considered here because of space limitations.

5-12.   GENERAL LUMPED-PARAMETER MODELS

In one example in Chapter 3 lumped-parameter models were treated as sequences of well-mixed pools in series.   This is not at all necessary, and tanks may be of unequal size, be in parallel, be in series, have recycles, or any combination thereof as illustrated by the network in Fig. 5-16.   Within a given pool, mass transfer, heat transfer, chemical reaction, or any other process may be occurring. In fact, one of the major applications is in modeling complex chemical reactors.

Often it is convenient to separate the development of the lumped parameter model into two parts.   First, an arrangement for the pools is selected by inspection of the system and from past experience. Second, numerical values for certain constants and parameters are determined from experimental data. These constants may range from sizes (time constants) of various pools to diffusivities or rate constants for the reactions. In any case, all one generally has at his disposal is an experimental record of an input and corresponding output record.

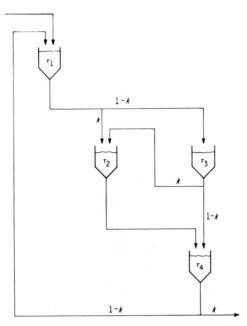

FIG. 5-16. Lumped-parameter model with split streams and recycle.

The determination of these numerical values is, in certains respects, a natural problem for the analog computer. Knowing (or assuming) a priori the arrangement or configuration for the pools, the analog patchboard can be assembled, with the unknown parameters appearing as pot settings. The experimental input record can be used as input to the computer to obtain the corresponding model output, which in turn can be compared to the experimental output to obtain adjustments for the pot settings. When a dual-trace oscilloscope and a "rep-op" or repetitive-operation feature (allowing the problem to be solved many times per second to obtain a continuous record on the scope) are available, a convenient arrangement is as shown in Fig. 5-17. With the comparison of the experimental and model responses in front of him, the operator can readily "knob twiddle" the unit until he has good numerical values (pot settings) for the unknown parameters.

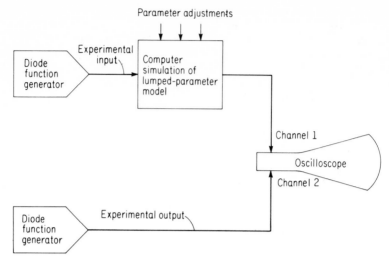

FIG. 5-17.   Manual implementation of model development.

When a hybrid computer (or analog computer with expanded logic) is available, this scheme can be automated even further. The first step is to obtain the difference between the model response and the experimental response, or the error $e(t)$ as shown in Fig. 5-18. This error can be integrated in one of at

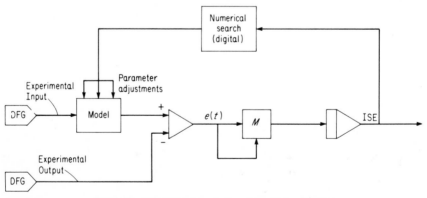

FIG. 5-18.   Hybrid implementation of model development.

least two ways to obtain a single numerical value representing the goodness of fit. Perhaps the most common integral is the integral of the square error (ISE), namely

$$\int_0^T [e(t)]^2 \, dt$$

Another common integral is the integral of the absolute error (IAE), which is

$$\int_0^T |e(t)| \, dt$$

In either case, the integral can be evaluated by the analog computer; it is illustrated in Fig. 5-18 for ISE. Thus at the end of each solution this value is available, and the optimum values of the parameters are those that minimize this integral. To obtain complete automation, a search procedure such as steepest descent, pattern search, random search, and the like is programmed on the digital portion of the hybrid to search for the minimum. Such optimization techniques are topics for Chapter 16.

5-13. SUMMARY

This chapter has served to introduce the subject of analog computers and to show their very useful application in the solution of differential equations. Entire books are devoted to this subject but the treatment presented here must, of necessity, be short. Where analog equipment is available, the engineer should strive to incorporate the analog computer into his program of study and analysis of mathematical models.

REFERENCES

1. H. Hamer, "Generation of Special Time Functions," Applications Report, Electronic Associates, Inc., West Long Branch, N. J.
2. M. L. James, G. M. Smith and J. C. Wolford, *Analog and Digital Computer Methods in Engineering Analysis,* International Textbook, Scranton, Pa., 1966.
3. C. L. Johnson, *Analog Computer Techniques,* McGraw-Hill, New York, 1956.
4. G. A. Korn and T. M. Korn, *Electronic Analog Computers,* 2d ed., McGraw-Hill, New York, 1956.
5. Thomas Matthews, "Analog Simulators," *Chemical Engineering,* Jan. 4, 1965, p. 79.
6. ——, "How To Analyze the Circuits and Mathematics of Analog Simulators," *Chemical Engineering,* Feb. 1, 1965, p. 79.
7. "Primer on Analog Computation," Applications Reference Library, Electronic Associates, Inc., West Long Branch, N. J., 1964.
8. C. L. Smith, "All-Digital Simulation for the Process Industries," *Instrument Society of America Journal,* Vol. 13, No. 7 (July 1966).
9. G. W. Smith and R. C. Wood, *Principles of Analog Computation,* McGraw-Hill, New York, 1959.

PROBLEMS

5-1. Determine the output of the circuit shown.

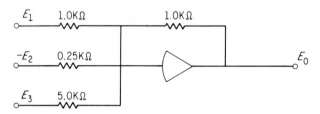

PROB. 5-1

5-2. Determine the output of the circuit shown.

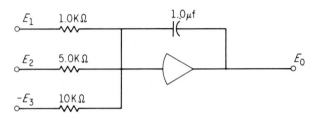

PROB. 5-2

5-3. Determine the output of the circuit shown.

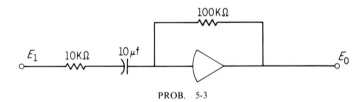

PROB. 5-3

5-4. Determine the output of the circuit shown.

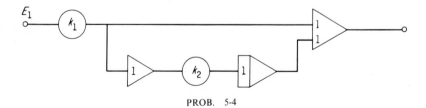

PROB. 5-4

5-5. Determine the output of the circuit shown.

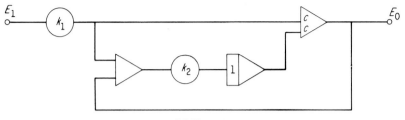

PROB.   5-5

5-6. Set up a circuit for the solution of the differential equations below.   Assume all initial conditions are zero and $r(t)$ is a unit step function.

(a) $(10p + 26)c(t) = r(t)$

(b) $(p^2 + 10p + 26)c(t) = r(t)$

(c) $(p^3 + 10p + 26)c(t) = r(t)$

5-7. Set up a circuit for the solution of the following nonlinear differential equation. Assume all initial conditions are zero and $r(t)$ is an arbitrary forcing function.

$$\frac{d^3[c(t)]}{dt^3} + 2\frac{d^2[c(t)]^2}{dt^2} + \left\{\frac{d[c(t)]}{dt}\right\}^2 = r(t)$$

5-8. Set up a circuit for the solution of the following nonlinear differential equation. Assume all initial conditions are zero and $r(t)$ is an arbitrary forcing function.

$$c(t)\frac{d^2[c(t)]}{dt^2} + [1 - c(t)]^2 = r(t)$$

5-9. Set up the circuit to solve the equation $(p^2 + b^2)c(t) = 0$ to generate a sinusoidal function for use as a forcing function or input to the analog-computer circuit for a separate differential equation.

5-10. The derived equation for an oven-heating system is

$$c(t) = \frac{K_s}{1 + \tau_s p}r(t)$$

Assume $K_s = 1$, the maximum expected value of τ_s is 20 min, and $r(t)$ is a step change of 100°F.

(a)   Set up the unscaled circuit to solve this.

(b)   Magnitude-scale this circuit for a $\pm$ 10-volt. computer.   Redo for a $\pm$ 100-volt. computer.

(c)   Time scale this circuit for $\beta = 0.5$.

5-11. A wetted wall column's descriptive differential equation was derived in Prob. 2-2. Prepare an analog-computer program to solve this to yield x_a as a function of z. How would this computer program have to be changed if operation were nonisothermal as in Prob. 2-3?

5-12. Program the tubular reactor results of Prob. 1-27 for an analog computer.

5-13. Prepare the analog-computer program to simulate the heat exchanger of Prob. 1-31.

5-14. Prepare the analog-computer program to simulate the tubular reactor of Prob. 2-7.

5-15. A nonisothermal absorption column was described in Prob. 2-12. Prepare the analog-computer program to solve the descriptive differential equations for this column.

5-16. Problem 2-23 is devoted to the use of mixed tanks in series to model an alkylation reactor. Develop the analog-computer program to solve the descriptive differential equations that yield the temperature and concentration profiles for the reactor.

5-17. The industrially important operation of leaching is described in Prob. 3-5. Develop the analog-computer program to solve the equations resulting from the three cases proposed in this problem.

5-18. Develop the analog-computer program to simulate the stirred-tank reactor of Prob. 3-13.

5-19. Develop the analog-computer program to simulate the distillation-columm reboiler of Prob. 3-16.

5-20. Set up an analog-computer program to simulate the reactor problem in Prob. 3-22 when $n = 4$.

5-21. Set up an analog-computer program to simulate the block diagram derived in Prob. 4-17.

chapter 6

Numerical Solutions of Ordinary Differential Equations

In earlier chapters of this book methods were presented to mathematically model systems that are described by ordinary differential equations, and this chapter will be devoted to a presentation of numerical methods to solve these ordinary differential equations. The treatment here is only intended to introduce the subject. Emphasis will be placed on techniques that can be readily implemented on a digital computer because the existence of modern high-speed digital computers often makes the numerical solution to a differential equation quite practical.

Throughout the presentation of this chapter the student might bear in mind the possibilities open to him in solving ordinary differential equations, i.e, analog-computer techniques as presented in the previous chapter, numerical techniques as presented in this chapter, or analytical solution methods. (The student using this book should have previously had an introduction to analytical methods, and this prior knowledge should prove sufficient for a book of this type.)

Our approach will be to introduce the subject by illustrating the numerical solution to an ordinary differential equation by approximating the derivatives in the equations with finite difference obtained from a Taylor series expansion. This will be extended to show the various forms of the finite-difference approximations to derivatives that can be obtained using a Taylor series expansion. Following this the numerical solution of two common types of ordinary differential equations will be presented. Integration of ordinary differential equations will then be discussed, using the method of Runge-Kutta integration formula and predictor-corrector techniques. Finally, solution of simultaneous sets of ordinary differential equations and higher-order differential equations will be developed.

6-1. TAYLOR SERIES EXPANSIONS [1,2,3,4,5]

A Taylor series expansion is an infinite power series representation of a function $f(x)$ in terms of x and about some point x_0. If the series converges it allows us to determine $f(x)$ at the point x based on the Taylor series about x_0. The Taylor series, therefore, allows us to extrapolate $f(x)$ to some point x based on its value at x_0. This is a basic tool of numerical analysis.

A Taylor series is expressed as

$$f(x) = \sum_{n=0}^{\infty} \frac{f^{(n)}(x_0)}{n!} (x - x_0)^n \qquad (6\text{-}1)$$

where

$$f^n(x_0) = \frac{d^n f}{dx^n}\bigg|_{x = x_0}$$

$$f^{(0)}(x_0) = f(x_0)$$

$$0! = 1$$

An equivalent form of Eq. 6-1 is also useful:

$$f(x_0 + x) = \sum_{n=0}^{\infty} \frac{f^{(n)}(x_0)}{n!} x^n \qquad (6\text{-}2)$$

where x is the distance between x_0 and the point at which the function is to be evaluated.

The use of a Taylor series can introduce us to numerical solutions via a simple example. Consider the solution of the simple equation

$$\frac{d^2 u}{dx^2} - Ku = 0 \qquad (6\text{-}3)$$

with the boundary conditions

$$u(0) = 1$$

$$u(1) = 0$$

This equation can be associated with the physical problem of describing the temperature profile in an uninsulated rod as shown in Fig. 6-1.

Both x and u are continuous variables as shown in Fig. 6-2a, but we will consider them as discrete quantities. This is illustrated for the rod in Fig. 6-2b where 21 discrete values of x are shown; and therefore, each increment of $x = 0.05$ unit. In general, if there are N increments then $\Delta x = 1/N$ and $x_i = i\Delta x$ for $0 \le i \le N$. For nomenclature we will assume

$$\frac{du}{dx} = u_x$$

$$\frac{d^2 u}{dx^2} = u_{xx}$$

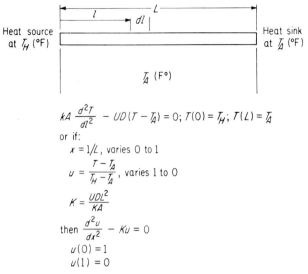

$$kA \frac{d^2T}{dl^2} - UD(T - T_A) = 0; \quad T(0) = T_H; \quad T(L) = T_A$$

or if:

$x = 1/L$, varies 0 to 1

$u = \dfrac{T - T_A}{T_H - T_A}$, varies 1 to 0

$K = \dfrac{UDL^2}{KA}$

then $\dfrac{d^2u}{dx^2} - Ku = 0$

$u(0) = 1$

$u(1) = 0$

FIG. 6-1.   Uninsulated-rod temperature problem.

$$u(x_i) = u_i$$

$$u(x_{i+1}) = u_{i+1}$$

Our problem is now to express u_{i+1} as a function of u_i. Writing the Taylor series about x_i for $u_{i+1} = u(x_i + \Delta x)$ we have

$$u_{i+1} = u_i + (u_x)_i \Delta x + \frac{(u_{xx})_i}{2!} (\Delta x)^2 \tag{6-4}$$

$$+ \frac{(u_{xxx})_i}{3!} (\Delta x)^3 + \frac{(u_{xxxx})_i}{4!} (\Delta x)^4 + \cdots$$

and writing the Taylor series about x_i for $u_{i-1} = u(x_i - \Delta x)$ we have

$$u_{i+1} = u_i + (u_x)_i (-\Delta x) + \frac{(u_{xx})_i}{2!} (-\Delta x)^2 \tag{6-5}$$

$$+ \frac{(u_{xxx})_i}{3!} (-\Delta x)^3 + \frac{(u_{xxxx})_i}{4!} (-\Delta x)^4 + \cdots$$

Adding Eqs. 6-4 and 6-5,

$$u_{i+1} + u_{i-1} = 2u_i + \frac{(2u_{xx})_i (\Delta x)^2}{2!} + \frac{2(u_{xxxx})_i}{4!} (\Delta x)^4 + \cdots \tag{6-6}$$

Solving Eq. 6-6 for $(u_{xx})_i$,

$$(u_{xx})_i = \frac{u_{i+1} - 2u_i + u_{i-1}}{(\Delta x)^2} + \frac{(u_{xxxx})_i (\Delta x)^2}{12} + \cdots \tag{6-7}$$

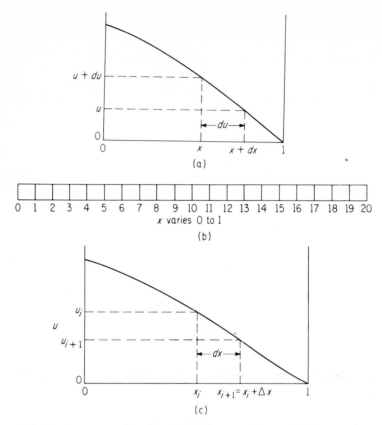

FIG. 6-2.   Two ways to view u(x).   (a) u and x are continuous.   (b) Discrete points
for x.   (c) u and x are discrete.

Equation 6-7 may be truncated after the first term and the approximation for the
second derivative $(u_{xx})_i$ will be "second-order" correct.

$$(u_{xx})_i = \frac{u_{i+1} - 2u_i + u_{i-1}}{(\Delta x)^2} \qquad (6\text{-}8)$$

Now our basic equation (Eq. 6-3) may be written

$$(u_{xx})_i - Ku_i = 0 \qquad (6\text{-}9)$$

and substituting Eq. 6-8 into Eq. 6-9:

$$\frac{u_{i+1} - 2u_i + u_{i-1}}{(\Delta x)^2} - Ku_i = 0$$

or

$$u_{i-1} - [2 + K(\Delta x)^2]\,u_i + u_{i+1} = 0 \qquad (6\text{-}10)$$

where the u's are unknowns, and therefore, we have one equation in three unknowns. But Eq. 6-10 may be written for $2 \leq i \leq N - 2$. Even considering this, however, there are still two more unknowns than equations, but the additional relationships are provided by the boundary conditions

$$u(0) = 1$$
$$u(1) = 0$$

The net result is $N - 1$ equations in $N - 1$ unknowns. In our case we have $N = 20$ (There are 21 points with one numbered zero) and our set of 19 equations in 19 unknowns is as shown in Fig. 6-3. The solution to this set of algebraic equations will be the set of temperatures along the rod which we will consider to be the numerical solution to the basic ordinary differential equation with which we started.

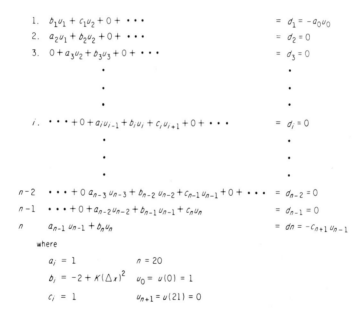

FIG. 6-3.   Set of algebraic equations to represent the basic differential equation.

The basic idea to be derived from this entire example is the use of finite-difference approximations for derivatives with the resultant transformation of the solution of the differential equation into the solution of a set of algebraic equations.

6-2.   DIFFERENTIATION[4]

In the previous example a finite-difference approximation was obtained for the first derivative by manipulation of the Taylor series written about x_i.   While, in general, this is the best approach to the approximation, it is instructive to approach the same problem from a geometric viewpoint.   Consider Fig. 6-4.

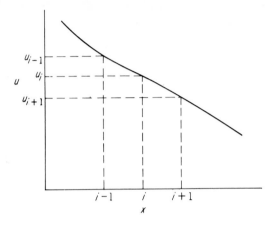

FIG. 6-4.   Geometric approach to approximation.

Now it can be seen that

$$\left(\frac{du}{dx}\right)_{i+\frac{1}{2}} \approx \frac{u_{i+1} - u_i}{\Delta x}$$

$$\left(\frac{d^2 u}{dx^2}\right)_{i-\frac{1}{2}} \approx \frac{u_i - u_{i-1}}{\Delta x}$$

Therefore

$$\left(\frac{d^2 u}{dx^2}\right)_i \approx \frac{\left(\dfrac{du}{dx}\right)_{i+\frac{1}{2}} - \left(\dfrac{du}{dx}\right)_{i-\frac{1}{2}}}{\Delta x} = \frac{u_{i+1} - 2u_i + u_{i-1}}{(\Delta x)^2}$$

which is the same as Eq. 6-8.

While this geometric approach helps to visualize the approximation being used, the Taylor series approach is by far the most useful.   In order to get a general idea of now numerical solution methods are developed, we will now consider the general problem of developing general numerical differentiation formulas from Taylor series.   Basically our motive is to develop these differentiation formulas, and whenever a finite-difference approximation is needed for a derivative term in a differential equation, we can consult our table for the appropriate term(s).

The points x_i are normally referred to as *pivotal points* and are normally evenly spaced by h (as is quite common, h is used here instead of Δx) and

$$x = x_0 \pm ih \qquad i = 0, 1, 2, 3, \ldots \tag{6-11}$$

At the pivotal points the function $f(x)$ will have the pivotal values $f_i = f(x_i)$. In general, one pivotal point x_j may be written in terms of some other pivotal point x_i:

$$x_j = x_{i \pm m} = x_i \pm mh \tag{6-12}$$

and the Taylor series (with $x_0 = x_i$ and $\Delta x = \pm mh$) is

$$f_j = f(x_i \pm mh) = f_i \pm mhf_i' + \frac{(mh)^2}{2!} f_i'' + \cdots \tag{6-13}$$

If we take $m = 1$ and 2 in Eq. 6-13 we have the following as the pivotal values on either side of x_i:

$$f_{i-2} = f_i - 2hf_i' + 2h^2 f_i'' - \frac{4h}{3} f_i''' + \cdots \qquad m = -2 \tag{6-14}$$

$$f_{i-1} = f_i - hf_i' + \frac{h^2}{2} f_i'' - \frac{h^3}{6} f_i''' + \cdots \qquad m = -1 \tag{6-15}$$

$$f_{i+1} = f_i + hf_i' + \frac{h^2}{2} f_i'' + \frac{h^3}{6} f_i''' + \cdots \qquad m = +1 \tag{6-16}$$

$$f_{i+2} = f_i + 2hf_i' + 2h^2 f_i'' + \frac{4h^3}{3} f_i''' + \cdots \qquad m = +2 \tag{6-17}$$

If we add Eq. 6-15 and Eq. 6-16 and solve for f_i'' we have

$$f_i'' = \frac{1}{h^2} (f_{i+1} - 2f_i + f_{i-1}) - \frac{h^2}{12} f_i^{iv} + \cdots \tag{6-18}$$

which is the same, of course, as Eq. 6-7. We can get other approximations for this second derivative, however. As an example, if we take Eq. 6-16 times 2 and subtract Eq. 6-17 we get

$$f_i'' = \frac{1}{h^2} (f_i - 2f_{i+1} + f_{i+2}) \tag{6-19}$$

This can go on and on. By taking more pivotal points into consideration and by combining the Taylor series representations for the pivotal values in various algebraic fashions, we can come up with an infinite number of approximations for f_i''. Any other derivative's approximation can be viewed in a similar fashion. A list of some of the possibilities is given in Fig. 6-5. This list

$h^n f^{(n)}$	Multiplier	Differentiation Formula Coefficients					Error in $f^{(n)}$
	1	**1**	1				$-(1/2)hf''$
	1/2	**-3**	4	-1			$(1/3)h^2 f'''$
	1/2	**-11**	18	-9	2		$-(1/4)h^3 f^{iv}$
hf' ←	1/2	**-25**	48	-36	16	-3	$(1/5)h^4 f^v$
	1/2	+1	**0**	-1			$-(1/6)h^2 f'''$
	1/6	-2	**-3**	6	-1		$(1/12)h^3 f^{iv}$
	1/12	-3	**-10**	18	-6	1	$(1/20)h^4 f^v$
	1/12	1	-8	**0**	8	-1	$(1/30)h^4 f^v$
	1	1	**-2**	1			$-hf'''$
	1	2	**-5**	4	-1		$(11/12)h^2 f^{iv}$
	1/12	**35**	-104	114	-56	11	$(5/6)h^3 f^v$
$h^3 f'''$	1	1	**-2**	1			$-(1/12)h^2 f^{iv}$
	1	1	**-2**	1	0		$(1/12)h^2 f^{iv}$
	1/12	11	**-20**	6	4	-1	$(1/12)h^3 f^v$
	1/12	-1	16	**-30**	16	-1	$(1/90)h^4 f^v$
	1	-1	**3**	-3	1		$-(3/2)hf^{iv}$
	1/2	**-5**	18	-24	14	-3	$(7/4)h^2 f^v$
$h^4 f'''$	1	-1	**3**	-3	1		$-(1/2)hf^{iv}$
	1/2	-3	**10**	-12	6	-1	$(1/4)h^2 f^v$
	1/2	-1	2	**0**	-2	1	$-(1/4)h^2 f^v$
	1	**1**	-4	6	-4	1	$-2hf^v$
$h^4 f^{iv}$	1	1	**-4**	6	-4	1	$-hf^v$
	1	1	-4	**6**	-4	1	$-(1/6)h^2 f^v$

FIG. 6-5. Differentiation formula. *Source*: J.M. McCormick and M.G. Salvadori, *Numerical Methods in Fortran Programming* (Englewood Cliffs, N.J.: Prentice-Hall, Inc, 1964). Used with permission.

considers from two to five pivotal points and is valid at the point with the *boldface* coefficient.

One way to view the numerical solution of ordinary differential equations is via a table of differentiation formulas as given in Fig. 6-5. The appropriate formula for each term in the differential equation is selected from the table and substituted into the differential equation. A set of algebraic equations will result, and the solution of this set will yield the solution to the differential equation. This general concept has various modifications involved in its actual implimentation.

6-3. SECOND-ORDER BOUNDARY-VALUE PROBLEMS

One of the most commonly encountered equations is the linear, second-order boundary-value problem of the form

$$f'' + y(x)f' + g(x)f = F(x) \qquad (6\text{-}20)$$

where $f(a) = f_0$ and $f(b) = f_n$.

In this case f is known at the points a and b, and it is to be determined at all intermediate points. This is illustrated in Fig. 6-6.

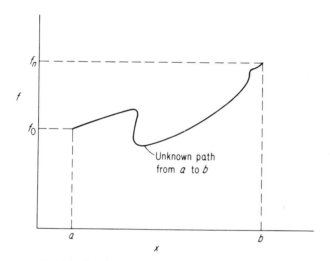

FIG. 6-6. Solution to second-order boundary-value problem.

The general numerical approach to this problem is[4]:

(a) Substitute approximate expressions for f' and f'' with the same order of error.

(b) Apply these linear-difference approximations at the $n - 1$ pivotal points where f_i is unknown.

(c) Solve the resulting set of $n - 1$ linear algebraic equations.

The results obtained in (c) are the values of f_i between $x = a$ and $x = b$.

The heat-conduction problem of Sec. 6-1 was an example of this general approach.

A convenient way to formulate this solution in terms of difference equations is to use three-point, centered formulas with errors of order h^2. Equation 6-20 can be written as

$$h^2 f'' + h y(x)[hf'] + h^2 g(x)f = h^2 F(x)$$

Substituting the appropriate finite-difference formulas from Fig. 6-5 gives

$$f_{i+1} - 2f_i + f_{i-1} + \frac{h}{2} y_i [f_{i+1} - f_{i-1}] + h^2 g_i f_i = h^2 F_i$$

Rearranging gives

$$(1 - hy_i/2)f_{i-1} - (2 - h^2 g_i)f_i + (1 + hy_i/2)f_{i+1} = h^2 F_i \qquad (6\text{-}20a)$$

This equation is applied for $i = 1, 2, 3... n - 1$. The result is a set of $n - 1$ unknowns, f_i. To illustrate the use of this finite-difference solution technique, , consider the following example.

Example 6-1

Solve the following ordinary differential equation

$$f'' - (1/x)f' + x^2 f = x$$

with the following boundary conditions of $f(0) = 1$, $f(1) = 0$ and a value of $h = 1/4$. Referring to Eq. 6-20,

$$y(x) = 1/x \qquad g(x) = x^2 \qquad F(x) = x$$

The corresponding finite-difference equation is

$$(1 - 1/8x_i)f_{i-1} - (2 - x_i^2/16)f_i + (1 + 1/8x_i)f_{i+1} = x_i/16$$

The matrix form of this equation for $i = 1, 2, 3$, is

$$\begin{bmatrix} -(2 - x_1^2/16) & (1 + 1/8x_1) & 0 \\ (1 - 1/8x_2) & -(2 - x_2/16) & (1 + 1/8x_2) \\ 0 & (1 + 1/8x_3) & -(2 - x_3^2/16) \end{bmatrix} \begin{bmatrix} f_1 \\ f_2 \\ f_3 \end{bmatrix} = \begin{bmatrix} x_1/16 - (1 - 1/8x_1)f_0 \\ x_2/16 \\ x_3/16 - (1 + 1/8x_3)f_4 \end{bmatrix}$$

With $f_0 = f(0) = 1$, $f_4 = f(1) = 0$, $x_1 = 1/4$, $x_2 = 1/2$, and $x_3 = 3/4$, the above set of equations can be solved and the results are

$$f_0 = 1 \qquad \text{(Boundary condition)}$$
$$f_1 = 0.310$$
$$f_2 = 0.091$$
$$f_3 = 0.015$$
$$f_4 = 0 \qquad \text{(Boundary condition)}$$

Higher-ordered, linear boundary-value problems are also sometimes encountered. In the usual case they are even ordered, and the same number of boundary conditions normally are available at each end of the integration range. The simplest approach to these problems involves an identical approach to that for second-order problems with one additional aspect. Fictitious pivotal points are assumed outside the range of integration in order to express the boundary conditions in terms of finite differences. This is illustrated with the following example.

Example 6-2

Solve the following differential equation:

$$f^{iv} - 16f = x$$

using $h = 1/4$, $f(0) = 0$, $f(1) = 1$, $f'(1) = 0$ and $f''(0) = 0$.
From Fig. 6.5 for a finite-difference approximation with an error of order h^2,
the resulting difference equation is

$$f_{i-2} - 4f_{i-1} + (6 - 16h^4)f_i - 4f_{i+1} + f_{i+2} = h^4 x$$

For $h = 1/4$ and $i = 1, 2, 3$ the set of difference equations to be solved is

$i = 1$ $\quad f_{-1} - 4f_0 + 5\dfrac{15}{16} f_1 - 4 f_2 + f_3$	$= x_1/256$
$i = 2$ $\qquad f_0 - 4 f_1 + 5\dfrac{15}{16} f_2 - 4 f_3 + f_4$	$= x_2/256$
$i = 3$ $\qquad\qquad f_1 - 4 f_2 + 5\dfrac{15}{16} f_3 - 4f_4 + f_5$	$= x_3/156$

There are seven unknowns, f_{-1} through f_5, and there are three equations.
Consequently four of the unknowns are specified by the boundary conditions.
The boundary conditions in finite-difference nomenclature are

1. $f_0 = 0$
2. $f_4 = .1$
3. $f_4' = 0$
4. $f_0'' = 0$

One and two above can be used directly in the solution to the difference
equation set to specify values of f_0 and f_5. Finite-difference approximations
from Fig. 6-5 with an error of the same order as used in the approximation for
the differential equation is used as follows:

$$f_4' = (f_3 - f_5)/2h = 0 \qquad \text{or} \qquad f_3 = f_5$$

and

$$f_0'' = (f_{-1} - 2f_0 + f_1)/h^2 = 0 \qquad \text{or} \qquad f_{-1} = f_1$$

Using these two relations the set of difference equations can be solved for the
values of f_1 through f_3, and this is left as an exercise for the student (Prob. 6-2).

6-4. INITIAL VALUE PROBLEMS BY EULER TYPE METHODS [4]

An initial value problem is an ordinary differential equation and a set of associated conditions, all of which are specified at one point. Forward-integration methods are most commonly employed in the solution of these equations. This implies that y_{i+1} is found in terms of preceding values of y such as y_i, y_{i-1} or other past pivotal point. One of the simplest of these forward-integration methods is Euler's method when applied to the first-order equation $y' = f(x,y)$ with $y(x_0) = y_0$ is

$$y_{i+1} = y_i + hf_i \qquad\qquad (6\text{-}21)$$

where f_i is the derivative of y at i. This method does not require even spacing of pivotal points, and it has on error the order of which is h^2 as shown in Fig. 6-5 for the finite-difference approximation to y'.

Example 6-3

Obtain the solution to the following differential equation by Euler's method and compare the results with the analytical solution to determine the accumulation of error in the numerical solution.

$$y' = (y + y^2)/x$$

with $y(1) = 1$ and $h = 0.1$ for $i = 1$, 2, and 3. The difference equation is

$$y_{i+1} = y_i + 0.1 [y_i + (y_i)^2]/x_i$$

The analytical solution to the differential equation is

$$y = x/(2 - x)$$

The following compares the results of evaluating the difference equation and the differential equation:

i	0	1	2	3	
x	1.0	1.1	1.2	1.3	
y_i	1.0	1.20	1.44	1.71	Numerical solution
y	1.0	1.222	1.50	1.86	Analytical solution

The error in the numerical solution can be seen to propagate as the solution is continued.

An improvement to the Euler method is the Euler-Richardson method which decreases the truncation error to the order of h^3. It first calculates $y_{i+1}^{(1)}$ and $y_{i+\frac{1}{2}}$ by the conventional Euler approach:

$$y_{i+1}^{(1)} = y_i + h_i f_i$$

where

$$f_i = f(x_i, y_i)$$

$$y_{i+\frac{1}{2}} = y_i + \frac{h_i}{2} f_i$$

Then for a better value of $y_{i+1} = y_{i+1}^{(2)}$,

$$y_{i+1}^{(2)} = y_{i+\frac{1}{2}} + h_i f_{i+\frac{1}{2}}/2 = y_i + h_i(f_i + f_{i+\frac{1}{2}}/2)$$

where

$$f_{i+\frac{1}{2}} = f(x_{i+\frac{1}{2}}, y_{i+\frac{1}{2}})$$

These two values are now combined to produce the Euler-Richardson formula. The combination is based on the Richardson $h^2/3$ extrapolation formula and yields:

$$y_{i+1} = \frac{4}{3} y_{i+1}^{(2)} - \frac{1}{3} y_{i+1}^{(1)} = y_i + \frac{h_i}{3}(f_i + 2f_{i+\frac{1}{2}}) \qquad (6\text{-}22)$$

Example 6-4

Solve Example 6-3 by the Euler-Richardson method. The difference equations for this method are

$$
\begin{aligned}
f_i &= (y_i + y_i^2)/x_i \\
x_{i+\frac{1}{2}} &= x_i + h_i/2 \\
y_{i+\frac{1}{2}} &= y_i + h_i f_i/2 \\
f_{i+\frac{1}{2}} &= [y_{i+\frac{1}{2}} + y_{i+\frac{1}{2}}^2]/x_{i+\frac{1}{2}} \\
y_{i+1} &= y_i + h_i(f_i + 2f_{i+\frac{1}{2}})/3
\end{aligned}
$$

The following table compares the results of the Euler-Richardson method with the Euler method and the analytical solution. The improvement of the numerical solution using the Euler-Richardson is significant.

i	0	1	2	3	
x	1.0	1.1	1.2	1.3	
y_i	1.0	1.20	1.44	1.71	Euler method
y_i	1.0	1.21	1.48	1.82	Euler-Richardson method
y	1.0	1.222	1.50	1.86	Analytical solution

The intermediate values in the solution are shown below.

i	x_i	y_i	f_i	$x_{i+\frac{1}{2}}$	$y_{i+\frac{1}{2}}$	$f_{i+\frac{1}{2}}$
0	1	1	2	1.05	1.1	2.20
1	1.1	1.213	2.440	1.15	1.335	2.711
2	1.2	1.475	3.042	1.25	1.627	3.419
3	1.3	1.804	–	–	–	–

The Euler-Gauss predictor-corrector method is another improvement of Euler's method. It is based on the use of Euler's method to find a predicted value of y_{i+1}, and then the Euler method is used repeatedly to obtain corrected values of y_{i+1} to any required degree of accuracy. The following equations are used to develop the Euler-Gauss predictor-corrector method.

$$y_{i+1}^{(1)} = y_i + h_i f_i$$
$$y_{i+\frac{1}{2}} = y_i + h_i f_i/2$$
$$x_{i+\frac{1}{2}} = x_i + h_i/2$$
$$f_{i+\frac{1}{2}} = f(x_{i+\frac{1}{2}}, y_{i+\frac{1}{2}})$$
$$y_{i+1}^{(2)} = y_{i+\frac{1}{2}} + h_i f_{i+1}/2 = y_i + h_i (f_i + f_{i+\frac{1}{2}})/2$$

An extrapolation formula, Richardson's method [4], is used to predict $y_{i+1}^{(1)}$ and $y_{i+1}^{(2)}$, and the result is

$$y_{i+1} = y_i + h_i (f_i + 2f_{i+\frac{1}{2}})/3 \qquad (6\text{-}23)$$

Example 6-5
Solve Example 6-3 by the Euler-Gauss method and compare the results with the Euler and the Euler-Richardson method.
To compute y_1 $(i = 0)$,

$$y_1 = y_0 + 0.1 (f_0 + 2f_{\frac{1}{2}})/3$$

where

$$y_0 = 1, \ x_0 = 1$$
$$f_0 = (y_0 + y_0^2)/x_0 = 2$$
$$x_{\frac{1}{2}} = 1 + 0.1/2 = 1.05$$
$$y_{\frac{1}{2}} = y_0 + 0.1 f_0/2 = 1.1$$
$$f_{\frac{1}{2}} = (y_{\frac{1}{2}} + y_{\frac{1}{2}}^2)/x_{\frac{1}{2}} = 2.20$$
$$y_1 = 1.0 + 0.1 (2 + (2)2.2)/3 = 1.213$$

The following table compares the results for the Euler-Gauss method with the two previous methods and the analytical solution.

i	0	1	2	3	
x	1	1.1	1.2	1.3	
y_i	1.0	1.21	1.475	1.79	Euler-Gauss method
y_i	1.0	1.21	1.44	1.80	Euler-Richardson method
y_i	1.0	1.20	1.44	1.71	Euler method
y	1.0	1.22	1.50	1.86	Analytical solution

In comparing the results of the three methods with the analytical solution, it is seen that more accurate results are obtained with the procedures that require more effort. This would be expected since the two modified Euler methods are approximating the quadratic term in the Taylor series expansion for improved accuracy.

These results also show the effect of truncation error on the solution of an initial value problem. If the solution was continued for a large value of i the error would propagate through the solution, and the results could be meaningless. After three steps there is a significant difference between the analytical solution (1.86) and the numerical solutions (1.71-1.80).

It is necessary to have the numerical solution converge to the true solution of the differential equation. The numerical solution is effected by the truncation error and the step size, h. With a digital computer the truncation error is usually determined by the number of significant figures used in the computations. Therefore to obtain an accurate numerical solution the step size is usually reduced until the values of the dependent variable do not change as a result of changing the step size. If this is the case the numerical integration procedure is said to be *stable* and to *converge*. It is beyond the scope of this chapter to discuss error analysis, and the reader is referred to Kunz[8] or Collatz[9] for a treatment of this topic.

6-5. RUNGE-KUTTA METHODS[1,4]

The Runge-Kutta methods are basically one-step techniques for numerically solving first-order differential equations of the type

$$y' = f(x,y) \tag{6-24}$$

Runge-Kutta methods are typically classed according to their order p, which is equivalent to (a) the order of h in term in the Taylor series through which the method approximates, (b) the number of points at which f must be evaluated over the interval, and (c) the error is of the order h^{p+1}.

First-Order Runge-Kutta

As this method must approximate the Taylor series through the term involving h, it can be derived directly from the Taylor series expansion

$$y_{i+1} = y_i + hy_i' = y_i + hf_i$$

This will be recognized as the Euler method presented earlier.

Second-Order Runge-Kutta

This formula must approximate the Taylor series through the h^2 term, or

$$y_{i+1} = y_i + hy_i' + \frac{h^2}{2} y_i''$$

$$= y_i + hf_i + \frac{h^2}{2} f_i'$$

As y is a function of x, f_i' can be expressed in terms of the partial derivatives by the chain rule

$$f' = \frac{df(x,y)}{dx} = \frac{\partial f}{\partial x} + \frac{\partial f}{\partial y}\frac{dy}{dx} = f_x + f_y f$$

Substituting in the previous relationship gives

$$y_{i+1} = y_i + hf_i + \frac{h^2}{2}[f_{x,i} + f_{y,i}f_i] \tag{6-25}$$

The general formula for a second-order Runge-Kutta is

$$y_{i+1} = y_i + h[a_1 f(x_i,y_i) + a_2 f(x_i + b_1 h, y_i + b_2 f_i h)]$$

Comparing to Eq. 6-25, it is seen that for this expression to approximate the Taylor series through the h^2 term, the constants a_1, a_2, b_1, and b_2 must be chosen so that following equality holds:

$$a_1 f_i + a_2 f(x_i + b_1 h, y_i + b_2 f_i h) = f_i + \frac{h}{2}[f_{x,i} + f_{y,i}f_i] \tag{6-26}$$

Expanding $f(x_i + b_1 h, y_i + b_2 f_i h)$ about $f(x_i,y_i)$ in a Taylor series gives

$$f(x_i + b_1 h, y_i + b_2 f_i h) = f_i + b_1 hf_{x,i} + b_2 hf_{y,i}f_i$$

Substituting into Eq. 6-26 gives

$$(a_1 + a_2)f_i + a_2 b_1 hf_{x,i} + a_2 b_2 hf_{y,i}f_i = f_i + \frac{h}{2}f_{x,i} + \frac{h}{2}f_{y,i}f_i$$

Equating coefficients yields

$$a_1 + a_2 = 1$$
$$a_2 b_1 = 1/2$$
$$a_2 b_2 = 1/2$$

Note that there are three equations and four unknowns, leaving one degree of freedom. It consequently follows that there is an infinite number of second-order Runge-Kutta formulas.

Perhaps the more common of these are

(1) Improved Euler method

$$a_1 = a_2 = 1/2$$
$$b_1 = b_2 = 1$$

(2) Modified Euler method

$$a_1 = 0 \quad a_2 = 1$$
$$b_1 = b_2 = 1/2$$

Fourth-Order Runge-Kutta

Although the algebra is quite more involved, fourth-order Runge-Kutta formulas can be obtained in a manner analogous to the previous derivation of second-order Runge-Kutta formulas. Although there is also an infinite number of possibilities for fourth-order Runge-Kutta formulas, the following is probably most frequently used:

$$y_{i+1} = y_i + h(K_1 + 2K_2 + 2K_4 + K_4)/6 \qquad (6\text{-}27)$$

where

$$K_1 = f(x_i, y_i)$$
$$K_2 = f(x_i + h/2, \quad y_i + hK_1/2)$$
$$K_3 = f(x_i + h/2, \quad y_i + hK_2/2)$$
$$K_4 = f(x_i + h, \quad y_i + hK_3)$$

Obviously any order Runge-Kutta method is possible, and the reader is referred to Refs. 6 and 7 for a through treatment.

Another point that should be noted is that all of the Runge-Kutta methods are "self-starting," that is, the only information required to initialize the solution is the value of y at some value of x. This is available for the first interval from the boundary condition and from the previous interval for all others. To illustrate the use of fourth-order Runge-Kutta method, consider the following example.

Example 6-6|5|

Use the fourth-order Runge-Kutta to solve the differential equation, $y' = x + y^2$; and to determine the value of $y(0.2)$ and $y(0.4)$. The boundary condition is $y(0) = 1$ and the interval $h = 0.2$.

$i = 1$

$$y_1 = y_0 + 0.2(K_1 + 2K_2 + 2K_3 + K_4)/6$$
$$y_0 = 1$$
$$K_1 = 0 + (1)^2 = 1$$
$$K_2 = 0.2/2 + 1 + 0.2(1)/2^2 = 1.310$$
$$K_3 = 0.2/2 + 1 + 0.2(1.31)/2^2 = 1.379$$
$$K_4 = 0.2 + 1 + 0.2(1.379)^2 = 1.828$$
$$y_1 = 1 + 0.2(8.206)/6 = 1.274$$

$i = 2$

$$y_2 = 1.274 + 0.2(1.274 + 2(2.264) + 2(2.551) + 2.651)/6$$
$$= 1.726$$

Fourth-order Runge-Kutta is a widely used numerical technique because of the accuracy that can be obtained with a straightforward application of the method. It can be extended for solution of second-order equations, and the following formula are given without proof [10].

Consider the second-order differential equation

$$y'' = f(x,y,y')$$

The fourth-order formulas are

$$y_{i+1} = y_i + hy'_i + (k_1 + k_2 + k_3)/6$$

$$y'_{i+1} = y'_i + (k_1 + k_2 + k_3 + k_4)/6$$

(6-28)

$$k_1 = hf(x_i, y_i, y'_i)$$
$$k_2 = hf(x_i + h/2, y_i + hy'_i/2 + hk_1/8, y'_i + k_1/2)$$
$$k_3 = hf(x_i + h/2, y_i + hy'_i/2 + hk_2/8, y'_i + k_2/2)$$
$$k_4 = hf(x_i + h, y_i + hy'_i + hk_3/2, y'_i + k_3)$$

The procedure for applying these is the same as illustrated in the example for the solution to the first-order differential equation. The only difference is that an additional boundary condition is necessary to start the solution and is used to compute k_1.

6-6. INTEGRATION FORMULAS[4]

In the last section several numerical methods were presented for solving equations of the form

$$y' = f(x,y)$$

Another approach for solving such problems is via integration of this equation in the following manner:

$$y_{i+n} = y_i + \int_{x_i}^{x_i + nh} y' \, dx$$

What is needed is an integration formula for evaluating this integral from the points at which f is available. Such integration formulas are analogous to the differentiation formulas presented in Sec. 6-2.

Suppose we consider evaluating the integral

$$I = \int_{x_i}^{x_i + nh} f(x) \, dx = \int_0^{nh} f(x_i + \xi) \, d\xi$$

Expanding $f(x_i + \xi)$ into a Taylor series and integrating gives

$$I = \int_0^{nh} [f(x_i) + f'(x_i) + f''(x_i)\frac{\xi^2}{2} + \cdots] \, d\xi$$

$$= nhf(x_i) + \frac{n^2 h^2}{2} f'(x_i) + \frac{n^3 h^3}{6} f''(x_i) + \cdots \tag{6-29}$$

The derivatives in this equation can now replaced by the differentiation formulas in Fig. 6-5. To illustrate this, let $n = 1$, and Eq. 6-29 becomes

$$I = hf_i + \frac{h^2}{2} f_i' + \frac{h^3}{6} f_i'' \tag{6-30}$$

Then using a two-point differentiation formula for f_i,

$$f_i' = \frac{f_{i+1} - f_i}{h} - \frac{1}{2} hf'' + \cdots$$

Equation 6-30 becomes

$$I = h\left[f_i + \frac{f_{i+1} - f_i}{2} \right] - \frac{h^3}{12} f_i'' + \cdots$$

Neglecting the terms of third-order and higher, the intergration formula becomes

$$I = \frac{h}{2} [f_i + f_{i+1}]$$

with error $\dfrac{-h^3 f''}{12}$. This is the familiar trapezoid rule and is the second entry in the table of integration formulas in Fig. 6-7.

For even values of n, this procedure is modified slightly by writing the Taylor series about the midpoint of the interval rather than the extreme. As an example, for $n = 4$, Eq. 6-29 becomes

$$I = \int_{x_0}^{x_1} f(x)\, dx = \int_{-2h}^{2h} f(x_2 + \xi)\, d\xi$$

$$= \int_{-2h}^{2h} \left[f(x_2) + \xi f'(x_2) + \frac{\xi^2}{2} f''(x_2) + \cdots \right] d\xi$$

$$= 4hf_2 + \frac{8}{3} h^3 f_2'' + \frac{8}{15} h^5 f_2^{iv} + \cdots$$

Then, using the three-point differentiation formula,

$$f_2'' = \frac{f_1 - 2f_2 + 3f_3}{h^2} - \frac{1}{12} h^2 f_2^{iv} + \cdots$$

to obtain

$$I = h\left[4f_2 + \frac{8f_1 - 16f_2 + 8f_3}{3} \right] + \left[\frac{8}{15} - \frac{8}{36} \right] h^5 f_2^{iv}$$

$I_{nm}(i)$	Multiplier	f_{i-1}	f_i	f_{i+1}	f_{i+2}	f_{i+3}	f_{i+4}	f_{i+5}	
$I_{11}(i)$	h		1	0					$(h^2/2)f'$
$I_{12}(i)$	$h/2$		1	1					$-(h^3/12)f''$
$I_{13}(i)$	$h/12$		5	8	-1				$(h^4/24)f'''$
$I_{14}(i)$	$h/24$		9	19	-5	1			$-(19/720)h^5f^{iv}$
$I_{15}(i)$	$h/720$		251	646	-264	106	-19		$(27/1440)h^6f^v$
$I_{14}(i)$	$h/24$	-1	13	13	-1				$(11/720)h^5f^{iv}$
$I_{15}(i)$	$h/720$	-19	346	456	-74	11			$-(11/1440)h^6f^v$
$I_{13}(i)$	$h/12$	-1	8	5					$-(h^4/24)f'''$
$I_{21}(i)$	$2h$		0	1	0				$(h^3/3)f''$
$I_{21}(i)$	2		1	0	0				$2h^2f'$
$I_{23}(i)$	$h/3$		1	4	1				$-(h^5/90)f^{iv}$
$I_{24}(i)$	$h/3$		1	4	1	0			$-(h^5/90)f^{iv}$
$I_{34}(i)$	$3h/8$		1	3	3	1			$-(3/80)h^5f^{iv}$
$I_{43}(i)$	$4h/3$		0	2	-1	2	0		$(14/45)h^5f^{iv}$
$I_{23}(i)$	$h/12$	4	-8	28	0				$(h^4/3)f'''$
$I_{22}(i)$	$h/2$		0	4	0				$(h^3/3)f''$
$I_{32}(i)$	$h/2$		-3	9	0	0			$(27/12)h^3f''$
$I_{42}(i)$	$h/2$		-8	16	0	0	0		$(80/12)h^3f''$
$I_{52}(i)$	$h/2$		-15	25	0	0	0	0	$(175/12)h^3f''$
$I_{31}(i)$	h		3	0	0	0			$(9/2)h^2f'$
$I_{41}(i)$	h		4	0	0	0	0		$8h^2f'$
$I_{51}(i)$	h		5	0	0	0	0	0	$(25/2)h^2f'$
$I_{33}(i)$	$h/12$		9	0	27	0			$(9/24)h^4f'''$
$I_{43}(i)$	$h/12$		32	-64	80	0	0		$(64/24)h^4f'''$
$I_{33}(i)$	$h/12$	27	-72	81	0	0			$(63/24)h^4f'''$
$I_{43}(i)$	$h/12$	80	-208	176	0	0	0		$(28/3)h^4f'''$

FIG. 6-7.   Integration formulas. *Source:* J.M. McCormick and M.G. Salvadori, *Numerical Methods in Fortran Programming* (Englewood Cliffs, N.J.: Prentice-Hall, Inc., 1964). Used with permission.

$$I = \frac{4h}{3}\left[2f_1 - f_2 + 2f_3\right] + \frac{14}{45}\,h^5f_2^{iv} + \cdots$$

Thus the formula is

$$I = \frac{4h}{3}\left[2f_1 - f_2 + 2f_3\right] \tag{6-31}$$

with error $14h_5f_2^{iv}/45$. This entry is I_{43} in Fig. 6-7.

The subscripts on I_{nm} in Fig. 6-7 have the following meanings

n = number of intervals over which the integral is taken

m = number of points in the differentiation formulation used to obtain the integration formula.

That is, I_{43} is the integral over four intervals using a three-point differentiation formula. The interval over which the integral is taken is given by the boldface entries in Fig. 6-7.

Probably the most widely used integration formula is Simpson's rule which is of a simple form and has an error of order of h^5. It is given in Fig. 6-7 as I_{23} and can be written as

$$I_{23} = (h/3)[f_i + 4f_{i+1} + f_{i+2}]$$ (6-32)

The integration formulas can be applied successively to a multiple of the basic interval h. For example Simpson's rule can be applied to any overall interval made up of an even number of subintervals of width h.

$$I_s = \frac{h}{3}\left[(f_0 + 4f_1 + f_2) + (f_2 + 4f_3 + f_4) + \cdots + (f_{n-2} + 4f_{n-1} + f_n)\right]$$ (6-33)

or

$$I_s = \frac{h}{3}\left[f_0 + 4f_1 + 2f_2 + 4f_3 + 2f_4 + \cdots + 2f_{n-2} + 4f_{n-1} + f_n\right]$$ (6-34)

This formula is readily implemented in FORTRAN using one DO loop.

6-7. PREDICTOR-CORRECTOR METHODS [1, 4]

Integration formula can be used in solving the following differential equation $y' = f(x,y)$. Suppose that values for y_3 and for three equidistantly spaced previous points, namely y_0, y_1, and y_2 are available. As at the start of the last section, the solution can be written as

$$y_4 = y_0 + \int_{x_0}^{x_4} y' dx$$ (6-35)

Using the integration formula of Eq. 6-31, namely,

$$I_{43} = \frac{4h}{3}(2f_1 - f_2 + 2f_3)$$

Equation 6-35 becomes

$$y_4 = y_0 + \frac{4h}{3}(2f_1 - f_2 + 2f_3)$$

This can be generalized (superscript o signifies the predicted value of y_{i+1}) as

$$y^0_{i+1} = y_{i-3} + \frac{4h}{3}(2f_{i-2} - f_{i-1} + 2f_i) \qquad (6\text{-}36)$$

which is known as *Milne's predictor*. Notice that it predicts the next point, y_{i+1} from *information already available,* namely, y_i and f_i at prior points.

To improve the accuracy of the results of Eq. 6-36, a *corrector* is used. Using Simpson's one-third rule, Eq. 6-32, the corrector is

$$y^{(j)}_{i+1} = y_{i-1} + \frac{h}{3}(f_{i-1} + 4f_i + f_{i+1}) \qquad (6\text{-}37)$$

where the superscript j denotes y_{i+1} after the jth application of the predictor formula. Typically Eq. 6-37, known as *Milne's corrector,* is applied only once. The error of Eq. 6-37 is of the order h^5.

Certainly a wide variety of predictor-corrector combinations could be developed using the integration formula in Fig. 6-7. Instead of presenting others, perhaps our space could be better devoted to comparing the predictor-corrector methods in general to the Runge-Kutta methods.

First, observe that the predictor-corrector is not "self-starting." That is, the present point and three previous points are required before the predictor-corrector can be applied. These points are typically obtained using Runge-Kutta methods.

Second, the predictor-corrector typically require fewer calculations. Note that the application of the corrector in Eq. 6-37 requires no additional evaluations of f, as f_i, f_{i-1}, and f_{i-2} are already available. Application of the corrector in Eq. 6-37 requires only one evaluation of f per interval. Contrast this with Runge-Kutta method with the same error (h^5). The fourth-order Runge-Kutta requires four evaluations of f per interval.

Example 6-7

Solve the differential equation $y' = x + y^2$ by the Milne predictor-corrector method for y_4. Extend the results from Example 6-6 to start the solution.

Predictor for $i = 3$ $\qquad y^0_4 = y_3 + \dfrac{4h}{3}(2f_1 - f_2 + 2f_3)$

Corrector for $j = 1$ $\qquad y^1_4 = y_2 + \dfrac{h}{3}(f_2 + 4f_3 + f_4)$

Extending the results from Example 6-6, using fourth-order Runge-Kutta gives

i	0	1	2	3	4
x_i	0	0.2	0.4	0.6	0.8
y_i	1	1.274	1.726	2.356	–
f_i	1	1.823	3.379	6.153	–

Predictor

$$y_4^0 = 2.356 + 4(0.2)[2(1.823) - 3.379 + 2(6.153)]/3$$
$$= 5.709$$
$$f_4 = 0.8 + (5.709)^2 = 33.393$$

Corrector

$$y_4^1 = 2.356 + (0.2)[3.379 + 4(6.153) + 33.393]/3$$
$$= 6.448$$

One could iterate to improve the value obtained from the corrector or the step-size could be reduced to obtain a more accurate value with one application of the corrector.

6-8. SIMULTANEOUS EQUATIONS

The Runge-Kutta methods and predictor-corrector methods presented above were developed specifically for the equation

$$y' = f(x,y)$$

Suppose that instead there are two simultaneous equations of the form

$$y' = f(x,y,z)$$
$$z' = g(x,y,z)$$

Consider the application of the fourth-order Runge-Kutta method in Eq. 6-27. The steps would be as follows:

1. Values are available for y_i and z_i.
2. Evaluate K_1 and G_1 as follows:

$$K_1 = f(x_i, y_i, z_i)$$
$$G_1 = g(x_i, y_i, z_i)$$

3. Evaluate K_2 and G_2 as follows:

$$K_2 = f(x_i + h/2, y_i + hK_1/2, z_i + hG_1/2)$$

4. Evaluate K_3, G_3, K_4, and G_4 similarly.
5. Calculate y_{i+1} and z_{i+1} as follows:

$$y_{i+1} = y_i + h(K_1 + 2K_2 + 2K_3 + 2K_4)/6$$
$$z_{i+1} = z_i + h(G_1 + 2G_2 + 2G_3 + 2G_4)/6$$

The basic procedure is to carry all the simultaneous equations through the solution in the same manner as one equation would be treated.

The predictor-corrector methods are treated similarly.   For example, Milne's predictor-corrector discussed above would be applied as follows:

1. Given values of y_i, z_i, and three previous points.
2. Use the predictor to calculate y_{i+1}^0 and z_{i+1}^0 as follows:

$$y_{i+1}^0 = y_{i-3} + \frac{4h}{3} (2f_{i-2} - f_{i-2} + 2f_i)$$

$$z_{i+1}^0 = z_{i-3} + \frac{4h}{3} (2g_{i-2} - g_{i-1} + 2g_i)$$

3. Apply the corrector:

$$y_{i+1}^{(j)} = j_{i-1} + \frac{h}{3} (f_{i-1} + 4f_i + f_{i+1})$$

$$z_{i+1}^{(j)} = z_{i-1} + \frac{h}{3} (g_{i-1} + 4g_i + g_{i+1})$$

Again, each equation is treated in the same manner as a single equation would be treated.

6-9. HIGHER-ORDER EQUATIONS

Recall that the Runge-Kutta methods and predicator-correctors presented above were all for first-order equations. Consider a second-order equation of the form

$$\frac{d^2y}{dx^2} + a_1 \frac{dy}{dx} + a_2 y = b(x) \qquad (6\text{-}38)$$

$$y(0) = C_1$$
$$y'(0) = C_2$$

This equation can be represented by defining a new variable equal to the derivative y—that is,

$$z = \frac{dy}{dx} \qquad (6\text{-}39)$$

Substituting into Eq. 6-33 gives

$$\frac{dz}{dx} + a_1 z + a_2 y = b(x)$$

This equation along with Eq. 6-34 forms a pair of simultaneous equations:

$$z' = b(x) - a_1 z - a_2 y = g(x,y,z)$$
$$y' = z \qquad\qquad = f(x,y,z)$$
$$z(0) = C_2$$
$$y(0) = C_1$$

This set of equations can be solved by the concepts presented in Sec. 6-8 and a high-order differential equation can be represented as a set of simultaneous first-order differential equations. Also in Sec. 6-5 the algorithm was presented for the Runge-Kutta solution of a second-order differential equation. Comparing the algorithms that one would use considering the higher-order equation as a set of first-order differential equations and the second-order algorithm, there is a difference in the two procedures. This difference arises in the development of the algorithm when the Taylor series expansion is substituted for the corresponding terms in the second-order differential equation. The result is a more accurate algorithm which requires less bookkeeping[11], and it may be advantageous to transform a set of first order equations into one higher-order equations. Collatz[12] tabulates the fourth-order Runge-Kutta algorithm for equations through fourth-order.

6-10. SUMMARY

This chapter has presented a review of some numerical methods adaptable to computer solutions of mathematical models. In no way should the contents be interpreted as the sole source of methods available to the reader, but rather a brief index of methods commonly applied in solving a variety of models.

The selection of one method over another is a function of a number of items:
1. The mathematical model under study
2. The computer hardware and software available
3. The individual's knowledge of the various methods
4. The time and money available to accomplish the task

Therefore, no standard procedure can be illustrated. It is important, however, to use a method in which the advantages and shortcomings are familar to the reader, since a model is only as good as the values generated by the numerical technique used.

REFERENCES

1. D. D. McCracken and W. S. Dorn, *Numerical Methods and Fortran Programming,* Wiley, New York, 1965.
2. R. W. Hamming, *Numerical Methods for Scientists and Engineers,* McGraw-Hill, New York, 1962.
3. James Singer, *Elements of Numerical Analysis,* Academic Press, New York, 1964.
4. J. M. McCormick and M. G. Salvadori, *Numerical Methods of Fortran,* Prentice-Hall, Englewood Cliffs, N.J., 1964.
5. A. L. Nelson, K. W. Foley, and M. Coral, *Differential Equations,* Heath, Boston, 1952.
6. Anthony Ralston, "Runge-Kutta Methods with Minimum Error Bounds," *Mathematics of Computers, Vol 16* 80, 1962, p. 431.

7. H. A. Luther, and H. P. Konen, "Some Fifth-order Classical Runge-Kutta Formulas," *S.I.A.M. Review,* Vol. 7, 4, (October 1965), p. 551.
8. K. A. Kunz, *Numerical Analysis,* McGraw-Hill, New York, 1957.
9. L. Collatz, *The Numerical Treatment of Differential Equations,* Springer-Verlag, New York, 1960.
10. M. Abramowitz and I. A. Stegun, *Handbook of Mathematical Functions,* National Bureau of Standards Applied Mathematics Series, 55, U. S. Government Printing Office, Washington, D.C., 1965.
11. Collatz, *op. cit.* p. 117.
12. Collatz, *op. cit.* p. 69.

PROBLEMS

6-1. Formulate the heat-conduction example of Sec. 6-1 as a second-order boundary-value problem. Solve the problem for the value of $K = 1$ and $h = 1/4$ and $h = 1/8$ using the boundary conditions $u(0) = 1$ and $u(1) = 0$.

6-2. Solve the difference equation set given in Example 6-2 for the values of the function at the intermediate points in the interval from 0 to 1.

6-3. Solve the differential equation for flow in a tubular reactor for a first-order chemical reaction with $C_{Af} = 1$, using the Euler method for $h = 0.1$, and compare the results with the analytical solution given in Sec. 1-9 for values of kV_s/vL of $0.1, 0.2$, and 0.3.

6-4. Solve Prob. 6-3 using the Euler-Richardson Method and compare the results with the analytical solution.

6-5. Develop the finite-difference algorithms for heat transfer the vapor-liquid heat exchanger, Sec. 1-6, for the Euler-Gauss method and the fourth-order Runge-Kutta method. Which would require the smaller interval size for the same accuracy?

6-6. A second-order ordinary differential equation is obtained from the model for heat conduction in a rod of Sec. 1-3. Develop the fourth-order Runge-Kutta equations for this differential equation, and compare with the set of fourth-order Runge-Kutta equations obtained from converting the second-order equation to a system of first-order equations.

6-7. Formulate the set of ordinary differential equations of Sec. 2-2 into a set of finite-difference equations by the Euler method and develop a digital-computer flow diagram to solve the set of equations.

6-8. Formulate the set of differential equations of Sec. 2-2 into a set of difference equations to be solved by the method of fourth-order Runge-Kutta. Compare the computer flow diagram for this problem with the one from Prob. 6-7, and discuss the difference in complexity of the two methods with the size of the interval that would be required to obtain an accurate solution.

6-9. Extend the computer flow diagram of Prob. 6-8 to use the fourth-order Runge-Kutta method to start the solution, then switch to the Milne predictor-corrector method.

6-10. Convert the following set of first-order ordinary differential equations into one higher-order ordinary differential equation. Give the fourth-order Runge-Kutta formulas for this case and compare it with the fourth-order Runge-Kutta formulation for the simultaneous solution of the equation set:

$$y' = xyz$$
$$z' = zy/z$$

chapter 7

Problems Requiring Partial Differential Equations

Although the approach to formulating models for all systems is basically the same, there are some slight differences between problems involving one independent variable and problems involving two or more independent variables. First, problems with only one independent variable are described by ordinary differential equations, whereas partial differential equations are required for problems with two or more independent variables. Second, the boundary conditions are also slightly different. In this chapter these differences are considered for representative problems with more than one independent variable. Only problem formulations are considered in this chapter; solutions are discussed in the next chapter.

7-1. REVIEW OF PARTIAL DIFFERENTIAL EQUATIONS

In the previous chapters two superficially different problems have been considered. The first type included steady-state problems in which the dependent variables were functions of position (one-dimensional problems). The second type included unsteady-state problems in which the dependent variables were constant throughout the system at any given instant of time (zero-dimensional problems). Each of these led to ordinary differential equations. If an unsteady-state problem is considered in which the dependent variables are also functions of position, two independent variables —time and position— must be included. Alternatively, a second position variable may be required, thus yielding two independent variables for position. The ultimate number of independent variables —four— is required for unsteady-state problems in three dimensions.

For such problems the basic principles are the same as for the problems considered in the previous chapters. The main difference is that partial derivatives must be used instead of ordinary derivatives.

For example, consider the case in which the unsteady-state temperature profiles in a bar are to be determined. The derivatives can be taken with respect to either the space (position) variable giving the temperature gradient at a given time, or with respect to the time variable which gives the time rate of temperature change at some position. The two derivatives are not the same and must be distinguished. Since more than one independent variable is present, partial derivatives must be used. In writing a partial derivative it is implied that other independent variables are treated as constants in the differentiation process. Different notations used are

$$\frac{\partial T}{\partial z} = \left(\frac{\partial T}{\partial z}\right)_t = T_z$$

The second of these, $\left(\frac{\partial T}{\partial z}\right)_t$, is used to indicate that the time variable t is held constant, which is implied in taking the partial derivative anyway. The third, T_z, is commonly used because of its ease in writing, but subscripts of this type are used in this text to indicate the directions for such quantities as velocity components. Consequently, the notation $\frac{\partial T}{\partial z}$ will be used for the partial derivative of T with respect to z.

7-2. BOUNDARY CONDITIONS FOR PARTIAL DIFFERENTIAL EQUATIONS [1,2]

Before treating an example in which partial differential equations are obtained, a word should be said concerning boundary conditions. A thorough discussion of the number and type of boundary conditions necessary and sufficient to insure a solution for a general partial differential equation is beyond the scope of this text. However, for most physical problems encountered in engineering, the following rules usually hold.

1. To have a sufficient boundary condition, the value of the dependent variable must be specified at some given value of one independent variable and at every value of the other independent variables.
2. The number of boundary conditions for one dependent variable in terms of one of the independent variables equals the order of the highest derivative with respect to the given independent variable.

For example, the vibration of a string stretched between two points on the x-axis as shown in Fig. 7-1 is described by the following partial differential equation [3] (the "wave" equation):

$$\frac{\partial^2 u(x,t)}{\partial x^2} = \alpha \, \frac{\partial^2 u(x,t)}{\partial t^2} \tag{7-1}$$

where u is the displacement from the x-axis. For this equation two boundary conditions are required for $u(x,t)$ at given values of x, and two boundary conditions are required for $u(x,t)$ at given values of t.

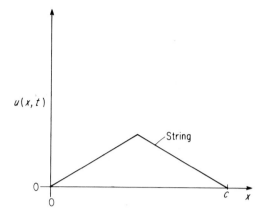

FIG. 7-1. The plucked string.

First, consider the boundary conditions for $u(x,t)$ at given values of x. If the two end points are held at fixed positions on the x-axis as shown in Fig. 7-1, the two boundary conditions would be

$$u(0,t) = 0$$
$$u(c,t) = 0$$

However, if the end of the string at $x = 0$ were moved with time, the boundary condition at $x = 0$ would become

$$u(0,t) = f(t)$$

where $f(t)$ describes the movement of the end point with time. It is perfectly acceptable for the boundary conditions at specified values of one independent variable to be functions of one or more *other* independent variables.

The two boundary conditions for Eq. 7-1 at given values of t follow the same rules as those on x. An interesting situation is that of the plucked string, i.e., where the string is given an initial displacement and released from rest at time zero. In this case one boundary condition describes the initial displacement and is of the form:

$$u(x,0) = f_1(x)$$

where $f_1(x)$ describes the initial displacement. The second boundary condition arises from the fact that the string is initially at rest, i.e., the velocity at time zero is zero. Since the velocity is the partial derivative of the displacement with respect to time, this boundary condition is

$$\frac{\partial u(x,0)}{\partial t} = 0$$

These are only a few of the possible combinations of boundary conditions that are sufficient to insure a solution of Eq. 7-1. The boundary conditions for problems involving other aspects of engineering work are usually equally easy to derive, and in the early examples in this chapter the boundary conditions are derived in detail. As the student becomes more familiar with the formulation of partial differential equations and their boundary conditions, such detailed derivations become unnecessary.

The boundary conditions for dependent variables at time equal zero are often referred to as initial conditions. In essence, they are the same as other boundary conditions, and no distinction will be made in this text.

7-3. FORMULATION FOR A ONE-DIMENSIONAL UNSTEADY-STATE PROBLEM

In Chapter 1 several one-dimensional steady-state problems were examined, all of which were described by ordinary differential equations. Recall that the accumulation term was zero in these examples, and the only independent variable was the position variable. If the unsteady-state equation instead of the steady-state equation were desired, the accumulation term would not equal zero, thus introducing the time variable in addition to the position variable. Thus a partial differential equation will result.

Although the unsteady-state formulation for one of the examples in Chapter 1 could be used as an example in this chapter, a completely different example will be used. In addition to illustrating the techniques for formulating partial differential equations, the following example also illustrates the use of spherical coordinates [7].

Consider the situation in which a hot ball initially at uniform temperature T_0 is dropped into a bucket of cool water. The temperature at a point inside the sphere is a function of both time t and distance r from the center. In this example, a partial differential equation is derived that can be solved for the temperature at a point within the sphere as a function of time. The following data are needed in the derivation:

T_w temperature of water, °F
c_p heat capacity of metal, Btu/lb-°F
ρ Density of metal, lb/ft³
k Thermal conductivity of metal, Btu/hr-ft²-°F/ft
R Radius of sphere, ft

Additional assumptions:

(a) Physical properties are constant.
(b) Heat is lost from the surface by convection.

(c) The convective heat-transfer coefficient h is assumed to be constant.

(d) The sphere is not large enough to change the water temperature significantly.

Quantity Conserved: Heat

System: A shell of thickness Δr at a distance r from the center of the sphere, as shown in Fig. 7-2.

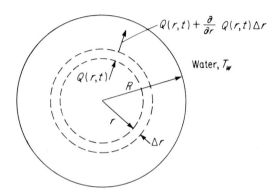

FIG. 7-2. Sphere considered in Sec. 7-3.

Independent Variables: Time t and distance r from the center,

Dependent Variables: Temperature T, conductive heat flow Q, and convective heat flux q_v. The temperature is a function of r and t, which is denoted by $T(r,t)$, and similarly for the rate of conductive heat flow, $Q(r,t)$. The heat flux by convection at the surface of the sphere is $q_v(t)$, since it is a function of time only.

Input: Heat is being conducted into the element at face r at rate $Q(r,t)$.

Output: Heat is being conducted from the element at face $r + \Delta r$ at rate $Q(r + \Delta r,t)$. In terms of $Q(r,t)$,

$$Q(r + \Delta r, t) = Q(r,t) + \frac{\partial}{\partial r}\,[Q(r,t)]\,\Delta r$$

In this case the partial derivative is used instead of the total derivative since Q is a function of two independent variables.

Accumulation: The rate of accumulation of heat within the element equals the rate of change of the enthalpy of the metal in the element. To determine the enthalpy, define a reference state at temperature T_R as discussed in Sec. 1-6. The accumulation term is

$$\frac{\partial}{\partial t}\,\{(4\pi r^2 \Delta r)\rho\, c_p\,[\,T(r,t) - T_R]\}$$

(The volume of a shell of inner radius r and thickness Δr is $4\pi r^2 \Delta r$.)   Again the partial instead of the total derivative is used, since T is a function of two independent variables.

Formulating the Equation:

$$\text{In} - \text{Out} = \text{Accumulation}$$

$$Q(r,t) - \left\{ Q(r,t) + \frac{\partial}{\partial r} [Q(r,t)] \Delta r \right\}$$

$$= \frac{\partial}{\partial t} \left\{ (4\pi r^2 \Delta r) \rho c_p [T(r,t) - T_R] \right\}$$

Recalling that $r, \Delta r, \rho, c_p$, and T_R are independent of time t,

$$-\frac{\partial Q(r,t)}{\partial r} = 4\pi r^2 \rho c_p \frac{\partial T(r,t)}{\partial t} \tag{7-2}$$

Recall that the surface area of a sphere of radius r is $4\pi r^2$.   By Fourier's law,

$$Q(r,t) = -k(4\pi r^2) \frac{\partial T(r,t)}{\partial r}$$

Substituting into Eq. 7-2,

$$\frac{k}{\rho c_p} \cdot \frac{1}{r^2} \frac{\partial}{\partial r} \left[r^2 \frac{\partial T(r,t)}{\partial r} \right] = \frac{\partial T(r,t)}{\partial t}$$

Expanding,

$$\frac{k}{\rho c_p} \left[\frac{\partial^2 T(r,t)}{\partial r^2} + \frac{2}{r} \frac{\partial T(r,t)}{\partial r} \right] = \frac{\partial T(r,t)}{\partial t}$$

The group $(k/\rho c_p)$ often appears in heat-transfer work and is consequently given a special name–*thermal diffusivity* α.   Therefore, the above equation becomes

$$\alpha \left[\frac{\partial^2 T(r,t)}{\partial r^2} + \frac{2}{r} \frac{\partial T(r,t)}{\partial r} \right] = \frac{\partial T(r,t)}{\partial t} \tag{7-3}$$

Boundary Conditions: For the above partial differential equation, three boundary conditions are needed–one on t and two on r.   The boundary condition for t is the initial temperature profile in the sphere.

$$T(r,0) = T_0 \tag{7-4}$$

This equation reads as follows: The temperature for all values of r and at time equal zero is T_0. To be a proper boundary condition the value of the independent variable must be specified at some value of one of the independent variables for all values of the remaining independent variable(s).

To develop the boundary conditions for r, the approach used is similar to that used in Sec. 1-5.   To develop one boundary condition, an element of radius Δr (as shown in Fig. 7-3) at the center of the sphere is examined.

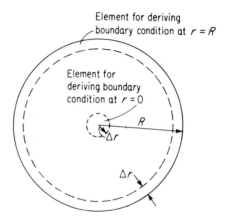

Element for deriving
boundary condition at $r = R$

Element for
deriving boundary
condition at $r = 0$

R

Δr

Δr

FIG. 7-3.   Special elements examined for deriving the boundary conditions for Eq. 7-3.

Input: No heat enters this element.

Output: Heat is being lost by conduction from the surface at $r = \Delta r$.   In terms of temperature gradients,

$$Q(0 + \Delta r, t) = -4\pi(\Delta r)^2 k \frac{\partial T(0 + \Delta r, t)}{\partial r}$$

Accumulation: The rate of change of enthalpy of the metal within this element is:

$$\frac{\partial}{\partial t}\left\{ \frac{4}{3}\pi(\Delta r)^3 \rho c_p [T(r,t) - T_R] \right\}$$

Formulating the Equation:

$$\text{In} - \text{Out} = \text{Accumulation}$$

$$0 + 4\pi(\Delta r)^2 k \frac{\partial T(0 + \Delta r, t)}{\partial r} = \frac{4}{3}\pi(\Delta r)^3 \rho c_p \frac{\partial T(r,t)}{\partial t}$$

Dividing by $(\Delta r)^2$ and taking the limit as $\Delta r \to 0$,

$$\frac{\partial T(0,t)}{\partial r} = 0 \qquad\qquad (7\text{-}5)$$

This boundary condition reads as follows: The partial of T with respect to r at r equals zero for all time t is zero.

The second boundary condition is formulated by examining an element of thickness Δr at the surface as also shown in Fig. 7-3.

Input: Heat is conducted into the element across the face $R - \Delta r$ at the following rate:

$$- k \, [4\pi \, (R - \Delta r)^2] \, \frac{\partial \, T(R - \Delta r, t)}{\partial r}$$

Output: The rate at which heat is lost by convection from the surface is given by:

$$h \, [4\pi R^2] \, [T(R, t) - T_w]$$

Accumulation: The rate of change of the enthalpy of the metal within the element is:

$$\frac{\partial}{\partial t} \, \{4\pi \, (R - \Delta r)^2 (\Delta r) \rho \, c_p [\, T(r, t) - T_R]\}$$

Formulating the Equation:

$$\text{In} - \text{Out} = \text{Accumulation}$$

$$- k \, [4\pi \, (R - \Delta r)^2] \, \frac{\partial \, T(R - \Delta r, t)}{\partial r} - h \, [4\pi \, R^2] \, [T(R, t) - T_w]$$

$$= \frac{\partial}{\partial t} \, \{4\pi \, (R - \Delta r)^2 \, (\Delta r) \rho \, c_p [\, T(r, t) - T_R]\}$$

Taking the limit as $\Delta r \rightarrow 0$ and simplifying,

$$\frac{\partial \, T(R, t)}{\partial r} = - \frac{h}{k} \, [\, T(R, t) - T_w] \tag{7-6}$$

In summary, the equation and boundary conditions for this example are as follows:

$$\alpha \left\{ \frac{\partial^2 \, T(r, t)}{\partial r^2} + \frac{2}{r} \, \frac{\partial \, T(r, t)}{\partial r} \right\} = \frac{\partial \, T(r, t)}{\partial t} \tag{7-3}$$

$$T(r, 0) = T_0 \tag{7-4}$$

$$\frac{\partial \, T(0, t)}{\partial r} = 0 \tag{7-5}$$

$$\frac{\partial \, T(R, t)}{\partial r} = - \frac{h}{k} \, [\, T(R, t) - T_w] \tag{7-6}$$

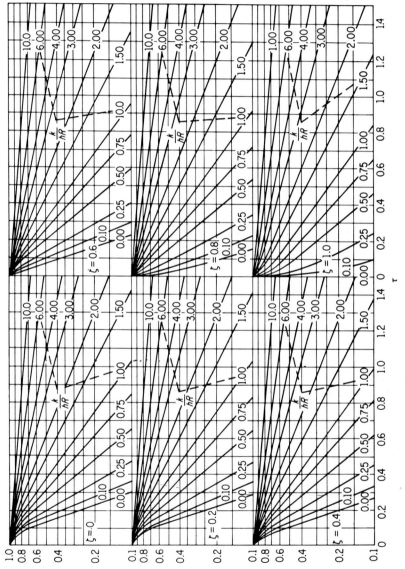

FIG. 7.4. Temperature profiles in sphere in terms of dimensionless variables and parameters. (Reprinted with permission from L.M.K. Boelter, V.H. Cherry, H.A. Johnson, and R.C. Martinelli. *Heat Transfer Notes*, U. of California Press, 1948.

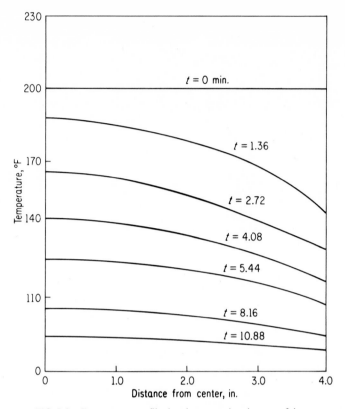

FIG. 7-5.   Temperature profiles in sphere at various instants of time.

Solution: The above set of equations can be solved analytically (separation of variables), numerically (Chapter 8), or by analogs (Chapter 10).   At this point let us consider not how to solve this equation, but only the nature of the solution.   A change of variable is often made prior to solving such equations for two reasons: (1) to obtain as many homogeneous boundary conditions as possible, and (2) to make the solution as general as possible.   For the above equation, the following changes of variable are useful.

$$\varphi = \frac{T - T_w}{T_i - T_w} \tag{7-7}$$

$$\zeta = \frac{r}{R} \tag{7-8}$$

$$\tau = \frac{\alpha t}{R^2} \tag{7-9}$$

In Chapter 9 a technique is presented for obtaining such changes of variables.
    Using these dimensionless variables, the general solution to Eq. 7-3 can be

expressed in terms of these variables and the dimensionless group k/hR. Using the technique of separation of variables, Mickley, Sherwood, and Reid[1] obtain a series solution. Although the details are of no interest here, the graphical solution is presented in Fig. 7-4 in such a manner that all possibilities are covered.

To illustrate the use of these general plots, consider the specific example of a steel ball 8 in. in diameter which is dropped into a water bath at 80 °F. The initial temperature of the ball was 200°F, and the convective heat transfer coefficient at the surface is 100 Btu/hr-ft²-°F. The thermal conductivity of steel is 25 Btu/hr-ft²-°F/ft, and its thermal diffusivity is about 0.49 ft²/hr. From this information and the graphs in Fig. 7-4, the temperature profile in the ball at various instants of time can be calculated. The first step is to calculate the value of k/hR, which is 0.75. Using this value and the graphs in Fig. 7-4, the temperature profiles in Fig. 7-5 can be obtained very easily.

7-4. A MULTIDIMENSIONAL PROBLEM: DERIVATION OF THE CONTINUITY EQUATION

In many engineering problems the dependent variables are functions of two or three space variables. Although it could be argued that essentially all realistic engineering problems are three-dimensional problems, the solutions are usually so much easier to obtain for one- or two-dimensional problems that the approximations are convenient whenever acceptable. The equations for two- and three-dimensional problems are formulated in a manner very similar to those for one-dimensional problems. The main difference is that the element for which the equations are written must be of infinitesimal length in the direction of each space variable required for the problem. In Fig. 7-6a are shown the elements for one-, two-, and three-dimensional problems in rectangular coordinates. The concepts are directly extended to cylindrical and spherical coordinates to obtain the three-dimensional elements shown in Fig. 7-6b.

Another added complexity in two- and three-dimensional problems above that for one-dimensional problems is the necessity of representing the direction in addition to the magnitude of fluxes. Since the flux is a vector quantity, the most convenient way to do this is in terms of the components of the vector in the direction of the independent variables [8,9]. For example, the velocity v of a fluid in motion can be represented by its component vectors v_x, v_y, and v_z in the direction of the axes. Recall that v_x is used to denote the component of velocity v in the x-direction, not $\partial v/\partial x$. A similar situation exists for spherical and cylindrical coordinates.

As the introductory example of multidimensional problems, the derivation of the continuity equation in three dimensions will be used [1,8,9].

In this example the law of conservation of mass is applied to an element in a compressible fluid in unsteady-state flow. The element examined is a rectan-

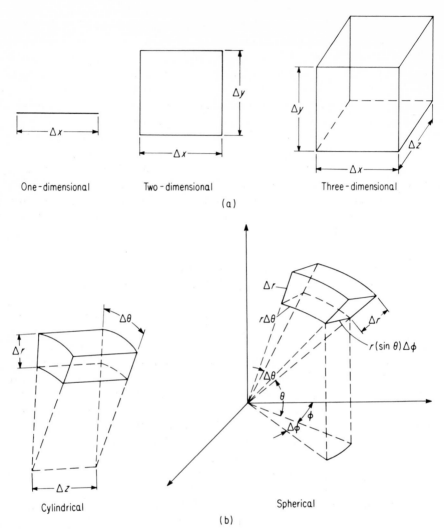

One-dimensional   Two-dimensional   Three-dimensional

(a)

Cylindrical   Spherical

(b)

FIG. 7-6.   Examples of differential elements.   (a) Elements in rectangular coordinates.   (b) Three-dimensional elements.

gular element of infinitesimal lengths Δx, Δy, and Δz as shown in Fig. 7-7.   The element is fixed in space, i.e., the Euler point of view instead of the Lagrange point of view is used.   At a point (x,y,z) and at time t, the velocity of the fluid is v, which can be expressed in terms of its components v_x, v_y, and v_z.   At time t the density at point (x,y,z) is ρ.   For this problem, v is a function of x, y, z, and t; that is, $v(x,y,z,t)$.   Thus, the components v_x, v_y, and v_z are also functions of x,y,z, and t.   However, these will not be shown as arguments for the sake of shorter notation.

If we consider v_x, v_y, and v_z as being positive in the positive x-, y-, and z-directions, the fluid then is entering the element in Fig. 7-7 at the faces perpendicular to the x-, y-, and z-axes at $x = x$, $y = y$, and $z = z$. Furthermore, the fluid is leaving this element at the faces perpendicular to the x-, y-, and z- axes at $x = x + \Delta x$, $y = y + \Delta y$, and $z = z + \Delta z$. This is also indicated by the arrows in Fig. 7-7. At the face perpendicular to the x-axis at $x = x$, the rate (lb/hr) at which fluid is entering the element equals $\rho v_x \Delta y \Delta z$. Since the components v_y and v_z are parallel to this face, they do not contribute anything at this face. Similarly, the rate at which the fluid is leaving the face at $x = x + \Delta x$ is $(\rho v_x \Delta y \Delta z)|_{x+\Delta x}$, which is given by

$$\rho v_x \Delta y \Delta z + \frac{\partial}{\partial x} (\rho v_x \Delta y \Delta z) \Delta x$$

Similarly, the rates at which fluid is entering the other faces are

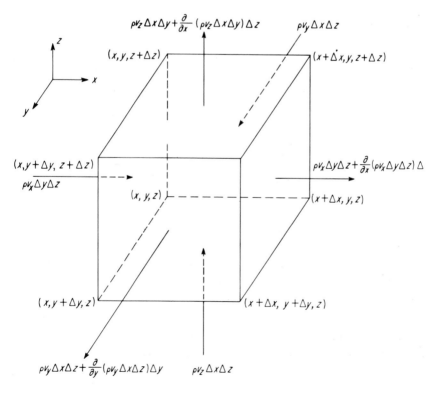

FIG. 7-7. Control volume used to derive the continuity equation.

Face at $y = y$

$$\rho v_y \Delta x \Delta z$$

Face at $y = y + \Delta y$

$$\rho v_y \Delta x \Delta z + \frac{\partial}{\partial y} (\rho v_y \Delta x \Delta z) \Delta y$$

Face at $z = z$

$$\rho v_z \Delta x \Delta y$$

Face at $z = z + \Delta z$

$$\rho v_z \Delta x \Delta y + \frac{\partial}{\partial z} (\rho v_z \Delta x \Delta y) \Delta z$$

The only remaining term is the accumulation term:

$$\frac{\partial}{\partial t} (\rho \Delta x \Delta y \Delta z)$$

Substituting these terms in a mass balance,

$$\rho v_x \Delta y \Delta z - \left[\rho v_x \Delta y \Delta z + \frac{\partial}{\partial x} (\rho v_x \Delta y \Delta z) \, \Delta x \right]$$

$$+ \rho v_y \Delta x \Delta z - \left[\rho v_y \Delta x \Delta z + \frac{\partial}{\partial y} (\rho v_y \Delta x \Delta z) \, \Delta y \right]$$

$$+ \rho v_z \Delta x \Delta y - \left[\rho v_z \Delta x \Delta y + \frac{\partial}{\partial z} (\rho v_z \Delta x \Delta y) \, \Delta z \right]$$

$$= \frac{\partial}{\partial t} (\rho \Delta x \Delta y \Delta z)$$

The equation can be simplified to

$$\frac{\partial \rho}{\partial t} + \frac{\partial (\rho v_x)}{\partial x} + \frac{\partial (\rho v_y)}{\partial y} + \frac{\partial (\rho v_z)}{\partial z} = 0 \qquad (7\text{-}10)$$

This is the continuity equation commonly used in fluid mechanics. For incompressible flow, the density ρ is constant, and Eq. 7-10 reduces to

$$\frac{\partial v_x}{\partial x} + \frac{\partial v_y}{\partial y} + \frac{\partial v_z}{\partial z} = 0 \qquad (7\text{-}11)$$

Alternate forms of Eq. 7-10 have been derived for other coordinate systems [8]. In cylindrical coordinates the continuity equation is

$$\frac{\partial \rho}{\partial t} + \frac{1}{r} \frac{\partial}{\partial r} (\rho r v_r) + \frac{1}{r} \frac{\partial}{\partial \theta} (\rho v_\theta) + \frac{\partial}{\partial z} (\rho v_z) = 0 \qquad (7\text{-}12)$$

In spherical coordinates the continuity equation becomes

$$\frac{\partial \rho}{\partial t} + \frac{1}{r^2} \frac{\partial}{\partial r}(\rho r^2 v_r) + \frac{1}{r \sin \theta} \frac{\partial}{\partial \theta}(\rho v_\theta \sin \theta)$$

$$+ \frac{1}{r \sin \theta} \frac{\partial}{\partial \varphi}(\rho v_\varphi) = 0 \tag{7-13}$$

These can be derived by either (1) a derivation similar to that used to derive Eq. 7-10, or (2) by making a transformation of axes on Eq. 7-10.

7-5. A TWO-DIMENSIONAL STEADY-STATE HEAT-CONDUCTION EXAMPLE

To illustrate the principles applicable in heat conduction problems, an idealized problem will be presented. Consider the block of length b in the x-direction and c in the y-direction as shown in Fig. 7-8. The faces at $x = b$ and $y = 0$ are insulated, the face at $x = 0$ is held at $0°$, and the face at $y = c$ is maintained such that the temperature is a prescribed function of x, namely $f(x)$. As heat conduction in the z-direction is neglected, this is a two-dimensional steady-state heat-transfer problem. This also allows us to develop the equations based on a unit length in the z-direction.

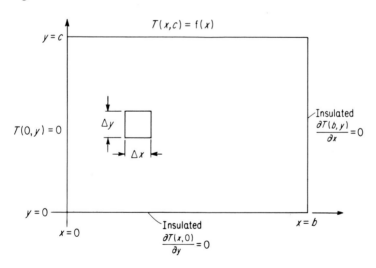

FIG. 7-8. Block considered in Sec. 7-5.

The derivation of the equations proceeds in much the same manner as for the continuity equation. Since this is a two-dimensional problem, the first step is to define an element of infinitesimal length Δx and Δy in the x- and y-directions, as shown in Fig. 7-8. The heat flux at a point is in reality a vector quantity—that is, the direction of flow must be specified in addition to the magnitude. Therefore, it can be represented by its components in the x- and y-directions. The heat flux is denoted as q, and its components in the x- and y-directions are denoted as q_x and q_y, respectively.

As in the one-dimensional problems, the components of the heat flux are taken to be positive when the flow is in the positive x- and y-directions. By Fourier's law the heat flux in the x-direction is related to the thermal gradient as follows:

$$q_x(x,y) = -k\frac{\partial T(x,y)}{\partial x} \tag{7-14}$$

Similarly the y-component is given by

$$q_y(x,y) = -k\frac{\partial T(x,y)}{\partial y} \tag{7-15}$$

To develop the equation for the temperature profiles in the block in Fig. 7-8, a heat balance must be made for the element. Since the heat flux is positive in the positive x-direction, the rate at which heat is conducted into the element across the face at x in Fig. 7-8 is $(\Delta y)q_x(x,y)$. The rate at which heat leaves the element across the face at $x + \Delta x$ is $(\Delta y)q_x(x + \Delta x,y)$. Similarly, the rate at which heat enters across the face at y is $(\Delta x)q_y(x,y)$, and the rate at which heat leaves the element across the face at $y +\Delta y$ is $(\Delta x)q_y(x,y +\Delta y)$.

Substituting these quantities into a heat balance indicates

$$\text{In} - \text{Out} = 0$$

$$\Delta y[q_x(x,y) - q_x(x + \Delta x,y)] + \Delta x[q_y(x,y) - q_y(x,y + \Delta y)] = 0$$

This equation can be simplified to

$$\frac{\partial q_x}{\partial x} + \frac{\partial q_y}{\partial y} = 0 \tag{7-16}$$

The next step is to substitute for q_x and q_y from Eqs. 7-4 and 7-5. If the thermal conductivity is constant, Eq. 7-16 becomes

$$\frac{\partial^2 T(x,y)}{\partial x^2} + \frac{\partial^2 T(x,y)}{\partial y^2} = 0 \tag{7-17}$$

This equation is commonly known as *Laplace's equation in two dimensions*.

The boundary conditions for this equation are readily derived from an inspection of the boundaries of the block. Since the faces at $x = b$ and $y = 0$ are insulated, two of the boundary conditions are

$$\frac{\partial T(b,y)}{\partial x} = 0 \qquad (7\text{-}18)$$

$$\frac{\partial T(x,0)}{\partial y} = 0 \qquad (7\text{-}19)$$

The face at $x = 0$ is maintained at a constant temperature of 0°, which is expressed as

$$T(0,y) = 0 \qquad (7\text{-}20)$$

Similarly, the face at $y = c$ is maintained at a constant temperature distribution represented by $f(x)$, which is expressed as

$$T(x,c) = f(x) \qquad (7\text{-}21)$$

The temperature distribution in essentially all problems involving two-dimensional steady-state heat conduction in solids is described by Eq. 7-17. The only major difference from one problem to the next is the boundary conditions. For the above boundary conditions an analytic solution is possible, but many problems require a numerical solution.

Using the technique of separation of variables [3,5,6], Eq. 7-17 and its boundary conditions can be solved to yield

$$T(x,y) = \frac{2}{b} \sum_{n=1}^{\infty} \frac{\sin \dfrac{(2n-1)\pi x}{2b} \cosh \dfrac{(2n-1)\pi y}{2b} \displaystyle\int_0^b f(\xi) \sin \dfrac{(2n-1)\pi \xi}{2b} \, d\xi}{\cosh \dfrac{(2n-1)\pi c}{2b}} \qquad (7\text{-}22)$$

Although this is the analytic solution, it is by no means simple. To evaluate such an expression a digital computer would be extremely helpful, and this is the reason that a numerical solution is used for many problems for which an analytic solution could be obtained.

As a specific example, consider the case in which b and c equal π and

$$f(x) = 10x$$

That is, at $x = 0$ the temperature at the upper edge of the block in Fig. 7-8 is 0°, and is 31.416° at $x = \pi$. With these values the integral in Eq. 7-22 can be evaluated to yield the final expression

$$T(x,y) = \frac{80}{\pi} \sum_{n=1}^{\infty} \frac{(-1)^{n+1} \sin \dfrac{(2n-1)x}{2} \cosh \dfrac{(2n-1)y}{2}}{(2n-1)^2 \cosh \dfrac{(2n-1)\pi}{2}}$$

The solution for various values of y is given in Fig. 7-9.

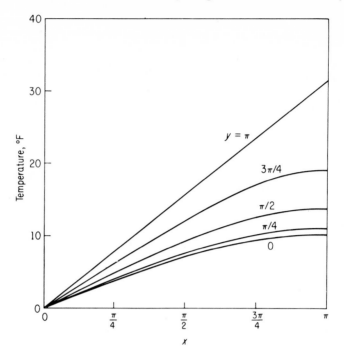

FIG. 7-9.   Temperature profiles in the block in Fig. 7-8.

7-6.   A PROBLEM REQUIRING SIMULTANEOUS PARTIAL DIFFERENTIAL EQUATIONS

Although no radically new concepts are required to formulate simultaneous partial differential equations, an example will be presented in order to illustrate the minor differences between this and previous examples.   In the following examples a tubular reactor in which the reactants are in laminar flow is considered. Since the heat of reaction is significant, an enthalpy balance as well as a component material balance must be made.

In Chapter 1 a tubular reactor was examined in which the velocity profile was considered to be flat, i.e., plug flow.   This reactor was described by an ordinary differential equation with the axial distance z being the independent variable.   In this example a similar reactor is considered but with two main differences: first, flow is assumed to be in the laminar region, in which case the velocity $v(r)$ is the parabolic function of the radial distance r as derived in Sec. 1-11; and second, the reaction is assumed to be exothermic with heat of reaction ΔH.   To provide cooling, the reactor tube is immersed in a fluid which is boiling at temperature T_b.   The overall heat-transfer coefficient between the

reacting fluid and the boiling liquid is U_i based on the inside area of the
reactor tube of radius R_i.   A sketch of the reactor is shown in Fig. 7-10.

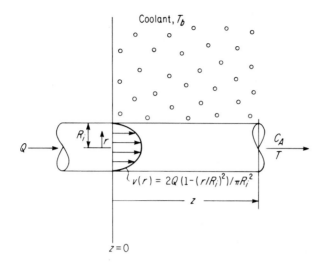

Coolant, T_b

$v(r) = 2Q\,(1-(r/R_i)^2)/\pi R_i^2$

z

$z = 0$

$$A \rightarrow B$$

with rate constant k_1.   The variation of k_1 with temperature obeys the Arrhenius
expression of the form $k_1 = k_0\,\exp(-a/T)$. The volume change on reaction is
negligible.   The fluid enters the reaction zone with an established velocity
profile, and at uniform temperature T_0 and concentration C_{A_0}.   The diffusivity
of A through B, D_{AB}, is assumed to be independent of temperature.   The
thermal conductivity of the liquid is k, also assumed independent of temperature.

For this problem steady-state partial differential equations are to be
developed for temperature T and concentration C_A as functions of r and z.
In the following discussion, the conduction and diffusion in the axial direction
are neglected.

The first step is to define an element of differential thickness in the r- and
z-directions.   The first equation to be developed is for the concentration profile,
and the element in Fig. 7-11 shows the terms in the balance for component A.
This component enters the element by bulk flow at face z, and this input term is

$$v(r)C_A\,(r,z)(2\pi r\Delta r)$$

At face $z + \Delta z$ the rate at which component A leaves with the bulk-flow stream is

$$v(r)C_A\,(r,z + \Delta z)\,(2\pi r\Delta r)$$

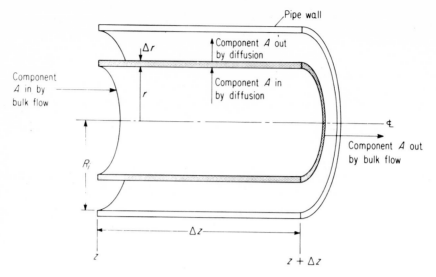

FIG. 7-11.   Control volume considered for the laminar-flow tubular reactor.

In the radial direction, component A enters and leaves by diffusion.   The rate at which A enters at face r is

$$(2\pi r \Delta z)\left[-D_{AB} \frac{\partial C_A(r,z)}{\partial r}\right]$$

At face $r + \Delta r$ the rate at which component A leaves is given by

$$2\pi (r + \Delta r)\left[-D_{AB} \frac{\partial C_A(r + \Delta r,z)}{\partial r}\right] \Delta z$$

The rate at which component A is formed by reaction within the element is

$$r_A (2\pi r\Delta r\Delta z)$$

Since component A is a reactant, the rate of formation of A is

$$r_A = -k_1 C_A(r,z)$$

Combining these terms into a component material balance yields

$$2\pi r(\Delta r)v(r)C_A(r,z)$$

$$-2\pi r(\Delta r)v(r)C_A(r,z + \Delta z)$$

$$-2\pi D_{AB}(\Delta z)r \frac{\partial C_A(r,z)}{\partial r}$$

$$+2\pi D_{AB}(\Delta z)(r + \Delta r) \frac{\partial C_A(r + \Delta r,z)}{\partial r}$$

$$-2\pi k_1 r(\Delta r)(\Delta z)C_A(r,z) = 0$$

This equation simplifies to

$$\frac{D_{AB}}{r} \frac{\partial}{\partial r} \left[r \frac{\partial C_A(r,z)}{\partial r} \right] - v(r) \frac{\partial C_A(r,z)}{\partial z} - k_1 C_A(r,z) = 0 \qquad (7\text{-}23)$$

For this equation, one boundary condition is required at a specified value of z and two boundary conditions are required at specified values of r. One boundary condition can be obtained from the inlet concentration profile:

$$C_A(r,0) = C_{A_0} \qquad (7\text{-}24)$$

Another boundary condition can be obtained by inspecting the conditions at the center of the tube. As no material diffuses across the center, the boundary condition is

$$\frac{\partial C_A(0,z)}{\partial r} = 0 \qquad (7\text{-}25)$$

This can be verified by making a component mass balance for a special infinitesimal element at the center. It is also obvious that there can be no diffusion in the r-direction at $r = R$, which is expressed mathematically by

$$\frac{\partial C_A(R_i,z)}{\partial r} = 0 \qquad (7\text{-}26)$$

This is the second boundary condition on r.

At this point an equation for $C_A(r,z)$ with $v(r)$ and k_1 as parameters has been developed. The velocity $v(r)$ is known to be parabolic from the momentum balance. However, the rate constant k_1 is a function of temperature T, and consequently it is a function of r and z. Equation 7-23 cannot be solved for $C_A(r,z)$ until the temperature T as a function of r and z is known, and thus an enthalpy balance must be introduced.

To formulate an equation for $T(r,z)$, an enthalpy balance is made for an element similar to the one in Fig. 7-11 for the component mass balance. The specific terms for the enthalpy balance are shown in Fig. 7-12.

Heat in by conduction:

$$- k(2\pi r \Delta z) \frac{\partial T(r,z)}{\partial r}$$

Heat out by conduction:

$$- 2\pi k(\Delta z)(r + \Delta r) \frac{\partial T(r + \Delta r, z)}{\partial r}$$

Enthalpy in with bulk flow:

$$v(r)(2\pi r \Delta r) \rho c_p [T(r,z) - T_R]$$

where T_R = reference temperature

ρ = density, independent of $T(r,z)$ and $C_A(r,z)$

c_p = heat capacity, independent of $T(r,z)$ and $C_A(r,z)$

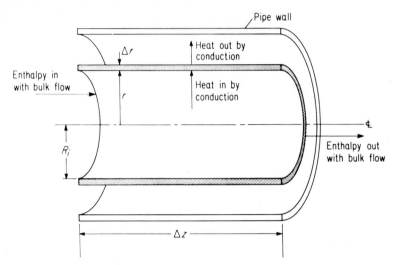

FIG. 7-12.   Control volume for making enthalpy balance. Heat is either absorbed or released by the reaction occurring in the control volume.

Enthalpy out with bulk flow:

$$v(r)\,(2\pi r\Delta r)\rho c_p \lfloor T(r,z+\Delta z) - T_R \rfloor$$

Heat of reaction: Assuming the reaction is exothermic, the rate of heat input is

$$(-\Delta H)\,k_1\,C_A(r,z)\,(2\pi r\Delta r\Delta z)$$

Collecting these terms into an overall enthalpy balance yields upon simplification

$$(-\Delta H)k_1 C_A(r,z) + k\,\frac{\partial^2 T(r,z)}{\partial r^2} + \frac{k}{r}\,\frac{\partial T(r,z)}{\partial r} - v(r)\rho c_p\,\frac{\partial T(r,z)}{\partial z} = 0$$

For this equation two boundary conditions on $T(r,z)$ are required at specified values of r and one boundary condition on $T(r,z)$ is required at a specified value of z.   This latter boundary condition can be the temperature profile at the entrance.   Since the fluid enters at a uniform temperature T_0, one boundary condition is

$$T(r,0) = T_0 \tag{7-28}$$

As for the component mass balance, one boundary condition can be specified at $r = 0$ and the other can be specified at $r = R_i$. These two boundary conditions can be determined by making enthalpy balances on special elements at the center and at the tube wall. However, at this point most readers probably have enough experience to write them from inspection. In either case, these boundary conditions are

$$\frac{\partial T(0,z)}{\partial r} = 0 \tag{7-29}$$

$$-\frac{\partial T(R_i,z)}{\partial r} = \frac{U_i}{k} \left[T(R_i,z) - T_b \right] \tag{7-30}$$

This last boundary condition essentially states that the heat transferred to the fluid at the wall by conduction must be transferred by convection to the coolant.

Recapitulating, two partial differential equations and their boundary conditions have been formulated, namely

$$D_{AB} \frac{\partial^2 C_A(r,z)}{\partial r^2} + \frac{D_{AB}}{r} \frac{\partial C_A(r,z)}{\partial r} - v(r) \frac{\partial C_A(r,z)}{\partial z} - k_1 C_A(r,z) = 0$$

$$C_A(r,0) = C_{A_0}$$

$$\frac{\partial C_A(0,z)}{\partial r} = 0$$

$$\frac{\partial C_A(R_i,z)}{\partial r} = 0$$

$$(-\Delta H) k_1 C_A(r,z) + k \frac{\partial^2 T(r,z)}{\partial r^2} + \frac{k}{r} \frac{\partial T(r,z)}{\partial r} - v(r)\rho c_p \frac{\partial T(r,z)}{\partial z} = 0$$

$$T(r,0) = T_0$$

$$\frac{\partial T(0,z)}{\partial r} = 0$$

$$\frac{\partial T(R_i,z)}{\partial r} = -\frac{U_i}{k} \left[T(R_i,z) - T_b \right]$$

Note that $C_A(r,z)$ must be known to solve for $T(r,z)$, and $T(r,z)$, and thus k_1, must be known to solve for $C_A(r,z)$. Consequently, these equations must be solved simultaneously.

7-7. LUMPED-PARAMETER MODEL OF A TWO-DIMENSIONAL SYSTEM[9]

In Sec. 3-8 a lumped-parameter model of a one-dimensional system was developed. In this section these concepts are extended to a two-dimensional system, and their futher extension to a three-dimensional system will be rather obvious.

To illustrate these principles, consider the chimney illustrated in Fig. 7-13. The gas inside the chimney is assumed to be at 200°F, and the outside gas is assumed to be at 100°F. The convective heat-transfer coefficient on the inside surface is 12 Btu/hr-ft²-°F, and is 3 Btu/hr-ft²-°F on the outside surface. The thermal conductivity, specific heat, and density are 0.6 Btu/hr-ft²-°F/ft, 0.20 Btu/lb-°F, and 144 lb/ft³, respectively. The steady-state temperature profiles for inside temperatures of 200°F and 300°F will be obtained, and the transition from an inside temperature of 200°F to 300°F will be determined also.

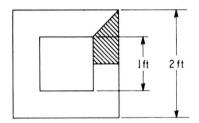

FIG. 7-13.   Cross section of chimney.

If the heat flow up the chimney is neglected, the problem involves heat flow in two directions, thus necessitating a two-dimensional grid. From the symmetry of the problem, only an eighth of the chimney must be considered. This section with its grid is shown in Fig. 7-14. The equations describing the elements for this grid are derived in the same manner as for the one-dimensional grid. Note from the symmetry that the temperature is the same for nodes 2 and 2', 5 and 5', etc. A unit distance in the z-direction is assumed for developing the equation.

For node 1 the equation describing the heat transfer in and out of this element is

$$\frac{k}{\Delta x}\frac{\Delta y}{2}[T_2(t) - T_1(t)] + \frac{k}{\Delta y}\frac{\Delta x}{2}[T_2(t) - T_1(t)]$$

$$+ h_0\frac{\Delta y}{2}[T_0 - T_1(t)] + h_0\frac{\Delta x}{2}[T_0 - T_1(t)]$$

$$= \rho c_p\frac{\Delta x\Delta y}{4}\frac{dT_1(t)}{dt}$$

Since Δx and Δy are equal, this equation becomes

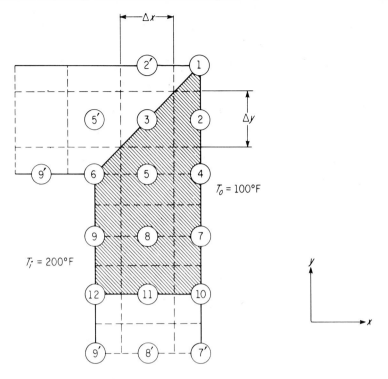

FIG. 7-14.   Eighth section of the chimney with its grid.

$$T_1(t) = -\frac{4k + 4h_0\,\Delta x}{\rho c_p(\Delta x)^2}\,T_1(t) + \frac{4k}{\rho c_p(\Delta x)^2}\,T_2(t) + \frac{4h_0}{\rho c_p\Delta x}\,T_0 \qquad (7\text{-}31)$$

where

$$\dot{T}_1(t) = \frac{dT_1(t)}{dt}$$

Let

$$a = \frac{k}{\rho c_p(\Delta x)^2}$$

and

$$c = \frac{h_0}{\rho c_p\Delta x}$$

Equation 7-31 reduces to

$$\dot{T}_1(t) = -4(a + c)\,T_1(t) + 4aT_2(t) + 4cT_0 \qquad (7\text{-}32)$$

The equations for the other eleven nodes are

$$\dot{T}_2(t) = -2(2a + c)T_2(t) + aT_1(t) + aT_4(t) + 2aT_3(t) + 2cT_0$$

$$T_3(t) = -4aT_3(t) + 2aT_2(t) + 2aT_5$$

$$\dot{T}_4(t) = -2(2a + c)T_4(t) + aT_2(t) + aT_7(t) + 2aT_5(t) + 2cT_0$$

$$\dot{T}_5(t) = -4aT_5(t) + aT_4(t) + aT_6(t) + aT_3(t) + aT_8(t)$$

$$\dot{T}_6(t) = -4(a + \frac{b}{3})T_6(t) + \frac{4}{3}aT_9(t) + \frac{8}{3}aT_5(t) + \frac{4}{3}bT_i$$

$$\dot{T}_7(t) = -2(2a + c)T_7(t) + aT_4(t) + aT_{10}(t) + 2aT_8(t) + 2cT_0$$

$$\dot{T}_8(t) = -4aT_8(t) + aT_7(t) + aT_9(t) + aT_5(t) + aT_{11}(t)$$

$$\dot{T}_9(t) = -2(2a + b)T_9(t) + aT_{12}(t) + aT_6(t) + 2aT_8(t) + 2bT_i$$

$$\dot{T}_{10}(t) = -2(2a + c)T_{10}(t) + 2aT_7(t) + 2aT_{11}(t) + 2cT_0$$

$$\dot{T}_{11}(t) = -4aT_{11}(t) + aT_{10}(t) + aT_{12}(t) + 2aT_8(t)$$

$$\dot{T}_{12}(t) = -2(2a + b)T_{12}(t) + 2aT_9(t) + 2aT_{11}(t) + 2bT_i$$

where

$$b = \frac{h_i}{\rho c_p \Delta x}$$

The procedures for solving the above set of simultaneous equations are the same as used in Chapter 3 for the one-dimensional case. In Table 7-1 the steady-state temperatures are given at each node in the grid when the inside temperature is 200°F.

To calculate the unsteady-state solution for a change in the inside temperature from 200°F to 300°F, the initial conditions to be used with the above equations are the temperatures at the respective nodes given in Table 7-1. The solutions for these equations are given in Fig. 7-15.

TABLE 7-1

Outside temperature	100	100
Inside temperature	200	300
Temperature of node		
1	106.1	112.1
2	113.6	127.3
3	131.0	161.9
4	120.7	141.3
5	148.3	196.5
6	186.1	272.2
7	124.1	148.1
8	155.4	210.8
9	192.1	284.2
10	125.0	150.0
11	157.1	214.3
12	192.7	285.5

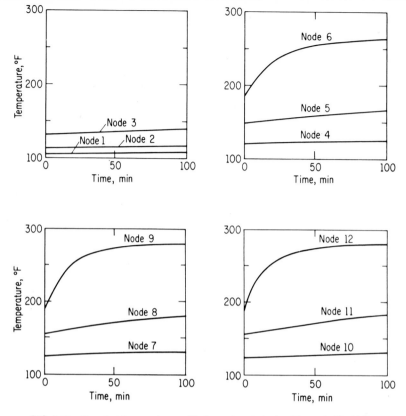

FIG. 7-15.   Transient temperature profile for each node in the chimney in Fig. 7-13.

7-8.   SUMMARY

Attention in this chapter has been focused on the formulation of partial differential equations.   The problems provide other exercises yielding such equations.   Chapter 8 considers their solution.

REFERENCES

1. H. S. Mickley, T. K. Sherwood and C. E. Reed, *Applied Mathematics in Chemical Engineering,* McGraw-Hill, New York, 1957.
2. B. G. Jenson and G. V. Jeffreys, *Mathematical Methods in Chemical Engineering,* Academic Press, New York, 1963.
3. R. V. Churchill, *Fourier Series and Boundary Value Problems,* McGraw-Hill, New York, 1963.
4. ———, *Operational Mathematics,* McGraw-Hill, New York, 1958.

5. D. E. Johnson and J. R. Johnson, *Mathematical Models in Engineering and Physics,* Ronald Press, New York, 1965.

6. C. R. Wylie, *Advanced Engineering Mathematics,* McGraw-Hill, New York, 1960.

7. F. Kreith, *Principles of Heat Transfer,* 2d ed., International Textbook, Scranton, Pa., 1965.

8. R. B. Bird, W. E. Stewart and E. N. Lightfoot, *Transport Phenomena,* Wiley, New York, 1960.

9. C. O. Bennett and J. E. Myers, *Momentum, Heat and Mass Transfer,* McGraw-Hill, New York, 1962.

PROBLEMS

7-1. Formulate the continuity equation in cylindrical coordinates.

7-2. Formulate the continuity equation in spherical coordinates.

7-3. Make an energy balance over the element used to derive the continuity equation in Sec. 7-4. Assume constant physical properties, and neglect heat generated by viscous dissipation.

7-4. Make a component material balance over the element used to derive the continuity equation in Sec. 7-4. The diffusivity of A through the fluid is D_{AB}, and the rate of formation of A by reaction is r_A mole A/unit volume.

7-5. Develop the unsteady-state equations describing the heat exchanger in Sec. 2-2. Let the cross-sectional area per tube pass be A_t, and the cross-sectional area for flow in the shell be A_s.

7-6. Suppose that the steam is off so that the temperature of the steam-line hanger considered in Sec. 1-3 is uniform at the ambient temperature. If at time = 0 the steam is turned on (temperature of steam line becomes T_s immediately), derive the equations and boundary conditions that can be solved for the transients.

7-7. Suppose that the assumption of uniform temperatures at a cross section were not made for the steam-line hanger in Sec. 1-3. Derive the equation and boundary conditions that can be solved for $T(x,y,z)$.

7-8. Suppose the steam-line hanger in Sec. 1-3 had been a round rod of radius R. Without assuming that the temperature is uniform over a cross section, derive the equations and boundary conditions that can be solved for $T(r,z)$.

7-9. Develop the equations and boundary conditions to give the temperature in the fin in Prob. 1-13 as a function of r and z (thickness variable).

7-10. Determine the concentration profile as a function of r and z in the wetted-wall column in Prob. 2-2 if the gas is in laminar flow. The diffusivity of A through B is D_{AB}, but diffusion in the direction of flow may be neglected. The effect of absorbed A on the velocity profile and gas rate may be neglected.

7-11. According to the penetration theory, mass transfer occurs by the following mechanism:

    (1) A turbulent eddy from the bulk fluid reaches the interface at time zero, displacing all previous fluid at the interface in the locality of the eddy.

    (2) Unsteady-state mass transfer by diffusion occurs while the eddy remains at the interface.

    (3) At time t_c, the contact time, another eddy displaces the previous eddy.

Consider a binary mixture whose interfacial concentration of component A is C_{a_i} and whose bulk-phase concentration is C_{a_b}. The diffusivity of A through B is D_{AB}.

    (a) Develop the partial differential equation and boundary conditions describing the concentration profile in the eddy while it is at the interface.

    (b) What is the expression for the total mass transferred while the eddy is at the interface?

    7-12. Air to a blast furnace is preheated in a Cowper stove or hot-blast stove. This unit operates on a cycle involving alternate heating and cooling periods as shown. Let the mass of ceramic material per unit height of column be M, the heat-transfer area per unit height be A, and the void volume per unit height be V. Assume constant physical properties, and neglect conduction in the axial direction. Let the heat-transfer coefficient between gas and ceramic be h. Develop the equations that give the temperature profiles in the gas and ceramic as functions of position and time. Assume constant temperature over a cross section.

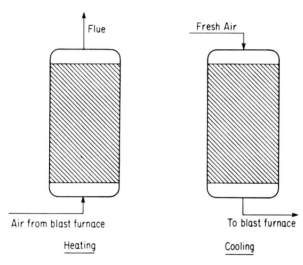

PROB. 7-12

    7-13. Consider the coffee pot shown in the drawing. Liquid rises up through the riser, is distributed uniformly over the bed of coffee grounds, passes through the bed taking up coffee extract, and falls back to the bottom. The properties of the system are:

    1. Cross-sectional area of the bed is A.

    2. Height of bed is L.

    3. No concentration gradients in radial direction.

    4. Holdup of liquid per unit height of bed is H_L lb H_2O/ft.

    5. Mass-transfer area per unit volume of bed is a.

    6. Mass-transfer rate (lb/sec-ft^2) is given by $k_m(C_s - C)$, where C_s is the saturated concentration (lb coffee/lb H_2O) of coffee and C is the concentration of coffee in the liquid.

    7. Density of the liquid is constant.

    8. Holdup of liquid in bottom of pot is H_t lb H_2O. This reservoir is well mixed.

9. Initial concentration of coffee in the liquid in the bed and reservoir is zero.
10. Circulation of liquid is w lb H_2O/min.

Develop the differential equations and boundary conditions that describe the operation of the coffee pot.

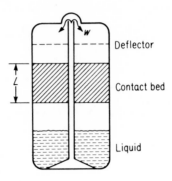

7-14. Suppose a long tube of radius R is filled with an incompressible Newtonian fluid. At time $t = 0$ a constant pressure gradient dP/dz is imposed on the fluid. Determine the equations and boundary conditions that describe the velocity profiles during the "start-up" of laminar flow.

7-15. Refer to Prob. 2-16. A tube of length L and cross-sectional area A contains a material of porosity ϵ and permeability k. Initially this tube is filled with water at a very high pressure P_1. The pressure gradient is related to the superficial velocity by Darcy's law, and the compressibility of the water is treated as in Prob. 2-16. If water is withdrawn from the end at $z = L$ at rate w, determine the necessary equations and boundary conditions that can be solved for the unsteady-state pressure profiles.

7-16. As all of the above problems can be approached via a lumped-parameter model, no separate exercises will be presented. It is suggested that the student examine several of the above problems using lumped-parameter models.

chapter 8

Solution of Partial Differential Equations

The previous chapter treated the derivation of partial differential equations for typical engineering systems. As the subject of solutions for these equations is indeed broad enough for entire texts, this chapter must begin by restricting the topics to be considered. The first restriction is that no analytic approaches will be discussed. Typical analytic techniques such as Laplace transforms[1], Fourier transforms [1], or separation of variables [2] can typically be applied to only certain linear partial differential equations. Even then they frequently yield a "solution" in the form of an infinite series, which is not always easy to evaluate.

Numerical techniques must usually be somewhat tailored to the specific partial differential equation to be solved. Although rather comprehensive texts are available on numerical solutions to such equations [3,4], the space in this text is probably insufficient even to briefly describe all the various approaches. Thus the techniques presented have been selected in order to devote primary attention to presenting *ideas* rather than long, complex equations and their idiosyncrasies.

Numerical techniques generally require significant storage space and running time on even large digital computers. As the advent of the hybrid computer has made previously "unsolvable" equations amenable to solution in a reasonable amount of time, it is felt that some space should also be devoted to introducing the basic concepts of this approach.

In all cases, this discussion will be restricted to partial differential equations involving only two independent variables. Extension to other cases is fairly straightforward.

Perhaps a reasonable starting point for the discussion of partial differential equations is with a classification of such equations.

8-1. TYPES OF PARTIAL DIFFERENTIAL EQUATIONS

It is surprising how many real, physical systems can be adequately represented by a partial differential equation of the following form:

$$A \frac{\partial^2 u(x,y)}{\partial x^2} + B \frac{\partial^2 u(x,y)}{\partial x \partial y} + C \frac{\partial^2 u(x,y)}{\partial y^2} + D \frac{\partial u(x,y)}{\partial x} + E \frac{\partial u(x,y)}{\partial y}$$

$$+ Fu(x,y) + G = 0 \tag{8-1}$$

Depending upon the coefficients, this partial differential equation is classified into one of the following categories.

1. Elliptic if $B^2 - 4AC < 0$. A typical example is Laplace's equation describing steady-state heat transfer in two dimensions as derived in Sec. 7-5:

$$\frac{\partial^2 u(x,y)}{\partial x^2} + \frac{\partial^2 u(x,y)}{\partial y^2} = 0$$

Note that $A = C = 1$, $B = 0$, and thus $B^2 - 4AC < 0$.

2. Parabolic if $B^2 - 4AC = 0$. A typical example is one-dimensional unsteady-state heat transfer described by the following equation, commonly called the *diffusion equation:*

$$\alpha \frac{\partial^2 u(x,t)}{\partial x^2} = \frac{\partial u(x,t)}{\partial t}$$

Note that $A = 1$, $B = C = 0$, and thus $B^2 - 4AC = 0$.

3. Hyperbolic if $B^2 - 4AC > 0$. The equation describing vibrations in a string falls into this category:

$$\frac{\partial^2 u(x,t)}{\partial x^2} = k \frac{\partial^2 u(x,t)}{\partial t^2}$$

Note that $A = 1$, $B = 0$, and $C = -k$, and thus $B^2 - 4AC > 0$.

As problems that typically produce hyperbolic equations are not treated to any extent in this text, their solution will not be considered either. Thus attention is restricted to parabolic and elliptic equations.

8-2. FINITE DIFFERENCES FOR PARTIAL DERIVATIVES

As most numerical methods for the solution of partial differential equations involve finite differences, perhaps a brief review is in order. Let $f(x)$ be a single valued, finite, and continuous function of x. Using Taylor's expansion theorem, the following two relationships can be obtained:

$$f(x + \Delta x) = f(x) + \frac{df(x)}{dx} \Delta x + \frac{d^2 f(x)}{dx^2} \frac{(\Delta x)^2}{2} + \frac{d^3 f(x)}{dx^3} \frac{(\Delta x)^3}{6} + \cdots \qquad (8\text{-}2)$$

$$f(x - \Delta x) = f(x) - \frac{df(x)}{dx} \Delta x + \frac{d^2 f(x)}{dx^2} \frac{(\Delta x)^2}{2} - \frac{d^3 f(x)}{dx^3} \frac{(\Delta x)^3}{6} + \cdots \qquad (8\text{-}3)$$

Note that truncation of Eq. 8-2 after two terms yields the following expression for $\dfrac{df(x)}{dx}$:

$$\frac{df(x)}{dx} = \frac{f(x + \Delta x) - f(x)}{\Delta x} \tag{8-4}$$

This is known as a *forward difference.* Truncation of Eq. 8-3 in a similar manner yields

$$\frac{df(x)}{dx} = \frac{f(x) - f(x - \Delta x)}{\Delta x} \tag{8-5}$$

This is called the *backward difference.* As both of the above expressions assume second and higher powers of Δx are negligible, these two expressions for $\dfrac{df(x)}{dx}$ are said to have errors of $O(\Delta x)$, i.e., of order Δx.

On the other hand, subtracting Eq. 8-3 from Eq. 8-2 gives

$$f(x + \Delta x) - f(x - \Delta x) = 2\frac{df(x)}{dx}\Delta x + \frac{d^3f(x)}{dx^3}\frac{(\Delta x)^3}{3} + \cdots$$

Now if the term involving $(\Delta x)^3$ and all higher powers are neglected, we have

$$\frac{df(x)}{dx} = \frac{f(x + \Delta x) - f(x - \Delta x)}{2\Delta x} \tag{8-6}$$

This is the *central difference,* and its error is $O(\Delta x^2)$.

Adding Eqs. 8-3 and 8-2 produces

$$f(x + \Delta x) + f(x - \Delta x) = 2f(x) + \frac{d^2f(x)}{dx^2}\Delta x^2 + \frac{d^4f(x)}{dx^4}\frac{\Delta x^4}{12} + \cdots$$

Truncating terms involving fourth and higher powers of Δx and solving for the second derivative yields

$$\frac{d^2f(x)}{dx^2} = \frac{f(x + \Delta x) - 2f(x) + f(x - \Delta x)}{(\Delta x)^2} \tag{8-7}$$

The error of this expression is $O(\Delta x^2)$.

Although the above discussion was for a function of one variable, it is readily extended to partial derivatives of a function of several variables.

To clarify the notation to be used in this chapter, consider a function $u(x,y)$. Now subdivide the x-y plane into a series of rectangles of sides Δx, Δy as shown in Fig. 8-1. This gives a *grid* of lines whose intersections are referred to as *nodes*

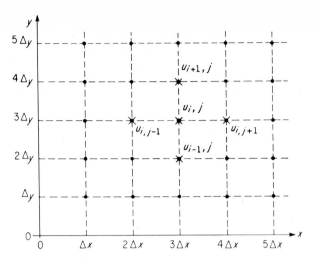

FIG. 8-1.   Nomenclature for nodal points.

(also called mesh points, grid points, lattice points, or pivotal points).   It is only at these nodes that the finite difference solution will give the value of the function.   The coordinates can be expressed in terms of the lengths of the sides of the rectangles

$$x = j\Delta x$$

$$y = i\Delta y$$

The function $u(x,y)$ now becomes $u(j\Delta x, i\Delta y)$, or shortened to $u_{i,j}$.   The reversing of subscripts here is a source of difficulty, the reason being an attempt to be consistent with matrix terminology ($a_{i,j}$ is the element on the ith row and jth column).   Thus, if the nodes are taken to be positions in a matrix, their representation would be $u_{i,j}$, not $u_{j,i}$.

Applying the previous expressions for derivatives to the function $u(x,y)$ at point (i,j) yields the following expressions:

Function	Difference	Expression	Error
$\dfrac{\partial u}{\partial x}$	Forward	$\dfrac{u_{i,j+1} - u_{i,j}}{\Delta x}$	(Δx)
	Backward	$\dfrac{u_{i,j} - u_{i,j-1}}{\Delta x}$	(Δx)
	Central	$\dfrac{u_{i,j+1} - u_{i,j-1}}{2\Delta x}$	$(\Delta x)^2$

$\dfrac{\partial^2 u}{\partial x^2}$	—	$\dfrac{u_{i,j+1} - 2u_{i,j} + u_{i,j-1}}{(\Delta x)^2}$	$(\Delta x)^2$
$\dfrac{\partial u}{\partial y}$	Forward	$\dfrac{u_{i+1,j} - u_{i,j}}{\Delta y}$	(Δy)
	Backward	$\dfrac{u_{i,j} - u_{i-1,j}}{\Delta y}$	(Δy)
	Central	$\dfrac{u_{i+1,j} - u_{i-1,j}}{2\Delta y}$	$(\Delta y)^2$
$\dfrac{\partial^2 u}{\partial y^2}$	—	$\dfrac{u_{i+1,j} - 2u_{i,j} + u_{i-1,j}}{(\Delta y)^2}$	$(\Delta y)^2$

These expressions will be used repeatedly in the ensuing discussion.

8-3. NUMERICAL SOLUTION OF PARABOLIC EQUATIONS

As the first application of finite differences to the solution of partial differential equations, suppose we consider parabolic equations. Specifically, the discussion will be oriented around the diffusion equation

$$\alpha \frac{\partial^2 T(x,t)}{\partial x^2} = \frac{\partial T(x,t)}{\partial t} \tag{8-8}$$

$$\alpha = \text{thermal diffusivity, ft}^2/\text{hr}$$

which describes one-dimensional, unsteady-state heat conduction. To obtain boundary conditions, suppose the specific problem involves unsteady-state heat conduction in a slender rod of length L. The temperatures at the ends are maintained at T_o and T_L. If the initial temperature profile is some known function $f(x)$, the boundary conditions for Eq. 8-8 are

$$T(0,t) = T_o, t > 0 \tag{8-9}$$

$$T(L,t) = T_L, t > 0 \tag{8-10}$$

$$T(x,0) = f(x), 0 \le x \le L \tag{8-11}$$

This equation can be solved analytically via the method of separation of variables[2], but the solution involves an infinite series.

As we have restricted our attention to numerical methods, we shall now apply finite differences to Eq. 8-8. We begin by constructing a lattice network in the x,t plane, and noting by x's the points at which the temperature is known, as illustrated in Fig. 8-2. Note that the temperatures for nodes $x=0$ are known from the boundary condition $T(0,t) = T_o$. This is expressed mathematically as

$$T_{i,o} = T_o, \qquad i = 1,2 \ldots$$

Similarly, the temperatures for the nodes at $x = L$ arise from the boundary condition $T(L,t) = T_L$, or

$$T_{i,n} = T_L, \qquad i = 1,2 \ldots$$

The temperatures for the nodes at $t = 0$ are derived from the boundary condition $T(x,0) = f(x)$, or

$$T_{o,j} = f(j\Delta x) \qquad j = 0,1, \ldots, n$$

Thus the boundary conditions specify the solution at several points, and the numerical method must be capable of extending the solution to unknown points.

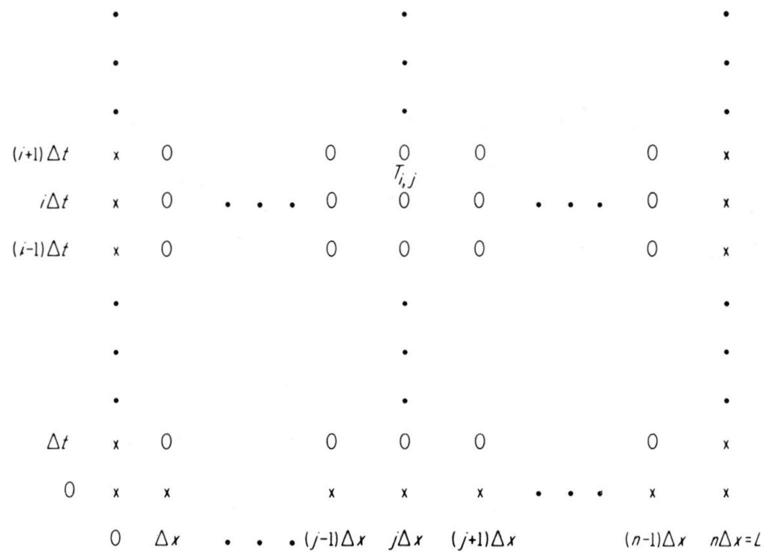

FIG. 8-2.   Lattice network illustrating known points.

Suppose that finite differences are used to represent Eq. 8-8 at node i,j. Using forward differences for the time partial, we obtain

$$T_{i+1,j} - T_{i,j} = \frac{\alpha \Delta t}{(\Delta x)^2} \left(T_{i,j+1} - 2T_{i,j} + T_{i,j-1} \right) \tag{8-12}$$

Suppose the temperatures of all nodes on row i are known, but only the boundary nodes are known on row $i + 1$ (obviously this condition exists for row 0 before the solution is started).   (See Fig. 8-2.)   Pictorially, this is represented as follows:

$$T_{i+1,j}$$

| Row $i+1$ | o | o | o | (unknown) |
| Row i | x | x | x | (known) |

$$T_{i,j-1} \qquad T_{i,j} \qquad T_{i,j+1}$$

Note that under these conditions Eq. 8-12 contains only one unknown, namely $T_{i+1,j}$, and can be solved for this unknown. This equation can be solved for nodes beginning with $j = 1$ and continuing through the node at $j = n - 1$. As the nodes at $j = 0$ and $j = n$ are known from the boundary conditions, the entire row $i + 1$ can be determined. Consequently, this procedure can be used to solve for all rows, beginning with $i = 1$ and working forward in time.

The above procedure is iterative in nature, and consequently is ideal for programming on a digital computer. It is indeed a simple task, which is a strong point in favor of this approach.

However, no consideration of a numerical technique is complete without some consideration of a stability analysis. Although space is insufficient here for a detailed analysis [Ref. 3 and 4], the following points are worthy of note.

1. The error involved in the finite difference for $\partial^2 T/\partial x^2$ is of order (Δx^2), but is of order Δt for the finite difference for $\partial T/\partial t$. Thus Δt must be much smaller than Δx.

2. Consider the specific problem in which initially the entire bar is maintained at $0°$ [i.e., $T(x,0) = 0$], the end at $x = L$ is maintained at $0°$ (i.e., $T(L,t) = 0$), but the end at $x = 0$ is maintained at $100°$ for $t > 0$ (i.e., $T(0,t) = 100°$, $t > 0$). Now suppose the technique suggested above is applied to generate the solution. Note in Eq. 8-12 that when $T_{i,j+1} = T_{i,j} = T_{i,j-1} = 0$, then $T_{i+1,j} = 0$. Applying this consideration to the case when $n = 8$ gives the following results (now denoting by x's nodes at which the temperature is *not zero*):

$i = 4$	x	x	x	x	x	o	o	o	o
$i = 3$	x	x	x	x	o	o	o	o	o
$i = 2$	x	x	x	o	o	o	o	o	o
$i = 1$	x	x	o	o	o	o	o	o	o
$i = 0$	x	o	o	o	o	o	o	o	o

Note that for each iteration of a row, the effect of the change in temperature propagates to one more node. Thus seven iterations (or time $= 7\Delta t$) are required before the temperature at node $n = 7$ changes. This is not what physically happens, as the temperature along the entire length of the bar begins to change as soon as the temperature at one end changes.

3. This procedure is stable only if

$$r = \frac{\alpha \Delta t}{(\Delta x)^2} \leq \frac{1}{2} \qquad (8\text{-}13)$$

The effect of this can be vividly seen in Fig. 8-3, where the solution when

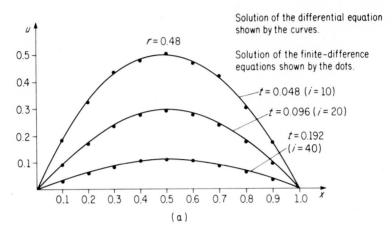

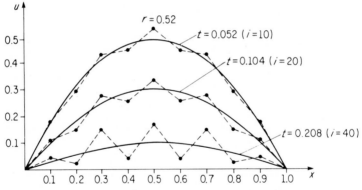

FIG. 8-3.   Effect of r on solution of diffusion equation.   (Reprinted by permission from Ref. 3, p. 16.)

$r = 0.48$ appears to be stable but is unstable when $r = 0.52$.   The proof of this statement for general equations can be found in Ref. 3.   However, for heat-conduction problems, it is interesting to note that whenever $r > 1/2$, the first law of thermodynamics is violated.   For example, suppose $T_{i,j-1} = T_{i,j+1} = 10°$ and $T_{i,j} = 0$.   Intuitively, we would suspect that $0 \leq T_{i+1,j} \leq 10°$.   However, for $r = 1$, Eq. 8-12 gives

$$T_{i+1,j} = 20°$$

For this to happen, heat must flow *against* the temperature gradient.

The effective results of the restriction $r \leq \frac{1}{2}$ is that the selection of Δx is critical. First, Δx must be small enough so that $\partial^2 T/\partial x^2$ is approximated reasonably well. Reducing Δx increases the number of nodes in the x-direction and thus the number of calculations required for each row. Furthermore, it *effectively increases the number of rows*. As Δt must satisfy the relationship

$$\Delta t \leq (\Delta x)^2/2\alpha$$

Δt must be decreased when Δx is decreased (also note the square in the expression). Thus, to proceed to a certain point in time, more rows must be calculated.

8-4. THE CRANK-NICHOLSON METHOD

The major disadvantage of the previous approach is the dependence of Δt on Δx. In 1947 Crank and Nicholson [5] proposed a technique by which the ratio r defined above can be any positive, finite value.

One approach to deriving their equations is to consider a node midway between node i,j and node $i + 1,j$, as follows:

$$T_{i+1,j-1} \qquad T_{i+1,j} \qquad T_{i+1,j+1}$$

Row $i+1$	o	o	o	(unknown)
		⊗		(intermediate node)
Row i	x	x	x	(known)

$$T_{i,j-1} \qquad T_{i,j} \qquad T_{i,j+1}$$

Using a central difference to approximate $\partial T/\partial t$ at the intermediate node yields

$$\frac{\partial T(x,t)}{\partial t} = \frac{T_{i+1,j} - T_{i,j}}{\Delta t}$$

This is correct to the order of $(\Delta t)^2$. Now $\partial^2 T/\partial x^2$ at this node is approximated by the average of $\partial^2 T/\partial x^2$ evaluated at the adjacent nodes:

$$\frac{\partial^2 T(x,t)}{\partial x^2} = \frac{1}{2(\Delta x)^2} \left[(T_{i+1,j+1} - 2T_{i+1,j} + T_{i+1,j-1}) + (T_{i,j+1} - 2T_{i,j} + T_{i,j-1}) \right]$$

This expression is correct to the order of (Δx^2). Substituting these into Eq. 8-8 yields

$$T_{i+1,j} - T_{i,j} = \frac{\alpha \Delta t}{2(\Delta x)^2} \left[(T_{i+1,j+1} - 2T_{i+1,j} + T_{i+1,j-1}) + (T_{i,j+1} - 2T_{i,j} + T_{i,j-1}) \right]$$

$$(8\text{-}14)$$

Note that application of this equation to each of the $n - 1$ interior points on one of the rows in Fig. 8-2 gives $n - 1$ equations (one for each node) and $n - 1$ unknowns (the temperature at each node).

The set of equations can be written generally as follows (the row subscript on T has been dropped):

$$
\begin{aligned}
b_1 T_1 - c_1 T_2 &= d_1 \\
- a_2 T_1 + b_2 T_2 - c_2 T_3 &= d_2 \\
&\vdots \\
- a_j T_{j-1} + b_j T_j - c_j T_{j+1} &= d_j \\
&\vdots \\
a_{n-1} T_{n-2} + b_{n-1} T_{n-1} &= d_{n-1}
\end{aligned}
\tag{8-15}
$$

where the a's, b's, c's, and d's are known.   This set of equations can be solved by matrix inversion, Gauss reduction, Thomas algorithm (see Prob. 8-1), or similar approaches.

As mentioned earlier, Δt and Δx are completely independent for this approach. Nor does this method suffer from the propagation phenomenon described earlier for the previous approach.   If one of the boundary temperatures changes, *all* nodes on the next row are affected simultaneously.

8-5.   SOLUTION OF NONLINEAR PARABOLIC EQUATIONS

In many practical problems, the coefficients in the partial differential equations depend upon the dependent variable itself.   For example, denoting the dependence of the thermal diffusivity upon temperature by $\alpha(T)$ and inserting into Eq. 8-8 yields

$$
\alpha(T)\frac{\partial^2 T(x,t)}{\partial x^2} = \frac{\partial T(x,t)}{\partial t}
\tag{8-16}
$$

which is a nonlinear partial differential equation.

If the Crank-Nicholson technique is used to solve this equation, a difficulty arises.   Recall that the left-hand side of this equation is approximated by the finite-difference expressions evaluated at row i (at which all temperatures are known) and at row $i + 1$ (where the temperatures are unknown).   Although equations analogous to the set in Eq. 8-15 can be derived, $\alpha(T)$ causes them to be nonlinear.   Solution of such sets of equations is typically a time-consuming trial-and-error task, and certainly not feasible to undertake for each row in the numerical solution.

For many nonlinear functions such as the dependence of thermal diffusivity on temperature, the value of the function changes only slightly for the small temperature change from one row to the next in the numerical solution.   In such cases, the value of the nonlinear function, in this case $\alpha(T)$, at the known row can also be used for the unknown row without introducing serious error.

For other functions which show a strong dependence on the dependent variable, an iterative procedure might be satisfactory. For example, a first estimate of the temperatures on row $i + 1$ might be calculated using $\alpha(T)$ evaluated at row i. Now these temperatures may be used to obtain a better value of $\alpha(T)$ on row $i + 1$, which is in turn used to evaluate a second estimate of the temperatures on this row. This procedure may be repeated as many times as desirable, but two iterations are often sufficient.

Other more sophisticated procedures are available for treating nonlinear equations, some of which can be found in the exercises.

8-6. NUMERICAL SOLUTION OF ELLIPTIC EQUATIONS

As a typical example of an elliptic partial differential equation, consider Laplace's equation

$$\frac{\partial^2 T(x,y)}{\partial x^2} + \frac{\partial^2 T(x,y)}{\partial y^2} = 0 \tag{8-17}$$

which describes the two-dimensional, steady-state flow of heat in a homogeneous conductor. For boundary conditions, consider the conductor as being a square block of width b on each side. Two of the adjoining sides are maintained at known temperature profiles, while the other two sides are insulated. This gives the following boundary conditions:

$$T(0,y) = f_2(y) \tag{8-18}$$

$$T(x,0) = f_1(x) \tag{8-19}$$

$$\frac{\partial T(z,b)}{\partial y} = 0 \tag{8-20}$$

$$\frac{\partial T(b,y)}{\partial x} = 0 \tag{8-21}$$

These conditions along with the finite-difference grid to be used are shown in Fig. 8-4.

From the boundary conditions in Eqs. 8-18 and 8-19, the temperatures for the nodes along the axes are known. Specifically, these are given by

$$T_{i,0} = f_2\left(\frac{ib}{4}\right) \qquad i = 1,2,3,4$$

$$T_{0,j} = f_1\left(\frac{jb}{4}\right) \qquad j = 1,2,3,4$$

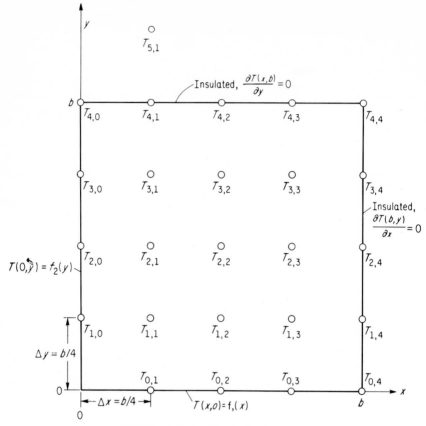

FIG. 8-4.   Grid for solving Laplace's equation.

The unknowns are the temperatures at the sixteen remaining nodes in the block (the node outside the block will be used for applying the boundary condition). At each of these nodes a finite-difference equation can be written. This gives sixteen equations which can be solved for the sixteen unknowns.

First, consider a typical interior node, say node 1,3. Equation 8-17 can be "finite-differenced" about this node to yield

$$\frac{T_{1,4} - 2T_{1,3} + T_{1,2}}{(\Delta x)^2} + \frac{T_{2,3} - 2T_{1,3} + T_{0,3}}{(\Delta y)^2} = 0 \tag{8-22}$$

Analogous equations can be written for the remaining interior nodes, giving a total of nine equations.

For the nodes on the boundaries at $x = b$ and $y = b$, the boundary conditions that the normal partial derivatives are zero must be incorporated. As an example, consider node 4,1. The external node 5,1 can be used to write the same finite-difference equation for this node as for the interior nodes, namely

$$\frac{T_{4,2} - 2T_{4,1} + T_{4,0}}{(\Delta x)^2} + \frac{T_{5,1} - 2T_{4,1} + T_{3,1}}{(\Delta y)^2} = 0 \qquad (8\text{-}23)$$

However, $T_{5,1}$ must be eliminated by using the boundary condition. Using central differences, the partial derivative at this node is as follows:

$$\frac{T_{5,1} - T_{3,1}}{2\Delta x} = \frac{\partial T(\Delta x, b)}{\partial y} = 0$$

or

$$T_{5,1} = T_{3,1}$$

(This equality does not hold in general, only when the partial derivative is zero). This equation can be substituted into Eq. 8-23 to give

$$\frac{T_{4,2} - 2T_{4,1} + T_{4,0}}{(\Delta x)^2} + \frac{2(T_{3,1} - T_{4,1})}{(\Delta y)^2} = 0 \qquad (8\text{-}24)$$

Similar equations can be developed for the remaining nodes on the boundary, giving a total of seven such equations.

The results of this are sixteen equations with sixteen unknowns. The solution of such equations can be undertaken by matrix inversion, Gauss reduction, or by some iterative procedure. Obviously one of the problems is that fine grids give a large number of such equations. Such large sets would usually be solved by iterative techniques, while smaller sets would be solved by direct methods. For large sets, it is also necessary to take advantage of the regularity of the coefficient matrix to reduce storage requirements. This and other considerations are discussed in Refs. 3 and 4.

For nonlinear elliptic partial differential equations, the above situation is complicated by the presence of nonlinear equations in the resulting set. One avenue for circumventing this problem is to use the relaxation method as described previously in Sec. 7-8. Kreith [6] gives a good discussion of such approaches for heat conduction problems.

8-7. SOLUTION BY ANALOG (HYBRID) TECHNIQUES

The solution of partial differential equations on the digital computer is normally rather expensive in terms of computer time. Thus it is natural to look toward the high-speed capabilities of the analog computer. The difficulty reduces to the simple fact that the analog computer utilizes only one independent variable, namely time.   This section is devoted to an introduction to approaches for circumventing this problem.

Instead of discussing problems in general, perhaps the discussion in this section should be confined to a simple example.   The isothermal, plug flow, tubular reactor in Fig. 8-5 is described by the following differential equation:

$$- v \frac{\partial C_A(z,t)}{\partial z} - k [C_A (z,t)]^n = \frac{\partial C_A(z,t)}{\partial t} \tag{8-25}$$

$$C_A(0,t) = C_{A_f}(t)$$

$$C_A(z,0) = f(z)$$

All other nomenclature is defined in Fig. 8-5.

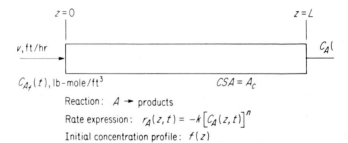

FIG. 8-5.   Isothermal, plug-flow reactor.

As the analog computer cannot handle the two independent variables in the above equation directly, obviously one of them must be eliminated in some manner.   One easy manner of accomplishing this is to use finite differences for *one* of the independent variables.   This is completely analogous to the use of finite differences for the numerical methods discussed earlier when *both* independent variables were replaced by finite differences.   However, we can choose to express either the partials with respect to time or the partials with respect to position as finite differences.   Both possibilities are considered below.

As the independent variable on the analog computer is time, suppose we first consider eliminating the position variable by finite differences.   Using the same techniques discussed earlier in this chapter, the function $C_A(z,t)$ can be represented as a series of functions $C_{A_f}(t)$, $C_{A_1}(t)$, $\cdots$, $C_{A_i}(t)$, $\cdots$, $C_{A_n}(t)$ which represent the values of the concentration at points along the length of the reactor. Note that each of these variables is a function of time.

The finite-difference equation representing the *i*th point in the reactor can be derived by using, for example, backward differences to represent the first partial. Of course, other approximations could be used.   This equation is

$$-\frac{v}{\Delta z}[C_{A_i}(t) - C_{A_{i-1}}(t)] - k[C_{A_i}(t)]^n = \frac{dC_{A_i}(t)}{dt}$$

Although the first equation must be modified slightly to account for the boundary condition, we now essentially have n simultaneous, ordinary differential equations similar to the above equation.

Of course, the analog computer is frequently used to solve simultaneous, ordinary differential equations. The schematic wiring diagram for the ith equation is shown in Fig. 8-6. Note that the only input to this circuit is the concentration at the previous node, and the output is the concentration at node i. When all the circuits are patched on the computer simultaneously, the speed of the analog computer produces a solution very rapidly. Of course, the accuracy depends upon the interval Δz used in the finite differences.

The limitation on this approach is rather obviously due to hardware availability. Note that the circuit in Fig. 8-6 requires one integrator, one summer, one inverter, and one function generator. For $n = 30$, the problem will require about 90 amplifiers, or in other words a rather large analog computer. This requirement can be reduced somewhat by multiplexing (i.e., one component being time-shared between various functions), but the solution speed usually must be reduced if this is used extensively.

Suppose that instead of finite-differencing the position variable, we finite-difference the time variable. Performing this manipulation to Eq. 8-25 yields

$$-v(\Delta t)\frac{dC_{A_i}(z)}{dz} - k(\Delta t)[C_{A_i}(z)]^n = C_{A_i}(z) - C_{A_{i-1}}(z) \qquad (8\text{-}26)$$

$$C_{A_i}(0) = C_{A_f}(i\Delta t)$$

$$C_{A_0}(z,0) = f(z)$$

This equation relates the concentration profile $C_{A_i}(z)$ at time $i\Delta t$ to the concentration profile at time $(i-1)\,\Delta t$ and the boundary condition. This is schematically represented as follows:

$t = i\Delta t$	o	o	o	o	o	o	o	o
$(i-1)\Delta t$	x	x	x	x	x	x	x	x
.	.	.	.	.	.	.	.	.
.	.	.	.	.	.	.	.	.
.	.	.	.	.	.	.	.	.
Δt	x	x	x	x	x	x	x	x
$t = 0$	x	x	x	x	x	x	x	x
	$z = 0$							$z = L$

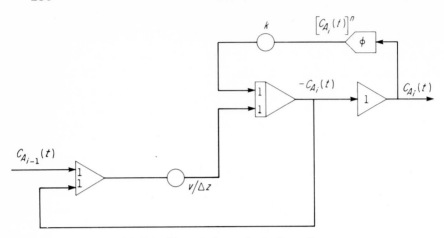

FIG. 8-6.   Schematic diagram for the ith equation.

Knowing the concentration profile at $(i-1)\,\Delta t$, the above equation can be used to generate the concentration profile at $i\Delta t$.   Thus we have a "recurrence relationship" for the successive concentration profiles.   To start the successive calculations, the concentration profile at time zero is known to be $f(z)$.

The schematic diagram for the analog circuit is shown in Fig. 8-7.   The solution procedure is as follows:

1. Begin the solution using $C_{A_0}(z) = f(z)$
2. Solve Eq. 8-26 using the circuit in Fig. 8-7 for $C_{A_1}(z)$.   Store these values as they are generated.
3. Reset the analog computer.
4. Solve Eq. 8-26 for $C_{A_2}(z)$.   Store the results as they are generated.
5. Reset the analog computer.
6. Repeat as long as desired.

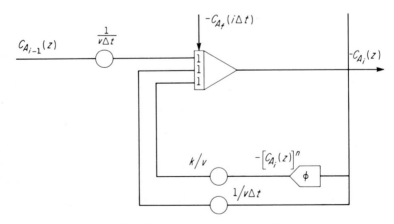

FIG. 8-7.   Analog schematic for Eq. 8-26.

Two things are important to note about this procedure. First, the analog computer repetitively solves the same set of equations, which it is capable of doing extremely rapidly. Second, the external storage device must be capable of function storage and playback. This device is usually a digital computer operated with the analog in a hybrid environment.

Although the above discussion is by no means an exhaustive treatment of the capabilities of the hybrid computer for solving partial differential equations, it is hoped that at least the basic ideas were covered. More detail is available in Ref. 7.

8-8. SUMMARY

This chapter has hopefully provided some insight into practical approaches for solving partial differential equations. Where insufficient information was presented for some specific item, the references listed should provide some assistance. A few additional points are developed in the problems.

REFERENCES

1. R. V. Churchill, *Operational Mathematics*. McGraw-Hill, New York, 1958.
2. ———, *Fourier Series and Boundary Value Problems*, 2d ed., McGraw-Hill, New York, 1963.
3. G. D. Smith, *Numerical Solution of Partial Differential Equations*, Oxford U. P., London, 1965.
4. G. E. Forsythe and W. R. Wasow, *Finite Difference Methods for Partial Differential Equations*, Wiley, New York, 1960.
5. J. Crank and P. Nicholson, "A Practical Method for Numerical Evaluation of Solutions of Partial Differential Equations of the Heat Conduction Type," *Proc. Cambridge Phil. Soc.*, Vol. 43 (1947), pp. 50-67.
6. F. Kreith, *Principles of Heat Transfer*, 2d ed., International Textbook, Scranton, Pa, 1965.
7. G. A. Bekey and W. J. Karplus, *Hybrid Computation*, Wiley, New York, 1968.

PROBLEMS

8-1. The Thomas algorithm was mentioned as one avenue for solving Eq. 8-15. Show that this set of equations can be solved by the following procedure:

1. Calculate the following:

$$B_i = b_i - \frac{a_i c_{i-1}}{B_{i-1}}, \qquad i = 2, ..., n-1$$

$$B_1 = b_1$$

$$\gamma_i = \frac{d_i + a_i \gamma_{i-1}}{B_i}, \qquad i = 2, ..., n-1$$

$$\gamma_1 = d_1 / b_1$$

2. Calculate the solution:

$$T_i = \gamma_i + \frac{c_i T_{i+1}}{B_i} \qquad i = 1, 2, \cdots, n - 2$$

$$T_{n-1} = \gamma_{n-1}$$

8-2. A more general form of the Crank-Nicholson method introduced in Sec. 8-4 is to consider the node between nodes i, j, and $i + 1, j$ to be at $(i + \theta)\Delta t, 0 \leq \theta \leq 1$. Note that for $\theta = 0$, this corresponds to the explicit method presented in Sec. 8-3. For $\theta = 1/2$, we have the Crank-Nicholson method as presented in Sec. 8-4. Derive the recurrence relationship for the diffusion equation considered in Sec. 8-4 when $\theta = 1$. Is the form of the resulting equations explicit or implicit?

8-3. Consider a partial differential equation of the following form:

$$-b \frac{\partial u}{\partial x} = \frac{\partial u}{\partial t}$$

Note that such an equation results if k in Eq. 8-25 is set equal to zero. An approach very similar to the Crank-Nicholson method can be applied to the equation. Central differences are written for a node centered in both the x-direction and the t-direction, as shown below:

$$
\begin{array}{cccccc}
 & & & & u_{i,j-1} & u_{i,j} \\
i\Delta t & \circ & \circ & & \circ & \circ \\
\\
 & & & & & + \\
\\
 & & & & u_{i-1,j-1} & u_{i-1,j} \\
(i-1)\Delta t & \circ & \circ & & \circ & \circ \\
\\
 & & & & (j-1)\Delta x & j\Delta x
\end{array}
$$

Using averaged central differences as for the Crank-Nicholson method, show that the following difference equation arises:

$$u_{i,j} = \frac{\dfrac{b}{\Delta x} - \dfrac{1}{\Delta t}}{\dfrac{b}{\Delta x} + \dfrac{1}{\Delta t}} \left[u_{i,j-1} - u_{i-1,j}\right] + u_{i-1,j-1}$$

Note that since $u_{i,0}$ and $u_{0,j}$ are known for all i and j from the boundary conditions, this equation is explicit ($u_{i,j}$ is only unknown).

8-4. For the partial differential equation in the above exercise, develop the following recurrence relationship using forward or backward differences:

$$\left[\frac{b}{\Delta x} + \frac{1}{\Delta t}\right] u_{i,j} = \frac{u_{i-1,j}}{\Delta t} + \frac{bu_{i,j-1}}{\Delta x}$$

This equation is not stable for all Δt and Δx, whereas the result of the previous exercise is.

8-5. Use only forward and backward differences to obtain the following recurrence relationship for Eq. 8-25 when the reaction is first order:

$$\left\{ \frac{v}{\Delta z} + k + \frac{1}{\Delta t} \right\} C_{A_{i,j}} = \frac{1}{\Delta t} C_{A_{i-1,j}} + \frac{v}{\Delta z} C_{A_{i,j-1}}$$

8-6. When the method in Prob. 8-3 is used for equations with distributed sources or sinks, the value of the dependent variable at the centered node is typically taken to be the average of the surrounding four nodes. Develop the resulting recurrence relationship for solving Eq. 8-25 if the reaction is first-order. Note that difficulties are encountered if non-linearities exist (reactions orders other than first).

8-7. Consider the single-pass counterflow heat exchanger shown. It can be readily shown that this exchanger is described by the following equations:

$$-\frac{\partial T_t(z,t)}{\partial z} + \frac{UA}{m_t c_{pt}} |T_s(z,t) - T_t(z,t)| = \frac{A_t \rho_t}{m_t} \frac{\partial T_t(z,t)}{\partial t}$$

$$\frac{\partial T_s(z,t)}{\partial z} - \frac{UA}{m_s c_{ps}} |T_s(z,t) - T_t(z,t)| = \frac{A_s \rho_s}{m_s} \frac{\partial T_s(z,t)}{\partial t}$$

$$T_s(L,t) = T_{s_0}$$
$$T_s(z,0) = f_s(z)$$
$$T_t(0,t) = T_{t_0}$$
$$T_t(z,0) = f_t(z)$$

Apply the procedure in Prob. 8-3 to obtain the recurrence relationships. Note that the split-boundary condition causes the resulting relationships to be implicit rather than explicit.

In deriving the recurrence relationships, note that T_t is known at $z = 0$. Thus, use a node centered between nodes $j - 1$ and j to obtain a recurrence relationship for $T_{t_{i,j}}$. On the other hand, T_s is known at $z = L$. Use a node centered between nodes j and $j + 1$ to obtain the recurrence relationship for $T_{s_{i,j}}$.

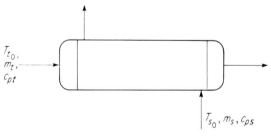

PROB. 8-7.

8-8. Repeat Prob. 8-7 using only forward or backward differences to obtain recurrence relationships for $T_{s_{i,j}}$ and $T_{t_{i,j}}$. Again consider the split-boundary condition in arranging the nodes.

8-9. The "start-up" of laminar flow considered in Prob. 7-14 is described by the following equation:

$$\mu \, \frac{1}{r} \, \frac{\partial}{\partial r} \left[r \, \frac{\partial v_z(r,t)}{\partial r} \right] - \frac{dP}{dz} = \rho \, \frac{\partial v_z(r,t)}{\partial t}$$

Boundary conditions:

$$v_z(r,0) = 0$$

$$v_z(R,t) = 0$$

$$\frac{\partial v_z(0,t)}{\partial r} = 0$$

Use the technique outlined in Sec. 8-3 to obtain a recurrence relationship. For the space derivatives, approximate the second partial as the central difference of the first partials about point $j(\Delta r)$, evaluating the first partials about $(j + \frac{1}{2}) \, \Delta r$ and $(j - \frac{1}{2}) \, \Delta r$. Then evaluate the first partials by central differences about these points. The final equation is

$$v_{i+1,j} = v_{i,j} + \frac{\mu \Delta t}{\rho (\Delta r)^2} \left[\left(\frac{2j+1}{2j} \right) v_{i,j+1} - 2v_{i,j} + \left(\frac{2j-1}{2j} \right) v_{i,j-1} \right] - \frac{\Delta t}{\rho} \, \frac{dP}{dz}$$

This equation can be applied whenever $j \neq 0$. How should the boundary condition $\dfrac{\partial v_z(0,t)}{\partial r} = 0$ be incorporated to give the relationship for $j = 0$?

8-10. Apply the Crank-Nicholson procedure to the previous exercise to obtain the recurrence relationships.

8-11. Consider the nonlinear partial differential equation

$$\alpha(T) \, \frac{\partial^2 T}{\partial x^2} = \frac{\partial T}{\partial t}$$

Recall that in the application of the Crank-Nicholson technique, the coefficient $\alpha(T)$ should be evaluated at $T_{i+\frac{1}{2},j}$. One manner of obtaining this value is to use the Taylor series approximation for $T_{i+\frac{1}{2},j}$ in terms of $T_{i,j}$, truncated after two terms. Show that this expression can be combined with the original partial differential equation to give

$$T_{i+\frac{1}{2},j} = T_{i,j} + \frac{(\Delta t) \alpha (T_{i,j})}{2 (\Delta x)^2} \left[T_{i,j-1} - 2T_{i,j} + T_{i,j+1} \right]$$

The coefficient $\alpha (T_{i+\frac{1}{2},j})$ is evaluated and used directly, thus allowing the Crank-Nicholson technique to be applied.

8-12. The equations describing the coffee pot in Prob. 7-13 are

$$- w \, \frac{\partial C(z,t)}{\partial z} + A a k_m [C_s - C(z,t)] = H_L \, \frac{\partial}{\partial t} [C(z,t)]$$

$$C(0,t) = C_r(t)$$

$$C(z,0) = 0$$

$$\frac{H_t}{w} \, \frac{dC_r(t)}{dt} + C_r(t) = C(L,t)$$

$$C_r(0) = 0$$

Finite-difference this equation in the z-direction as discussed in Sec. 8-7 and develop the analog wiring diagram if only three nodes are taken for the bed.

8-13. Now finite-difference both equations in the previous exercise for t instead of z. Develop the necessary wiring diagram for the analog and the recurrence relationship for the second equation (assuming a digital computer will be performing the function storage and playback, and thus would be available for numerical calculations also).

8-14. If the thermal diffusivity is unity, the diffusion equation becomes

$$\frac{\partial^2 u(x,t)}{\partial x^2} = \frac{\partial u(x,t)}{\partial t}$$

$$u(0,t) = 0$$
$$u(1,t) = 1$$
$$u(x,0) = 0$$

Finite-difference this equation in the x-direction and develop the analog wiring circuit for a typical node.

8-15. Finite-difference the diffusion equation in the above exercise in the t-direction. Develop the analog wiring diagram. What difficulties are encountered?

chapter 9

Transformations

In the previous chapters changes of variable were used in order to reduce the number of groups of parameters in the equations describing the physical system. The most important result from such changes of variable is that the solution to the differential equations could be easily represented in such a fashion as to be applicable to general types or classes of problems rather than for specific examples. In this chapter the details of the techniques for obtaining the appropriate changes of variable are presented.

9-1. A SYSTEMATIC APPROACH TO CHANGES OF VARIABLE

Although the techniques for making appropriate changes of variable come with experience, a step-by-step technique is presented in this chapter for two reasons. First, those inexperienced in the field can use this technique to derive the proper changes of variable. Second, this systematic technique can be used for problems in which the proper changes of variable are not obvious. Normally, the changes of variable are made for the dependent and independent variables in order to reduce the number of groups of parameters required to adequately describe the problem. However, in a few cases the changes of variable are even more helpful. Sometimes the number of independent variables required in the equations can be reduced, such changes in the independent variables being known as *similarity transforms*. The derivation of similarity transforms is a degree more difficult than the derivation of changes of variable, but the technique to be presented will obtain such similarity transforms, if they exist, in addition to changes of variable.

Although similarity transforms are very useful, they exist for only a few problems. However, changes of variable are made in almost all practical problems. Consequently, the technique is first developed to obtain changes of variable for problems for which no similarity transform exists. Finally, it is extended to derive similarity transforms whenever possible.

The terms *dimensionless variable* and *normalized variable* are often associated with the process of introducing a change of variable into equations. The term *dimensionless* is used because the final equations will be in terms of dimensionless variables and dimensionless groups of parameters. The term *normalized* is used to denote that the range of a variable is restricted to the region between zero and one. In normalization, the variable is made dimensionless, but in many cases it is impossible to normalize a variable even though it can be made dimensionless.

An example of such a variable is the time variable for situations in which time can increase to infinity. In this case a dimensionless variable can be substituted for time, but this variable will not be normalized.

The techniques for reducing the number of parameters or dimensionless groups required to mathematically describe a physical problem have been investigated for many years. The description of problems using the minimum number of parameters is the subject of such apparently diversified fields as dimensional analysis, inspectional analysis, and scale-up theory. However, these techniques typically consider only changes of variable, and not similarity transforms. Since the following technique applies to similarity transforms as well, it would seem to be preferable.

The basic steps in this method of analysis are as follows [1]:

1. Introduce arbitrary reference quantities in order to place the dependent and independent variables in dimensionless form. This typically takes the following form:

$$\frac{\text{Variable} - \text{arbitrary reference value}}{\text{Arbitrary reference difference}}$$

For example, the temperature T may be nondimensionalized as $(T - T^*)/\Delta T^*$, where T^* is some arbitrary reference value and ΔT^* is some arbitrary characteristic temperature difference in the problem. This will become clearer in the examples.

2. Introduce these changes of variable into the differential equations and their boundary conditions, rearranging as necessary so that the parameters appear in dimensionless groups.

3. Equate each dimensionless parameter containing an arbitrary reference quantity to either zero or one until all arbitrary constants have been defined.

In the third step, a system of algebraic equations is obtained that relate the arbitrary constants to the original parameters of the problem. There are three possiblities for this set of algebraic equations:

1. More equations than unknowns, i.e., an overdetermined set of equations. In this case there will be groups of parameters or dimensionless groups remaining in the transformed equations.

2. Number of equations equals number of unknowns. For such situations no dimensionless groups of parameters remain in the transformed equations.

3. Fewer equations than unknowns, i.e., an underdetermined set. In this case a similarity transform is possible.

Such techniques as this are best illustrated in terms of examples. The following examples have already been considered previously in the text, and are repeated here to illustrate the derivation of the changes of variable used.

9-2. EXAMPLE—THE STEAM-PIPE HANGER

In Sec. 1-3 the equations and boundary conditions were derived for a metal bar used as a pipe hanger. The differential equation and boundary conditions for this problem are as follows:

$$\frac{d^2 T(z)}{dz^2} = \frac{4h}{Wk}[T(z) - T_a] \qquad\qquad [1\text{-}9]$$

$$T(0) = T_s \qquad\qquad [1\text{-}10]$$

$$T(L) = T_a \qquad\qquad [1\text{-}11]$$

where the parameters in these equations were defined in Sec. 1-3.

First, the following dimensionless variables are defined:

$$\varphi(\zeta) = \frac{T(z) - T^*}{\Delta T^*} \qquad\qquad (9\text{-}1)$$

$$\zeta = \frac{z - Z^*}{\Delta Z^*} \qquad\qquad (9\text{-}2)$$

where

$$
\begin{aligned}
T^* &= \text{Reference temperature, }^\circ\text{F} \\
\Delta T^* &= \text{Reference temperature difference, }^\circ\text{F} \\
Z^* &= \text{Reference distance, ft} \\
\Delta Z^* &= \text{Reference difference in distances, ft}
\end{aligned}
$$

Making these changes of variable in Eq. 1-9 yields

$$\frac{\Delta T^*}{(\Delta Z^*)^2}\frac{d^2\varphi(\zeta)}{d\zeta^2} = \frac{4h}{Wk}[\varphi(\zeta)\Delta T^* + T^* - T_a]$$

This equation can be rearranged to

$$\frac{d^2\varphi(\zeta)}{d\zeta^2} = \frac{4h(\Delta Z^*)^2}{Wk}\left[\varphi(\zeta) - \frac{T_a - T^*}{\Delta T^*}\right] \qquad\qquad (9\text{-}3)$$

The same transformations must also be made in the boundary conditions:

$$\varphi\left(\frac{-Z^*}{\Delta Z^*}\right) = \frac{T_s - T^*}{\Delta T^*} \qquad\qquad (9\text{-}4)$$

$$\varphi\left(\frac{L - Z^*}{\Delta Z^*}\right) = \frac{T_a - T^*}{\Delta T^*} \qquad\qquad (9\text{-}5)$$

The groups appearing in Eqs. 9-3 through 9-5 are

$$\frac{4h(\Delta Z^*)^2}{Wk}, \quad \frac{T_a - T^*}{\Delta T^*}, \quad \frac{T_s - T^*}{\Delta T^*}, \quad \frac{-Z^*}{\Delta Z^*}, \quad \text{and} \quad \frac{L - Z^*}{\Delta Z^*}$$

Since there are five groups, five algebraic equations could be obtained if all these groups were set equal to either zero or one. Since this would be an overdetermined set, it would not have a solution. Thus only four of the groups can be selected. To obtain the simplest changes of variable, the last four groups are selected and set equal to zero or one as follows:

$$\frac{T_a - T^*}{\Delta T^*} = 0 \qquad (9\text{-}6)$$

$$\frac{T_s - T^*}{\Delta T^*} = 1 \qquad (9\text{-}7)$$

$$\frac{-Z^*}{\Delta Z^*} = 0 \qquad (9\text{-}8)$$

$$\frac{L - Z^*}{\Delta Z^*} = 1 \qquad (9\text{-}9)$$

These equations can be solved to yield the following values:

$$T^* = T_a$$
$$\Delta T^* = T_s - T_a$$
$$Z^* = 0$$
$$\Delta Z^* = L$$

Substituting these values and the values of the groups in Eqs. 9-6 though 9-9 into Eqs. 9-3 though 9-5 yields

$$\frac{d^2\varphi(\zeta)}{d\zeta^2} = \frac{4hL^2}{Wk}\varphi(\zeta) \qquad (9\text{-}10)$$

$$\varphi(0) = 1 \qquad (9\text{-}11)$$

$$\varphi(1) = 0 \qquad (9\text{-}12)$$

Thus all but one dimensionless group are eliminated.

An alternative procedure would be to reverse the order in which the groups in Eqs. 9-8 and 9-9 are set equal to zero and one; that is,

$$\frac{-Z^*}{\Delta Z^*} = 1$$

$$\frac{L - Z^*}{\Delta Z^*} = 0$$

These two equations can be solved to yield

$$Z^* = L$$
$$\Delta Z^* = L$$

Substituting these values and the previous values of T^* and ΔT^* into Eqs. 9-3 though 9-5 yields

$$\frac{d^2\varphi(\zeta)}{d\zeta^2} = \frac{4hL^2}{Wk}\,\varphi(\zeta) \tag{9-13}$$

$$\varphi(0) = 0 \tag{9-14}$$

$$\varphi(1) = 1 \tag{9-15}$$

Comparing these equations to Eqs. 9-10 though 9-12 reveals that the only difference is that the boundary conditions are reversed.

The solutions for this set and the previous set of equations are compared graphically in Fig. 9-1. Although these two are different, both will yield the same profile for T vs. z.

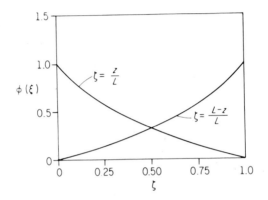

FIG. 9-1.   Comparison of solutions to Eqs. 9-10 through 9-12 and Eqs. 9-13 through 9-15.

9-3.   COOLING A SPHERE—A PROBLEM WITH PARTIAL DIFFERENTIAL EQUATIONS

In Sec. 7-3 the following partial differential equation and boundary conditions were derived for the cooling of a sphere:

$$\alpha \left[\frac{\partial^2 T(r,t)}{\partial r^2} + \frac{2}{r}\frac{\partial T(r,t)}{\partial r} \right] = \frac{\partial T(r,t)}{\partial t} \tag{7-3}$$

$$T(r,0) = T_0 \tag{7-4}$$

$$\frac{\partial T(0,t)}{\partial r} = 0 \tag{7-5}$$

$$\frac{\partial T(R,t)}{\partial r} = -\frac{h}{k}[T(R,t) - T_w] \tag{7-6}$$

The principles for deriving changes of variable for partial differential equations is basically the same as for ordinary differential equations. The first step is to make the following changes of variable:

$$\varphi(\zeta,\tau) = \frac{T - T^*}{\Delta T^*} \tag{9-16}$$

$$\zeta = \frac{r - r^*}{\Delta r^*} \tag{9-17}$$

$$\tau = \frac{t - t^*}{\Delta t^*} \tag{9-18}$$

Making these substitutions into Eqs. 7-3 through 7-6 yields

$$\alpha \left[\frac{\Delta T^* \partial^2 \varphi(\zeta,\tau)}{(\Delta r^*)^2 \partial \zeta^2} + \frac{2\Delta T^*}{(\Delta r^*\zeta + r^*)\Delta r^*} \frac{\partial \varphi(\zeta,\tau)}{\partial \zeta} \right]$$

$$= \frac{\Delta T^*}{\Delta t^*} \frac{\partial \varphi(\zeta,\tau)}{\partial \tau} \tag{9-19}$$

$$\varphi\left(\zeta, \frac{-t^*}{\Delta t^*} \right) = \frac{T_0 - T^*}{\Delta T^*} \tag{9-20}$$

$$\frac{\partial \varphi\left(-\dfrac{r^*}{\Delta r^*}, \tau \right)}{\partial \zeta} = 0 \tag{9-21}$$

$$\left(\frac{\Delta T^*}{\Delta r^*} \right) \frac{\partial \varphi\left(\dfrac{R - r^*}{\Delta r^*}, \tau \right)}{\partial \zeta} = -\frac{h}{k} \left\{ \Delta T^* \varphi\left(\frac{R - r^*}{\Delta r^*}, \tau \right) + T^* - T_w \right\} \tag{9-22}$$

Equations 9-19 and 9-22 can be rearranged to

$$\frac{\partial^2 \varphi(\zeta,\tau)}{\partial \zeta^2} + \frac{2}{\left(\zeta + \dfrac{r^*}{\Delta r^*} \right)} \frac{\partial \varphi(\zeta,\tau)}{\partial \zeta} = \frac{(\Delta r^*)^2}{\alpha \Delta t^*} \frac{\partial \varphi(\zeta,\tau)}{\partial \tau} \tag{9-23}$$

$$\frac{\partial \varphi\left(\dfrac{R - r^*}{\Delta r^*}, \tau \right)}{\partial \zeta} = \frac{\Delta r^* h}{k} \left\{ \varphi\left(\frac{R - r^*}{\Delta r^*}, \tau \right) - \frac{T_w - T^*}{\Delta T^*} \right\} \tag{9-24}$$

Thus the following dimensionless groups of parameters result:

$$\left(\frac{r^*}{\Delta r^*} \right) \qquad\qquad \left(\frac{T_w - T^*}{\Delta T^*} \right)$$

$$\left(\frac{R - r^*}{\Delta r^*} \right) \qquad\qquad \left(\frac{T_0 - T^*}{\Delta T^*} \right)$$

$$\left(\frac{(\Delta r^*)^2}{\alpha \Delta t^*} \right) \qquad\qquad \left(\frac{h \Delta r^*}{k} \right)$$

$$\left(\frac{t^*}{\Delta t^*} \right)$$

Note that there are seven groups with six arbitrary constants. Consequently, all but one dimensionless groups of parameters can be eliminated from the dimensionless equations.

Although there is some freedom in selecting which groups are set equal to zero, the following reasoning was used to obtain the changes of variable suggested in Sec. 7-3:

1. In the second term in Eq. 9-23 note that this term is much simpler if the group $(r^*/\Delta r^*)$ is set equal to zero. This also simplifies the boundary condition in Eq. 9-21. Thus:

$$r^* = 0$$

2. In Eq. 9-22 the group $(R/\Delta r^*)$ remains. If this group were set equal to unity, the new variable ζ will be normalized. Consequently

$$\Delta r^* = R$$

3. As written, the boundary condition in Eq. 9-24 is in a rather complex form. Some simplification can be obtained if the group $(T_w - T^*)/\Delta T^*$ is set equal to zero. Therefore

$$T^* = T_w$$

4. Since the temperature must be within the range between T_0 and T_w, the variable $\varphi(\zeta,\tau)$ can be normalized by setting the group $(T_0 - T^*)/\Delta T^*$ equal to unity. Consequently,

$$\Delta T^* = T_0 - T_w$$

5. The reference time t^* only appears in the group $(t^*/\Delta t^*)$. Setting this group equal to zero simplifies the boundary condition in Eq. 9-20. If $(t^*/\Delta t^*)$ equals zero, then

$$t^* = 0$$

6. The group $(\Delta r^*)^2/\alpha \Delta t^*$ will not explicitly appear in the final expressions if it is set equal to one. Recall that Δr^* equals R, which yields

$$\Delta t^* = R^2/\alpha$$

With these values for the arbitrary constants, the changes of variable are

$$\varphi(\zeta,\tau) = \frac{T(r,t) - T_w}{T_0 - T_w}$$

$$\zeta = \frac{r}{R}$$

$$\tau = \frac{\alpha t}{R^2}$$

Inserting these simplifications into Eqs. 9-20, 9-21, 9-23, and 9-24 yields

$$\frac{\partial^2 \varphi(\zeta,\tau)}{\partial \zeta^2} + \frac{2}{\zeta}\frac{\partial \varphi(\zeta,\tau)}{\partial \zeta} = \frac{\partial \varphi(\zeta,\tau)}{\partial \tau}$$

$$\varphi(\zeta,0) = 1$$

$$\frac{\partial \varphi(0,\tau)}{\partial \zeta} = 0$$

$$\frac{\partial \varphi(1,\tau)}{\partial \zeta} = -\frac{Rh}{k}\,\varphi(1,\tau)$$

Note that the only group remaining in these expressions is (Rh/k). In Fig. 7-4 is a set of graphs representing the solution of this set of equations in terms of the three dimensionless variables and the dimensionless group (Rh/k).

9-4. INTRODUCTION TO SIMILARITY TRANSFORMS

In the above examples the changes of variable have resulted in a simpler set of equations since fewer groups of parameters are present. However, no independent or dependent variables have been eliminated. In some cases the above procedure can lead to the reduction in the number of independent variables. The remainder of this chapter is devoted to this subject.

Before beginning this discussion, perhaps a review of a few principles of calculus is in order. The principal item of interest is the so-called *chain rule*, which can be expressed as follows: Let $f(\xi,\eta)$ be a function with continuous first partial derivatives, and let ξ and η in turn be functions of x and y. If ξ and η are continuous and have first partial derivatives at the point (x,y), then

$$\frac{\partial}{\partial x}\Big[f(\xi,\eta)\Big] = \frac{\partial f(\xi,\eta)}{\partial \xi}\frac{\partial \xi}{\partial x} + \frac{\partial f(\xi,\eta)}{\partial \eta}\frac{\partial \eta}{\partial x}$$

$$\frac{\partial}{\partial y}\Big[f(\xi,\eta)\Big] = \frac{\partial f(\xi,\eta)}{\partial \xi}\frac{\partial \xi}{\partial y} + \frac{\partial f(\xi,\eta)}{\partial \eta}\frac{\partial \eta}{\partial y}$$

In this section, a special case of this rule will be considered. The initial form of the equations contains a function $g(x,y)$ of two variables, but a similarity transform will be used to find a new function $f_1(\eta)$ in which η is a function of x and y. Thus a partial differential equation can be reduced to an ordinary differential equation, provided that such a function exists. For those cases in which it does exist, its partial derivatives can be obtained as follows:

$$\frac{\partial f_1(\eta)}{\partial x} = \frac{df_1(\eta)}{d\eta}\frac{\partial \eta}{\partial x} \tag{9-25}$$

$$\frac{\partial f_1(\eta)}{\partial y} = \frac{df_2(\eta)}{d\eta}\frac{\partial \eta}{\partial y} \tag{9-26}$$

In this usage, $df_1(\eta)/d\eta$ is usually represented by $f_1'(\eta)$.

9-5. SIMILARITY TRANSFORM FOR ONE-DIMENSIONAL UNSTEADY-STATE HEAT CONDUCTION

The concepts applicable in making similarity transforms are best presented by illustrating with an example. Therefore, consider the problem of heat conduction in the semi-infinite slab shown in Fig. 9-2. In Sec. 7-5 the equation for steady-state heat conduction in two dimensions was derived. By analogous procedures, the following equation for unsteady state can be obtained:

$$\frac{\partial T(x,t)}{\partial t} = \alpha \frac{\partial^2 T(x,t)}{\partial x^2} \qquad (9\text{-}27)$$

This equation is commonly called the *diffusion equation* for the one-dimensional case. Equation 9-27 is in rectangular coordinates, whereas Eq. 7-3 is the diffusion equation in spherical coordinates.

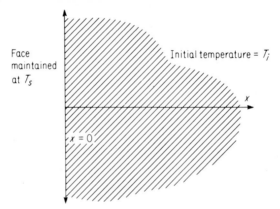

Face maintained at T_s

Initial temperature = T_i

x

$x = 0$

FIG. 9-2.   Heat conduction in a semi-infinite slab.

The boundary conditions for Eq. 9-27 can be obtained by inspection of Fig. 9-2. Since the initial temperature of the entire slab is T_i, the first boundary condition is

$$T(x,0) = T_i \qquad (9\text{-}28)$$

The temperature of the face at $x = 0$ is kept at T_s, which is expressed mathematically as

$$T(0,t) = T_s \qquad (9\text{-}29)$$

Equation 9-27 requires three boundary conditions, two of which were obtained above. The other boundary condition could be at $x = \infty$, but such boundary conditions are difficult to formulate and difficult to use. Instead, it is sufficient to know that the temperature $T(x,t)$ is a well-behaved physical function which has a finite value at every point in the slab. This can be expressed by

$$T(x,t) < M \qquad (9\text{-}30)$$

where M is some unspecified positive number. This restriction essentially states that the value of $T(x,t)$ cannot be infinitely large for any value of x and t.

To derive a similarity transform we start in the same manner as in the last section; i.e., by introducing appropriate changes of variable. As in the previous chapter, first make changes of variable with arbitrary reference values. For this problem, the following variables are introduced:

$$\varphi(\zeta,\tau) = \frac{T(x,t) - T^*}{\Delta T^*}$$

$$\zeta = \frac{x - x^*}{\Delta x^*}$$

$$\tau = \frac{t - t^*}{\Delta t^*}$$

where T^*, ΔT^*, x^*, Δx^*, t^*, and Δt^* are arbitrary constants.

From the results of the previous examples in this chapter, one could easily predict that t^* and x^* are both zero. Substituting the above dimensionless variables with x^* and t^* equal zero into Eq. 9-27 and its boundary conditions yields

$$\frac{\partial \varphi(\zeta,\tau)}{\partial \tau} = \frac{\alpha \Delta t^*}{(\Delta x^*)^2} \frac{\partial^2 \varphi(\zeta,\tau)}{\partial \zeta^2} \tag{9-31}$$

$$\varphi(\zeta,0) = \frac{T_i - T^*}{\Delta T^*} \tag{9-32}$$

$$\varphi(0,\tau) = \frac{T_s - T^*}{\Delta T^*} \tag{9-33}$$

$$\varphi(\zeta,\tau) < M' \tag{9-34}$$

where M' is another arbitrary positive number. In effect, this last boundary condition says that since $T(x,t)$ is bounded, $\varphi(\zeta,\tau)$ must be bounded also.

From these equations the following groups are obtained:

$$\left(\frac{\alpha \Delta t^*}{(\Delta x^*)^2} \right)$$

$$\left(\frac{T_i - T^*}{\Delta T^*} \right)$$

$$\left(\frac{T_s - T^*}{\Delta T^*} \right)$$

Note that there are three dimensionless groups with four arbitrary constants. Since one arbitrary constant can be specified for each group, one of these arbitrary constants will remain after each group is eliminated. Thus, a similarity transform is possible.

The last two groups in the above set of groups will specify the values of T^* and ΔT^*. One of these groups must be set equal to one and the other to zero. In this case it makes little difference which group is set equal to zero, so the first group will be selected arbitrarily. Setting $(T_i - T^*)/\Delta T^*$ equal to zero indicates

$$T^* = T_i$$

Now setting the group $(T_s - T^*)/\Delta T^*$ equal to one reveals

$$\Delta T^* = T_s - T_i$$

Thus the new dependent variable is defined as

$$\varphi(\zeta,\tau) = \frac{T(x,t) - T_i}{T_s - T_i} \qquad (9\text{-}35)$$

Although setting the group $\alpha\Delta t^*/(\Delta x^*)^2$ equal to one does not specify either Δt^* or Δx^*, it will provide the following relationship between them:

$$\Delta t^* = (\Delta x^*)^2/\alpha \qquad (9\text{-}36)$$

If this group were set equal to zero, Eq. 9-31 would reduce to

$$\partial\varphi/\partial\tau = 0$$

which does not describe the physical system. Letting Δx^* remain arbitrary and substituting this value for Δt^* into the expression for the independent variable τ yields

$$\tau = \frac{t\alpha}{(\Delta x^*)^2} \qquad (9\text{-}37)$$

Recall that the independent variable ζ is defined as

$$\zeta = \frac{x}{\Delta x^*}$$

Therefore, an arbitrary constant appears in the relationships between the dimensionless independent variables and the original independent variables.

In such cases a similarity transform can be made in order to reduce the number of independent variables. The new independent variable can be determined by eliminating the arbitrary constant between the expressions for the two dimensionless independent variables. This can be accomplished by defining a new independent variable η as follows:

$$\eta = \frac{\zeta}{\tau^{\frac{1}{2}}} = \frac{x/\Delta x^*}{t^{\frac{1}{2}}\alpha^{\frac{1}{2}}/\Delta x^*} = \frac{x}{t^{\frac{1}{2}}\alpha^{\frac{1}{2}}} \qquad (9\text{-}38)$$

Consequently, the dependent variable φ can be expressed as some function of η, or

$$\frac{T(x,t) - T_i}{T_s - T_i} = \varphi(\zeta,\tau) = f(\eta) \qquad (9\text{-}39)$$

At this point it is not clear why η was defined as $\zeta/\tau^{\frac{1}{2}}$ as in Eq. 9-38. Clearly, η could be defined as ζ^2/τ, τ/ζ^2, etc., all of which would result in an expression that does not contain Δx^*. However, note that when η is defined as in Eq. 9-38, then

$$\frac{\partial^2 \eta}{\partial x^2} = 0$$

This does not hold for any of the other possible definitions of η. Note that in Eq. 9-27, the second partial of $T(x,t)$ with respect to x appears. In expressing this partial derivative in terms of $f(\eta)$, the term $\partial^2 \eta/\partial x^2$ appears. Defining η such that $\partial^2 \eta/\partial x^2$ is zero eliminates one term when $\partial^2 T/\partial x^2$ is evaluated in terms of $f(\eta)$. The evaluation of $\partial^2 T(x,t)/\partial x^2$ will be presented in detail shortly.

As an insurance against possible errors in deriving the similarity transform, it is usually preferable to introduce the transformation into the original differential equation and its boundary conditions. Introducing the changes of variable in Eqs. 9-38 and 9-39 into Eq. 9-27 and its boundary conditions is accomplished as follows:

1. Solve Eq. 9-39 for $T(x,t)$:

$$T(x,t) = (T_s - T_i)\, f(\eta) + T_i$$

2. Calculate $\partial T(x,t)/\partial t$:

$$\frac{\partial T(x,t)}{\partial t} = \frac{\partial\, [(T_s - T_i)f(\eta) + T_i]}{\partial t}$$

$$= (T_s - T_i)\frac{\partial f(\eta)}{\partial t}$$

$$= (T_s - T_i)\frac{df(\eta)}{d\eta}\frac{\partial \eta}{\partial t}$$

$$= -\frac{(T_s - T_i)x}{2\alpha^{\frac{1}{2}} t^{3/2}} f'(\eta)$$

3. In a similar manner, $\partial^2 T(x,t)/\partial^2 x$ is determined:

$$\frac{\partial^2 T(x,t)}{\partial x^2} = \frac{\partial}{\partial x}\left[\frac{\partial T(x,t)}{\partial x}\right] = \frac{\partial}{\partial x}\left[(T_s - T_i)\frac{df(\eta)}{dn}\frac{\partial \eta}{\partial x}\right]$$

$$= (T_s - T_i)\frac{\partial}{\partial x}\left[\frac{df(\eta)}{d\eta}\left(\frac{\partial \eta}{\partial x}\right)\right]$$

$$= (T_s - T_i)\left[\frac{d^2f(\eta)}{d\eta^2}\left(\frac{\partial \eta}{\partial x}\right)^2 + \frac{df(\eta)}{d\eta}\frac{\partial^2 \eta}{\partial x^2}\right]$$

If η is chosen such that $\partial^2 \eta/\partial x^2$ is zero, then the last term in this expression vanishes. Evaluating the partial derivative in the above expression yields

$$\frac{\partial^2 T(x,t)}{\partial x^2} = \frac{f''(\eta)}{t\alpha} (T_s - T_i)$$

4. Substituting into Eq. 9-27 and simplifying,

$$-\frac{(T_s - T_i)x}{2\alpha^{\frac{1}{2}} t^{3/2}} f'(\eta) = \alpha \frac{f''(\eta)}{t\alpha} (T_s - T_i)$$

or

$$f''(\eta) + \frac{\eta}{2} f'(\eta) = 0 \tag{9-40}$$

5. Substituting into the boundary condition in Eq. 9-28,

$$f(0) = 0 \tag{9-41}$$

(Note that $\eta = 0$ when $t = 0$).

6. Substituting into the boundary condition in Eq. 9-29,

$$f(\infty) = 1 \tag{9-42}$$

(Note that $\eta = \infty$ when $x = 0$)

By this procedure, the solution to the original partial differential equation can be obtained from the solution of the following ordinary differential equation and its boundary condition:

$$f''(\eta) + \frac{\eta}{2} f'(\eta) = 0 \tag{9-40}$$

$$f(0) = 0 \tag{9-41}$$

$$f(\infty) = 1 \tag{9-42}$$

As a comparison the solution to Eq. 9-40 and its boundary conditions is given in Fig. 9-3, and the solution to the original partial differential equation is given in

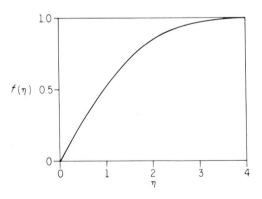

FIG. 9-3.   Solution to Eqs. 9-40 through 9-42.

Fig. 9-4. As a rule the solutions to ordinary differential equations are much simpler to obtain than for partial differential equations. However, for this case the partial differential equation has an analytic solution, namely,

$$T(x,t) = T_i + (T_s - T_i)\cdot\left[1 - \frac{2}{\sqrt{\pi}}\int_0^{\frac{x}{2\sqrt{\alpha t}}} e^{-\eta^2}\,d\eta\right]$$

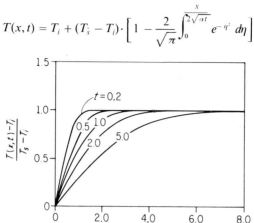

FIG. 9-4. Solution to original partial differential equation.

9-6. A PROBLEM IN BOUNDARY-LAYER THEORY

Although similarity transforms can be developed for a relatively small number of problems, they have been used extensively in the area of boundary-layer theory[2]. Probably the main reason for their popularity in this area is that it developed in an era (the 1930's) before computers became available. Thus the only practical way to solve problems was analytically. Since the similarity transform could reduce the equation describing the problem from a partial differential equation to an ordinary differential equation, it was quite useful.

The general principles and ideas of boundary-layer theory are illustrated in the following example[3].

Consider the hypothetical case in which an infinitely wide flat plate is immersed in an infinitely wide and thick stream of fluid. The temperature of the fluid is uniform at T_1 before the leading edge of the plate, and the temperature of the plate is maintained constant at T_0. See Fig. 9-5 for a schematic representation of the process. Let x be the distance from the leading edge of the plate, and y be the distance from the surface of the plate. The velocity distribution in the fluid can be approximately represented as

$$v_x = ay \qquad (9\text{-}43)$$
$$v_y = 0$$
$$v_z = 0$$

where a is a constant. The rate of heat transfer by conduction is assumed to be important only in the y-direction.   In addition the fluid properties are assumed to be constant.

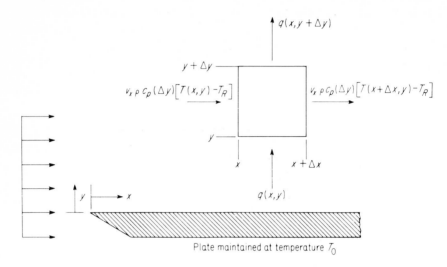

FIG. 9-5.   Flat plate in uniform fluid stream.

Although some of these assumptions are not very accurate, the results will be at least a first approximation to the actual case.   A significant improvement can be obtained by formulating a momentum balance to give the velocity profile. However, this adds so much complexity to the problem that the general approach would be obscured.

To formulate the differential equation to describe the temperature as a function of x and y, a heat balance is applied to the element shown in Fig. 9-5.   The rate of heat input through the face at x by bulk flow is $v_x \rho c_p (\Delta y) |\, T(x,y) - T_R\,|$ where T_R is the reference temperature for computing enthalpy.   The rate of heat output at face $x + \Delta x$ is

$$v_x \rho c_p (\Delta y) |\, T(x + \Delta x, y) - T_R\,|$$

Since the rate of heat transfer by conduction is assumed to be negligible in all directions except the y-direction, no other inputs or outputs occur in the x-direction.   Through the face at y the rate of heat input by conduction is

$$q(x,y) = -k(\Delta x)\frac{\partial T(x,y)}{\partial y}$$

Similarly, the rate of heat output at face $y + \Delta y$ is $q(x, y + \Delta y)$.   If the steady-state temperature profile is desired, the accumulation term must be set equal to zero. Combining the above terms into an energy balance yields

$$v_x \rho c_p (\Delta y) | T(x,y) - T_R | + q(x,y) - v_x \rho c_p (\Delta y) | T(x + \Delta x,t) - T_R |$$
$$- q(x, y + \Delta y) = 0$$

This equation can be simplified to

$$\frac{\partial}{\partial x} \left\{ v_x \rho c_p [T(x,y) - T_R] \right\} = \frac{\partial}{\partial y} \left[k \frac{\partial T(x,y)}{\partial y} \right]$$

Recalling that v_x is a function of y only and that all physical properties were assumed to be constant, this equation becomes

$$\frac{\partial T(x,y)}{\partial x} = \frac{k}{v_x \rho c_p} \frac{\partial^2 T(x,y)}{\partial y^2}$$

Substituting for v_x from Eq. 9-43 yields

$$\frac{\partial T(x,y)}{\partial x} = \frac{k}{a \rho c_p} \frac{1}{y} \frac{\partial^2 T(x,y)}{\partial y^2} \tag{9-44}$$

The boundary conditions for this equation are

$$T(0,y) = T_1 \tag{9-45}$$
$$T(x,\infty) = T_1 \tag{9-46}$$
$$T(x,0) = T_0 \tag{9-47}$$

In order to derive a similarity transform, the following dimensionless variables will be substituted into Eq. 9-44 and its boundary conditions:

$$\varphi(\zeta,\xi) = \frac{T(x,y) - T^*}{\Delta T^*}$$

$$\zeta = \frac{x}{\Delta x^*}$$

$$\xi = \frac{y}{\Delta y^*}$$

Note that the reference quantities x^* and y^* have been set equal to zero. Substituting these variables into Eq. 9-44 and its boundary conditions yields

$$\frac{\partial \varphi(\zeta,\xi)}{\partial \zeta} = \frac{k \Delta x^*}{(\Delta y^*)^3 a \rho c_p} \frac{1}{\xi} \frac{\partial^2 \varphi(\zeta,\xi)}{\partial \xi^2}$$

$$\varphi(0,\xi) = \frac{T_1 - T^*}{\Delta t^*}$$

$$\varphi(\zeta,\infty) = \frac{T_1 - T^*}{\Delta T^*}$$

$$\varphi(\zeta,0) = \frac{T_0 - T^*}{\Delta T^*}$$

In these equations, note that the following three dimensionless groups appear:

$$\frac{T_1 - T^*}{\Delta T^*}$$

$$\frac{T_0 - T^*}{\Delta T^*}$$

$$\frac{k\Delta x^*}{(\Delta y^*)^3 a\rho c_p}$$

It makes little difference which of the first two groups is set equal to zero or to one. Arbitrarily setting the group $(T_1 - T^*)/\Delta T^*$ equal to zero yields

$$T^* = T_1$$

Setting the group $(T_0 - T^*)/\Delta T^*$ equal to one yields

$$\Delta T^* = T_0 - T_1$$

With these values for T^* and ΔT^*, the dimensionless variable $\varphi(\zeta,\xi)$ is related to $T(x,y)$ as follows:

$$\varphi(\zeta,\xi) = \frac{T(x,y) - T_1}{T_0 - T_1} \tag{9-48}$$

The group $\dfrac{k\Delta x^*}{(\Delta y^*)^3 a\rho c_p}$ is set equal to one to obtain the following relationship between Δx^* and Δy^*:

$$\Delta x^* = \frac{(\Delta y^*)^3 a\rho c_p}{k}$$

Expressing the dimensionless variables ζ and ξ both in terms of Δy^* yields

$$\zeta = \frac{xk}{(\Delta y^*)^3 a\rho c_p}$$

$$\xi = \frac{y}{\Delta y^*}$$

Since both of these dimensionless variables contain the arbitrary constant Δy^*, a new variable η can be defined such that Δy^* is eliminated. Therefore, define η as

$$\eta = \frac{\xi}{\zeta^{1/3}} = \frac{y(a\rho c_p/k)^{1/3}}{x^{1/3}} \tag{9-49}$$

Consequently the solution to Eq. 9-44 and its boundary conditions can be expressed as a function of η, or

$$\frac{T(x,y) - T_1}{T_0 - T_1} = \varphi(\zeta,\xi) = f(\eta) = f\left(\frac{y(a\rho c_p/k)^{1/3}}{x^{1/3}}\right) \tag{9-50}$$

Note that in Eq. 9-44 the second partial of T with respect to y appears. Therefore η was chosen in such a manner that $\dfrac{\partial^2 \eta}{\partial y^2}$ equals zero. Since the details of substituting the expression into the original partial differential equation are the same as in the previous section, they will not be repeated for this example. Substitution of the above expression for $f(\eta)$ and the expression for η in Eq. 9-49 into Eq. 9-44 and its boundary conditions yields

$$f''(\eta) + \frac{\eta^2}{3} f'(\eta) = 0 \tag{9-51}$$

$$f(0) = 1 \tag{9-52}$$

$$f(\infty) = 0 \tag{9-53}$$

To obtain a solution, Eq. 9-51 can be integrated once to yield

$$f'(\eta) = C_1 \exp(-\eta^3/9)$$

Integrating again from ∞ to η yields

$$\int_{\infty}^{\eta} f'(\eta)d\eta = C_1 \int_{\infty}^{\eta} |\exp(-\lambda^3/9)| \, d\lambda$$

or

$$f(\eta) - f(\infty) = C_1 \int_{\infty}^{\eta} |\exp(-\lambda^3/9)| \, d\lambda$$

However, the boundary condition in Eq. 9-53 indicates that $f(\infty)$ is zero. Therefore this equation reduces to

$$f(\eta) = C_1 \int_{\infty}^{\eta} [\exp(-\lambda^3/9)| \, d\lambda \tag{9-54}$$

Evaluating this expression at $\eta = 0$ and applying the boundary condition in Eq. 9-52 yields

$$f(0) = 1 = C_1 \int_{\infty}^{0} \exp(-\lambda^3/9) \, d\lambda$$

or

$$C_1 = -\frac{1}{\displaystyle\int_{0}^{\infty} \exp(-\lambda^3/9) \, d\lambda} \simeq 0.43$$

Substituting yields

$$f(\eta) = 0.43 \int_{\eta}^{\infty} \exp(-\lambda^3/9) \, d\lambda = \frac{T(x,y) - T_1}{T_0 - T_1}$$

9-7. SUMMARY

In this chapter the techniques for making a change of variable and, when possible, a similarity transform have been presented. Although a systematic approach was presented for the benefit of novices in the field, changes of variables will become more and more obvious as the practitioner gains experience. Perhaps the most useful area for this systematic approach is in making similarity transforms. Although a competitive technique is to guess a general expression for the new independent variable in terms of the present ones, the more systematic technique presented in this chapter seems preferable especially for those inexperienced in the art of solving such problems.

REFERENCES

1. J. D. Hellums, and S. W. Churchill, "Simplification of the Mathematical Description of Boundary and Initial Value Problems," *AIChE Journal,* vol. 10, no. 1 (January 1969), pp. 110-114.
2. H. Schlichting, *Boundary Layer Theory,* McGraw-Hill, New York, 1955.
3. H. S. Mickley, T. S. Sherwood, and C. E. Reed, *Applied Mathematics in Chemical Engineering,* McGraw-Hill, New York, 1957.

PROBLEMS

9-1. The concept of a time constant was introduced in Chapter 3. For example, the temperature of the billet in Prob. 3-1 is given by

$$\frac{Mc_p}{hA}\frac{dT(t)}{dt} + T(t) = T_w$$

$$T(0) = T_0$$

Derive the dimensionless variables for this equation, noting that the dimensionless time variable is simply the real time divided by the time constant.

9-2. Heat flow in the radial direction in a cylindrical conductor is described by the following differential equation and boundary conditions:

$$\frac{d}{dr}\left[r\frac{dT(r)}{dr} \right] = 0$$

$$T(R_i) = T_i$$

$$T(R_0) = T_0$$

Derive appropriate changes of variable.

9-3. The rate of diffusion from the mothball considered in Prob. 1-5 is given by

$$\frac{d}{dr}\left[\frac{r^2}{1 - x_A(r)} \frac{dx_A(r)}{dr}\right] = 0$$

$$x_A(R_0) = x_{A_0}$$

$$\lim_{r \to \infty} x_A(r) = 0$$

Derive the changes of variable

$$\zeta = \frac{r}{R_0}$$

$$\Psi(\zeta) = \frac{1 - x_A(r)}{1 - x_{A_0}}$$

and then substitute $\xi = \dfrac{1}{\zeta}$ so that $0 \le \xi \le 1$

9-4. A single-pass counterflow heat exchanger is described by the following differential equations (see Prob. 8-7):

$$\frac{dT_t(z)}{dz} - \frac{UA}{m_t c_{p_t}} [T_s(z) - T_t(z)] = 0$$

$$\frac{dT_s(z)}{dz} - \frac{UA}{m_s c_{p_s}} [T_s(z) - T_t(z)] = 0$$

$$T_s(0) = T_{s_0}$$

$$T_t(L) = T_{t_0}$$

Derive the appropriate changes of variable.

9-5. A tubular reactor in which a series reaction $A \xrightarrow{k_1} B \xrightarrow{k_2} C$ is occurring is described by the following equations:

$$V\frac{dC_A(z)}{dz} + A_r k_1 C_A(z) = 0$$

$$V\frac{dC_B(z)}{dz} + A_r k_2 C_B(z) - A_r k_1 C_A(z) = 0$$

$$C_A(0) = C_{A_0}$$

$$C_B(0) = 0$$

Derive the appropriate changes of variable.

9-6. The concentration profiles in a wetted-wall column (if the gas is in laminar flow) are given by the following partial differential equation:

$$D_{AB} \frac{\partial}{\partial r}\left[r\frac{\partial C_A(r,z)}{\partial r}\right] - rv(r)\frac{\partial C_A(r,z)}{\partial z} = 0$$

$$C_A(r,0) = 0$$

$$\frac{\partial C_A(0,z)}{\partial r} = 0$$

$$\frac{\partial C_A(R,z)}{\partial r} = \frac{k}{D_{AB}} [C_{A_i} - C_A(R,z)]$$

Derive the appropriate changes of variable. Recall that $v(r)/v_{\max}$ is dimensionless and normalized.

9-7. The equations describing the coffee pot in Prob. 7-13 are

$$- w \frac{\partial C(z,t)}{\partial z} + A a k_m [C_s - C(z,t)] = H_L \frac{\partial C(z,t)}{\partial t}$$

$$\frac{H_t}{w} \frac{dC_r(t)}{dt} + C_r(t) = C(L,t)$$

$$C(0,t) = C_r(t)$$

$$C(z,0) = 0$$

$$C_r(0) = 0$$

Derive the appropriate changes of variable.

9-8. The partial differential equation describing the "start-up" of laminar flow was derived in Prob. 7-14 to be

$$\mu \frac{1}{r} \frac{\partial}{\partial r} \left[r \frac{\partial v_z(r,t)}{\partial r} \right] - \frac{dP}{dz} = \rho \frac{\partial v_z(r,t)}{\partial t}$$

$$v_z(r,0) = 0$$

$$v_z(R,t) = 0$$

$$\frac{\partial v_z(0,t)}{\partial r} = 0$$

Derive the appropriate changes of variable.

9-9. The unsteady-state equation describing the temperature profile in the exchanger in Sec. 1-6 is

$$\rho c_p \frac{\partial T(z,t)}{\partial t} = - \frac{w c_p}{\pi R^2} \frac{\partial T(z,t)}{\partial z} + \frac{2U_i}{R} [T_s - T(z,t)]$$

$$T(0,t) = T_i$$

$$T(z,0) = T_i$$

Derive the appropriate changes of variable.

9-10. In developing the equations for mass transfer using the penetration theory (see Prob. 5-12), the following equation and boundary conditions are encountered:

$$\frac{\partial C_A(x,t)}{\partial t} = D_{AB} \frac{\partial^2 C_A(x,t)}{\partial x^2}$$

$$C_A(x,0) = C_{A_0}$$
$$C_A(0,t) = C_{A_s}$$
$$C_A(\infty,t) = C_{A_0}$$

Develop a similarity transform for this set, substitute, and solve.

9-11. According to boundary-layer theory flow near the leading edge of a flat plate is described by

$$v_x \frac{\partial v_x}{\partial x} - \left(\int \frac{\partial v_x}{\partial x} \, dy \right) \frac{\partial v_x}{\partial y} = v \frac{\partial^2 v_x}{\partial y^2}$$

$$v_x = 0 \quad \text{at} \quad y = 0$$
$$v_x = v_0 \quad \text{at} \quad y = \infty$$
$$v_x = v_0 \quad \text{at} \quad x = 0$$

Derive the following changes of variable:

$$\varphi(\eta) = \frac{v_x}{v_0}$$

$$\eta = \frac{y}{\sqrt{\dfrac{vx}{v_0}}}$$

Substitute into the original equation.

chapter 10

Physical Analogies

To extend our knowledge of familiar systems to those less familiar, an analogy between the two is often very helpful. If two systems are described by the same differential equations, these two systems are said to be *analogs*. This is the basic principle behind the use of analog computers for investigating system behavior —an analog to the physical system can be developed from the components of the analog computer. This same idea extends to distributed-parameter systems, except that the analog is often more specialized. Such practical utilization of the analogies between systems will be discussed later in this chapter, but it will first be discussed in connection with obtaining the differential equations for distributed-parameter systems.

10-1. THE SIGNIFICANCE OF PHYSICAL SIMILARITY

One of the amazing aspects of science is that essentially all physical systems, no matter how diverse, are described by the same general differential equation. From one system to another only the physical connotation of the variables differs. The reasons for this are not clear, with some writers preferring to attach a religious significance, some contending that it is a matter of semantics, and others proposing different reasons and explanations. Here these various philosophical reasons are not important, only that such a similarity exists, and that it can serve a very useful purpose.

In this section the main objective is to show that such a similarity exists. Attention is devoted primarily to distributed-parameter or "field" problems, as contrasted to lumped-parameter problems. In a lumped-parameter model the physical system is regarded as being composed of a network of distinct elements interconnected in some specific manner. In electrical circuit theory a resistor is characterized by a lumped-parameter R which relates the voltage drop across the resistor to the current through it. All the characteristics of the resistor have been lumped into this one parameter R, and it is quite satisfactory for many purposes.

On the other hand, if the internal behavior of the resistor is being investigated, this lumped-parameter model is not acceptable. For example, in designing the resistor it may be necessary to know at what point the current density is largest, and the lumped-parameter model is useless for this purpose.

To investigate this problem, each small segment of the resistor must be treated as contributing some resistance to the flow of current, thus introducing the concept that the resistivity is distributed throughout the resistor. Consequently, the voltage and current flow must be treated as being functions of position in the resistor. Any problems in which the spatial dimensions enter the analysis are considered to be field problems. In formulating differential equations, one or more independent variables describing position enter the problem, whereas time is the only independent variable in a lumped-parameter model.

At this point, two different types of independent variables have been recognized: those for position and the one for time. Similarly, there are two different types of dependent variables. The first of these is commonly called an *across* variable, primarily because it relates the condition at one point in the field to that at some other point in the field. Across variables are seldom measured in absolute units, but as the difference in the magnitudes of the across variable at two different points in the field. In an electrical system, the across variable is the voltage, which can be measured in two different ways:

1. Difference in electrical potential between two points in the system.
2. Difference in potential between a point in the system and some reference point, typically a "ground," i.e., ground potential.

Note that the across variable is measured as the difference between the scalar potential at two points and is thus a scalar quantity.

On the other hand, a *through* variable represents the flux of some quantity traversing an elemental cross section in the field. In an electrical system, the through variable would be the current flux (density), i.e., flow rate of electrons per unit area. This quantity is specified by measuring at only one point in the field. Note that flux is typically measured in terms of a magnitude and a direction of flow, and is consequently a vector quantity.

10-2. GENERALIZED SYSTEM PARAMETERS [1,2]

The relationship between the across variables and the through variables is expressed in terms of the system parameters. These parameters for field systems can be divided into three different types, the distinction being the part they play in the transfer mechanisms associated with the system characteristics. The three different types are discussed below:

Dissipators

The system can dissipate energy, or in the thermodynamic interpretation, the entropy of system and surroundings is increasing. An example of a dissipative system is one with electrical resistivity. The relationship between the across and through variables for a dissipative system is analogous to the relationship for lumped-parameter systems that dissipate energy. For an electrical resistor the equation relating voltage drop ΔV and current I is

$$\Delta V = IR$$

However, for a distributed-parameter system the voltage drop is replaced by a voltage gradient, and the equation for an electrical system is

$$\frac{\partial V}{\partial x} = -\frac{1}{\sigma} I$$

where σ is the electrical conductivity and x is the position variable for the field. For systems in general, this equation takes the form

$$\frac{\partial p}{\partial x} = -Df \tag{10-1}$$

where D = dissipativity per unit length
$\quad\quad f$ = flux or through variable
$\quad\quad p$ = potential or across variable
$\quad\quad x$ = position variable

The minus sign appears because the flux is in the direction of decreasing potential.

Reservoirs of Flux

In many cases the field acts as a reservoir of flux or energy. Examples are the inductance of an electrical system and the inertia of a mechanical or fluid system in motion. For example, the pressure gradient is proportional to the rate of change of velocity or mass flux with respect to time. Assuming constant density, this is expressed by

$$\frac{\partial P}{\partial x} = -\frac{\rho}{g} \frac{\partial v}{\partial t}$$

where P = pressure, lb_f/ft^2

$\quad\quad g_c$ = gravitational constant, $32.2 \dfrac{ft\text{-}lb_m}{lb_f\,sec^2}$
$\quad\quad t$ = time, sec
$\quad\quad v$ = velocity, ft/sec
$\quad\quad \rho$ = fluid density, lb_m/ft^3

For systems in general, this equation is expressed as

$$\frac{\partial p}{\partial x} = -E_f \frac{\partial f}{\partial t} \tag{10-2}$$

where E_f is a measure of capacity of the flux reservoir per unit length of the field.

Reservoirs of Potential

The third and last type of parameter is for the case in which the field acts as a reservoir of potential. Examples are the capacitance of an electrical system or the heat capacity of a thermal system. For conductive heat transfer it can be derived that the gradient of the flux is related to the rate of change of the temperature with time as follows:

$$\frac{\partial q}{\partial x} = - c_p \rho \frac{\partial T}{\partial t}$$

where c_p = heat capacity, Btu/lb-$^\circ$F
q = heat flux, Btu/hr-ft^2
T = temperature, $^\circ$F

For general systems this equation is expressed as follows:

$$\frac{\partial f}{\partial x} = - E_P \frac{\partial p}{\partial t} \qquad (10\text{-}3)$$

where E_P is a measure of the capacity of the potential reservoir per unit length of the field.

TABLE 10-1

Across and Through Variables for Specific Systems. (Reprinted by permission from W. J. Karplus, *Analog Simulation*, McGraw-Hill Book Co., © 1958.)

Physical Area	Across Variable, p	Through Variable, f	PARAMETERS		
			Dissipators, D	Reservoirs of Flux, E_f	Reservoirs of Potential, E_P
Electrodynamics	Voltage	Current	Resistivity	Inductivity	Capacivity
Magnetics	Potential, MMf	Flux	Reluctance	Permeability	—
Electromagnetics	Potential, EMf	Flux	Conductivity	Permeability	Dielectric permittivity
Translational mechanics	Displacement or velocity	Force	Viscous damping	Spring constant	Mass (inertia)
Rotational mechanics	Angular displacement or velocity	Torque	Viscous damping	Spring constant	Inertia
Fluid dynamics	Velocity potential (Pressure)	Flow rate	Viscosity	Inertia (density)	Compressibility
Mass transfer	Concentration	Mass flux	Diffusivity	Inertial forces	Compressibility
Heat transfer	Temperature	Heat flux	Thermal resistance	—	Thermal capacitance

These three basic types of parameters can characterize practically all physical systems. The presence or absence of one or more of these basic types of parameters determines the general form of the differential equation that describes the system. Not all physical systems have parameters of all three basic types. For example, there is no reservoir of flux for thermal systems. In Table 10-1 are listed the physical meanings of the parameters for various types of physical systems.

In the ensuing sections, the techniques for developing partial differential equations from these concepts are presented.

10-3. GENERAL TYPES OF EQUATIONS

Fortunately, only a relatively small number of different types of partial differential equations appear. In fact, practically all of these are special cases of the very general partial differential equation given below for two independent variables ζ and ξ:

$$a_1 \frac{\partial^2 p}{\partial \zeta^2} + 2a_2 \frac{\partial^2 p}{\partial \zeta \partial \xi} + a_3 \frac{\partial^2 p}{\partial \xi^2}$$

$$= a_4 \frac{\partial p}{\partial \zeta} + a_5 \frac{\partial p}{\partial \xi} + a_6 p + a_7 \tag{10-4}$$

where $a_1, \ldots, a_7$ are constants. Fortunately, it is seldom necessary to solve this equation without making some simplifications. Analogous forms of this equation are available for three and four independent variables.

In the ensuing sections several specific situations are examined in order to derive partial differential equations commonly encountered in engineering work.

Laplace's Equation (Parabolic Equation)

Consider the case in which the system contains neither reservoirs of flux nor reservoirs of potential, thus leaving only the dissipative parameter D. Such conditions often arise in describing potential field problems under steady-state conditions, e.g., the temperature profile in a slab or the electrical field inside or around a conductor.

In the ensuing discussion, Laplace's equation in one-dimension is derived for the Cartesian coordinate system. The results are readily extended to two- or three-dimensional systems and to cylindrical or spherical coordinate systems. To derive Laplace's equation in one dimension, consider the case in which a flux (the through variable) is passing through the system in Fig. 10-1. As in preceding chapters, a balance is made over an infinitesimal element for which the flux is considered to be the quantity conserved. As is customary, a positive flux is considered to be flowing in the positive direction of the independent variable.

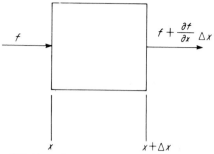

FIG. 10-1. Typical one-dimensional element.

Thus if the flux flowing in at face x is defined as f, the flux flowing out at face $x + \Delta x$ is $f + \dfrac{df}{dx}\,\Delta x$. Since there are no reservoirs of potential within the element, the rate of flux in must equal the rate of flux out. Therefore

$$f = f + \frac{df}{dx}\,\Delta x$$

or

$$\frac{df}{dx} = 0 \tag{10-5}$$

Note that this is Eq. 10-3 with E_P equal zero. For dissipative systems, the flux f is related to the potential gradient by Eq. 10-1, which can be rearranged to

$$f = -\frac{1}{D}\frac{dp}{dx}$$

Substituting into Eq. 10-5 yields

$$\frac{d}{dx}\left[\frac{1}{D}\frac{dp}{dx}\right] = 0 \tag{10-6}$$

This form of Laplace's equation is for cases in which the dissipativity varies with x. When the dissipativity is constant, Eq. 10-6 reduces to

$$\frac{d^2 p}{dx^2} = 0 \tag{10-7}$$

It is interesting to note that the dissipativity parameter D does not appear in this equation.

For a three-dimensional system, Laplace's equation can be derived in a manner very similar to that used above. For systems with variable dissipativity the result is

$$\frac{\partial}{\partial x}\left(\frac{1}{D}\frac{\partial p}{\partial x}\right) + \frac{\partial}{\partial y}\left(\frac{1}{D}\frac{\partial p}{\partial y}\right) + \frac{\partial}{\partial z}\left(\frac{1}{D}\frac{\partial p}{\partial z}\right) = 0 \tag{10-8}$$

If D is constant, this equation reduces to

$$\frac{\partial^2 p}{\partial x^2} + \frac{\partial^2 p}{\partial y^2} + \frac{\partial^2 p}{\partial z^2} = 0 \qquad (10\text{-}9)$$

Again, note that the diffusivity does not appear. The above forms of the equations are so common that a shorthand notation has been introduced. Equation 10-9 is often written as

$$\nabla^2 p = 0 \qquad (10\text{-}10)$$

where ∇^2 (del square) is the Laplacian operator defined as

$$\nabla^2 = \frac{\partial^2}{\partial x^2} + \frac{\partial^2}{\partial y^2} + \frac{\partial^2}{\partial z^2} \qquad (10\text{-}11)$$

Similarly, Eq. 10-8 is often written

$$\nabla \cdot \left(\frac{1}{D} \nabla p \right) = 0 \qquad (10\text{-}12)$$

where ∇ is

$$\nabla = \left(\frac{\partial}{\partial x}, \frac{\partial}{\partial y}, \frac{\partial}{\partial z} \right) \qquad (10\text{-}13)$$

The notation in Eq. 10-12 is the dot product of two vectors, i.e.,

$$\left(\frac{\partial}{\partial x}, \frac{\partial}{\partial y}, \frac{\partial}{\partial z} \right) \cdot \left(\frac{1}{D} \frac{\partial p}{\partial x}, \frac{1}{D} \frac{\partial p}{\partial y}, \frac{1}{D} \frac{\partial p}{\partial z} \right)$$

$$= \frac{\partial}{\partial x} \left(\frac{1}{D} \frac{\partial p}{\partial x} \right) + \frac{\partial}{\partial y} \left(\frac{1}{D} \frac{\partial p}{\partial y} \right) + \frac{\partial}{\partial z} \left(\frac{1}{D} \frac{\partial p}{\partial z} \right)$$

These types of expressions appear throughout this chapter, and the above shorthand notation is used because of its convenience.

Laplace's equation can also be derived in spherical and cylindrical coordinates by examining elements of the appropriate configuration. For cylindrical coordinates Laplace's equation for systems with constant dissipativity becomes

$$\frac{1}{r} \frac{\partial}{\partial r} \left(r \frac{\partial p}{\partial r} \right) + \frac{1}{r^2} \frac{\partial^2 p}{\partial \theta^2} + \frac{\partial^2 p}{\partial z^2} = 0 \qquad (10\text{-}14)$$

In this case the operator ∇^2 is defined as

$$\nabla^2 = \frac{1}{r} \frac{\partial}{\partial r} \left(r \frac{\partial}{\partial r} \right) + \frac{1}{r^2} \frac{\partial^2}{\partial \theta^2} + \frac{\partial^2}{\partial z^2} \qquad (10\text{-}15)$$

An expression for systems in which the dissipativity varies can also be developed. For spherical coordinates Laplace's equation for constant D is

$$\frac{1}{r^2}\left(\frac{\partial}{\partial r}\,r^2\,\frac{\partial p}{\partial r}\right) + \frac{1}{r^2\sin^2\theta}\,\frac{\partial^2 p}{\partial\varphi^2} + \frac{1}{r^2\sin^2\theta}\,\frac{\partial}{\partial\varphi}\left[\sin\theta\,\frac{\partial p}{\partial\varphi}\right] = 0$$

$$(10\text{-}16)$$

The expression for ∇^2 in spherical coordinates can be readily deduced from the above.

Recall that the above reasoning was for systems in which the dissipative parameter was the only significant parameter. In cases in which only one parameter, whether it be D, E_f, or E_P, is important, the system is still described by Laplace's equation.

The Diffusion Equation (Elliptic Equation)

If (1) the dissipativity D is significant, (2) the capability to act as a reservoir of potential E_P is significant, and (3) the capability to act as a reservoir of flux E_f is insignificant, the system is described by the diffusion equation. Such conditions arise in the description of unsteady-state conductive heat-transfer problems, compressible flow problems, and electrical problems in which the capacitance and resistance of the system are important.

As for Laplace's equation, the diffusion equation will also be derived for a one-dimensional system in Cartesian coordinates. The system to be examined is shown in Fig. 10-1, which is the same system examined in deriving Laplace's equation. As in the previous section, the flux entering the element through the face at x is denoted by f, and the flux leaving the element at the face at $x + \Delta x$ is $f + \dfrac{\partial f}{\partial x}\,\Delta x$. For a system which can act as a reservoir of potential, the difference between the flux in and the flux out is related to the capability of the system to act as a reservoir of potential. Since E_P is defined as the capability of the system to act as a potential reservoir per unit length for a one-dimensional system, the flux in and flux out of the system are related by

$$f - \left\{f + \frac{\partial f}{\partial x}\,\Delta x\right\} = E_P\Delta x\,\frac{\partial p}{\partial t}$$

or

$$\frac{\partial f}{\partial x} = -E_P\,\frac{\partial p}{\partial t} \qquad (10\text{-}17)$$

Recall that the flux f is related to the potential gradient by

$$f = -\frac{1}{D}\,\frac{\partial p}{\partial x}$$

Substituting into Eq. 10-19 yields

$$\frac{\partial^2 p}{\partial x^2} = E_P D\,\frac{\partial p}{\partial t} \qquad (10\text{-}18)$$

when D is constant. This is the diffusion equation in one dimension. In three dimensions, it becomes

$$\frac{\partial^2 p}{\partial x^2} + \frac{\partial^2 p}{\partial y^2} + \frac{\partial^2 p}{\partial z^2} = E_P D \frac{\partial p}{\partial t} \tag{10-19}$$

This equation is often written using the Laplacian operator ∇^2 as follows:

$$\nabla^2 p = E_P D \frac{\partial p}{\partial t} \tag{10-20}$$

This equation can be expressed in cylindrical and spherical coordinates by using the definition of the Laplacian operator in these coordinate systems. Note that the diffusion equation reduces to Laplace's equation at steady-state conditions.

It can also be shown that the diffusion equation applies to cases in which (1) the dissipativity of the system is significant, (2) the capability of the system to act as a reservoir of flux E_f is significant, but (3) the capability of the system to act as a reservoir of potential is insignificant. Proof of this is left to the problems at the end of the chapter. Therefore, the diffusion equation applies to systems in which the dissipative parameter is important and the capability to act as *one* but not both reservoirs is present.

The Wave Equation (Hyperbolic Equation)

The third general partial differential equation applies to systems which act as both a reservoir of flux and a reservoir of potential, but do not have a significant dissipativity parameter. Examples of such systems are the perfectly elastic vibrating string and an electrical system in which only capacitance and inductance are significant.

To derive this partial differential equation, start with Eqs. 10-2 and 10-3 for reservoirs of flux and reservoirs of potential, respectively:

$$\frac{\partial p}{\partial x} = -E_f \frac{\partial f}{\partial t} \tag{10-2}$$

$$\frac{\partial f}{\partial x} = -E_P \frac{\partial p}{\partial t} \tag{10-3}$$

Taking the partial of Eq. 10-2 with respect to x and the partial of Eq. 10-3 with respect to t yields

$$\frac{\partial^2 f}{\partial x \partial t} = -\frac{1}{E_f} \frac{\partial^2 p}{\partial x^2}$$

$$\frac{\partial^2 f}{\partial x \partial t} = -E_P \frac{\partial^2 p}{\partial t^2}$$

Equating these two equations gives the wave equation in one dimension:

$$\frac{\partial^2 p}{\partial x^2} = E_f E_P \frac{\partial^2 p}{\partial t^2} \qquad (10\text{-}21)$$

In three dimensions this equation becomes

$$\frac{\partial^2 p}{\partial x^2} + \frac{\partial^2 p}{\partial y^2} + \frac{\partial^2 p}{\partial z^2} = E_f E_P \frac{\partial^2 p}{\partial t^2}$$

Using the Laplacian operator ∇^2, this is shortened to

$$\nabla^2 p = E_f E_P \frac{\partial^2 p}{\partial t^2}$$

Note that this equation also reduces to Laplace's equation at steady-state. However, systems described by this equation do not reach a steady-state, i.e., a state in which p is not a function of time. Instead, any motion in them continues indefinitely without damping.

10-4.   AN EXAMPLE OF SIMILAR SYSTEMS

In the above sections the three fundamental equations of physics were derived for general types of systems. In a later section some common modifications of these equations are considered. Before considering these further extensions, a few examples of systems described by these equations are discussed.

Steady-State Temperature Profile in a Solid

At steady-state the reservoir of potential, i.e., the heat capacity, of the solid is not important. Furthermore, there is no reservoir of flux for thermal systems. Consequently, the dissipativity parameter is the only significant parameter, and Laplace's equation describes the temperature profile in the solid. The dissipativity parameter in this case equals the reciprocal of the thermal conductivity. Laplace's equation for solids whose thermal conductivity is constant is

$$\nabla^2 T = 0$$

where T is the temperature.

Electrical Field in a Conductor at Steady-State

Since the capacitive and inductive effects are not important at steady-state, the only significant parameter in this case is the electrical conductivity σ. Hence, Laplace's equation is again applicable. Thus the voltage V at a point in the conductor is described by

$$\nabla^2 V = 0$$

for conductors in which σ is constant.

Concentration Profile in a Diffusing Medium at Steady-State

Again, the reservoirs of flux or potential are not important, and the concentration C_A in the diffusing medium is given by

$$\nabla^2 C_A = 0$$

for systems in which the molecular diffusivity is constant.

Steady-State Flow Through a Porous Media

For laminar flow through a porous media the flow velocity (through variable) is related to the pressure gradient by Darcy's law, which is

$$v_s = -\frac{\kappa}{\mu}\frac{\partial P}{\partial x} \tag{10-22}$$

where v_s = superficial velocity
  κ = permeability of the porous media
  μ = fluid viscosity
  P = Pressure

At steady-state, the reservoirs of energy are unimportant, leaving the only significant parameter being the dissipativity as defined by Darcy's law. Thus the pressure profile in the media is given by Laplace's equation:

$$\nabla^2 P = 0$$

for beds in which the ratio κ/μ is constant.

Although all the above systems are quite different, note that all are described by Laplace's equation. This forms the basis for the construction of analog models, which will be described in a later section. A similar treatment is possible for a variety of systems described by the diffusion equation (in fact, these same systems at unsteady-state) and by the wave equation.

10-5. FIELDS WITH DISTRIBUTED SOURCES OR SINKS

Although this modification can be applied to all three of the above equations, it will be illustrated for Laplace's equation, from which it can be readily extended to the other equations. Consider a bar through which current is flowing. If the current were not flowing, the steady-state temperature profile would be described by Laplace's equation. However, when current flows through a conductor, the resistance of the conductor causes some of the electric energy to be converted to thermal energy. Since the resistance of the bar is a distributed parameter, this means that heat is generated at every point in the bar. Thus there is a distributed source of thermal energy which is not accounted for by Laplace's equation. Other situations in which this phenomenon occurs are heat transfer from the surface of a fin or rod, mass transfer by molecular diffusion accompanied by chemical reaction, heat liberated by viscous dissipation in flow systems, and a variety of others.

To develop a modified form of Laplace's equation, consider the one-dimensional element shown in Fig. 10-2. As before, the flux entering the element

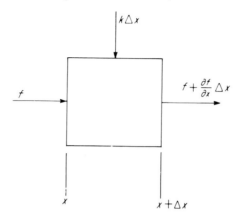

FIG. 10-2. One-dimensional element with distributed or internal source.

across the face at x is f, and the flux leaving the element across the face at $x + \Delta x$ is $f + \dfrac{df}{dx}\,\Delta x$. If the element has neither the capability to act as a reservoir of flux nor as a reservoir of potential, the flux into the element must equal the flux out of the element. If the rate at which flux enters a unit length of the system is denoted by k, a balance on the element in Fig. 10-2 reveals

$$f + k\Delta x = f + \frac{df}{dx}\,\Delta x$$

or

$$\frac{df}{dx} = k$$

Substituting the relationship between f and p for a dissipative element gives

$$\frac{d^2p}{dx^2} = -kD$$

when D is independent of x. This equation is readily extended to higher-dimensional systems, which in terms of the Laplacian operator ∇^2 is

$$\nabla^2 p = -kD$$

This is commonly known as *Poisson's equation*. In this equation k may be a function of p, which will be the case for reacting systems.

If the above modification is made to the diffusion equation, it becomes

$$\nabla^2 p = DE_P \frac{\partial p}{\partial t} - kD$$

Similarly, the wave equation becomes

$$\nabla^2 p = C_3 \frac{\partial^2 p}{\partial t^2} - C_4$$

where C_3 and C_4 are functions of the system parameters. A case where this is important is a vibrating elastic string under the influence of gravity.

10-6. THE DAMPED-WAVE EQUATION

If an actual string is excited, these excitations gradually die out with time. Since the wave equation was derived for the case in which no damping occurs, the excitation never ceases. To account for damping effects, the dissipative parameter must be considered in developing the wave equation. Although the details of the derivation will be omitted, the damped-wave equation is

$$\nabla^2 p = C_5 \frac{\partial^2 p}{\partial t^2} + C_6 \frac{\partial p}{\partial t} + C_7$$

where C_5, C_6, and C_7 depend upon the system parameters. This equation appears in fluid-dynamics problems in which viscous, compressibility, and inertial effects are appreciable; in electrodynamics problems in which resistance, inductance, and capacitance are significant; and in any other problem in which all three parameters $(D, E_P, \text{ and } E_J)$ appear.

10-7. SYSTEMS IN MOTION–THE SUBSTANTIAL DERIVATIVE

In the derivation of the diffusion equation, it was assumed that the system was not in motion. However, for systems such as a fluid in motion the partial derivative term $\partial/\partial t$ in the diffusion equation must be replaced by the substantial derivative D/Dt defined as

$$\frac{D}{Dt} = \frac{\partial}{\partial t} + v_x \frac{\partial}{\partial x} + v_y \frac{\partial}{\partial y} + v_z \frac{\partial}{\partial z}$$

where v_x, v_y, and v_z are the velocity components in the x, y, and z directions, respectively. The reason for this distinction is that the real time must be replaced by an effective time which includes the effects of fluid motion, as discussed in Sec. 2-9. Consequently, the diffusion equation for systems in motion becomes

$$\nabla^2 p = \frac{Dp}{Dt}$$

Examples in which this is important include heat transfer in a flowing stream and diffusion in a flowing stream.

10-8. EQUATIONS ENCOUNTERED IN THEORY OF ELASTICITY

Probably the majority of systems which the previous equations cannot describe occur in the theory of elasticity. However, since such problems have not been considered previously in this text, they are not discussed here either. Interested readers are suggested to consult the texts by Karplus[1,3].

10-9. GENERAL BOUNDARY CONDITIONS

The possible forms and general rules for boundary conditions for these problems are given in Table 10-2. Although this table is valid for the majority of systems, there are exceptions.

10-10. LAMINAR-FLOW TUBULAR REACTOR

In Sec. 7-6 the equations describing the concentration and temperature profiles in a laminar-flow tubular reactor were derived by inspecting a general differential element or control volume in the system, which was the technique used in the first six chapters. However, the concepts of the general partial differential equations presented in this chapter can also be used.

In connection with these concepts, the following points could be made concerning the partial differential equation for the mass balance on component A derived in Sec. 7-6:

1. The dissipativity parameter is significant, but in the r-direction only. (Recall that mass transfer by diffusion in the z-direction has been neglected.) In this problem the dissipativity parameter is the reciprocal of the molecular diffusivity D_{AB}.

2. The system also has a significant ability to act as a reservoir of potential, and the general parameter E_p in this case becomes $1/\rho$, where ρ is the molar density.

3. Since inertial effects are not significant in this case, the system does not have the capability to act as a reservoir of flux.

4. The substantial derivative instead of the partial derivative with respect to time must be used because the fluid is in motion.

5. As component A is being consumed by reaction, a constant term as in Poisson's equation must be included.

The general differential equation describing this situation is the diffusion equation with a distributed source as in Poisson's equation, namely,

TABLE 10-2

Boundary Conditions. (Reprinted by permission from W. J. Karplus, *Analog Simulation*, McGraw-Hill Book Co., © 1958.)

Equation	Mathematical Form	Number of Boundary Conditions	Types of Boundary Conditions	Number of Initial Conditions	Types of Initial Conditions	Number of Parameter Types	Types of Parameters
Laplace's	$\nabla^2 = 0$	Two for each space variable	$p = k$ $\dfrac{\partial p}{\partial n} = k$	None		One	Dissipator or *PE* reservoir or *KE* reservoir
Diffusion	$\nabla^2 p = k \dfrac{\partial p}{\partial t}$	Two for each space variable	$p = k$ $\dfrac{\partial p}{\partial n} = k$	One	At $t = 0$: $p(x,y,z)$	Two	Dissipator and *PE* reservoir or *KE* reservoir
Wave	$\nabla^2 p = k \dfrac{\partial^2 p}{\partial t^2}$	Two for each space variable	$p = k$ $\dfrac{\partial p}{\partial n} = k$	Two	At $t = 0$: $p(x,y,z)$ and $\dfrac{\partial p}{\partial t}(x,y,z)$	Two	*PE* reservoir and *KE* reservoir
Damped wave	$\nabla^2 p = k_1 \dfrac{\partial^2 p}{\partial t^2}$ $+ k_2 \dfrac{\partial p}{\partial t} + k_2 p$	Two for each space variable	$p = k$ $\dfrac{\partial p}{\partial n} = k$	Two	At $t = 0$: $p(x,y,z)$ and $\dfrac{\partial p}{\partial t}(x,y,z)$	Three	Dissipator and *PE* reservoir and *KE* reservoir
Diffusion with moving coordinate system	$\nabla^2 p = k_1 \dfrac{\partial p}{\partial t}$ $+ k_2 \dfrac{\partial p}{\partial x} + k_2 \dfrac{\partial p}{\partial y}$ $+ k_4 \dfrac{\partial p}{\partial z}$	Two for each space variable	$p = k$ $\dfrac{\partial p}{\partial n} = k$	One	At $t = 0$: $p(x,y,z)$	Two	Dissipator and *PE* reservoir or *KE* reservoir
Poisson's	$\nabla^2 p = -ki$	Two for each space variable	$p = k$ $\dfrac{\partial p}{\partial n} = k$	None		One	Dissipator or *PE* reservoir or *KE* reservoir

$$\nabla^2 p = DE_p \frac{\partial p}{\partial t} - kD$$

Recalling that the system is in cylindrical coordinates and that the diffusivity is significant only in the radial direction, the previous equation becomes

$$\frac{1}{r} \frac{\partial}{\partial r} \left[r \frac{\partial C_A}{\partial r} \right] = \frac{\rho}{D_{AB}} \frac{DC_A}{Dt} - r_A / D_{AB}$$

where r_A is the rate of appearance of component A by chemical reaction. Note that all the terms in the above equation have the units moles A /hr-ft^3.

The right-hand side does not include the term $D_{AB} \; \partial^2 T/\partial z^2$ because the molecular diffusion in the z-direction is being neglected. Expanding the substantial derivative yields

$$\frac{DC_A}{Dt} = \frac{\partial C_A}{\partial t} + \frac{v_r}{r} \frac{\partial (rC_A)}{\partial r} + \frac{v_\theta}{r} \frac{\partial C_A}{\partial \theta} + v_z \frac{\partial C_A}{\partial z}$$

Since the reactor is in steady-state operation, the term $\partial C_A/\partial t$ is zero. Furthermore, v_r and v_θ, the components of the velocity in the r- and θ-directions, are zero, thus reducing the substantial derivative to

$$\frac{DC_A}{Dt} = v_z \frac{\partial C_A}{\partial z}$$

The reaction-rate term is the rate of appearance of component A by reaction, which is

$$r_A = -k_1 C_A$$

Substituting these terms into the original equation yields

$$\frac{D_{AB}}{r} \frac{\partial}{\partial r} \left[r \frac{\partial C_A}{\partial r} \right] = \rho v_z \frac{\partial C_A}{\partial r} + k_1 C_A$$

Expanding the partial and rearranging,

$$D_{AB} \frac{\partial^2 C_A}{\partial r^2} + \frac{D_{AB}}{r} \frac{\partial C_A}{\partial r} - \rho v_z \frac{\partial C_A}{\partial z} - k_1 C_A = 0$$

This equation is identical to Eq. 7-23 derived earlier, and of course, the boundary conditions are also the same.

As for the partial differential equation for T as a function of r and z, it can be developed in a similar manner. Therefore, note these five points about the system:

1. Since the conduction of heat has been assumed to be significant in the r-direction only, the system has a significant dissipativity parameter. In this case the diffusivity parameter is the reciprocal of the thermal conductivity k.

2. This system has a significant capability to act as a reservoir of potential. In this case the general parameter E_P is $1/\rho c_P$.
3. Recall that thermal systems never act as reservoirs of flux.
4. Since the fluid is in motion, the substantial derivative must be used instead of the partial derivative with respect to time.
5. Since the reaction has a nonzero heat of reaction, a constant term as in Poisson's equation must be included.

These points indicate that the following form of the diffusion equation should be used:

$$\frac{1}{r}\frac{\partial}{\partial r}\left[r\frac{\partial T}{\partial r}\right] = \frac{\rho c_p}{k}\frac{DT}{Dt} - \frac{Q}{k}$$

where Q is the heat liberated as the reaction proceeds. All terms in this equation have units Btu/hr-ft^3. Since conduction of heat in the z-direction is being neglected, the left-hand side does not include the term $k\partial^2 T/\partial z^2$. The substantial derivative may be expanded as follows:

$$\frac{DT}{Dt} = \frac{\partial T}{\partial t} + \frac{v_r}{r}\frac{\partial(rT)}{\partial r} + \frac{v_\theta}{r}\frac{\partial T}{\partial\theta} + v_z\frac{\partial T}{\partial z}$$

Since $\partial T/\partial t$, v_r, and v_θ are zero, this equation reduces to

$$\frac{DT}{Dt} = v_z\frac{\partial T}{\partial z}$$

The heat liberated as component A is consumed by reaction is

$$Q = (-\Delta H)k_1 C_A$$

Substituting these terms into the original equation, expanding the partial, and rearranging gives

$$k\frac{\partial^2 T}{\partial r^2} + \frac{k}{r}\frac{\partial T}{\partial r} + (-\Delta H)k_1 C_A - \rho c_p v_z\frac{\partial T}{\partial z} = 0$$

Note that this equation is identical to Eq. 7-27 and the boundary conditions are the same as developed earlier.

10-11. INTRODUCTION TO ANALOG MODELS–CONTINUOUS ANALOGS

Although it is possible to develop the mathematical equations describing engineering problems using the preceding concepts, very few engineers seem to find this approach easy to use. On the other hand, physicists and mathematicians find this approach most enlightening. Although an engineer may not need to use this approach to describe his problem, he may find it very useful in obtaining a solution.

For example, suppose an engineer has obtained by some means an equation or set of equations along with the appropriate boundary conditions that describe his problem. The next logical step would be to solve these equations. Unfortunately, the solution of partial differential equations is not an easy task. Presently, there are four major avenues to the solution:

1. Analytic solution
2. Numerical techniques
3. Analog or hybrid computer techniques
4. Analog models

For practical engineering problems, analytic solutions are rarely possible. Numerical techniques are long and tedious, and can only be undertaken with a digital computer. Although analog computers readily solve most ordinary differential equations, the solution of partial differential equations requires special techniques and special equipment, preferably a hybrid machine. Analog models require considerable time and expense for construction, but once available they will yield the solution almost instantaneously.

Discussion of analytic techniques has been avoided throughout this text. As numerical and analog techniques have already been considered, it seems reasonable to devote the remainder of this chapter to analog models. Although these models are not used as much today as they were once, they are still common. Perhaps more important, the techniques for developing an analog model are often used in obtaining a numerical or analog-computer solution.

Basically, there are two types of analog models–the continuous analog and the network analog. The continuous analog is discussed in this section, and the network analog in the next.

The previous sections illustrated that two completely dissimilar systems may be described by the same differential equation. Thus the solution to the differential equation describing one of the systems may be obtained from the other system by making experimental measurements. To illustrate, consider the cross section of a furnace support as shown in Fig. 10-3. If only the steady-state temperature profile is required, this can be obtained by solving Laplace's equation in two dimensions:

$$\frac{\partial^2 T}{\partial x^2} + \frac{\partial^2 T}{\partial y^2} = 0$$

As dictated by the geometry of the support, the boundary conditions would be quite difficult to formulate. Thus an analytic solution would be very difficult to obtain, if possible at all.

Although the temperature profile could be obtained by numerical or analog-computer techniques, the continuous analog approach can be readily applied and will be discussed in this section. The first objective is to find another system

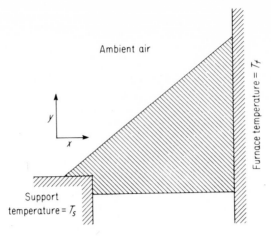

FIG. 10-3.   Furnace support.

which is described by the same differential equation, i.e., Laplace's equation.   In
Sec. 10-4 three other examples of systems described by Laplace's equation were
presented.   Although many other systems are also described by Laplace's equation, electrical systems such as the conductor discussed in Sec. 10-4 are so convenient to use that they are seldom rejected in favor of another continuous analog.

Consider the electrical conductive material shown in Fig. 10-4.   Note that
the shape of this conductor is the same as that of the furnace support shown in
Fig. 10-3.   It is not necessary for the size of the analog to be the same as the size
of the system.   The only requirement is that the ratios of the length of the analog

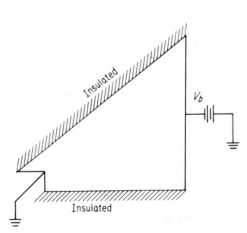

FIG. 10-4.   Electrical conductor analog for furnace
support.

to the respective length of the system must be constant. If rectangular coordinates are used for both system and analog, the ratio of the distances must satisfy the following relationships:

$$x = K_x x_a \qquad (10\text{-}23)$$

$$y = K_y x_a \qquad (10\text{-}24)$$

where x = x-distance in system
y = y-distance in system
x_a = x-distance in analog
y_a = y-distance in analog
K_x = scale factor for x-distance
K_y = scale factor for y-distance

Although it is not always necessary for K_x to equal K_y, it will be shown later to be necessary in this example.

To complete the relationship between the model and the system, an equation is needed to relate the temperature in the support to the voltage potential in the conductor. Note that where the temperature in the support is T_s, the voltage in the model is the ground or zero voltage. Furthermore, the voltage in the model is V_b where the temperature in the system is T_f. Therefore, the voltage and temperatures are related by an equation of the form

$$K_1(T - T_s) = V \qquad (10\text{-}25)$$

where V = voltage
K_V – scale factor

If the voltage is V_b where T is T_f, the constant K_1 equals $V_b/(T_f - T_s)$. Substituting into Eq. 10-25 yields

$$\frac{T - T_s}{T_f - T_s} = \frac{V}{V_b} \qquad (10\text{-}26)$$

Although some heat is lost by convection from the surface of the support, this is assumed to be negligible for this case. Thus the corresponding parts of the analog are electrically insulated. For systems in which these effects are important, Kayan [4] has presented techniques for including them in the model.

If the electrical resistance is the only significant parameter in the conductive material, the model is described by Laplace's equation in two dimensions:

$$\frac{\partial^2 V}{\partial x_a^2} + \frac{\partial^2 V}{\partial y_a^2} = 0 \qquad (10\text{-}27)$$

Solving Eqs. 10-23, 10-24, and 10-26 for x_a, y_a, and V and substituting the results into Eq. 10-27 yields

$$\frac{V_b K_x^2}{(T_f - T_s)} \frac{\partial^2 T}{\partial x^2} + \frac{V_b K_v^2}{T_f - T_s} \frac{\partial^2 T}{\partial y^2} = 0$$

This equation can be simplified to

$$\left(\frac{K_x}{K_y}\right)^2 \frac{\partial^2 T}{\partial x^2} + \frac{\partial^2 T}{\partial y^2} = 0$$

If the coefficient $(K_x/K_y)^2$ is equal to unity, this equation reduces to Laplace's equation

$$\frac{\partial^2 T}{\partial x^2} + \frac{\partial^2 T}{\partial y^2} = 0 \tag{10-28}$$

which is the equation that describes the temperature profile in the support. Requiring K_x to equal K_y means that the ratio of lengths in the x-direction equals the ratio of lengths in the y-direction, which in turn is the requirement for geometrical similarity.

Although this is not the only possible analog model for heat transfer systems, the conductive material analog is very common [5]. Extensive work has been done in the conductive analog area, and many types of conductive sheets and papers are available, along with many special devices for measuring the voltage at a point on the sheet. Another popular model is an electrolytic tank [1], which has the advantage that the physical properties of the fluid in the tank are more uniform than the physical properties of the papers or sheets.

Analogs can also be developed for systems described by the diffusion equation, wave equation, Poisson's equation, etc. Since the fundamental principles for these equations are basically the same as for Laplace's equation, no further examples of continuous analogs are presented.

10-12. NETWORK ANALOGS

The continuous analogs have several disadvantages, many of which are obvious to the reader. One main disadvantage is that a three-dimensional system can be described only by a three-dimensional analog, but the voltage at the center of the analog would be very difficult to measure. Another disadvantage is that the capacitance of electrical analogs is so low that transients are so fast that they cannot be measured accurately if at all. This is mainly because the same material must satisfy both resistive and capacitive requirements, and few materials will satisfy both. This situation is even more complicated for systems described by the damped-wave equation, where all three parameters are important.

The first step in developing a network analog for a system is to develop a lumped-parameter model of the system under consideration. This may be accomplished in one of two ways. First, the differential equation describing the system may be expressed in terms of finite differences. This is equivalent to the procedure used in initiating a numerical solution as discussed in Chapter 8. Alternatively, the lumped-parameter model may be developed from inspection of an element of finite size within the system as was done in Chapter 3. Using either approach the end result will be the same. Both of these techniques will be illustrated in this section.

Perhaps one of the most common uses of analog models is in reservoir engineering[6]. The structure of a simplified oil reservoir is shown in Fig. 10-5. Since oil is less dense than water, it will accumulate in the high points of the reservoir. The oil and usually a vastly larger amount of water, called the *aquifer*, are contained between two impermeable layers of rock. The formation between these two impervious layers is porous, and the oil and water can slowly pass through to the wells. Since the reservoir is usually some distance below the earth's surface, the water is under a very high pressure. When a well is drilled through the impervious layer into the reservoir, the pressure at this point is substantially reduced as the oil begins to flow out. This decrease in the volume of the oil in the reservoir is offset by the increase in the volume of the water accompanied by the reduction in pressure. Thus the compressibility of the water cannot be neglected.

Although most actual reservoirs must be described by a two-dimensional model, a few are amenable to representation by a one-dimensional model. Since this latter case is the simpler, it will be used for illustrative purposes. The reservoir shown in Fig. 10-5 will be the subject of the following analysis.

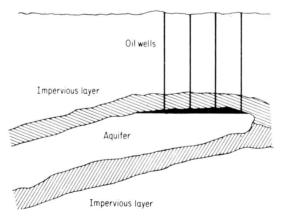

FIG. 10-5. One-dimensional example of an aquifer.

First using the lumped-parameter approach, consider the three elements of the aquifer in the reservoir shown in Fig. 10-6. Since this is a lumped-parameter system the distributed properties are considered to be lumped at the center. The

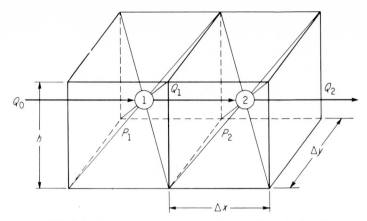

FIG. 10-6.   Lumped-parameter elements for a one-dimensional aquifer.

approach will be to first relate the volumetric flow rate Q between two centers or nodes to the pressure drop.   Next, the difference in volumetric flow rates into and out of a node will be related to the change in pressure within the node with time.

Since the water in the aquifer is flowing through a porous media, the volumetric flow rate is related to the pressure gradient by the following form of Darcy's law:

$$Q = -\frac{kA}{\mu}\frac{\partial P}{\partial x}$$

(10-29)

where Q =   volumetric flow rate, ft³/hr
k =   permeability
A =   area for flow, ft²
μ =   viscosity, lb/ft-hr
$\partial P/\partial x$ =   pressure gradient in x-direction (lb/ft²)/ft

If the pressure gradient $\partial P/\partial x$ is expressed in finite difference form, Eq. 10-29 when applied to determine Q_1 in Fig. 10-5 becomes

$$Q_1 = \frac{k}{\mu}\,h(\Delta y)\,\frac{(P_1 - P_2)}{\Delta x}$$

(10-30)

Now note the similarity between this equation and the equation for the voltage drop across the resistor in Fig. 10-7a:

$$V_1 - V_2 = I_1 R$$

(10-31)

where I_1 = current, amperes
R = resistance, ohms
$V_1 - V_2$ = difference in electric potential, volts

Thus the resistor may be used to represent the element in Fig. 10-6 if the voltage drop across the resistor corresponds to the pressure drop across the element, the current through the resistor corresponds to the volumetric flow rate of water between the elements, and the resistance R of the resistor corresponds to the term $\mu \Delta x/(kh \Delta y)$.

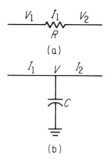

FIG. 10-7. Network components. (a) Electrical analog of Eq. 10-31. (b) Electrical analog of Eq. 10-35.

Next, consider relating the differences in flow in and out of an element to the change in pressure. If a mass balance is made for element 2 in Fig. 10-6 the results are

$$\text{In} - \text{Out} = \text{Accumulation}$$

$$w_1 - w_2 = \frac{d}{dt}(\epsilon \rho_2 h \Delta x \Delta y) \qquad (10\text{-}32)$$

where w = mass flow rate = $Q\rho$
     ϵ = porosity

Now an equation of state is required to relate the density ρ to the pressure P. A common equation of state for a compressible fluid is

$$\rho = \rho_0 \exp(\kappa P) \qquad (10\text{-}33)$$

where ρ_0 = density at $P = 0$
     κ = compressibility factor

Substituting Eq. 10-33 into Eq. 10-32 yields (for constant ϵ),

$$w_1 - w_1 = \epsilon(\Delta x \Delta y)h \frac{d}{dt}(\rho_0 e^{\kappa P_2})$$

$$= \epsilon(\Delta x \Delta y)h\rho_0 \kappa (e^{\kappa P}) \frac{dP_2}{dt}$$

$$= \epsilon(\Delta x \Delta y)h\rho \kappa \frac{dP_2}{dt}$$

Since $Q = w/\rho$, this equation becomes

$$Q_1 - Q_2 = \epsilon(\Delta x \Delta y)h\kappa \frac{dP_2}{dt} \qquad (10\text{-}34)$$

Now note the similarity between this equation and the equation for the capacitor shown in Fig. 10-7b:

$$I_1 - I_2 = C\frac{dV}{dt_E} \qquad (10\text{-}35)$$

where C = capacitance, farads
$\quad t_E$ = time for model

The similarity between these two equations requires that I correspond to Q, V correspond to P, and C correspond to $\epsilon(\Delta x \Delta y)h\kappa$.

Since each node in the aquifer can be represented by the capacitor and resistor analogs, the network analog for a one-dimensional aquifer is shown in Fig. 10-8. The only question remaining is the size of the resistors and capacitors in this network. To determine this, the following scale factors are needed:

$$V = LP \qquad (10\text{-}36)$$

$$I = MQ \qquad (10\text{-}37)$$

$$t_E = Nt \qquad (10\text{-}38)$$

The scale factor L relates the pressure, typically in pounds per square inch, to the electric potential (typically in volts). Similarly, the scale factor M relates the flow (typically in barrels/day), to the current (usually in milliamperes). The scale factor N represents the relationship for elapsed time (typically in years) for the reservoir to the elapsed time (typically in seconds or minutes) for the model.

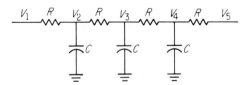

FIG. 10-8.   Electrical analog for a one-dimensional aquifer.

To determine the values of the resistances and capacitances for the model, Eqs. 10-36 through 10-37 are substituted into the equations for the model, namely Eqs. 10-31 and 10-35, yielding

$$L(P_1 - P_2) = Q_1 R$$

$$M(Q_1 - Q_2) = C\frac{d(LP_2)}{d(Nt)}$$

or

$$Q_1 = \frac{L}{R} (P_1 - P_2) \tag{10-39}$$

$$Q_1 - Q_2 = \frac{CL}{MN} \frac{dP_2}{dt} \tag{10-40}$$

Comparison of Eq. 10-39 with Eq. 10-30 indicates

$$R = \frac{L\mu\Delta x}{kh\Delta y}$$

Similarly, comparison of Eq. 10-40 with Eq. 10-34 reveals

$$C = \frac{MN\epsilon h\kappa\Delta x\Delta y}{L}$$

Thus the values of the resistances and capacitances for the network analog are specified.

As mentioned previously, a finite-difference approach may also be used. However, the equation describing the system must be derived first. Since only the dissipativity and the capability of the system to act as a reservoir of potential are important, the system will be described by the diffusion equation. Thus D and E_P must be evaluated in terms of the system parameters. Recall that D is defined as follows:

$$\frac{\partial p}{\partial x} = -Df \tag{10-1}$$

Rearrangement of Eq. 10-29 and substitution of $h\Delta y$ for A gives

$$\frac{\partial P}{\partial x} = -\frac{\mu}{kh\Delta y} Q$$

Thus the dissipativity is

$$D = \frac{\mu}{kh\Delta y} \tag{10-41}$$

Similarly, recall that E_P is

$$\frac{\partial f}{\partial x} = -E_P \frac{\partial p}{\partial t} \tag{10-3}$$

Expressing Eq. 10-34 in differential form yields

$$\frac{\partial Q}{\partial x} = -\epsilon h\kappa(\Delta y) \frac{\partial P_2}{\partial t}$$

Thus the value of E_P is

$$E_P = \epsilon h\kappa\Delta y \tag{10-42}$$

For systems for which D and E_P are the only significant parameters, the diffusion equation for a one-dimensional field in which the across variable is the pressure P is

$$\frac{\partial^2 P}{\partial x^2} = DE_P \frac{\partial P}{\partial t} \qquad (10\text{-}43)$$

where D and E_P are given by Eqs. 10-41 and 10-42.   Expressing the field variable in Eq. 10-43 in finite-difference form yields

$$\frac{P_1 + P_3 - 2P_2}{(\Delta x)^2} = DE_P \frac{dP_2}{dt}$$

or

$$\frac{P_1 - P_2}{D\Delta x} + \frac{P_3 - P_2}{D\Delta x} = E_P \Delta x \frac{dP_2}{dt}$$

Now compare this equation to the equation for the node in Fig. 10-8:

$$\frac{V_1 - V_2}{R} + \frac{V_3 - V_2}{R} = C \frac{dV_2}{dt}$$

Thus the same network analog for the aquifer is obtained as in the preceding section.   The values of R and C are determined as before.

For a two-dimensional reservoir, the aquifer can be represented by the two-dimensional network in Fig. 10-9.   A common addition to such a network is a device for representing the location of an oil well.   Since a volume of oil is being removed at each well, a device is added to the network which removes a prescribed amount of current at the node corresponding to the well.

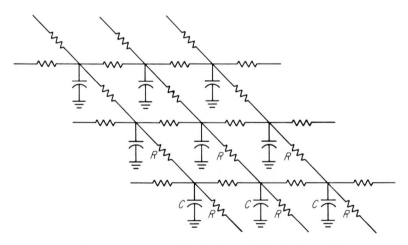

FIG. 10-9.   Electric network analog for a two-dimensional reservoir.

In practice it is not possible to calculate precise values for the resistances and capacitances in the network. Instead they are determined to give the best representation of the past history of the field. With these values of the resistances and capacitances, the future behavior of the field can be predicted with some reliability.

Although it is possible to construct a three-dimensional network of resistances and capacitances to represent the aquifer more accurately, these are seldom used because of their complexity.

10-13. ANALOGS FOR LUMPED-PARAMETER SYSTEMS

Up to this point distributed-parameter or field problems have been the primary subject, although in the last section a lumped-parameter approximation to a distributed-parameter system was considered. In this section lumped-parameter systems will receive the attention.

Generally there are five types of lumped-parameter systems commonly considered. These are translational systems, rotational systems, electrical systems, fluid systems, and thermal systems. For each of these systems there are three basic elements corresponding to the dissipative, capability to act as a reservoir of potential, and capability to act as a reservoir of flux parameters for a distributed-parameter system. For example, these three elements for an electrical system are a resistor, a capacitor, and an inductor. For a translational system these elements are a viscous damper, a mass, and a spring. For other systems these elements are shown schematically in Table 10-3.

As for the distributed-parameter systems, there are two types of variables–across and through variables. For an electrical system the across variable is the electrical potential, whereas for a translational system it is the velocity. The through variable for electrical systems is the current, and is the force for translational systems. Even for lumped-parameter systems the through variable retains its vector properties, since its direction of flow is important even though there are only two possibilities.

To illustrate the development and use of network analogs, consider the pumping system in Fig. 10-10. The pump removes liquid from a reservoir and delivers it through about 5,000 ft of 4 in. schedule 40 pipe to a process unit. From steady-state consideration of the pressure drop, the line size and pump size can be calculated to give a required flow. Suppose the system should deliver 175 gpm at essentially zero discharge pressure. This corresponds to 0.389 ft^3/sec or a liquid velocity of 4.4 ft/min. For simplicity, the liquid is assumed to be water. Using the equation

$$f = 0.00140 + \frac{0.125}{N_{Re}^{0.32}}$$

TABLE 10-3

Typical Lumped-Parameter Elements. (Reprinted by special permission from "Mathematical Models of Dynamic Systems," *Chemical Engineering,* January 17, 1966, p. 129, © 1966, by McGraw-Hill, Inc, New York.)

Physical Element	Diagram	Equation for Ideal Element
Translational spring		$v_{21} = \frac{1}{k}\frac{dF}{dt}$
Rotational spring		$\omega_{21} = \frac{1}{k}\frac{d\tau}{dt}$
Inductance		$v_{21} = L\frac{di}{dt}$
Fluid inertance		$p_{21} = I\frac{dQ}{dt}$
Thermal inductance	Does Not Exist	
Translational mass	$v_1 = $ Constant	$F = m\frac{dv_2}{dt}$
Rotational mass	$\omega_1 = $ Constant	$\tau = J\frac{d\omega_2}{dt}$
Electrical capacitance		$i = C\frac{dv_{21}}{dt}$
Fluid capacitance	$p_1 = $ Constant	$Q = C_f\frac{dp_2}{dt}$
Thermal capacitance	$T = $ Constant	$q = C_t\frac{dT_2}{dt}$
Translational damper		$F = bv_{21}$
Rotational damper		$\tau = B\omega_{21}$
Electrical resistance		$i = \frac{1}{R}v_{21}$
Fluid resistance		$Q = \frac{1}{R_f}p_{21}$
Thermal resistance		$q = \frac{1}{R_t}T_{21}$
Across-variable source		
Through-variable source		

Through variables: $F = $ force, $\tau = $ torque, $i = $ current, $Q = $ fluid flow
         $q = $ heat flow
Across variables: $v = $ translational velocity, $\omega = $ angular velocity
         $v = $ voltage, $p = $ pressure, $T = $ temperature
Inductance-type proportionality constants:
    $1/k = $ reciprocal translational stiffness, $1/K = $ reciprocal rotational stiffness, $L = $ inductance, $I = $ fluid inertance
Capacitance-type proportionality constants:
    $m = $ mass, $\zeta = $ moment of inertia, $C = $ capacitance, $C_f = $ fluid capacitance, $C_t = $ thermal capacitance
Resistance-type proportionality constants:
    $1/b = $ reciprocal translational damping, $1/B = $ reciprocal rotational damping, $r = $ resistance, $R_f = $ fluid resistance $R_t = $ thermal resistance

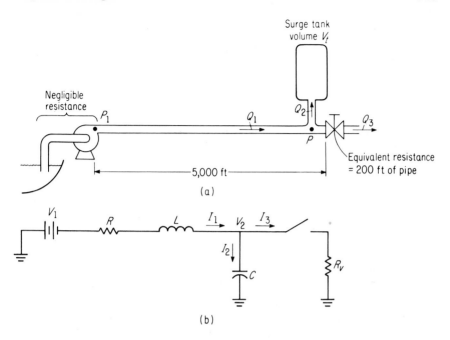

FIG. 10-10. (a) Simple pumping system. (b) Its electrical analog.

to compute the friction factor[8], this gives a pressure drop of 6.58 psi/1,000 ft. Thus the pressure drop from pump to valve is 32.91 psi, and is 1.317 psi through the valve, requiring the pump to produce a head of 34.23 psi. Assume that a pump is available to do just this, and will not vary significantly with flow.

So far, the surge tank has not affected the analysis. In fact, it is completely unnecessary in many cases. However, if it is necessary to close the valve practi-cally instantaneously, the surge tank is needed to prevent excessive pressures arising from the inertia of the fluid. Thus the sizing of the surge tank is strictly a dynamic problem, requiring a set of equations to describe the dynamics of the system. To standardize, let the pressure be the across variable and volumetric flow rate be the through variable.

To derive the equations describing this unit, first apply Newton's law ($F = ma$) to the 5,000-ft section of pipe. The force exerted at the pump discharge is $P_1 A$ where A is the cross-sectional area of the pipe. Similarly, the force at the valve is $P_2 A$ in the opposite direction. There is also a force in the opposite direction due to the friction loss, and is given by $\dfrac{2fL_c \rho V^2 A}{Dg_c}$ Substituting these terms into Newton's law gives

$$P_1 A - P_2 A - \frac{2fL_c \rho V^2 A}{Dg_c} = \frac{m}{g_c}\frac{dV}{dt}$$

Substituting $m = \rho L_c A$ and $V = Q_1/A$ yields after rearrangement,

$$(P_1 - P_2) = \left(\frac{2fL_c\rho Q_1}{A^2 Dg_c}\right) Q_1 + \frac{\rho L_c}{g_c A} \frac{dQ_1}{dt} \qquad (10\text{-}44)$$

Now consider a resistor and an inductor connected in series as follows:

The equation describing this circuit is

$$(V_1 - V_2) = RI + L\frac{dI}{dt} \qquad (10\text{-}45)$$

Note the similarity between this equation and Eq. 10-44 above. Obviously V corresponds to P, I to Q_1, L to $\rho L_c/g_c A$, and R to $\dfrac{2fL_c\rho Q_1}{A^2 Dg_c}$. However, this latter term is a function of Q_1, which requires that R be proportional to Q_1 for an exact analogy. As this is inconvenient to construct, R will be taken as constant and evaluated at the value of Q_1 when the valve is open. The effect of this approximation will be shown later; note, however, that the approximation is linear whereas the true representation is nonlinear.

Now consider the surge tank. Making a volumetric balance on the vessel yields

$$-Q_2 = \frac{d}{dt} V_a \qquad (10\text{-}46)$$

where V_a is the volume of the air space in the vessel. Assume that the tank is initially filled with air at atmospheric pressure. Assuming the ideal gas law to hold, V_a is related to P_2 as follows:

$$V_a = \frac{V_t P_0}{P_2 + P_0}$$

where P_0 is atmospheric pressure and P_2 is the gauge pressure at the tank. Substituting into Eq. 10-46 and expanding the derivative, we get

$$Q_2 = \frac{V_t P_0}{(P_2 + P_0)^2} \frac{dP_2}{dt} \qquad (10\text{-}47)$$

Next consider the equation describing a capacitor:

$$I_2 = C\frac{dV_2}{dt} \qquad (10\text{-}48)$$

Again, comparing to Eq. 10-47 indicates that I_2 corresponds to Q_2, V_2 to P_2, and C to $\dfrac{V_t P_0}{(P_2 + P_0)^2}$. However, this latter term is a function of P_2, requiring C to be a function of V_2 for exact analogy. As this is also inconvenient from a practical standpoint, C will be taken as constant and evaluated at the value of P_2 when the valve is open.

This type of reasoning can be continued to construct the complete electrical analog shown in Fig. 10-10b. The battery represents the pump, and the resistance R_v represents the resistance of the valve.

The equations describing the electrical circuit in Fig. 10-10b could be solved analytically to give the currents and voltages as functions of time. Furthermore, they could be differentiated and solved to give the peak voltage as a function of the capacitance. However, most pumping systems encountered in practice are considerably more complex than the one in Fig. 10-10a, thus making analytic solutions impractical. Thus we abandon the analytic approach in favor of other avenues.

One avenue immediately open is to construct the electrical circuit for the analog. As it is composed of such common elements as batteries, switches, resistors, inductors, and capacitors, this would be quite straightforward. Except for the approximations involved in using constant values for the resistors and capacitances the solution would represent the true solution quite well. As it is an electrical circuit, recorders and other peripheral equipment would be easy to obtain. Thus, even for large, complex systems, the construction of an electrical analog would be feasible.

However, the electrical analog so constructed would be specific to the system under consideration. Any changes would require modifications of the analog, and once the analog had passed its period of usefulness, its value drops to the salvage value of the components. For this reason a more feasible approach might be to simulate the system on an analog computer. This requires constructing a wiring diagram, patching the board for the computer, and debugging the results. The solution of the problem could be obtained repetitively (in rep-op) at a rate of several times a second, and the solutions displayed on an oscilloscope to give a continuous curve. The engineer could then "knob twiddle" until he obtained the desired results.

Clearly it would be difficult to account for the nonlinearities using the electrical analog network. Although analog computers normally contain nonlinear components such as multipliers and function generators, these are often in limited supply and are not used whenever they can be avoided. Thus a practical point to consider is the approximation involved in using constant values for the resistance and inductance. In Fig. 10-11 is shown the exact and two approximate solutions for a 50 gallon surge tank. The nonlinearities associated with the capacitance contributes heavily to the discrepancies, as evidenced by the change in results when the capacitance is evaluated at a pressure of 1.317 psi (pressure

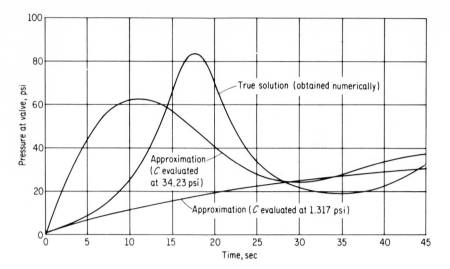

FIG. 10-11.   Comparison of solutions for transient pressures at the valve.

at valve under steady flow) and 34.23 psi (pressure at valve when closed).   Another nonlinearity not considered for either case is the large resistance to backward flow through the pump.

This example was chosen to illustrate the potential pitfalls to using approximations too freely.   It is certainly acceptable to use approximations; in fact, a good engineer will wisely take advantage of approximations when they do not seriously degrade the usefulness of the results.   Generally, approximations should be checked.   The clue to the poor approximation in the previous example is the drastic change in results when the capacitance is varied.

10-14.   SUMMARY

This chapter has sought to introduce the subject of physical similarity, and to show how it can be used in practice.   However, perhaps the main justification for serious study is the insight it can give to the systems around us and the mathematics we try to use to describe them.

REFERENCES

1.   W. J. Karplus, *Analog Simulation*, McGraw-Hill, New York, 1958.
2.   W. W. Soroyka, *Analog Methods in Computation and Simulation*, McGraw-Hill, New York, 1954.
3.   W. J. Karplus and W. E. Soroyka, *Analog Methods*, 2d ed., McGraw-Hill, New York, 1959.

4. C. F. Kayan, "Heat-Transfer Temperature Patterns of a Multicomponent Structure by Comparative Methods," *Trans. ASME,* Vol. 71 (1949).

5. F. Kreith, *Principles of Heat Transfer,* 2d ed., International Textbook, Scranton.

6. B. C. Craft and M. F. Hawkins, *Applied Petroleum Reservoir Engineering,* Prentice-Hall, Englewood Cliffs, N. J., 1959.

7. A. T. Murphy and D. L. Wise, "Mathematical Models of Dynamic Systems," *Chemical Engineering,* Jan. 17, 1966, pp. 125-132; Feb. 14, 1966, pp. 161-264, March 14, 1966, pp. 167-172; May 9, 1966, pp. 165-172.

8. W. L. McCabe and J. C. Smith, *Unit Operations of Chemical Engineering,* McGraw-Hill, New York, 1956, p. 67.

PROBLEMS

10-1. In Sec. 10-3 it was stated that Laplace's equation describes any system in which only *one* of the fundamental parameters (D, E_P, or E_f) is significant. This exercise will illustrate this by considering three network analogs.

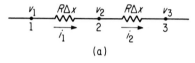

(a)

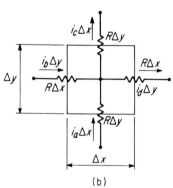

(b)

PROB. 10-1

(a) Only D is significant. In this case the electrical network analog to the one-dimensional element in Fig. 10-1 is as shown. The equation describing this network is derived as follows:

1. Let $\Delta i = i_2 - i_1$ and $\Delta v = v_2 - v_1$. As Δx of differential size, so are Δi and Δv.

2. Applying Ohm's law,

$$v_2 - v_1 = \Delta v = i_1 R \, \Delta x$$

3. Summation of currents about node 2:

$$i_2 - i_1 = \Delta i = 0$$

or

$$\frac{\Delta i}{\Delta x} = 0$$

4. Substituting for i from Ohm's law,

$$\frac{\Delta}{\Delta x} \left[\frac{\Delta v}{R \Delta x} \right] = 0$$

5. Taking the limit as $\Delta x \to 0$,

$$\frac{d}{dx} \left[\frac{1}{R} \frac{dv}{dx} \right] = 0$$

This is Laplace's equation in one dimension.

(b) For a two-dimensional system, the electrical analog for and element shown in part (b) of the accompanying sketch.

Note that $i_a, i_b, \cdots, i_d$ are current *fluxes*, and are multiplied by the area of the respective face to give current flow.   From this network, derive Laplace's equation in two dimensions.

(c) Construct the element and derive Laplace's equation in three dimensions when only D is significant.

10-2 (a).   When only the parameter E_f is significant the electrical analog of a finite difference element is as shown.

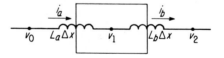

PROB.   10-2

Show that Laplace's equation also describes this system.

(b) Develop the network analog and derive Laplace's equation for a two-dimensional system.

10-3.   From Sec. 1.5, the equation describing the temperature profile in a cylindrical conductor with internal heat generation is

$$\frac{1}{r} \frac{d}{dr} \left[r \frac{dT}{dr} \right] = - \frac{\dot{q}}{k}$$

This is Poisson's equation in cylindrical coordinates.   Show that a similar equation could be derived by considering the element shown in the figure.   The current i_f is the analog to $\dot{q}$.

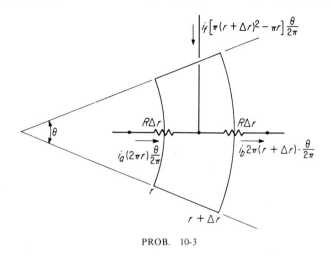

PROB. 10-3

10-4. In Sec. 1-3, the equation describing the steam hanger was derived to be

$$\frac{d^2 T(z)}{dz^2} = \frac{4h}{Wk} \left[T(z) - T_a \right]$$

Show that the network in the sketch is equivalent:

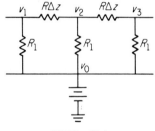

PROB. 10-4

10-5. Systems in which D and E_P are significant but E_f is not are described by the diffusion equation. (a) Using the one-dimensional network analog shown in the figure, derive the diffusion equation. (b) Develop a two-dimensional network and derive the diffusion equation in two dimensions.

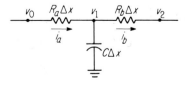

PROB. 10-5

10-6. Show that the network analog in the drawing represents the diffusion equation in cylindrical coordinates. Assume constant physical properties.

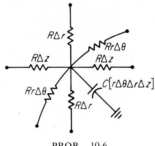

PROB. 10-6

10-7. Systems for which D and E_f are significant but E_p is not also obey the diffusion equation. Show that the network analog of the sketch represents the diffusion equation.

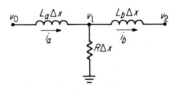

PROB. 10-7

10-8. In Sec. 7-4 the continuity equation was derived to be (two dimensions):

$$\frac{\partial(\rho v_x)}{\partial x} + \frac{\partial(\rho v_y)}{\partial y} = -\frac{\partial \rho}{\partial t}$$

Show that this equation is equivalent to the circuit in the sketch. The current flux is the analog to ρv (the mass flux), and the voltage is the analog to ρ.

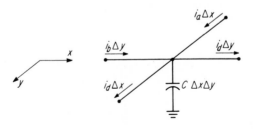

PROB. 10-8

10-9. Consider the circular fin frequently used on the tubes for air-cooled exchangers and radiators as shown. The temperature of the tube is T_t; the thermal conductivity of the metal in the fin is k: the convective heat-transfer coefficient from the surface is h; and the ambient temperature is T_A. Neglect heat transfer from the outer perimeter. Divide the fin into four elements and develop the network analog for steady-state conditions. Relate the resistances to the parameters of the problem.

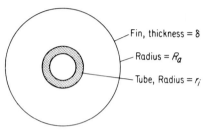

PROB. 10-9

10-10. The heat exchanger considered in Sec. 1-6 was described by the following differential equation:

$$wc_p \frac{dT(z)}{dz} - UA[T_s - T(z)] = 0$$

$$T(0) = T_f$$

Consider developing a network analog for this system. (a) Show that the analog for one node is as shown. (b) The difficulty arises when connecting the individual nodes. They must be isolated by an amplifier or cathode follower of some type as shown, where $\Delta z = L/4$. This amplifier must not draw a significant grid current. (c) Develop the circuit describing the unsteady-state behavior of this exchanger. If 10 volts represent 100°F and 1 min real time should be 1 sec on the model, relate R_1, R_2, and C to the parameters of the problem.

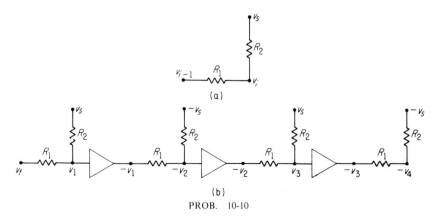

PROB. 10-10

10-11. Develop an electric network analog of the heat exchanger in Sec. 2-2. Let $\Delta z = L/4$. Recall that when these equations are solved on the analog computer, a trial-and-error solution is necessary. Does the network analog require similar procedures?

10-12. Using the generalized-parameter approach, develop the *unsteady-state* equation describing the temperature profile in the uninsulated steam pipe hanger in Sec. 1-3. (a) Is D significant? If so, what is D? (b) Is E_P significant? If so, what is E_P? (c) Is E_f significant? If so, what is E_f? (d) Is there a distributed source or sink (that is, is there a term as in Poisson's equation)? If so, what is it? (Recall that this term should be the heat *received* per unit *volume*.) (e) What is the final equation?

$$\frac{\partial^2 T}{\partial z^2} = \frac{c_p \rho}{k} \frac{\partial T}{\partial t} + \frac{4h(T - T_a)}{Wk} \quad Ans.$$

10-13. Using the generalized-parameter approach, develop the *unsteady-state* equation describing the fin in Prob. 10-9. (a) Is D significant? If so, what is D? (b) Is E_P significant? If so, what is E_P? (c) Is E_f significant? If so, what is E_f? (d) Is there a distributed source or sink (that is, is there a term as in Poisson's equation)? If so, what is its value? (Recall that this term should be the heat *received* per unit *volume*.) (e) What is the final equation?

$$\frac{1}{r} \frac{\partial}{\partial r}\left[k \frac{\partial T}{\partial r} \right] = c_p \rho \frac{\partial T}{\partial t} + \frac{2h}{\delta} [T - T_a] \quad Ans.$$

10-14. Using the generalized-parameter approach, formulate the *unsteady-state* equation describing laminar flow in circular conduits as in Sec. 1.10. (a) Is D significant? If so, what is D? (b) Is E_p significant? If so, what is E_p? (c) Is E_f significant? If so, what is E_f? (d) Is there a distributed source or sink? If so, what is it? (e) What is the final equation?

$$\frac{\mu}{gc} \cdot \frac{1}{r} \frac{\partial}{\partial r}\left[r \frac{\partial v_z}{\partial r} \right] = \frac{\rho}{g_c} \frac{\partial v_z}{\partial t} - \frac{P_L - P_0}{L} - \frac{\rho g}{g_c} \quad Ans.$$

10-15. Using the generalized-parameter approach, formulate the steady-state equation describing the temperature profile in the steam-jacketed exchanger in Sec. 1-6. (a) Is D significant? If so, what is D? (b) Is E_p significant? If so, what is E_p? (c) Is E_f significant? If so, what is E_f? (d) Is there a distributed source or sink? If so, what is it? (e) What is the final equation?

$$wc_p \frac{dT(z)}{dz} + 2\pi RU_i[T(z) - T_s] = 0 \quad Ans.$$

10-16. Develop a translational mechanical network analog of the system in Fig. 10-10.

10-17. Develop a rotation network analog of the system in Fig. 10-10.

10-18. Develop an electrical network analog of the control system shown in Figs. 4-9 and 4-15.

10-19. Develop an electrical network analog of the control system shown in Prob. 4-9.

10-20. Develop an electrical network analog of the control system shown in Prob. 4-13.

chapter 11

Large-Scale Systems

Previous chapters of this text have concentrated on the development of models for individual components (reactors, heat exchangers) frequently encountered in large-scale plants. The purpose of this chapter is to introduce the techniques for combining these individual models to construct the model of a large system.

Three problems will be analyzed in this chapter. The first example will be a simple steady-state problem to illustrate the principles concerning steady-state models and their solution, and the other two will be the steady- and unsteady-state formulations of the same problem.

First, a word about models in general.

11-1. TYPES OF MODELS

Models fall into two broad categories–steady-state models or unsteady-state models. As the name implies, the steady-state model only represents the conditions in the process when it is at, steady-state, i.e., the process variables are unchanging with time. The unsteady-state model also represents the variation of the process variables with time. Certainly any model that represents a changing or dynamic unit also represents the unit when under static or steady-state conditions. Consequently, it follows that the steady-state model is a special case of the dynamic model.

The type of model best suited to a particular application is often dictated by the application for which the model is developed. As the steady-state model does not represent the dynamic behavior of the process, an unsteady-state model is required for problems concerning control, start-up, operator training, etc. On the other hand, the steady-state model is usually sufficient for economic analysis, feasibility studies, and plant design. However, the current trend is to consider more closely many of the dynamic aspects of plant operation at the design stage, and consequently the unsteady-state model is finding some use here.

Another contrast between the two models is the types of equations encountered and the techniques for solving them. An unsteady-state model must contain one or more differential equations involving derivatives of the process variables with time. The steady-state model may be composed entirely of algebraic equations (typically nonlinear), although some steady-state models include differential equations involving derivatives of the process variables with position. An example of the latter case is a flow reactor in which the variables are functions of the position in the reactor.

As the types of equations found in each type of model differ, the techniques and mechanisms for finding the solutions differ. Unsteady-state models invariably contain a differential equation, often making the analog computer the best mechanism for solving the problem. Sometimes the application, e.g., operator training, virtually dictates its use. For other applications a numerical integration technique with a digital computer is sufficient. The increasing speed of digital machines is making this very feasible, and digital simulation languages have evolved to make the programming quite easy [1].

Some steady-state models also include differential equations, and consequently must be solved by the above techniques. However, many are composed exclusively of algebraic equations. The analog computer is of no use for these equations, and no numerical integration procedures are required. For reasonable processes these algebraic equations will be nonlinear, and there will be several of them. Solving large sets of linear algebraic equations may not be easy, and large sets of nonlinear algebraic equations are almost always difficult to solve. The solution is usually trial-and-error, and considerable thought must be devoted to the calculation procedure. Although attempts to relegate this to the digital computer have met with considerable success [2,3,4,5,6], a good understanding of the problem by the engineer is often necessary.

At this point it seems preferable to introduce a specific example for the ensuing discussion. Subsequently, additional points concerning the calculation procedure can be illustrated in detail.

11-2. THE MODULAR APPROACH

To illustrate ensuing points the evaporator system in Fig. 11-1 will be used. This is a reasonably simple unit, should be easy to understand, and will be sufficient to illustrate several interesting concepts. The equations developed in the ensuing discussion are for steady-state operation.

This system is basically a triple-effect evaporator unit with backward feed. The heat to each effect is supplied via an external heat exchanger operated under sufficient head so that no vaporization occurs in the exchanger. It will also be assumed that only latent heat is transferred from the vapor in the exchanger, all superheat having been lost in the vapor lines. The feed is a caustic solution, selected primarily to illustrate the role of physical properties.

The distinction between model development and design must be recognized. A model can only be developed for a specific unit or proposed configuration, whereas one of the objectives of a design is to develop such configurations. However, once a certain amount of information is available, a model can be

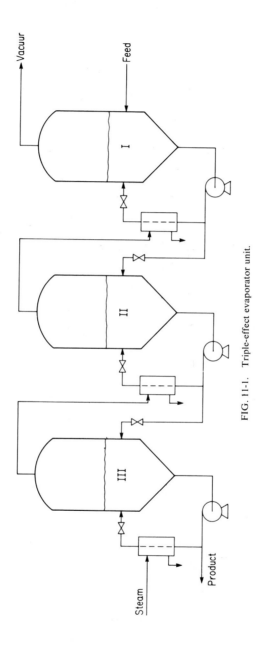

FIG. 11-1. Triple-effect evaporator unit.

developed to augment the design. For example, the model developed can be subsequently used to determine from economic considerations alone the optimum size of the heat exchangers and the optimum flow through the pumps. Similarly, the model could be extended to additional effects to ascertain the economic return. In other cases two widely different schemes to obtain the same objectives may both be modeled to obtain a good comparison.

A recent trend in developing models of units is to use the modular approach. This permits only one part of the overall unit to be examined at a time, thus transforming a large, complex problem into several smaller, hopefully simpler problems. This often also effectively reduces the magnitude of the overall problem. For example, the triple-effect evaporator system in Fig. 11-1 is composed of three individual units of the type shown in Fig. 11-2. Using the modular approach, a mathematical model is developed for this individual unit, and three of these models are properly combined to form the triple-effect evaporator system.

Actually this unit could be subdivided even further. For example, the pump, the heat exchanger, the valve, and the vessel itself could each be examined individually. There are certain advantages for this, especially at the equation-writing stage. However, these individual models must be subsequently recombined in a computational structure to obtain a numerical solution. It seems to be a personal preference as to how much subdivision is advantageous. Certainly subdivide the unit until each subsystem can be understood, but further subdivision should be left to a personal choice.

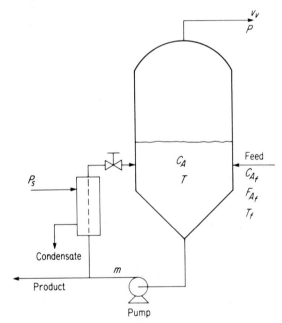

FIG. 11-2. Individual unit.

Another important step in the model development is to ascertain what quantities are specified and what quantities must be determined. Consider the case in which the evaporator in Fig. 11-2 is operated as a single effect. The following quantities would typically be specified:

C_{A_f} feed concentration
T_f feed temperature
F_f feed flow rate
P pressure in the effect

These quantities should be evaluated:

C_A concentration of liquor in the effect
T temperature of liquor in effect
F product flow rate
v_v vapor flow rate
T_s steam temperature
v_s steam rate
T_e temperature of exit stream from exchanger
Q rate of heat transfer

The following quantities are parameters:

m flow rate through pump
A area for heat transfer

In addition, relationships are needed to determine all physical properties and the heat-transfer coefficient. Figure 11-3 gives several physical properties of aqueous caustic solutions as functions of temperature and concentration.

Now for the question of subdividing further. Suppose the exchanger is examined individually. Then the following quantities are either known or should be specified by the equations for the remainder of the effect:

C_A inlet concentration
$(m - F)$ inlet flow
T inlet temperature
P_s pressure in the steam chest

The following parameters must be determined from the equations for the heat exchanger:

T_e temperature of the exit stream
Q rate of heat transfer
v_s steam rate

The heat-transfer area A is a parameter.

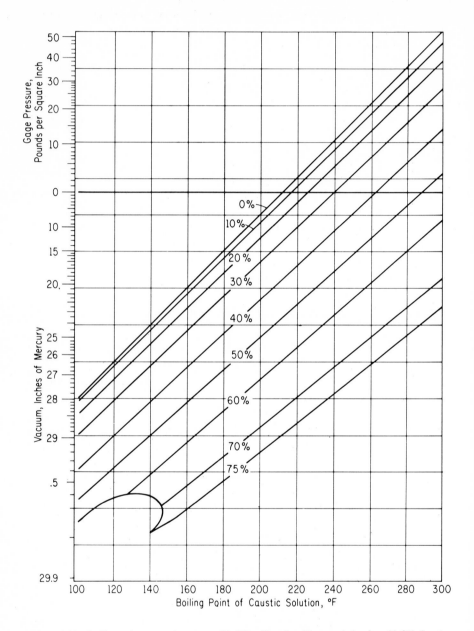

FIG. 11-3.(a)   Boiling-point curves for aqueous NaOH.   (Reprinted by permission from NaOH *Caustic Soda*, © 1953 by PPG Industries, Inc., Pittsburgh, Pa.)

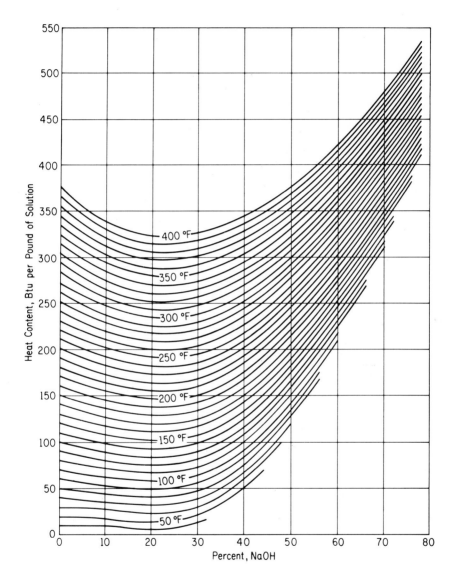

FIG. 11-3.(b) Enthalpy of aqueous NaOH solutions. (Reprinted by permission from NaOH *Caustic Soda*, © 1953 by PPG Industries, Inc., Pittsburgh, Pa.)

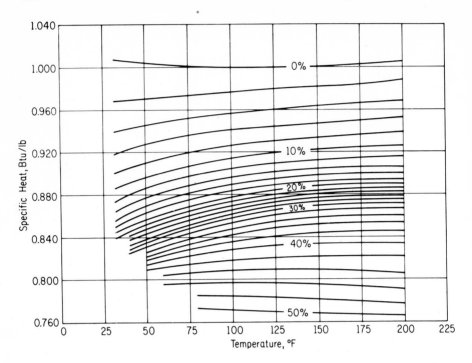

FIG. 11-3.(c) Specific heat of aqueous NaOH solutions. (Reprinted by permission from NaOH *Caustic Soda,* © 1953 by PPG Industries, Inc., Pittsburgh, Pa.)

A convenient way to visualize the input/output quantities for a model is by a block representation of the type shown.   The model itself consists of all necessary equations to relate the outputs to the inputs and parameters.

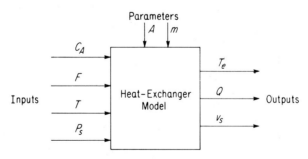

The rate of heat transfer Q is given by each of the following relationships:

$$Q = UA \frac{(T_s - T_e) - (T_s - T)}{\ln \dfrac{(T_s - T_e)}{(T_s - T)}} \tag{11-1}$$

$$Q = (m - F) c_p (T_e - T) \qquad (11\text{-}2)$$

$$Q = v_s \lambda_s \qquad (11\text{-}3)$$

In addition, the physical property relationships are needed for c_p, λ_s, T_s, and U, and as these must be included in the set of equations, they will be represented as follows:

$$c_p = f_1 \left(\frac{T_e + T}{2}, C_A \right) \qquad (11\text{-}4)$$

where c_p is evaluated at the averate temperature using Fig. 11-3c;

$$\lambda_s = f_2(P_s) \qquad (11\text{-}5a)$$
$$T_s = f_3(P_s) \qquad (11\text{-}5b)$$

where f_2, f_3 is really a set of steam tables;

$$U = f_4 (T, T_e, T_s, m, F, v_s, \text{etc.}) \qquad (11\text{-}6)$$

where f_4 represents the computational procedure for calculating U from the various quantities.

In this set of equations, the specified input quantities and parameters are

$$C_A, F, T, P_s, A, m$$

or a total of six. As a total of thirteen variables appear in the above equations, seven are unknown, namely

$$Q, v_s, T_e, c_p, \lambda_s, U, T_s$$

As there are seven equations and seven unknowns, there are no degrees of freedom (= number of unknowns – number of equations) remaining and the model is specified.

Next consider the vessel itself. The input/output relationships for this model are represented as follows:

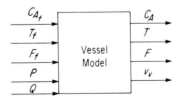

The model consists basically of a total mass balance, a component mass balance, and an enthalpy balance. Equating inputs to outputs in each case yields the following equations:

Total mass balance

$$F_f = F + v_v \qquad (11\text{-}7)$$

Component mass balance

$$C_{A_f}F_f = C_A F \tag{11-8}$$

Enthalpy balance

$$F_f H_f + Q = FH + v_v H_v \tag{11-9}$$

In addition to these the temperature in the effect is determined from the boiling-point curves in Fig. 11-3, and the enthalpies are determined from Fig. 11-3 or from steam tables. As for the heat exchanger, it can be shown that these relationships leave no degrees of freedom.

Although the equations for the two individual models could be combined to give a model for the complete system, this usually offers no advantage. Any further manipulations of the equations should be undertaken to facilitate the computational procedure for obtaining a solution.

In general the development of the mathematical model via the modular approach is reasonably simple. The fundamental relationships—heat balances, mass balances, force balances, and the like—are applied to each individual module. Each module is a "unit operation"—a heat exchanger, a mixer (this is really all the vessel is), a flash unit, a distillation column.

11-3.   COMPUTATIONAL PROCEDURE FOR NONLINEAR ALGEBRAIC EQUATIONS

The above section treated only the derivation of the model, with no attention devoted to techniques for solving the equations.

Eliminating Q between Eqs. 11-1 and 11-2 and solving for T_e gives

$$T_e = T_s - (T_s - T)\exp\left(-UA/(m - F)c_p\right) \tag{11-10}$$

Recall that the heat-transfer coefficient U and the heat capacity c_p are functions of T_e. Consequently, the right-hand side of Eq. 11-10 is a function of T_e, the quantity to be calculated. Although this equation cannot be solved explicitly for T_e, it has a solution. The problem is to find it.

The dependence of U and c_p on T_e makes Eq. 11-10 a nonlinear algebraic equation, which typically must be solved by trial-and-error methods or by other iterative procedures. Assuming U to be constant for the moment, consider the following procedure:

1. Assume T_e
2. Calculate the average temperature.
3. Evaluate c_p.
4. Evaluate T_e by Eq. 11-10.
5. Adjust assumed value of T_e and repeat calculations until assumed value matches calculated value.

The evolution of the values for a specific example is given in Table 11-1. Obviously two iterations suffice if the assumed value of T_e is simply the value of T_e calculated in the previous iteration.

TABLE 11-1

Progress of Exchanger Calculations

Constant Values

    Heat-transfer area: 200 ft²
    Heat-transfer coefficient: 200 Btu/hr-ft²-°F
    Concentration: 25% NaOH
    Steam temperature: 250°F (30 psia)
    Flow: 60,000 lb/hr

Inlet Temperature, °F	Assumed Outlet Temperature, °F	Heat Capacity, Btu/lb °F	Calculated Outlet Temperature, °F
130.0	130.0	0.874	194.0
130.0	194.0	0.878	193.8
130.0	193.8	0.878	193.8

Now consider the case in which U also varies. Without writing the necessary equations, recall from the film theory that U depends upon the unknown quantity v_s as well as the physical properties. Now consider the following calculation procedure:

1. Obtain a reasonable estimate of Q as follows:
    (a) Assume a reasonable value for U and evaluate c_p at T.
    (b) Calculate T_e by Eq. 11-10.
    (c) Calculate Q by Eq. 11-2. Assume this value for Q.
2. Calculate v_s by Eq. 11-3.
3. Calculate T_e by Eq. 11-2. Note that a slight difficulty arises since c_p is dependent upon T_e. However, this is overcome as before.
4. Now U can be evaluated.
5. Q is calculated from Eq. 11-1.
6. If the calculated value for Q does not equal the assumed value, correct the assumption and repeat.

Note that this procedure is somewhat more involved than in the previous case.

From the above discussion it should be obvious that the calculational procedure for solving the equations is often more involved than developing the equations themselves. This will be borne out more clearly as the discussion progresses.

Another noteworthy point is that the computational procedure for the heat exchanger has also been examined independently of the remainder of the system. Both the calculation procedure and the equations are often easier to grasp when considered alone, a definite advantage of the modular approach.

Now consider the remainder of the evaporator unit.   Note that the equation

$$F_j H_f + Q = FH + v_v H_v \qquad [11\text{-}9]$$

is quite nonlinear, as the enthalpies as well as Q are functions of C_A and/or T. Consequently an explicit solution for C_A is highly unlikely and a trial-and-error or iterative procedure is again in order.

Consider the following procedure:

1. Assume C_A.
2. Calculate T from C_A and P via the boiling-point curves in Fig. 11-3.
3. Evaluate all quantities for Eq. 11-9.
   - (a) Evaluate F from Eq. 11-8.
   - (b) Evaluate v_v from Eq. 11-7.
   - (c) Evaluate Q for the heat exchanger.
   - (d) Evaluate enthalpies.
4. Now both sides of Eq. 11-9 could be evaluated and checked for equality. However, the following procedure will facilitate making a new assumption:
   - (a) Using all values determined above except F, solve Eq. 11-9 for F. This gives a new estimate.   Even better, eliminate v_v between Eqs. 11-7 and 11-9 and then solve for F.
   - (b) From Eq. 11-8 determine a new estimate for C_A, i.e.,

$$C_A = C_{A_f} F_f / F.$$

5. If the calculated value for C_A does not agree with the assumed value, make another assumption and recalculate.

Note that the above procedure is iterative.   Also note that on each iteration the calculations for the heat exchanger are required, which are also iterative. Thus the calculations involve an iteration within a iteration—a task for a computer.

But if these calculations are to be relegated to the computer, some technique must be devised to make the new assumption.   An equation or algorithm is needed to determine a new value from the previously assumed value(s) and the calculated value.   Note in this case the calculations begin with an assumed value for the concentration, say C_{A_i}.   Upon completion of an iteration, a new value, say C_A^*, for the concentration is obtained.   The objective is to determine the assumed value, say $C_{A_{i+1}}$, for the next iteration.   One common technique is to use an equation of the following type:

$$C_{A_{i+1}} = C_{A_i} + h\,(C_A^* - C_{A_i}) \qquad (11\text{-}11)$$

This equation causes the original assumption to be increased by some factor times the difference between the calculated value and the original assumption.   If h equals one, the new assumption is the calculated value.   However, h may be greater than one or less than one, positive or negative.

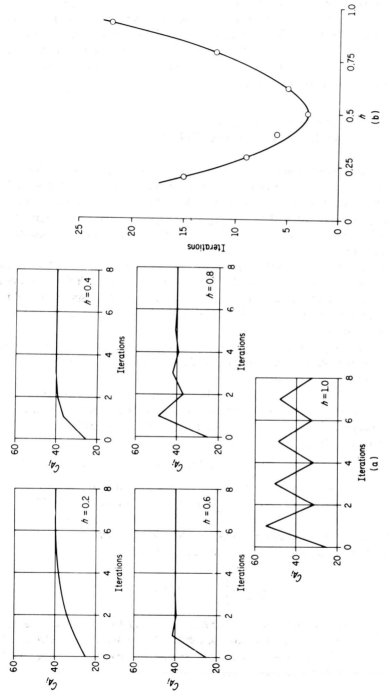

FIG. 11-4. The convergence problem. (a) Progress of convergence for specific cases. (b) Dependence of iterations required on h.

The selection of h for the general case is often difficult. Equation 11-11 is analogous to the reset (integral) only controller, and selecting h is analogous to tuning the controller. The effect of h on the convergence for a few specific cases is shown in Fig. 11-4a. Note that low values cause slow convergence, whereas high values cause oscillations. As shown in Fig. 11-4b, the optimum value of h is reasonably close to 0.5 *for this case.*

11-4. THE DESIGN PROBLEM VS. THE SIMULATION PROBLEM

In recent years a distinction has developed between design problems and simulation problems. A simulation is the mathematical modeling of an "existing" unit—either in fact or on paper. All parameters are specified, and all output quantities must be determined. The design problem differs in that one or more of the parameters are unspecified but one or more of the output quantities have been specified.

In the above problem the parameters m and A were specified and the output concentration is to be determined. Consider the case in which the output concentration C_A is specified, A (the size of the exchanger) is specified, and the flow m must be determined. All other specifications remain as before. This is a simple design problem.

One technique is to use the computational procedure for the simulation problem in the following manner, often called "design by simulation."

1. Assume a value for A.
2. Letting C_A be unknown, the simulation problem can be solved for C_A.
3. If C_A is not equal to the desired value C_{A_d}, adjust A and recalculate.

One immediate problem is how to adjust A. A slightly different form of Eq. 11-11 may be used, namely,

$$A_{i+1} = A_i + h(C_{A_i} - C_{A_d}) \tag{11-11a}$$

In this case h is not dimensionless, its value is not at all obvious, and it may be positive or negative. However, it can sometimes be surmounted, as illustrated below.

1. Note from the previous simulation that when $A = 400$ ft², $C_A = 40.6$.
2. As the feed was 20%, a pseudo-gain can be calculated as

$$\frac{400 \text{ ft}^2}{(40.6 - 20)\%} = 19.4 \text{ ft}^2/\%$$

This is interpreted to mean that it requires 19.4 ft² of heat-transfer area to raise the outlet concentration 1.0%. Due to the nonlinear characteristics of the unit, this value cannot be extrapolated over wide ranges. However, it is at least approximate.

3. Note that h is a factor to give the change in area to give a desired change in outlet concentration. Thus h can be chosen as this value. In practice it would probably be decreased somewhat.

Fig. 11-5 shows the progression of the calculations for the specific case of $h = (0.75)(19.4) = 14.6$

The main disadvantage of this approach is that it has three nested trial-and-error loops to calculate the following quantities:

1. The exit temperature of the exchanger.
2. The exit concentration from the evaporator. This loop includes the above loop.
3. The area of the exchanger. This loop includes the above two loops.

Sometimes the calculational procedure can be modified to avoid this, in many cases it cannot. In this case it can.

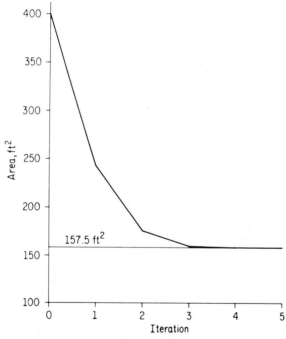

FIG. 11-5. Calculation of appropriate heat-transfer area.

As the specified quantities have changed, the computational procedure could also change. Having specified C_A permits the following quantities to be calculated directly:

1. Calculate F from Eq. 11-8.
2. Calculate v_y from Eq. 11-7.

3. Determine T from Fig. 11-3.
4. All enthalpies appearing in Eq. 11-9 can be evaluated.
5. Equation 11-9 can be solved for Q.
6. Equation 11-2 can be solved for T_e. Although the dependence of c_p on T_e causes a one or two iterations here, it is not serious.
7. Equation 11-1 is solved for A.

Carrying out these calculations for the previous example also gives 157.5 ft² for A. Note how simple this is as compared with the previous approach. Unfortunately, very few problems behave in this manner.

A more reasonable design problem would be to select both m and A so that the cost is minimized. As the value of A corresponding to a given m can be calculated by the above procedure, the line marked "Area" in Fig. 11-6 can be constructed. The problem now reduces to selecting either m or A.

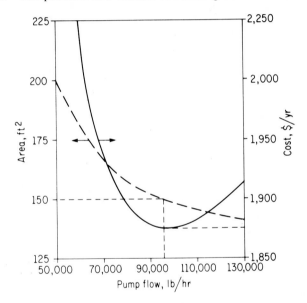

FIG. 11-6. Selection of optimum recirculation rate.

The following factors contribute to the economic picture:
1. The cost of the pump, whose size is determined by the flow rate m and the head. The head is a function of T_e, since no vaporization should occur in the exchanger. Knowing T_e and C_A, the vapor pressure of the caustic can be determined. The minimum allowable head is this pressure minus the pressure in the effect. A factor of 10% or more is often added for safety.
2. The power required for the pump.
3. The cost of the exchanger, a direct function of its size.

In adding economics to the model, the following factors are assumed:

1. Cost of electricity: $ 0.01/kwhr
2. Hours of operation: 8,000 hr/yr
3. Cost of pump: $ 20/gpm, installed.
4. Cost of electric driver: _ $ 200/hp, installed.
5. Efficiency of pump and driver: 40%.
6. Cost of exchanger: $ 100/ft^2, installed.
7. Depreciation: 10%/year, straight-line, no salvage value.

Using these values the "cost" curve in dollars per year can be constructed as shown in Fig. 11-6. As the most economic operation is desired, the minimum on this curve should be the operating point.

11-5. MULTIPLE-EFFECT EVAPORATOR PROBLEM

The previous paragraphs have illustrated several concepts in terms of a single-effect evaporator. At the beginning of the article a triple-effect unit was proposed, and multiple-effect units will be considered in this section.

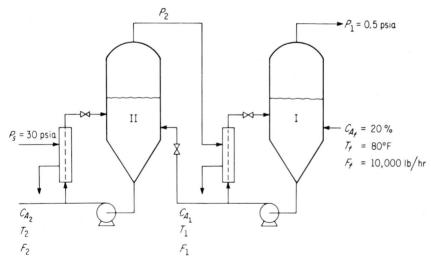

FIG. 11-7. Double-effect evaporator unit.

First, consider the double-effect unit shown in Fig. 11-7. Using the model for a single effect developed above, the model for Effect I can be represented schematically as shown in the first sketch.

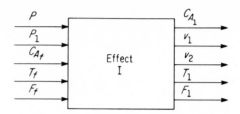

Similarly, the model for Effect II is as shown in the second sketch.

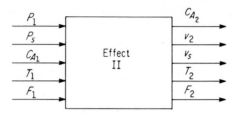

Now consider combining these to form the complete model, as shown in the third sketch.

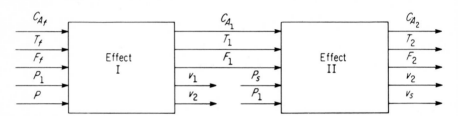

These models combine very nicely except for two difficulties:

1. The variable P_1 (the pressure in Effect I) is an input to *both* effects, but does not appear as an output.
2. The variable v_2 (the vapor flow from Effect I) is an output from both effects.

Other than P_1, all other inputs are specified variables.

Thus, the problem basically reduces to finding P_1. Clearly P_1 could be assumed, and each of the models solved. The order of solution would be to solve Effect I first, as all inputs are known (having assumed P_1). This yields values for C_{A_1}, T_1, and F_1, making all inputs to Effect II known. However, P_1 has been assumed, and this must be checked. For any value of P_1, there is no assurance that the two *calculated* values for v_2 will be equal. However, physically this stream can have only one value. Thus, the *correct* assumption for P_1 is the value that makes the two calculated values of v_2 equal.

Now the computational procedure is as follows:

1. Assume P_1.
2. Perform the calculations for Effect I. Recall that this is a trial-and-error calculation involving the assumption of C_{A_1}. Let the calculated value of v_2 be $(v_2)_I$.
3. Perform the calculations for Effect II. This is also a trial-and-error calculation. Let the calculated value of v_2 be $(v_2)_{II}$.
4. If $(v_2)_I \neq (v_2)_{II}$, assume a new value for P_1 and repeat calculations.

As this is clearly a task for the computer, again a systematic procedure must be devised for updating the estimate for P_1. As in previous examples, the following equation, similar to Eq. 11-11, can be used:

$$(P_1)_{i+1} = (P_1)_i + h[(v_2)_I - (v_2)_{II}]$$

Again evaluating h is difficult. (11-11b)

One way to accomplish this is to obtain the first solution by a "strong-arm" approach, as illustrated in Figure 11-8. Here the values of $(v_2)_I$ and $(v_2)_{II}$ are plotted vs. the assumed value of P_1. It is obvious that the solution is $P_1 = 3.8$ psia and $v_2 = 1650$ lb/hr. If only one solution is needed, then we have it. However if many solutions will be needed (e.g., if the unit is to be optimized), then there is significant incentive to evaluate h for the convergence algorithm.

Suppose that we assume P_1 equal 8 psia. As given in Figure 11-8, $(v_2)_I$ and

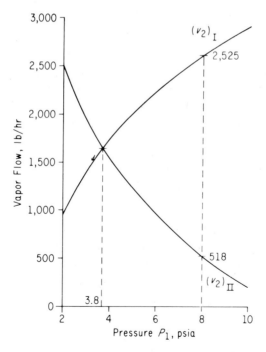

FIG. 11-8. Progression of calculations for double effect.

$(v_2)_{II}$ are 2625 lb/hr and 518 lb/hr. The ultimate would be for the next assumed value for P_1 to be the solution, namely 3.8 psia. Substituting these values into Eq. 11-11b gives

$$3.8 = 8 + h(2625 - 518)$$

or

$$h = -0.00199 \text{ psia}/(1 \text{b}/\text{hr})$$

In practice, this value should be reduced somewhat to be on the safe side.

Now consider the design problem. Suppose we again require the outlet concentration, in this case C_{A_2}, to be 30%. Again m is set at a given value, say 60,000 lb/hr, and A will be determined. As in typical installations, the same value of A is used in each effect.

For the single effect, recall that two approaches were used:

1. Assume A, calculate C_{A_2} (by above procedure), and correct assumption as necessary. Note that this approach involved three nested trial-and-errors, not counting the exchanger.
2. Modification of computational procedure.

Clearly the first procedure will work, but would require considerable calculations. As the second procedure was so obviously superior for the single effect, consider its application here.

Unfortunately knowing C_{A_2} does not permit any immediate calculations to be performed for Effect II, as neither the pressure P_1 nor the conditions of the input stream are known. However, consider the following procedure.

1. Assume C_{A_1}.
2. Now F_1, T_1, v_1, and Q_1 (heat-transfer rate in Exchanger I) can be calculated by Eqs. 11-7, 11-8, 11-9, and the boiling-point curves in Fig. 11-3.
3. As C_{A_2} is known, F_2 and v_2 can be calculated.
4. Now λ_1 (latent heat of vaporization of steam in Exchanger I) can be calculated from $\lambda_1 = Q_1/v_2$.
5. Thus P_1 can be determined from steam tables. Also, T_{s_1}, the condensation temperature, can be determined.
6. This permits the solution of Eqs. 11-1 and 11-2 for A_1, the area of Exchanger I.
7. From P_1 and C_{A_2} the temperature T_2 in Effect II can be calculated.
8. Now all enthalpies can be evaluated and Eq. 11-9 solved for Q_2.
9. Equations 11-1 and 11-2 can be solved for A_2.
10. As the exchangers should have equal areas, C_{A_1} must be reassumed until $A_1 = A_2$.

Note that this procedure requires only one trial-and-error.

The above approach has two disadvantages:

1. It destroys the modular approach, thus requiring the entire system to be examined simultaneously. This is not easy for large units.

2. One step in this procedure could be a source of numerical problems. Consider the case of the x-y plot in Fig. 11-9. Given a value of x, say x_1, the corresponding value of y can be read with greater certainty since a large change in x would entail a small change in y. That is, δ_x is much greater than δ_y. However, given y the value of x cannot be determined so exactly. Only a small error in the value of y_1 would produce a large error in x_1. This is analogous to the step in the above procedure in which the pressure P_1 is determined from λ_1. A small error in λ_1 would be considerably amplified in the resulting value of P_1. Situations such as this should be avoided whenever possible.

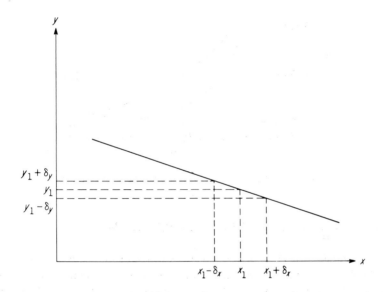

FIG. 11-9. The look-up problem.

Using either of the above approaches, the value of A can be calculated for the corresponding value of m to yield the curve "Area" in Fig. 11-10. As for the single-effect system, A and m should be chosen so as to minimize the cost. Using the same economics as before, the curve "Cost" in Fig. 11-10 can be calculated. Of course, A and m are selected to minimize the cost.

This procedure could be extended to also determine the optimum number of effects. From the cost of steam and the cost of the vessel itself, the cost of operation for one and two effects can be readily ascertained from previous calculations. These ideas can be extended to three, four, or however many additional effects can be added to further reduce the operating cost.

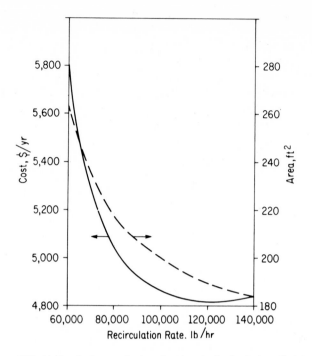

FIG. 11-10. Optimum selection of exchanger size for double-effect.

11-6.   STEADY-STATE REACTOR PROBLEM

The problem in the preceding sections was kept simple in order to illustrate the fundamental points.   In this section the somewhat more complex unit in Fig. 11-11 is considered.   Fresh feed, composed of only A and D, enters the unit at flow rate w_f, concentration C_{A_f} of component A, and temperature T_f.   This feed is combined with a recycle stream and fed to a well-mixed, jacketed reactor of volume V_r.   Water enters the jacket at flow rate m and temperature T_w, and the recirculation of the jacket water is assumed sufficient to consider the jacket water to be at a uniform temperature T_j.   The overall heat-transfer coefficient is U and the heat-transfer area is A.   In the reactor the reaction $2A \rightarrow B + C$ occurs.   This reaction is second-order, and the dependence of the rate constant with temperature is given by:

$$k = k_0 e^{-a/T}$$

where T is the temperature in degrees Rankine.   The heat of reaction ΔH is based on one mole of A reacting.   Table 11-2 gives numerical values of all parameters.

The discharge stream from the reactor is w_1 lb/min, and is composed of

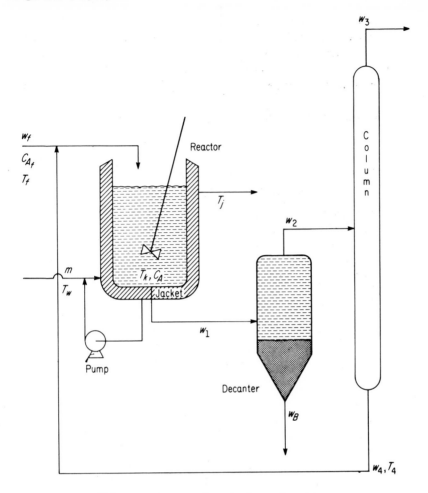

FIG. 11-11.   Reactor and associated separation equipment.

components A, B, C, and D.   The moles/min of each component in this stream
are denoted as W_{1_A}, W_{1_B}, W_{1_C}, and W_{1_D} (lower case w denotes units in lb/min,
upper case W denotes units in mole/min).   The reactor effluent is a two-phase
mixture, the heavy phase being pure B and the light phase containing all other
components plus 10 mole percent (the equilibrium concentration) of component B.

The light phase is separated into the various components by distillation.   Let
us assume that this unit has only reached the preliminary design stage, and the
primary concern is the feasibility of the reactor.   From the molecular weights of
the components in Table 11-2, note that component D is the light key and
component A is the heavy key.   Assume that the column is to be designed so that

no more than 5% of the heavy key in the feed appears in the overheads and no more than 5% of the light key appears in the bottoms.   Assume all of the lighter-than-light-key components (component D) appears in the overheads and all heavier-than-heavy-key components (component B) appear in the bottoms.   Also assume that the bottoms temperature is essentially the boiling point of component A, namely T_4.

The physical-property data are also given in Table 11-2, and are assumed to be constant.

TABLE 11-2

Numerical Values for Unit in Fig. 11-11.

Molecular weight of component A	46.0
Molecular weight of component B	76.0
Molecular weight of component C	16.0
Molecular weight of component D	26.0
Volume of reactor, V_r	100.0 gal
Heat-transfer area, A	29.9 ft²
Heat-transfer coefficient, U	74.5 Btu/hr-ft²-°F
Feed rate, w_f	73.5 lb/min
Concentration of component A in feed, C_{A_f}	1.0 lb mole/ft³
Temperature of feed, T_f	80.0°F
Density of feed and reacting mass, ρ	55.0/ft³
Heat capacity of feed and reacting mass, c_p	0.9 Btu/lb-°F
Water flow rate, m	74.3 lb/min
Inlet water temperature, T_w	75.0°F
Heat capacity of water, c_p'	1.0 Btu/lb-°F
Coefficients in Arrhenius expression	
k_0	0.566 mole B/hr-ft³
a	1000. °R
Heat of reaction, ΔH	− 5000. Btu/mole A
Bottoms temperature, T_4	190.0°F

Again the best way to approach this problem is with a modular approach, considering the reactor, decanter, and column as individual units.

First, consider the decanter.   Making material balances around the unit indicates

$$W_{1_A} \qquad - W_{2_A} = 0 \tag{11-12}$$

$$W_{1_B} - W_B \ - W_{2_B} = 0 \tag{11-13}$$

$$W_{1_C} \qquad - W_{2_C} = 0 \tag{11-14}$$

$$W_{1_D} \qquad - W_{2_D} = 0 \tag{11-15}$$

The equilibrium between the phases gives the relationship

$$\frac{W_{2_B}}{W_{2_A} + W_{2_B} + W_{2_C} + W_{2_D}} = 0.10 \qquad (11\text{-}16)$$

Assuming the inputs (W_{1_A}, W_{1_B}, W_{1_C}, and W_{1_D}) will be specified independently, there are five unknowns in this set of five equations, leaving no degrees of freedom.

Knowing W_{1_A}, W_{1_B}, W_{1_C}, and W_{1_D}, the solution of these equations is trivial. Equation 11-12 is solved for W_{2_A}, Eq. 11-14 for W_{2_C}, and Eq. 11-15 for W_{2_D}. Equation 11-16 is rearranged to

$$W_{2_B} = (W_{2_A} + W_{2_C} + W_{2_D})/9$$

As this equation gives W_{2_B} directly from quantities previously calculated, Eq. 11-13 is solved for W_B to complete the solution.

Material balances on the distillation column yield the following expressions:

$$W_{3_B} = 0 \qquad (11\text{-}17)$$

$$W_{3_A} = 0.05\ W_{2_A} \qquad (11\text{-}18)$$

$$W_{4_C} = 0 \qquad (11\text{-}19)$$

$$W_{4_D} = 0.05\ W_{2_D} \qquad (11\text{-}20)$$

$$W_{3_C} = W_{2_C} - W_{4_C} \qquad (11\text{-}21)$$

$$W_{3_D} = W_{2_D} - W_{4_D} \qquad (11\text{-}22)$$

$$W_{4_A} = W_{2_A} - W_{3_A} \qquad (11\text{-}23)$$

$$W_{4_B} = W_{2_B} - W_{3_B} \qquad (11\text{-}24)$$

Considering W_{2_A}, W_{2_B}, W_{2_C}, and W_{2_D} as known, this leaves eight unknowns in this set of eight equations, leaving zero degrees of freedom. The solution is also simple and straightforward.

Now for the reactor. Making an enthalpy balance on the jacket yields after simplification

$$mc'_p\,(T_w - T_j) + UA\,(T_k - T_j) = 0 \qquad (11\text{-}25)$$

Making an enthalpy balance on the reacting mass yields after simplification

$$w_f\, c_p\, T_f + w_4 c_p\, T_4 - w_1\, c_p\, T_k - UA\,(T_k - T_j)$$
$$- 2\,kC_A^2\, V_r\,(\Delta H) = 0 \qquad (11\text{-}26)$$

A total material balance gives

$$w_1 = w_f + w_4 \qquad (11\text{-}27)$$

Also note that w_4 can be calculated from W_{4_A}, W_{4_B}, W_{4_C}, and W_{4_D} and the molecular weights. Material balances on the individual components gives

$$
\begin{aligned}
W_{4_A} + W_{f_A} - W_{1_A} - 2kC_A^2V_r &= 0 & (11\text{-}28)\\
W_{4_B} \qquad\quad - W_{1_B} + kC_A^2V_r &= 0 & (11\text{-}29)\\
W_{4_C} \qquad\quad - W_{1_C} + kC_A^2V_r &= 0 & (11\text{-}30)\\
W_{4_D} + W_{f_D} - W_{1_D} \qquad\qquad\ &= 0 & (11\text{-}31)
\end{aligned}
$$

The concentration C_A is given by

$$
C_A = \frac{W_{1A}\,\rho}{w_1} \tag{11-32}
$$

Also k as a function of temperature must be included:

$$
k = k_0 \exp\left[-a/(T_k + 460.)\right] \tag{11-33}
$$

This set of nine equations contains nine unknowns (T_j, T_k, W_{1_A}, W_{1_B}, W_{1_C}, W_{1_D}, w_1, k, C_A), thus leaving no degrees of freedom.

This set of equations for the reactor is not especially easy to solve. One approach is as follows:

1. Calculate w_1 from Eq. 11-27.
2. Assume T_k.
3. Calculate T_j from Eq. 11-25.
4. Calculate $r_A = -kC_A^2V_r$ from Eq. 11-26.
5. Calculate W_{1_A} by Eq. 11-28.
6. Calculate C_A by Eq. 11-32.
7. As r_A is known, k can be calculated.
8. Equation 11-33 is solved for T_k, giving a check on the assumption.

This scheme is not without its disadvantages (negative concentrations and similar impossible phenomena appear during the search), but it does work. Another peculiarity is that when the equation $T_k|_{i+1} = T_k|_i + h\lfloor T_k|_{calc} - T_k|_i \rfloor$ is used to update the estimates, a good value of h is $- 0.02$. Table 11-3 gives the sequence of values for the iteration when there is no recycle to the reactor.

In the above paragraphs, the equations for each of the modules (units) were developed, and when given the inputs, a technique for obtaining the output was devised. Now these must be combined to form a model of the entire unit, and a solution technique must be developed. Suppose the following procedure is used.

1. Assume W_{1_A}, W_{1_B}, W_{1_C}, and W_{1_D}.
2. This provides all inputs to the decanter, and its equations can be solved to give w_2 and all its components.

TABLE 11-3

Progress of Iterations for Reactor

Iteration	Assumed Temperature	Calculated Temperature	C_A	T_j
1	100.0000	− 124.0665	0.7094	83.3297
2	104.4813	− 76.3592	0.6484	84.8228
3	108.0981	− 28.5513	0.5992	86.0279
4	110.8311	15.9790	0.5620	86.9385
5	112.7282	52.9110	0.5362	87.5706
6	113.9245	79.5154	0.5199	87.9692
7	114.6127	96.2178	0.5106	88.1985
8	114.9806	105.6176	0.5056	88.3211
9	115.1679	110.5367	0.5030	88.3835
10	115.2605	113.0046	0.5018	88.4143
11	115.3056	114.2152	0.5011	88.4294
12	115.3274	114.8024	0.5008	88.4366
13	115.3379	115.0855	0.5007	88.4401
14	115.3429	115.2218	0.5006	88.4418
15	115.3454	115.2872	0.5006	88.4426
16	115.3465	115.3186	0.5006	88.4430
17	115.3471	115.3337	0.5006	88.4432
18	115.3474	115.3409	0.5006	88.4433

3. As w_2 and its components are known, the equation for the column can be solved for w_3, w_4, and all component flows.
4. Now all inputs to the reactor are known, and the set of equations for it can be solved for w_1 and its components.
5. This provides a check on the assumption, and new assumed values for each of the components is obtained by the equation

$$W_{1j}\Big|_{i+1} = W_{1j}\Big|_i + h\,[\,W_{1j}\Big|_{calc} - W_{1j}\Big|_i\,], \qquad j = A, B, C, D$$

The progression of the calculations for a specific case is shown in Table 11-4. The value of h used to construct the table was 0.8.

TABLE 11-4

Iterative Calculations for Complete Unit

Iteration		W_{1_A}	W_{1_B}	W_{1_C}	W_{1_D}	T_k
1	Assumed	0.5000	0.2500	0.2500	0.1000	
	Calculated	0.9977	0.5013	0.4068	0.4676	138.3127
2	Assumed	0.8982	0.4510	0.3755	0.3941	
	Calculated	1.2849	0.6377	0.4524	0.4823	149.8204
3	Assumed	1.2075	0.6003	0.4370	0.4646	
	Calculated	1.5051	0.7235	0.4892	0.4858	156.2417
4	Assumed	1.4456	0.6989	0.4787	0.4816	
	Calculated	1.6763	0.7840	0.5167	0.4867	160.2508
5	Assumed	1.6302	0.7670	0.5091	0.4857	
	Calculated	1.8106	0.8289	0.5372	0.4869	162.9338
6	Assumed	1.7745	0.8165	0.5316	0.4866	
	Calculated	1.9166	0.8631	0.5528	0.4869	164.8159
7	Assumed	1.8882	0.8537	0.5485	0.4869	
	Calculated	2.0008	0.8895	0.5647	0.4869	166.1825
8	Assumed	1.9783	0.8824	0.5614	0.4869	
	Calculated	2.0679	0.9102	0.5739	0.4869	167.1987
9	Assumed	2.0500	0.9046	0.5714	0.4869	
	Calculated	2.1215	0.9265	0.5811	0.4869	167.9692
10	Assumed	2.1072	0.9221	0.5792	0.4869	
	Calculated	2.1645	0.9394	0.5868	0.4869	168.5612
11	Assumed	2.1531	0.9360	0.5853	0.4869	
	Calculated	2.1991	0.9497	0.5914	0.4869	169.0210
12	Assumed	2.1899	0.9470	0.5902	0.4869	
	Calculated	2.2268	0.9579	0.5950	0.4869	169.3811
13	Assumed	2.2194	0.9558	0.5940	0.4869	
	Calculated	2.2492	0.9645	0.5978	0.4869	169.6651
14	Assumed	2.2432	0.9628	0.5971	0.4869	
	Calculated	2.2672	0.9698	0.6001	0.4869	169.8905
15	Assumed	2.2624	0.9684	0.5995	0.4869	
	Calculated	2.2817	0.9741	0.6020	0.4869	170.0695
16	Assumed	2.2779	0.9729	0.6015	0.4869	
	Calculated	2.2935	0.9775	0.6034	0.4869	170.2124
17	Assumed	2.2903	0.9766	0.6030	0.4869	
	Calculated	2.3029	0.9802	0.6046	0.4869	170.3271
18	Assumed	2.3004	0.9795	0.6043	0.4869	
	Calculated	2.3106	0.9824	0.6056	0.4869	170.4188
19	Assumed	2.3085	0.9818	0.6053	0.4869	
	Calculated	2.3168	0.9842	0.6064	0.4869	170.4924
20	Assumed	2.3151	0.9837	0.6062	0.4869	
	Calculated	2.3218	0.9857	0.6070	0.4869	170.5520
21	Assumed	2.3204	0.9853	0.6068	0.4869	
	Calculated	2.3258	0.9868	0.6075	0.4869	170.5997

11-7. STEADY-STATE SOLUTION FROM DYNAMIC EQUATIONS

The unsteady-state problem may be considered for two reasons: (1) to obtain and investigate the true transient response of the unit, or (2) as an avenue to obtain the steady-state solution. It is this last reason that will receive the attention in this section.

In the previous sections of this chapter, the techniques for writing steady-state equations and methods to solve these equations were discussed. In each case, these equations were nonlinear, and required trial-and-error procedures to solve them. Not only were trial-and-error solutions encountered, but trial-and-errors within trial-and-errors (nested trial-and-errors) were necessary. Convergence to a solution was always a problem, and no matter how good the technique, there is always a chance that the scheme will not converge.

On the other hand, convergence is not a problem with unsteady-state solutions. Given enough time, and barring such phenomena as instabilities in numerical intergration schemes, the solution of the unsteady-state equations will approach the steady-state solution. Thus assured convergence is obtained, but at a price. Although the calculations associated with the unsteady-state solution are simple, the sheer number of calculations is generally reflected in longer computer times to obtain the solution. Even so, this trade-off is very lucrative, and deserves proper perspective.

If only the steady-state solution is needed, then there is no need to require that the equations accurately represent the transient conditions. Thus it is possible to mix steady and unsteady-state equations. The unsteady-state equations only provide a sure route to the steady-state solution.

In the previous problem, suppose that the steady-state equations for the decanter and for the column are still used. For the reactor, unsteady-state equations will be used for T_k, W_{1_A}, W_{1_B}, W_{1_C} and W_{1_D}. To obtain an expression for T_k, an unsteady-state enthalphy balance for the reactor in Fig. 11-11 yields after simplifications:

$$w_f c_p T_f + w_4 c_p T_4 - w_1 c_p T_k - UA(T_k - T_j)$$

$$- 2k C_A^2 V_r(\Delta H) = V_r \rho c_p \frac{dT_k}{dt} \qquad (11\text{-}34)$$

Next, unsteady-state component material balances are developed for the reactor.

$$W_{4_A} + W_{f_A} - W_{1_A} - 2kC_A^2 V_r = \frac{d}{dt}\left(\frac{V_r \rho W_{1_A}}{w_1}\right)$$

$$W_{4_B} \qquad - W_{1_B} + kC_A^2 V_r = \frac{d}{dt}\left(V_r \rho \frac{W_{1_B}}{w_1}\right)$$

$$W_{4C} \quad - W_{1C} + kC_A^2 V_r = \frac{d}{dt}\left(V_r \rho \frac{W_{1C}}{w_1}\right)$$

$$W_{4D} + W_{fD} - W_{1D} = \frac{d}{dt}\left(V_r \rho \frac{W_{1D}}{w_1}\right)$$

Even though w_1 is a function of time, it will be treated as a constant so that the derivatives in the above equation can be expanded as follows:

$$W_{4A} + W_{fA} - W_{1A} - 2kC_A^2 V_r = \frac{V_r \rho}{w_1}\frac{dW_{1A}}{dt} \qquad (11\text{-}35)$$

$$W_{4B} \quad - W_{1B} + kC_A^2 V_r = \frac{V_r \rho}{w_1}\frac{dW_{1B}}{dt} \qquad (11\text{-}36)$$

$$W_{4C} \quad - W_{1C} + kC_A^2 V_r = \frac{V_r \rho}{w_1}\frac{dW_{1C}}{dt} \qquad (11\text{-}37)$$

$$W_{4D} + W_{fD} - W_{1D} = \frac{V_r \rho}{w_1}\frac{dW_{1D}}{dt} \qquad (11\text{-}38)$$

The justification for this is that it does not affect the steady-state solution, which is all that is desired.

For these unsteady-state equations, initial conditions are required. The values used will not affect the final results, but values closer to the steady-state solution should be used whenever possible to reduce the computer time required for a solution.

The numerical solution procedure now becomes

1. Begin with the initial conditions for W_{1A}, W_{1B}, W_{1C}, W_{1D}, and T_k.
2. Solve the steady-state equations for the decanter to obtain the outlet streams.
3. Solve the steady-state equations for the column to obtain w_3, w_4, and their components.
4. Solve Eq. 11-25 for T_j (recall that T_k is known from the initial conditions).
5. Calculate w_1 by Eq. 11-27.
6. Calculate C_A by Eq. 11-32.
7. Calculate k by Eq. 11-33.
8. Now the left-hand sides of the unsteady-state equations, i.e., Eqs. 11-34 through 11-38, can be evaluated.
9. A numerical integration procedure, finite differences, or other scheme can be used to obtain values of T_k, W_{1A}, W_{1B}, W_{1C}, and W_{1D} after some time increment Δt.
10. These calculations, beginning with step 2, can be repeated to obtain the "time response" of the unit. After sufficient time, steady-state conditions are approached.

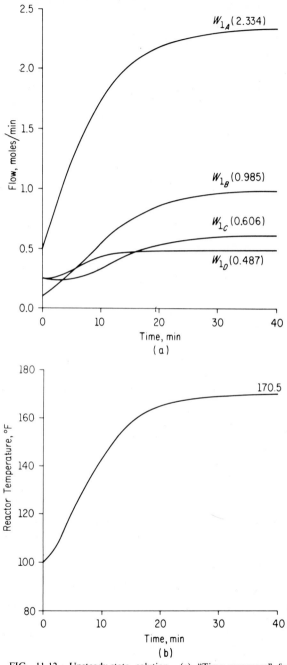

FIG. 11-12. Unsteady-state solution. (a) "Time responses" for component flows. (b) "Time response" for reactor temperature.

Although the above procedure is numerically oriented, the analog computer can be used in a similar fashion.

The "time response" for a specific case is shown in Fig. 11-12. Quotations are used because the "time response" is not the true response, but only a route to the steady-state solution. The solution was arbitrarily started at $T_k = 100°F$, $W_{1_A} = 0.5$ mole/min, $W_{1_B} = 0.25$ mole/min, $W_{1_C} = 0.25$ mole/min, and $W_{1_D} = 0.1$ mole/min. For comparison, construction of these curves required more than ten times as much computer time than required to construct Table 11-3. Even so, it still required less than a minute on an IBM 7040.

11-8. SUMMARY

This chapter has really only introduced the subject of modeling of plant-scale systems. Hopefully some insight has been gained into some of the problems of formulating such models and obtaining a solution. This procedure can be facilitated to a large extent by using one of the computer-aided process design programs that have appeared in recent years. Detailed discussion of these is beyond the scope of this text, and the reader is referred to Ref. 6 for further discussion and a good bibliography.

REFERENCES

1. C. L. Smith, "All-Digital Simulation for the Process Industries." *ISA Journal,* Vol. 13, No. 7 (July, 1966), pp. 53-60.
2. R. L. Norman, "A Matrix Method for Location of Cycles of a Directed Graph", *A.I.Ch.E.,* Vol. 11, No. 3 (May 1965), pp. 450–2.
3. W. Lee, and D. F. Rudd, "Ordering of Recycle Calculations," *A.I.Ch.E. Journal,* Vol. 12, No. 6 (November, 1966), pp. 1184-1190.
4. W. Lee, J. H. Christensen, and D. F. Rudd, "Design Variable Selection to Simplify Process Calculations, *A.I.Ch.E. Journal,* Vol. 12, No. 6 (November, 1966), pp. 1104-1110.
5. P. T. Shannon, et al., "Digital Computer Simulation of a Contact Sulphuric Acid Plant," presented at the 58th Annual A.I.Ch.E. Meeting, Philadelphia, Dec. 5-9, 1965.
6. L. B. Evans, D. G. Steward, and C. R. Sprague, "Computer-Aided Chemical Process Design," *Chemical Engineering Progress,* Vol. 64, No. 4 (April, 1968), pp. 39-46.

PROBLEMS

11-1. One industrial process for the production of benzene is by the reforming of cyclohexane. A simplified schematic of this process is shown in the figure. The cyclohexane feed (350 lb/min) and recycle are combined with a hydrogen recycle to form a stream whose composition is 90% H_2, 10% C_6H_{12} by volume. This stream is heated to 950°F and fed to a reactor operated at 610 psia. Two reversible reactions

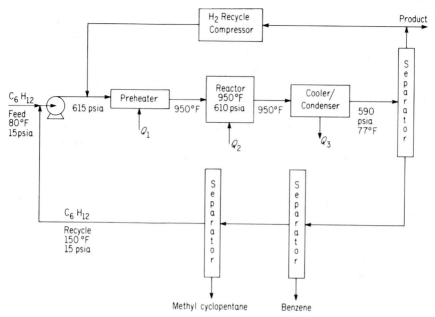

PROB. 11-1.   Process for reforming cyclohexane.

are important:

The reactor effluent is substantially at equilibrium, the equilibrium constants at 950°F being $K_1 (= 10^6$ atm^3) and K_2 ($= 12$). The products are cooled (condensing all but the hydrogen) and separated. Part of the hydrogen is recycled, the rest being a salable product. The liquid is separated into a benzene product stream, a methyl cyclopentane product stream, and a cyclohexane recycle stream. Assume all separations are perfect.

Using the modular approach, develop the necessary unit models for this process. Be sure that the following variables are calculated: (a) work by feed pump; (b) Q_1; (c) Q_2; (d) Q_3; (e) hydrogen recycle; (f) hydrogen product; (g) work input at compressor (assume isentropic with $n = 1.4$);   (h) compressor outlet temperature;   (i) benzene product; (j) methyl cyclopentane product.   Suggest an iterative procedure for solving this problem.

11-2.   A "depropanizer" is a distillation column used to remove propane from streams containing higher-molecular-weight paraffins.   Suppose the feed to such a column consists of 20% (vol) propane, 17% butane, and 63% higher-molecular-weight hydrocarbons that will be treated as pentane.   The feed flow is F mole/hr at 250° F.   The overheads product must be at least 98% propane, and the remaining 2% is assumed to be butane.   The bottoms is no more than 2% propane.   The column pressure is regulated so that the overheads temperature is 130°F.   The pressure differential between top and bottom of the column is 35 psi.   The water enters the condenser at 80°F and leaves at 110°F.   Assume the steam pressure to the reboiler will be 20°F higher than the reboiler temperature.   The reflux/ overhead product ratio is 8.   Assume the condenser subcools the overheads by 15°.

Formulate the equations and outline a procedure for calculating the water flow to the condenser and steam flow to the reboiler.

11-3.   Suppose the unit shown is used to recover compound A in the following isomerization reaction:

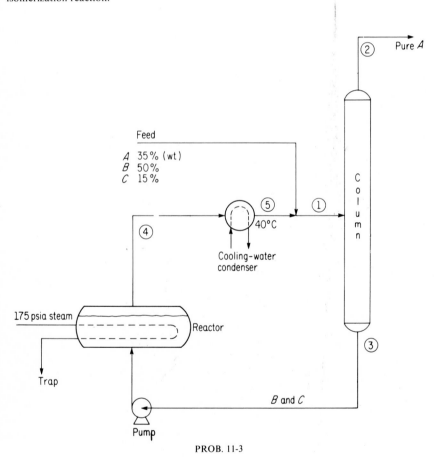

PROB. 11-3

$$B \underset{k_1}{\overset{k_2}{\rightleftharpoons}} A \underset{k_4}{\overset{k_3}{\rightleftharpoons}} C$$

This reaction is carried out in the presence of a catalyst in solvent D. The physical properties of the compounds are as shown in the table.

Property	A	B	C	D
Molecular weight	125	125	125	108
Heat capacity, pcu/lb-°C	0.41	0.45	0.42	0.5
Latent heat of vaporization, pcu/lb	75	75	75	75
Vapor pressure at 80°C, mm Hg	170	50	60	1.4 (110°C)
Vapor pressure at 120°C, mm HG	670	240	280	2.3

The reactor (volume = 2,000 gal) is operated isothermally at 105°C and contains 30% (wt) catalyst-solvent solution. At 105°C, the reaction rates are

$$k_1 = 2.2/hr$$
$$k_2 = 0.82/hr$$
$$k_3 = 0.21/hr$$
$$k_1 = 0.06/hr$$

The heats of reaction are neglegibly small. As the vapor pressure of D is so low, assume no D is vaporized from the reactor. The density of the reacting mixture at 105°C is 35 lb/ft³.

The pressure at the top of the column is about 300 mm Hg, with a 100 mm Hg differential from top to bottom. The overheads is essentially pure A, and the bottoms is composed of only B and C.

The cooling water to the condenser enters at 27°C and leaves at 35°C.

Develop the necessary models and suggest a calculation procedure to determine the following: (a) flow from bottom of the column; (b) bottoms temperature; (c) work input at pump; (d) reactor pressure; (e) steam requirements; (f) cooling-water requirements.

11-4. Coke deposited on a solid catalyst is burned by hot, recirculating gases containing a low percentage of O_2. Low oxygen concentrations are necessary to prevent overheating of the metallic catalyst. The total recirculation time is on the order of 30 hrs, steady-state (except inside the reactor) being obtained after about 1 hr. Use the schematic shown and the following data on the unit.

Reactor
Assume all O_2 in inlet gas is consumed by combustion of the coke.
Exchanger
$U_R = 60$ Btu/hr-ft²-°F
$A_e = 8,000$ ft²
Cooler
$U_c = 50$ Btu/hr-ft²-°F
$A_c = 4,000$ ft²
Cooling-water inlet = $T_{W_i} = 90°$F

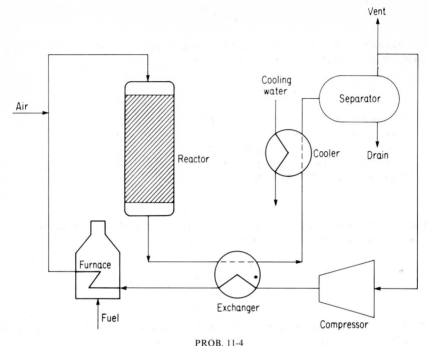

PROB. 11-4

Separator
   Removes any entrained particles in the gas to the compressor.
   Little condensation occurs, and it can be treated as a wide spot in the line.

Compressor
   Isentropic, $n = 1.2$
   Discharge Flow, 30,000 scfm

Furnace
   Efficiency $= 70\%$
   Heating value of fuel $= 18,000$ Btu/lb

Pressures
   Compressor discharge pressure $= 100$ psig
   ΔP for exchanger $= 5$ psi
   ΔP for reactor $= 35$ psi
   ΔP for cooler $= 5$ psi
   ΔP for furnace $= 5$ psi

Given values for the inlet air flow rate (inlet temperature $= 80°$F), flow ra of fuel to the
furnace, and flow rate of cooling water to the cooler, determine the equations than can be
solved for the O_2 concentration in the inlet gas to the reactor and the temperatures at all
points. Suggest a computational procedure. Is any data missing?

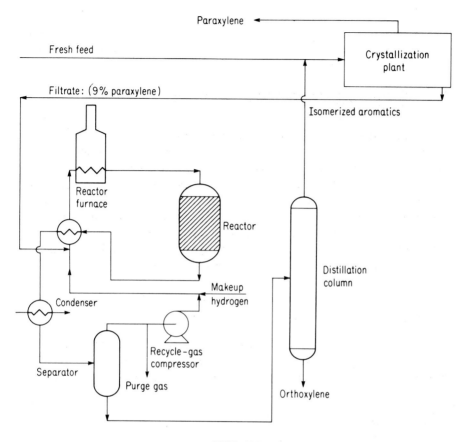

Paraxylene

Fresh feed

Crystallization plant

Filtrate: (9% paraxylene)

Isomerized aromatics

Reactor furnace

Reactor

Distillation column

Makeup hydrogen

Condenser

Recycle-gas compressor

Separator

Purge gas

Orthoxylene

PROB. 11-5

11-5. Consider the simplified flow diagram in the figure for the production of para-xylene. The feed composition is 5% (by weight) ethylbenzene, 23% paraxylene, 50% metaxylene, and 22% orthoxylene. This feed is combined with the tower overheads and sent to a low-temperature crystallization unit where essentially pure paraxylene is removed. The filtrate containing 9% paraxylene is mixed with a hydrogen stream before entering the preheater and reactor furnace. The xylene mixture is converted in the reactor to its equilib-rium composition of 24% paraxylene, 23% orthoxylene, and 53% metaxylene. After being cooled, the products enter a separator where the hydrogen is removed (neglect dissolved hydrogen). The orthoxylene is then removed by distillation, the remaining liquid (the column overheads) being sent to the crystallization plant. Determine the necessary equations that can be solved for the production rate of paraxylene.

The main simplification is the neglecting of side reactions producing benzene, toluene, etc. (Adapted by special permission from "Unique Catalyst is Key to Paraxylene Production," *Chemical Engineering* (October 7, 1968). Copyright 1968 by McGraw-Hill Inc., New York, N. Y. 10036.)

chapter 12

Matrix-Oriented Models

Some mathematical models tend to be cumbersome because of the sheer number of equations involved in the model itself, and there exists a number of instances when the model of a system can be very conveniently developed in the format of matrix mathematics. It will be the purpose of this chapter to illustrate a number of these situations. For completeness and terminology reasons a very brief introduction will be given to some elementary matrix mathematics.

This particular chapter will not be directly concerned with the use of matrix techniques in the numerical solution of differential equations, in optimization applications, or in sophisticated matrix mathematics applications.

12-1. ELEMENTARY MATRIX CONCEPTS

A matrix is defined as a two-dimensional *array* of numbers. These numbers are arranged in *rows* and *columns* and the size of a matrix with m rows and n columns is designed as $(m \times n)$. In general, the following will be used to indicate matrices:

$$\mathbf{A} = [a_{ij}] = \begin{bmatrix} a_{11} & a_{12} & \cdot & \cdot & \cdot & a_{1n} \\ a_{21} & a_{22} & \cdot & \cdot & \cdot & a_{2n} \\ \cdot & \cdot & & \cdot & & \cdot \\ a_{m1} & a_{m2} & \cdot & \cdot & \cdot & a_{mn} \end{bmatrix} \qquad (12\text{-}1)$$

A matrix is normally denoted by a boldface capital letter such as $\mathbf{A}$. Many times it is desirable to refer to one of the individual *elements* which go together to make up the array, and the symbol a_{ij} would refer to the element located at the intersection of the ith row and the jth column. Sometimes the symbol $[a_{ij}]$ is assumed to imply the matrix itself when it is inconvenient to list all of the individual elements.

A matrix with one row $(1 \times n)$ or one column $(m \times 1)$ is termed a *row vector* or a *column vector*. Normally vectors are referred to by lowercase, boldface symbols such as $\mathbf{u}$.

It should be noted that the use of *vector* to denote row and column matrices is not in conflict with the conventional usage of vector in the physical sense to denote a quantity having magnitude and direction. A physical vector is uniquely determined by its three components along three coordinate axes and any ordered set of three elements (either a row or column matrix) can be used to denote the physical vector. A physical vector therefore is an example of a matrix vector.

Matrix mathematics is different from conventional mathematics. For convenience, it is often desirable, therefore, to set aside a term to describe all those quantities which follow the laws of ordinary algebra. These quantities we will henceforth refer to as *scalars*.

Once the concept of matrices or arrays is established, it is then possible to develop the definitions, identities, and manipulations necessary to handle these arrays in mathematical operations. It is not the purpose of this chapter to outline this development; instead, a summary of the most useful items is given. For complete details on the development of this matrix mathematics the reader should consult any of the typical textbooks on engineering mathematics such as Refs. 2-10. The Problems at the end of this chapter provide a brief review of elementary matrix mathematics and many of the aspects of the usage of matrix mathematics will be illustrated in the context of the remaining sections of this chapter. As a start, however, a few elementary items will be summarized.

A matrix in which all elements are zero is referred to as a *null matrix* or as a *zero matrix* $\mathbf{0}$. It is possible to form an $n \times m$ matrix from an $m \times n$ matrix by interchanging rows and columns. This new matrix is called the *transpose* and will be referred to as $\mathbf{A}^T$. A matrix in which $n = m$, i.e., the same number of rows as columns, is called a *square matrix*. (A determinant of order n may be thought of as a function of a square $n \times n$ matrix.)

A square matrix in which $a_{ij} = a_{ji}$ for all values of i and j is a *symmetric matrix*. A square matrix in which $a_{ij} = -a_{ji}$ and $a_{ii} = 0$ is a *skew-symmetric matrix*. The *principal diagonal* of a matrix is composed of the elements a_{ii}. A matrix in which all of the elements on one side of the principal diagonal are zero is a *triangular matrix*. A square matrix in which all of the elements not on the principal diagonal are zero is referred to as a *diagonal matrix*. A diagonal matrix in which all of the elements on the principal diagonal are 1 is a *unit matrix* $\mathbf{I}$.

The sum or difference of two matrices is the sum or difference of their individual elements:

$$\mathbf{C} = \mathbf{A} + \mathbf{B} \qquad \text{where } c_{ij} = a_{ij} + b_{ij} \qquad (12\text{-}2)$$

and

$$\mathbf{D} = \mathbf{A} - \mathbf{B} \qquad \text{where } d_{ij} = a_{ij} - b_{ij} \qquad (12\text{-}3)$$

If $\mathbf{A}$ and $\mathbf{B}$ do not each have the same number of rows and each have the same number of columns, addition and subtraction are not defined.

The product of a scalar w and a matrix $\mathbf{A}$ is a new matrix $\mathbf{B}$ the elements of which are obtained by multiplying each of the elements of $\mathbf{A}$ by w;

$$\mathbf{B} = w\,\mathbf{A} \qquad \text{where } b_{ij} = wa_{ij} \qquad (12\text{-}4)$$

The product of two matrices is normally defined in terms of the "dot" product:

$$\mathbf{C} = \mathbf{A} \cdot \mathbf{B} \qquad \text{where } c_{ij} = \sum_{k=1}^{n} a_{ik} b_{kj} \tag{12-5}$$

A and **B** must be *conformable* before this product is defined. The two matrices are conformable if the number of columns in **A** equals the number of rows in **B**. The product **C** will have the number of rows in **A** and the number of columns in **B**. An example of matrix multiplication might be as follows:

$$\begin{bmatrix} 1 & 1 & 1 \\ 2 & -1 & 0 \\ -1 & 0 & 2 \end{bmatrix} \begin{bmatrix} 1 & 2 & 3 & -1 \\ 3 & -1 & 1 & 0 \\ 0 & 0 & -2 & 1 \end{bmatrix}$$

$$= \begin{bmatrix} 4 & 1 & 2 & 0 \\ -1 & 5 & 5 & -2 \\ -1 & -2 & -7 & 3 \end{bmatrix}$$

A matrix has no numerical value but associated with any square matrix is a determinant which does have a numerical value. A very convenient way to determine this numerical value for a determinant is to expand the determinant into its *minors* by the Laplace expansion. Given a determinant $|A|$, each element a_{ij} in the determinant has a minor $|M_{ij}|$ which is a new determinant which is formed from $|A|$ by deleting the ith row and the jth column. Each element also has a *cofactor* $|A_{ij}|$ which is defined as

$$|A_{ij}| = (-1)^{i+j} |M_{ij}| \tag{12-6}$$

where $|A_{ij}|$ is the cofactor of element a_{ij} and $|M_{ij}|$ is the minor of element a_{ij}. The Laplace expansion to determine the value of a determinant is then defined in terms of these cofactors:

$$\sum_{j=1}^{n} a_{ji} |A_{jk}| = \sum_{j=1}^{n} a_{ij} |A_{kj}| = \begin{cases} |A| & \text{if } i=k \\ 0 & \text{if } i \neq k \end{cases} \tag{12-7}$$

If the determinant $|A|$ of a square matrix **A** of order n does not vanish, **A** is said to be *nonsingular* and possesses a *reciprocal or inverse matrix*. In order to determine this inverse matrix $\mathbf{A}^{-1}$ of a matrix **A** it is possible to calculate a *cofactor matrix*, each element of which is the numerical value of the cofactor for the corresponding element in **A**. The transpose $\mathbf{A}^T$ of this cofactor matrix is the *adjoint matrix* adj**A** of **A**. The reciprocal or inverse matrix is then defined by

$$\mathbf{A}^{-1} = \frac{\text{adj } \mathbf{A}}{|A|} \tag{12-8}$$

This inverse matrix has the property

$$\mathbf{A}^{-1} \cdot \mathbf{A} = \mathbf{I} \tag{12-9}$$

where $\mathbf{I}$ is the unit matrix.

These tools make it possible to use matrices to solve large systems of simultaneous equations of the form

$$
\begin{aligned}
a_{11}x_1 + a_{12}x_2 + \cdots + a_{1n}x_n &= b_1 \\
a_{21}x_1 + a_{22}x_2 + \cdots + a_{2n}x_n &= b_2 \\
&\ \ \vdots \\
a_{n1}x_1 + a_{n2}x_2 + \cdots + a_{nn}x_n &= b_n
\end{aligned}
\tag{12-10}
$$

These may be expressed in matrix format as

$$\mathbf{Ax} = \mathbf{b} \tag{12-11}$$

where

$$
\mathbf{A} = \begin{bmatrix}
a_{11} & a_{12} & \cdot & \cdot & \cdot & a_{1n} \\
a_{21} & a_{22} & \cdot & \cdot & \cdot & a_{2n} \\
\cdot & \cdot & \cdot & \cdot & \cdot & \cdot \\
\cdot & \cdot & \cdot & \cdot & \cdot & \cdot \\
\cdot & \cdot & \cdot & \cdot & \cdot & \cdot \\
a_{n1} & a_{n2} & \cdot & \cdot & \cdot & a_{nn}
\end{bmatrix}
$$

$$
\mathbf{x} = \begin{bmatrix}
x_1 \\
x_2 \\
\cdot \\
\cdot \\
\cdot \\
x_n
\end{bmatrix}
\qquad
\mathbf{b} = \begin{bmatrix}
b_1 \\
b_2 \\
\cdot \\
\cdot \\
\cdot \\
b_n
\end{bmatrix}
$$

The solution vector $\mathbf{x}$ may then be determined:

$$\mathbf{Ax} = \mathbf{b} \tag{12-12}$$
$$\mathbf{A}^{-1}\mathbf{Ax} = \mathbf{A}^{-1}\mathbf{b}$$
$$\mathbf{Ix} = \mathbf{A}^{-1}\mathbf{b}$$
$$\mathbf{x} = \mathbf{A}^{-1}\mathbf{b} \tag{12-13}$$

It is also possible to carry out the operations of differentiation and integration of matrices. If the elements of a matrix $[a_{ij}(t)]$ are all functions of a scalar variable t, the matrix is said to be a *matrix function* of t. The elements of the *derivative matrix* $\dfrac{d[a_{ij}(t)]}{dt}$ are the derivatives of the corresponding elements in $[a_{ij}(t)]$.

Similarly, the integral $\int |a_{ij}(t)| \, dt$ is obtained by integrating each of the corresponding elements.

It is sometimes convenient to regard matrices as being constructed from elements that are submatrices or minor elements of matrices. As an example:

$$\mathbf{A} = \begin{bmatrix} & \vdots & \\ & \vdots & \\ & \vdots & \\ 19 & \vdots & 3 \\ & \vdots & \\ 28 & \vdots & 4 \\ & \vdots & \\ \cdots & \cdots \cdots & \cdots \\ & \vdots & \\ 62 & \vdots & 7 \\ & \vdots & \\ & \vdots & \end{bmatrix}$$

A may then be considered as

$$\mathbf{A} = \begin{bmatrix} \mathbf{P} & \mathbf{Q} \\ \mathbf{R} & \mathbf{S} \end{bmatrix}$$

where

$$\mathbf{P} = \begin{bmatrix} 19 \\ 28 \end{bmatrix} \qquad \mathbf{Q} = \begin{bmatrix} 3 \\ 4 \end{bmatrix} \qquad \mathbf{R} = [62] \qquad \mathbf{S} = [7]$$

This property of matrices is especially helpful for inverting large matrices. Consider the matrix **A** of order n whose inverse $\mathbf{A}^{-1}$ is desired. Partition this into four submatrices:

$$\mathbf{A} = \begin{bmatrix} \mathbf{A}_{11}(r,r) & \mathbf{A}_{12}(r,s) \\ \mathbf{A}_{21}(s,r) & \mathbf{A}_{22}(s,s) \end{bmatrix}$$

where the letters in parentheses refer to the number of rows and columns in each submatrix. (Note that $n = s + r$.) If we define $\mathbf{A}^{-1}$ as **B** and partition **B**

$$\mathbf{A}^{-1} = \mathbf{B} = \begin{bmatrix} \mathbf{B}_{11}(r,r) & \mathbf{B}_{12}(r,s) \\ \mathbf{B}_{21}(r,s) & \mathbf{B}_{22}(s,s) \end{bmatrix}$$

Then

$$\mathbf{B} \cdot \mathbf{A} = \mathbf{I}$$

and

$$\mathbf{B}_{11} \mathbf{A}_{11} + \mathbf{B}_{12} \mathbf{A}_{21} = \mathbf{I}_r$$
$$\mathbf{B}_{11} \mathbf{A}_{12} + \mathbf{B}_{12} \mathbf{A}_{22} = \mathbf{O}$$
$$\mathbf{B}_{21} \mathbf{A}_{11} + \mathbf{B}_{22} \mathbf{A}_{21} = \mathbf{O}$$
$$\mathbf{B}_{21} \mathbf{A}_{12} + \mathbf{B}_{22} \mathbf{A}_{22} = \mathbf{I}_s$$

Define

$$\mathbf{X} = \mathbf{A}_{11}^{-1} \mathbf{A}_{12} \qquad \mathbf{Y} = \mathbf{A}_{21} \mathbf{A}_{11}^{-1}$$
$$\mathbf{Z} = \mathbf{A}_{22} - \mathbf{Y} \mathbf{A}_{12} = \mathbf{A}_{22} - \mathbf{A}_{21} \mathbf{X}$$

Then

$$\mathbf{B}_{11} = \mathbf{A}_{11}^{-1} + \mathbf{X} \mathbf{Z}^{-1} \mathbf{Y}$$
$$\mathbf{B}_{12} = - \mathbf{X} \mathbf{Z}^{-1}$$
$$\mathbf{B}_{21} = - \mathbf{Z}^{-1} \mathbf{Y}$$
$$\mathbf{B}_{22} = \mathbf{Z}^{-1}$$

These then determine $\mathbf{B}$ which is $\mathbf{A}^{-1}$.

Another convenient way to determine inverse matrices (especially for pencil and paper calculations) is to take $\mathbf{A}$ and perform the necessary operations to transform $\mathbf{A}$ into the unit matrix $\mathbf{I}$. If these same operations are applied to the unit matrix $\mathbf{I}$ they will convert $\mathbf{I}$ into $\mathbf{A}^{-1}$.

Many other matrix operations are also possible and some of these will be further illustrated in subsequent sections of this chapter and in the problems at the end of the chapter.

12-2. SIMPLE ARRAY MODELS

As a simple example of how a physical system's math model might be presented in matrix format, consider the material balance on a plant to produce nitric acid by the oxidation of ammonia:

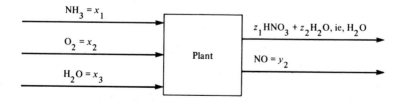

z is in mole fractions and x and y are expressed in moles per unit time. The material balance on oxygen and hydrogen may be expressed[1] as

$$
\begin{bmatrix} y_1 \\ \\ \\ y_2 \end{bmatrix} = \begin{bmatrix} 0 & \dfrac{2}{3z_1 + z_2} & \dfrac{1}{3z_1 + z_2} \\ \\ \dfrac{3}{z_1 + 2z_2} & 0 & \dfrac{2}{z_1 + 2z_2} \end{bmatrix} \begin{bmatrix} x_1 \\ \\ x_2 \\ \\ x_3 \end{bmatrix} \qquad (12\text{-}14)
$$

or $y = Ax$. The matrix A represents the plant operation. All chemical processing plants may be expressed in this simple matrix equation which an operator matrix A, a product vector y, and a feed vector x. Energy balances and efficiencies also may be included.

Another simple example of the use of matrices is in terms of helping to analyze large amounts of data such as in bubble-point calculations [11]. The calculations are of the form

$$
y_i = \sum_{j=1}^{m} K_{ij} x_j = \frac{1}{P} \sum_{j=1}^{m} f_{ij} x_j \qquad (12\text{-}15)
$$

where x_j = mole fraction of component j in liquid phase
      f_{ij} = fugacity function of component j at temperature i
      K_{ij} = vapor liquid equilibrium constant of component j at temperature i
      P = pressure
      y_i = vapor-phase mole fractions at temperature i

These calculations may be placed in the matrix format

$$
P\,Y = FX \qquad (12\text{-}16)
$$

where for normal hydrocarbons the matrices would be defined as in Fig. 12-1 for k cases (in this example $k = 3$) of assumed composition X, at three assumed temperatures, and with $P = 17$ atm. The product $P\,Y$ may be calculated and the elements Py_{ik} in each column in $P\,Y$ can be plotted versus the temperature of each row to obtain the bubble point where $y_{ik} = 1$ for case k.

These examples illustrate the use of matrix models for simple systems. It is now desirable to extend these concepts more fully to some complex physical examples.

Fugacity matrix @ $P = 17$ atm, **F**

Temperature, °F (i)	Component (j)											
	C_1	C_2^-	C_2	C_3^-	C_3	iC_4	nC_4	iC_4^-	nC_4^-	$C_4^=$	C_5	C_6
150	—	72.5	54.0	24.0	21.0	11.5	9.0	10.3	8.5	9.9	3.8	1.7
200	—	89.0	68.0	33.5	29.5	17.5	14.3	15.9	13.7	15.7	7.0	3.4
250	—	102	83.0	43.0	39.0	24.0	20.3	22.1	19.5	22.3	11.0	6.0

Composition, matrix **X**

Component (i)	Case No., (k)		
	1	2	3
C_1	0	0	0
C_2^-	0	0	0
C_2	0.001	0.003	0.005
C_3^-	0.010	0.025	0.040
C_3	0.025	0.042	0.060
iC_4 •	0.364	0.430	0.405
nC_4	0.350	0.370	0.300
iC_4^-	0	0	0
nC_4^-	0	0	0
$C_4^=$	0	0	0
C_5	0.250	0.130	0.190
C_6	0	0	0

Temperature, °F (i)	Case No., (k)		
	1	2	3
150	9.11	10.4	10.6
= 200	14.3	16.0	16.2
250	20.1	22.2	22.4

FIG. 12·1. Typical bubble-point calculation matrices [11].

12-3. MULTICOMPONENT DISTILLATION[11,14]

Multicomponent distillation calculations is one area in which extensive work has been done in matrix-type formulations.   For the steady state case the problem can be visualized as shown in Fig. 12-2 and 12-3.   The nomenclature is as follows:

j = stage number
i = component number
m = total number of components
x_{ij} = liquid mole fraction for ith component leaving jth stage
y_{ij} = vapor mole fraction for ith component leaving jth stage
z_{ij} = mole fraction of ith component entering jth stage
T_j = temperature of jth stage, °F
V_j = vapor stream leaving jth stage, mole/hr
L_j = liquid stream leaving jth stage, mole/hr
F_j = feed stream at jth stage, mole/hr
W_j = vapor side stream leaving jth stage, mole/hr
U_j = liquid side stream from jth stage, mole/hr
Q_j = heat removed from jth stage, Btu/hr
B = bottom product, mole/hr
D = overhead distillate product, mole/hr
Q_B = heat duty of reboiler, Btu/hr
Q_D = heat duty of condenser, Btu/hr
K_{ij} = equilibrium ratio for component i on jth stage, $K_{ij} = y_{ij}/x_{ij}$

The component material balance equations for each stage are then as follows:

$$L_{j-1}x_{ij-1} + F_j z_{ij} + V_{j+1}y_{ij+1} - (V_j + W_j)y_{ij} - (L_j + U_j)x_{ij} = 0 \qquad (12\text{-}17)$$

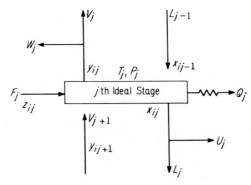

FIG. 12-2.   Ideal multicomponent equilibrium stage [12].

The equilibrium expression is

$$K_{ij} = y_{ij}/x_{ij} \tag{12-18}$$

The overall material balance of all stages from the condenser through the jth stage is

$$L_j = V_{j+1} + \sum_{k=2}^{j} (F_k - W_k - U_k) - D \qquad 2 \le j \le n-1 \tag{12-19}$$

where $D = V_1 + U_1$.

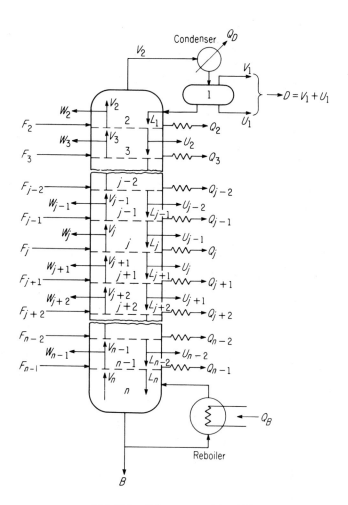

FIG. 12-3. Column nomenclature [12].

Equations 12-17, 12-18, and 12-19 may be combined:

$$B_1 x_{i1} + C_1 x_{i2} = D_1 \tag{12-20}$$

$$A_j x_{i,j-1} + B_j x_{ij} + C_j x_{i,j+1} = D_j \qquad 2 \le j \le n-1 \tag{12-21}$$

$$A_n x_{i,n-1} + B_n x_{in} = D_n \tag{12-22}$$

where

$$B_1 = -(V_1 K_{i1} + U_1)$$

$$C_1 = V_2 K_{i2}$$

$$D_1 = 0$$

$$A_j = L_{j-1} = V_j + \sum_{k=2}^{j-1} (F_k - W_k - U_k) - D \qquad 2 \le j \le n-1$$

$$B_j = -[(V_j + W_j)K_{ij} + (L_j + U_j)] \qquad 2 \le j \le n-1$$

$$= -[(V_j - W_j)K_{ij} + V_{j+1} + \sum_{k=2}^{j} (F_k - W_k - U_k) - D + U_j]$$

$$C_j = V_{j+1} K_{i,j+1} \qquad 2 \le j \le n-1$$

$$A_n = V_n + B$$

$$B_n = -(V_n K_{in} + B)$$

$$D_n = 0$$

In matrix notation Eqs. 12-20 through 12-22 become

$$
\begin{bmatrix}
B_1 & C_1 & & & & & \\
A_2 & B_2 & C_2 & & & & \\
 & & \cdot & \cdot & & & \\
 & & \cdot & \cdot & & & \\
 & & \cdot & \cdot & & & \\
 & & A_j & B_j & C_j & & \\
 & & & \cdot & \cdot & & \\
 & & & \cdot & \cdot & & \\
 & & & \cdot & \cdot & & \\
 & & & & A_n & B_n
\end{bmatrix}
\begin{bmatrix}
x_{i1} \\
x_{i2} \\
\cdot \\
\cdot \\
\cdot \\
x_{ij} \\
\cdot \\
\cdot \\
\cdot \\
x_{in}
\end{bmatrix}
=
\begin{bmatrix}
D_1 \\
D_2 \\
\cdot \\
\cdot \\
\cdot \\
D_j \\
\cdot \\
\cdot \\
\cdot \\
D_n
\end{bmatrix}
\tag{12-23}
$$

Inherent in the above formulation are summation equations of the form

$$\sum_{i=1}^{m} y_{ij} = 1.0 \tag{12-24}$$

$$\sum_{i=1}^{m} x_{ij} = 1.0 \tag{12-25}$$

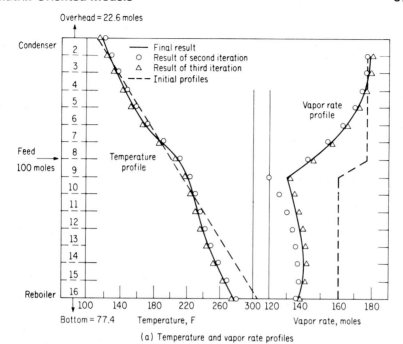

(a) Temperature and vapor rate profiles

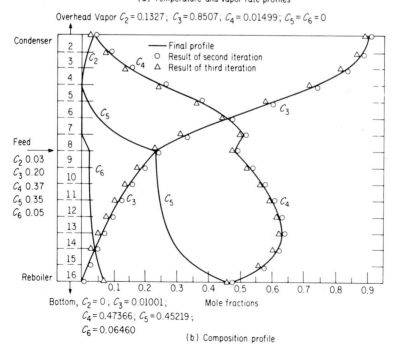

(b) Composition profile

FIG. 12-4.   Some typical results from solution of distillation model [12].   (a) Temperature
and vapor rate profiles.   (b) Composition profile.

It is also possible to formulate an enthalpy balance for each stage, but this is left as an exercise for the student.   The sum total of Eqs. 12-17, 12-18, 12-24, 12-25, and these enthalpy equations represents $n(m + 2)$ equations and $n(m + 2)$ independent variables.   An iterative solution technique for these is proposed in Ref. 12 and some typical results are shown in Fig. 12-4.

This general approach to model formulation can be extended to many similar stagewise examples, but it is better to consider, instead, the possibilities in formulating dynamic models.

12-4.   MATRIX MODEL OF A DYNAMIC DISTILLATION SYSTEM

Considerable work has been done using various lumped-parameter models of a plate-distillation column to yield calculated responses that closely approximate the internal and external dynamic response of an actual column [16-21].   From this work, a reasonable model seems to be one that considers each plate to be a perfectly mixed pool, each downcomer to be a perfectly mixed pool, and both the reboiler and condenser to be perfectly mixed also.   In Fig. 12-5 a simple column is illustrated to show the numbering scheme.   Note that each pool is numbered in the direction of liquid flowing from the top of the column.

In addition to the above assumptions, the following are also made for simplicity and not because of limitations of the general method:

1. The feed liquid falls on the plate below the feed point, and the vapor portion enters the plate above.
2. No vapor-liquid contact occurs in the downcomer.
3. A partial condenser is used.
4. The pressure is constant throughout the column.
5. Delay times in the lines to and from the condenser and reboiler are negligible.
6. Each plate, the reboiler, and the condenser are assumed to be ideal stages.
7. No heat is lost to the surroundings.
8. The components have equal latent heats of vaporization.
9. Mixtures are ideal.

Any method can be used to calculate the vapor-liquid equilibrium constants, and the Antoine equation is convenient for doing this because the coefficients for the equation are readily available for a variety of substances.   For simplicity, the equations describing the system are developed for binary mixtures.   The concentration of the light component is calculated from the equations, and the concentration of the other component is determined by difference.   When more than

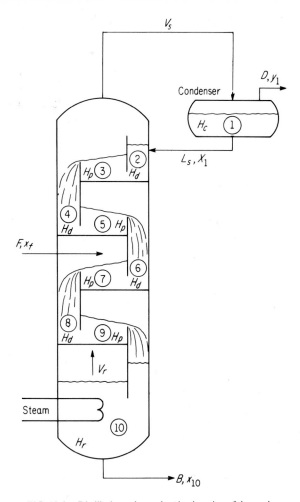

FIG. 12-5. Distillation column showing location of the pools.

two components are present, the concentration of each component is calculated individually and usually must be adjusted subsequently so that the sum of mole fractions is unity for every pool.

To develop the equations to describe the column in Fig. 12-5 the condenser is considered first. Making an unsteady-state material balance for the light component,

$$V_s y_3 - L_s x_1 - D y_1 = d(H_c x_1)/dt \qquad (12\text{-}26)$$

Since the vapor leaving the condenser is in equilibrium with the liquid,

$$y_1 = K_1 x_1 \qquad (12\text{-}27)$$

If the holdup of the condenser is considered to be constant,

$$L_s = V_s - D \qquad (12\text{-}28)$$

Substituting Eqs. 12-27 and 12-28 into Eq. 12-26 and rearranging;

$$x_1 = [(D - DK_1 - V_s)/H_c]x_1 + V_s K_3 x_3 \qquad (12\text{-}29)$$

Considering pool 2, an unsteady-state material balance around the light component yields

$$L_s x_1 - L_s x_2 = d(H_a x_2)/dt \qquad (12\text{-}30)$$

Rearranging and substituting Eq. 12-28 into Eq. 12-30,

$$\dot{x}_2 = [(V_s - D)/H_d]x_1 - [(V_s - D)/H_d]x_2 \qquad (12\text{-}31)$$

In Fig. 12-6 the ten differential equations required to describe the column in Fig. 12-5 are presented.   Note that the first two equations are identical to Eqs. 12-29 and 12-31, respectively.   All the other equations are developed in a similar manner as these two.

Upon examination of Fig. 12-6, these equations may be expressed by the following vector-matrix differential equation:

$$\dot{\mathbf{x}} = \mathbf{A}\,\mathbf{x} + \mathbf{b} \qquad (12\text{-}32)$$

In this equation, $\mathbf{A}$ is a 10×10 matrix, and $\mathbf{x}$ and $\mathbf{b}$ are 10×1 vectors.   The elements of $\mathbf{A}$ and $\mathbf{b}$ are defined in Fig. 12-7, where the a's and b's correspond to the analogous coefficients in Fig. 12-6.   The first subscript of each a corresponds

$$\dot{x}_1 = \left[(D - K_1 D - V_s)/H_c\right] x_1 + (V_s K_3/H_c)\, x_3$$

$$\dot{x}_2 = \left[(V_s - D)/H_d\right] x_1 - \left[(V_s - D)/H_d\right] x_2$$

$$\dot{x}_3 = \left[(V_s - D)/H_p\right] x_2 + \left[(D - V_s - K_3 V_s)/H_p\right] x_3 + (V_s K_5/H_p)\, x_5$$

$$\dot{x}_4 = \left[(V_s - D)/H_d\right] x_3 - \left[(V_s - D)/H_d\right] x_4$$

$$\dot{x}_5 = \left[(V_s - D)/H_p\right] x_4 + \left[(D - V_s - K_5 V_s)/H_p\right] x_5 + (V_f K_7/H_p)\, x_7$$
$$\qquad + (x_f V_f K_f/H_p)$$

$$\dot{x}_6 = \left[(V_s - D)/H_d\right] x_5 - \left[(V_s - D)/H_d\right] x_6$$

$$\dot{x}_7 = \left[(V_s - D)/H_p\right] x_6 - \left[(B + V_r + V_r K_7)/H_p\right] x_7 + (V_r K_9/H_p) x_9$$
$$\qquad \left[+ (F - V_f) x_f H_p\right]$$

$$\dot{x}_8 = \left[(V_r + B)/H_d\right] x_7 - \left[(V_r + B)/H_d\right] x_8$$

$$\dot{x}_9 = \left[(V_r + B)/H_p\right] x_8 - \left[(B + V_r + K_9 V_r)/H_p\right] x_9 + (V_r K_{10}/H_p) x_{10}$$

$$\dot{:} \ = \left[(V_r + B)/H_r\right] x_9 - \left[(B + V_r K_{10})/H_r\right] x_{10}$$

FIG. 12-6. Set of differential equations that describe the column
in Fig. 12-5.

$$\dot{x}_1 = a_{1,1} \; x_1 + a_{1,3} \; x_3$$

$$\dot{x}_2 = a_{2,1} \; x_1 + a_{2,2} \; x_2$$

$$\dot{x}_3 = a_{3,2} \; x_2 + a_{3,3} \; x_3 + a_{3,5} \; x_5$$

$$\dot{x}_4 = a_{4,3} \; x_3 + a_{4,4} \; x_4$$

$$\dot{x}_5 = a_{5,4} \; x_4 + a_{5,5} \; x_5 + a_{5,7} \; x_7 + b_5$$

$$\dot{x}_6 = a_{6,5} \; x_5 + a_{6,6} \; x_6$$

$$\dot{x}_7 = a_{7,6} \; x_6 + a_{7,7} \; x_7 + a_{7,9} \; x_9 + b_7$$

$$\dot{x}_8 = a_{8,7} \; x_7 + a_{8,8} \; x_8$$

$$\dot{x}_9 = a_{9,8} \; x_8 + a_{9,9} \; x_9 + a_{9,10} \; x_{10}$$

$$\dot{x}_{10} = a_{10,9} \; x_9 + a_{10,10} \; x_{10}$$

FIG. 12-7. Definition of nonzero elements in the matrices in Eq. 12-32.

to its row number in the **A** matrix, and the second subscript corresponds to its column number. For both **A** and **b** matrices, all elements not defined in Fig. 12-7 are zero. The **x** matrix is defined as

$$\mathbf{x} = \begin{bmatrix} x_1 \\ x_2 \\ \cdot \\ \cdot \\ \cdot \\ x_{10} \end{bmatrix}$$

Column Characteristics

Number of plates	10
Location of feed plate	5 from top
Condenser holdup	10 moles
Reboiler holdup	5 moles
Plate holdup	0.5 moles
Downcomer holdup	0.25 moles
Holdup at reflux addition point	0.25 moles

Components

The feed is a mixture of propane and butane. The constants for the Antoine equation are taken from Ref. 22.

Initial Conditions

Reflux ratio (external)	1.9
Fraction vapor in feed	0.33 (mole basis)
Overheads rate	60 mole/hr
Feed rate (total)	100 mole/hr
Mole fraction C_3 in feed	0.6

FIG. 12-8. Description of the column and the initial conditions for which the response for various changes is determined.

Using a matrix differential equation analogous to the one in Eq. 12-32, the response of the column whose size and initial conditions are given in Fig. 12-8 can be calculated.   Figures 12-9 and 12-10 show the response to a change in the reflux ratio and feed composition, respectively.

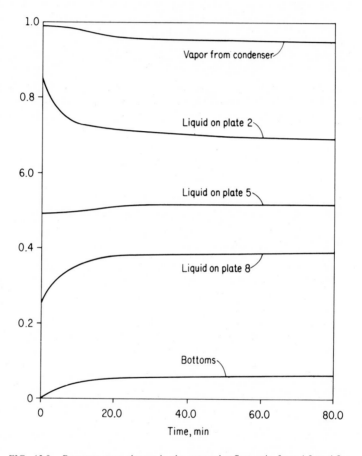

FIG. 12-9.   Response to a change in the external reflux ratio from 1.9 to 1.0.

As revealed in the exercises several different systems can be described by the vector-matrix differential equation in 12-32, thus permitting some continuity of solution procedures from one system to the next.

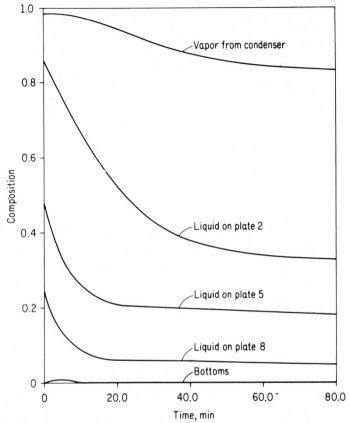

FIG. 12-10. Response to a change in feed composition from 0.6 mole fraction of more volatile component to 0.5.

12-5. SOLUTION TECHNIQUES FOR MATRIX DIFFERENTIAL EQUATIONS

In the previous section and in the Problems, matrix differential equations are developed to describe physical systems. In this section solution techniques applicable to such equations are described briefly. Specifically, attention is focused on the solution of the differential equation

$$\dot{\mathbf{x}} = \mathbf{A}\mathbf{x} + \mathbf{b} \tag{12-33}$$

First, consider the steady-state solution, in which case Eq. 12-33 reduces to

$$\mathbf{A}\mathbf{x} = -\mathbf{b}$$

Solving for $\mathbf{x}$,

$$\mathbf{x} = -\mathbf{A}^{-1}\,\mathbf{b} \tag{12-34}$$

where A^{-1} is the inverse of matrix A. Thus, the steady state solution can be obtained simply by matrix inversion. However, complications arise when the elements of A are functions of x, which is the case for a distillation column. To treat this situation an iterative procedure is required.

Now consider the unsteady-state solution of Eq. 12-33. This equation is the vector-matrix equivalent of the ordinary differential equation

$$\dot{x}(t) = ax(t) + b(t) \tag{12-35}$$

which is known to have the solution

$$x(t) = x(0)e^{at} + \int_0^t e^{a(t-\tau)} b(\tau)d\tau \tag{12-36}$$

We will show that the vector-matrix differential equation has a similar solution.

Although there are more rigorous approaches to the solution of Eq. 12-33, we shall use Laplace transforms to justify our conclusions. As we have shown that the integral of a matrix is simply the integral of each element, it follows that an integral transform (specifically the Laplace transform) of a matrix (or vector) is the Laplace transform of the elements. Taking the Laplace transform of Eq. 12-33 gives

$$s \, \mathbf{x} \, (s) - \mathbf{x} \, (0) = \mathbf{A} \mathbf{x} \, (s) + \mathbf{b} \, (s)$$

Solving for $\mathbf{x}$ (s):

$$\mathbf{x}(s) = [s \, \mathbf{I} - \, \mathbf{A}]^{-1} \mathbf{x} \, (0) + [s \, \mathbf{I} - \, \mathbf{A}]^{-1} \mathbf{b} \, (s)$$

To take the inverse transform we need the inverse Laplace transform of $[s\mathbf{I} - \mathbf{A}]^{-1}$. This inverse is customarily called the *transition matrix*, $\varphi \, (t)$, whose properties we will investigate shortly. Now taking the inverse transform of the above equation,

$$\mathbf{x}(t) = \varphi \, (t)\mathbf{x}(0) + \int_0^t \varphi \, (t - \tau)\mathbf{b}(\tau)d\tau \tag{12-37}$$

Note the similarity to Eq. 12-36 when e^{at} is the counterpart to $\varphi \, (t)$.

By analogy to the scalar case, the transition matrix is customarily devoted by the matrix exponential $e^{\mathbf{A}t}$, that is,

$$\varphi \, (t) = e^{\mathbf{A}t}$$

This function retains many of the properties of the scalar exponential such as
    (a) Exponential of the null matrix $\mathbf{O}$

$$e^{\mathbf{O}t} = \mathbf{I}$$

    (b) Derivative

$$\frac{d}{dt} \, e^{\mathbf{A}t} = \mathbf{A}e^{\mathbf{A}t}$$

(c) Product

$$e^{\mathbf{A}t_1} \, e^{\mathbf{A}t_2} = e^{\mathbf{A}(t_1 + t_2)}$$

(d) Series representation

$$e^{\mathbf{A}t} = \mathbf{I} + \mathbf{A}t + \frac{\mathbf{A}^2 t^2}{2} + \cdots = \sum_{k=0}^{\infty} (\mathbf{A}t)^k / k!$$

There are other important properties of the matrix exponential, but these are sufficient for our purposes.

The numerical computation of $e^{\mathbf{A}t}$ should not be taken lightly. One approach is to evaluate the series given above, and this is computationally feasible. For small matrices the inverse Laplace transform of $[s\mathbf{I} - \mathbf{A}]^{-1}$ can be obtained, but this is not very amenable to machine computation. A third way which we shall only mention is to use Sylvester's expansion theorem [5]. A fourth approach that we shall discuss in some detail is via the reduction to *canonical form* using *eigenvectors*.

Before discussing the method, we must first define *eigenvalues*. Let $\mathbf{x}$ and $\mathbf{y}$ be nth-order vectors and $\mathbf{M}$ be an nth-order square matrix (called a *transformation matrix*) which are related as follows:

$$\mathbf{y} = \mathbf{M} \, \mathbf{x}$$

Now suppose we require that $\mathbf{y}$ be a scalar multiple of $\mathbf{x}$, or

$$\mathbf{y} = \lambda \, \mathbf{x}$$

where λ is a scalar. Substituting gives

$$\lambda \, \mathbf{x} = \mathbf{M} \, \mathbf{x} \qquad (12\text{-}38)$$

For a specified matrix $\mathbf{M}$, there will generally be only n values of λ, say λ_i, and r corresponding values of $\mathbf{x}$, say $\mathbf{x}_i$, for which this equation will hold. These values of λ_i are called the *eigenvalues,* and the corresponding vectors $\mathbf{x}_i$ are called the *eigenvectors.*

To calculate the λ_i's and corresponding $\mathbf{x}_i$'s, first rearrange Eq. 12-38 to obtain

$$[\lambda \mathbf{I} - \mathbf{M}] \, \mathbf{x} = \mathbf{0}$$

For this relationship to hold the determinant of $[\lambda \mathbf{I} - \mathbf{M}]$ must equal zero, or

$$\left| \lambda \mathbf{I} - \mathbf{M} \right| = 0$$

Expanding this determinant gives a polynomial in λ, i.e., $f(\lambda)$, which is called the *characteristic function* of $\mathbf{M}$. The roots of this characteristic function are, of course, the eigenvalues.

In calculating the transition matrix, the eigenvalues and eigenvectors determined above will be used to diagonalize matrix **A** in Eq. 12-33. Introducing the transformation $\mathbf{x} = \mathbf{P}\,\mathbf{y}$ into Eq. 12-33 gives

$$\mathbf{P}\,\dot{\mathbf{y}} = \mathbf{A}\,\mathbf{P}\,\mathbf{y} + \mathbf{b}$$

or

$$\dot{\mathbf{y}} = \mathbf{P}^{-1}\mathbf{A}\mathbf{P}\mathbf{y} + \mathbf{P}^{-1}\mathbf{b} = \lambda\mathbf{y} + \mathbf{c} \qquad (12\text{-}39)$$

We now need to find the matrix **P** such that the product $\mathbf{P}^{-1}\mathbf{A}\mathbf{P} = \lambda$ gives a diagonal matrix.

If no two eigenvalues are equal, the matrix **P** that accomplishes this is simply composed of the eigenvectors as columns, i.e.,

$$\mathbf{P} = \left[\begin{array}{cccc} \mathbf{x}_1 & \vdots & \mathbf{x}_2 & \cdots & \mathbf{x}_n \end{array}\right] \qquad (12\text{-}40)$$

Under this transformation, the matrix $\lambda = \mathbf{P}^{-1}\mathbf{A}\mathbf{P}$ is a diagonal matrix with the eigenvalues on the diagonal, i.e.,

$$\lambda = \begin{bmatrix} \lambda_1 & & & & 0 \\ & \lambda_2 & & & \\ & & \cdot & & \\ & & & \cdot & \\ & & & & \cdot \\ 0 & & & & \lambda_n \end{bmatrix} \qquad (12\text{-}41)$$

The matrix exponential $e^{\lambda t}$ is also a diagonal matrix with exponentials of the eigenvalues on the diagonal, i.e.,

$$e^{\lambda t} = \begin{bmatrix} e^{\lambda_1 t} & & & & 0 \\ & e^{\lambda_2 t} & & & \\ & & \cdot & & \\ & & & \cdot & \\ & & & & \cdot \\ 0 & & & & e^{\lambda_n t} \end{bmatrix} \qquad (12\text{-}42)$$

It is left as an exercise to show that

$$e^{\mathbf{A}t} = \mathbf{P}e^{\lambda t}\mathbf{P}^{-1} \qquad (12\text{-}43)$$

This calculates the matrix exponential.

Equation 12-39 is called the *canonical form* of the state equation. To show why this form of the equation is of special importance, consider writing Eq. 12-39 for $n = 2$:

$$\frac{d}{dt}\begin{bmatrix} y_1 \\ y_2 \end{bmatrix} = \begin{bmatrix} \lambda_1 & 0 \\ 0 & \lambda_2 \end{bmatrix}\begin{bmatrix} y_1 \\ y_2 \end{bmatrix} + \begin{bmatrix} c_1(t) \\ c_2(t) \end{bmatrix} \qquad (12\text{-}44)$$

Note that the two states y_1 and y_2 are completely independent, and the two first-order scalar equations obtained from the canonical form of the equation are completely independent:

$$\dot{y}_1 = \lambda_1 y_1 + c_1(t) \tag{12-45}$$

$$\dot{y}_2 = \lambda_2 y_2 + c_2(t) \tag{12-46}$$

Note that only the vector **c** influences the states. If any element of **c** is zero, then this state variable is said to be *uncontrollable*.

Many characteristics, e.g., stability of systems, can be determined by examination of the state equation, but the purpose of this portion of the text has been only to introduce the subject. An excellent discussion can be found in Ref. 29.

12-6. MATRIX FORMULATION FROM PARTIAL DIFFERENTIAL EQUATIONS [25]

A large number of chemical process devices are characterized by partial differential equation, few of which are analytically solvable. Extensive work has been done on approximate solution methods and numerical methods, the more difficult problem being finite-boundary situations. Problems without analytic solutions have been approximated; e.g., by cascaded tank approximations, but these methods are limited and provides no a priori measure of error. This approach has been applied to a tubular device with longitudinal dispersion [26], which is represented by the Taylor diffusion model:

$$\frac{\partial x}{\partial \theta} = \frac{1}{Pe_L} \frac{\partial^2 x}{\partial z^2} - \frac{\partial x}{\partial z} \tag{12-47}$$

x = dimensionless normalized potential
θ = dimensionless time
z = dimensionless space $0 \leq z \leq 1.0$
Pe_L = Peclet number for longitudinal dispersion = $\dfrac{(v)L}{D_1}$

This equation is a practical example for which the solution technique to be developed is applicable. However it will be developed for a more general equation, and applied to the above equation subsequently.

The partial differential equation to be examined is

$$a \frac{\partial^2 x}{\partial z^2} + b \frac{\partial x}{\partial z} + cx = \frac{\partial x}{\partial t} \tag{12-48}$$

defined on

$$0 \leq z \leq 1.0$$
$$0 \leq x \leq 1.0$$
$$0 \leq t$$
$$a, b, c \text{ real}$$

The equation is transformed to a set of n ordinary differential difference equations by means of the usual Taylor expansions:

$$\frac{\partial x}{\partial z} = \frac{x_i - x_{i-1}}{\Delta z} + 0(\Delta z)$$

$$\frac{\partial^2 x}{\partial z^2} = \frac{x_{i+1} - 2x_i + x_{i-1}}{\Delta z^2} + 0(\Delta z^2)$$

Before deleting the error term, the implications of its contribution will be examined.

$$0(\Delta z) \sim \frac{\partial^2 x}{\partial z^2} \frac{\Delta z}{2!}$$

$$\sim \frac{1}{a}\left(\frac{\partial x}{\partial t} - b\frac{\partial x}{\partial z} - cx\right)\frac{\Delta z}{2!} \tag{12-49}$$

The error will rise rapidly with decreasing a, which can only be compensated by decreasing Δz. It appears then that maintaining $\Delta z/a$ constant should provide a relative measure of error in moving from one system to another, and should also indicate the number of increments required.

The error term is dropped and the respective substitutions made. The matrix differential equations that result will contain the physical boundary conditions. In this case, it is assumed that the value of a is zero when outside the range $0 \le z \le 1$. In the Taylor dispersion model this corresponds to no diffusion into or from the system at $z = 0$ and $z = 1$, a difficult boundary condition to formulate analytically. If we let x_0 be the value of x for $z < 0$, the first finite-difference equation for Eq. 12-48 is

$$a\left(\frac{x_2 - x_1}{\Delta z^2}\right) + b\left(\frac{x_1 - x_0}{\Delta z}\right) + cx_1 = \frac{dx_1}{dt}$$

If we define n as $1/\Delta z$, then

$$n^2 ax_2 - n^2 ax_1 + nbx_1 - nbx_0 + cx_1 = \frac{dx_1}{dt}$$

This equation can be rearranged to

$$n[\alpha x_2 - (\alpha - \beta - \delta)x_1] - n\beta x_0 = \frac{dx_1}{dt} \tag{12-50}$$

where $\alpha = an$
$\quad\quad\quad \beta = b$
$\quad\quad\quad \delta = c/n$

Similarly, the second difference equation is

$$a\left(\frac{x_3 - 2x_2 + x_1}{\Delta z^2}\right) + b\left(\frac{x_2 - x_1}{\Delta z}\right) + cx_2 = \frac{dx_2}{dt}$$

Making the same substitutions as above yields

$$n[\alpha x_3 - (2\alpha - \beta - \delta)x_2 + (\alpha - \beta)x_1] = \frac{dx_2}{dt} \qquad (12\text{-}51)$$

For the ith element, this equation becomes

$$n[\alpha x_{i+1} - (2\alpha - \beta - \delta)x_i + (\alpha - \beta)x_{i-1}] = \frac{dx_i}{dt} \qquad (12\text{-}52)$$

For the nth or final element, the finite-difference equation is (recall that a is zero for $z > 1$):

$$a\left(\frac{x_{n-1} - x_n}{\Delta z^2}\right) + b\left(\frac{x_n - x_{n-1}}{\Delta z}\right) + cx_n = \frac{dx_n}{dt}$$

This equation can be rearranged to

$$n[-(\alpha - \beta - \delta)x_n + (\alpha - \beta)x_{n-1}] = \frac{dx_n}{dt} \qquad (12\text{-}53)$$

Equations 12-50 through 12-53 can be readily expressed by the following vector-matrix differential equation:

$$\frac{d}{dt}\begin{bmatrix} x_1 \\ x_2 \\ \cdot \\ \cdot \\ \cdot \\ x_n \end{bmatrix}$$

$$= n\begin{bmatrix} (\alpha - \beta - \delta) & \alpha & & & & \\ (\alpha - \beta) & -(2\alpha - \beta - \delta) & \alpha & & & \\ & (\alpha - \beta) & -(2\alpha - \beta - \delta)\,\alpha & & & \\ & & & \cdot & & \\ & & & & \cdot & \\ & & & (\alpha - \beta) & -(2\alpha - \beta - \delta) & \alpha \\ & & & & -(\alpha - \beta) & (\alpha - \beta - \delta) \end{bmatrix}$$

$$\begin{bmatrix} x_1 \\ x_2 \\ \cdot \\ \cdot \\ \cdot \\ x_n \end{bmatrix} - nb\begin{bmatrix} x_0 \\ 0 \\ 0 \\ \cdot \\ \cdot \\ \cdot \end{bmatrix} \qquad (12\text{-}54)$$

Defining $\mathbf{A}$ and γ appropriately, this equation may be expressed as follows:

$$\frac{d\mathbf{x}}{dt} = n\,\mathbf{A}\,\mathbf{x} + \gamma\,\mathbf{x}_0 \qquad (12\text{-}55)$$

At this point, the partial differential equation in Eq. 12-55 has been transformed into a vector-matrix differential equation. Solution of such an equation was considered in Sec. 12-5.

12-7. AN EXAMPLE INVOLVING CHEMICAL REACTIONS

As an example of a matrix model of a chemically reacting system, we will consider a case of reacting flow in a porous medium. The particular example chosen is designed to predict accurately the heat phenomena associated with the reentry of a space capsule. For proper heat shield design it is necessary to model the char zone of a charring ablator where the porous medium is subjected to a steep temperature gradient and in which chemical reactions are occurring in the gas and between the gas and the porous medium (typically phenolic-nylon composite).

Some of the basic equations are the continuity equations for the individual species:

$$\frac{dn_i}{dz} = r_i \qquad i = 1, 2, \cdots k \qquad (12\text{-}56)$$

where i = individual component subscript
z = distance
r_i = reaction of component i
n_i = mass flux of component i

Also important is an equation that describes the energy transfer in the solid and gas:

$$W_g \epsilon C_p \frac{dT}{dz} = \frac{d}{dz}\left(k_e \frac{dT}{dz}\right) - \sum_{i=1}^{n} H_{xi} R_i \qquad (12\text{-}57)$$

where k_e = effective thermal conductivity
$\displaystyle\sum_{i=1}^{n} H_{xi} R_i$ = heat absorbed or released by chemical reactions
ϵ = porosity
T = temperature
W_g = gas mass flux
C_p = gas heat capacity

The chemical reactions of importance in a recent study [27,28] are given in Fig. 12-11. Note that there are 8 simultaneous reactions involving 11 chemical

1. $CH_4 = 1/2C_2H_6 + 1/2H_2$

2. $C_2H_6 = C_2H_4 + H_2$

3. $C_2H_4 = C_2H_2 + H_2$

4. $C_2H_2 = 2C + H_2$

5. $CH_4 = 1/2C_2H_2 + 3/2H_2$

6. $CH_4 = C + 2H_2$

7. $C + H_2O = CO + H_2$

8. $NH_3 = 1/2N_2 + 3/2H_2$

FIG. 12-11. Important chemical reactions taking place in the char zone during ablation in the temperature range from 500 to 2500 °F.

species. Matrix formulations can lead to an easy way to structure this problem. The reactions can be stated in the form:

$$\sum_{i=1}^{n} r_{ji} A_i \overset{k_{fi}}{\underset{k_{ri}}{\rightleftarrows}} \sum_{i=1}^{n} P_{ji} A_i \qquad (12\text{-}58)$$

where n = number of chemical species

i = species under consideration

j = chemical reaction under consideration

r_{ji} = stoichiometric coefficient of reactant A_i

p_{ji} = stoichiometric coefficient of product A_i

k_{fi} = forward reaction rate constant

k_{ri} = reverse reaction rate constant

In Fig 12-12 is shown the matrix formulation of the reactions in Fig. 12-11.

$$
\begin{bmatrix}
0 & 1 & 0 & 0 & 0 & 0 & 0 & 0 & 0 & 0 & 0 \\
0 & 0 & 1 & 0 & 0 & 0 & 0 & 0 & 0 & 0 & 0 \\
0 & 0 & 0 & 1 & 0 & 0 & 0 & 0 & 0 & 0 & 0 \\
0 & 0 & 0 & 0 & 1 & 0 & 0 & 0 & 0 & 0 & 0 \\
0 & 1 & 0 & 0 & 0 & 0 & 0 & 0 & 0 & 0 & 0 \\
0 & 1 & 0 & 0 & 0 & 0 & 0 & 0 & 0 & 0 & 0 \\
0 & 0 & 0 & 0 & 0 & 1 & 0 & 0 & 0 & 0 & 0 \\
0 & 0 & 0 & 0 & 0 & 0 & 0 & 0 & 1 & 0 & 0
\end{bmatrix}
\cdot
\begin{bmatrix}
H_2 \\ CH_4 \\ C_2H_6 \\ C_2H_4 \\ C_2H_2 \\ C \\ CO_2 \\ CO \\ NH_3 \\ N_2 \\ H_2O
\end{bmatrix}
=
\begin{bmatrix}
\frac{1}{2} & 0 & \frac{1}{2} & 0 & 0 & 0 & 0 & 0 & 0 & 0 & 0 \\
1 & 0 & 0 & 1 & 0 & 0 & 0 & 0 & 0 & 0 & 0 \\
1 & 0 & 0 & 0 & 1 & 0 & 0 & 0 & 0 & 0 & 0 \\
1 & 0 & 0 & 0 & 0 & 2 & 0 & 0 & 0 & 0 & 0 \\
\frac{3}{2} & 0 & 0 & 0 & \frac{1}{2} & 0 & 0 & 0 & 0 & 0 & 0 \\
2 & 0 & 0 & 0 & 0 & 1 & 0 & 0 & 0 & 0 & 0 \\
1 & 0 & 0 & 0 & 0 & 0 & 0 & 1 & 0 & 0 & 0 \\
\frac{3}{2} & 0 & 0 & 0 & 0 & 0 & 0 & 0 & 0 & \frac{1}{2} & 0
\end{bmatrix}
\cdot
\begin{bmatrix}
H_2 \\ CH_4 \\ C_2H_6 \\ C_2H_4 \\ C_2H_2 \\ C \\ CO_2 \\ CO \\ NH_3 \\ N_2 \\ H_2O
\end{bmatrix}
$$

FIG. 12-12. Matrix formulation of the chemical reactions given in Fig. 12-11 [27,28].

The reaction rate expressions for the ith species, R_i, for m chemical reactions can also be expressed by

$$R_i = \sum_{j=1}^{m} (p_{ji} - r_{ji}) \left[k_{fj} \prod_{k=1}^{n} c_k^{r_{ji}} - k_{rj} \prod_{k=1}^{n} c_k^{p_{ji}} \right] \tag{12-59}$$

where c_k is the concentration of component k.

This equation has the powers on the composition terms the same as the stoichiometric coefficients. However, it is not necessary to do this. If this is not the case for some of the reactions being used, it is only necessary to include two additional matrices besides r_{ji} and p_{ji} in the computer implementation.

To illustrate the use of this rate equation, the rate of reaction of methane (component 2) is given by the following:

$$R_2 = \sum_{j=1}^{8} (p_{j2} - r_{j2}) \left[k_{fj} \prod_{k=1}^{11} c_k^{r_{ji}} - k_{rj} \prod_{k=1}^{11} c_k^{p_{ij}} \right] \tag{12-60}$$

or expanding,

$$R_2 = (0 - 1) \left[k_{f1} c_2 - k_{r1} c_1^{1/2} c_3^{1/2} \right]$$

$$+ (0 - 1) \left[k_{f5} c_2 - k_{r5} c_1^{3/2} c_5^{1/2} \right]$$

$$+ (0 - 1) \left[k_{f6} c_2 - k_r c_1^2 c_6 \right]$$

The above equation contains five other terms in the expanded form, but these are not included since the coefficients are zero. Of course this equation is not exactly correct since the powers on the compositions are different in some cases for the actual rate expressions. For example, the carbon-water reaction involves the surface area of carbon, and not a "concentration" of carbon. To take the surface area of the char into account the reaction rate constant was modified, and the exponent on the carbon "concentration" was taken equal zero.

This example illustrates one way in which matrices can be used to simplify the accounting and computational problems associated with complex physical and chemical systems. There are, of course, many other ways in which matrices are useful in formulating models of chemical reacting systems, but this one example does at least illustrate their usefulness. Other examples will be developed in the Problems at the end of this chapter.

12-8.   SUMMARY

This chapter has given elementary insight into matrix mathematics and its usefulness in model formulation and, to some extent, in model solution. This has been accomplished by looking at a broad—but by no means complete—spectrum of the use of matrices in mathematical model applications. This spectrum is widely amplified by the Problems at the end of the chapter.

REFERENCES

1. Luh C. Tao, "Principles of Matrix Mathematics," *Chemical Engineering,* March 1, 1965, pp. 77-82.

2. C. R. Wylie, *Advanced Engineering Mathematics,* McGraw-Hill, New York, 1960.

3. L. A. Pipes, *Matrix Methods for Engineering,* Prentice-Hall, Englewood Cliffs, N.J., 1963.

4. R. Bellman, *Introduction to Matrix Analysis,* McGraw-Hill, New York, 1960.

5. T. L. Wade, *The Algebra of Vectors and Matrices,* Addison-Wesley, Reading, Mass., 1957.

6. L. Lapidus, *Digital Computation for Chemical Engineers,* McGraw-Hill, New York, 1962.

7. V. G. Jensen, and G. V. Jeffreys, *Mathematical Methods in Chemical Engineering,* Academic Press, New York, 1963.

8. N. R. Admundson, *Mathematical Methods in Chemical Engineering,* Prentice-Hall, Englewood Cliffs, N.J., 1966.

9. E. Kreyszig, *Advanced Engineering Mathematics,* 2nd ed., Wiley, New York, 1967.

10. Louis A. Pipes, *Applied Mathematics for Engineers and Physicists,* 2nd ed., McGraw-Hill, New York, 1958.

11. G. T. Westbrook, "A Powerful Tool-Matrix Algebra," *Hydrocarbon Processing and Petroleum Refiner,* Vol. 42, No. 6 (June 1963), p. 173.

12. J. C. Wang, G. E. Henke, "Tridiagonal Matrix for Distillation," *Hydrocarbon Processing,* Vol. 45, No. 8 (August 1966), p. 155.

13. N. R. Amundson, and A. J. Pontinen, "Multicomponent Distillation Calculations on a Large Digital Computer," *Industrial and Engineering Chemistry,* Vol. 50, No. 5 (May 1958), p. 730.

14. A. Acrivos, and N. R. Amundson, "Applications of Matrix Mathematics to Chemical Engineering Problems," *Industrial and Engineering Chemistry,* Vol. 47, No. 8 (August 1955), p. 1533.

15. L. C. Tao, "Linear Matrix Differential Equations," *Chemical Engineering,* May 24, 1965, p. 125.

16. J. H. Duffin, and J. D. Gamen, "Dynamic Model of a Multistage-Multicomponent Column Including Tray Hydraulics and Boundary Lags and Delays," presented at the 58th National Meeting of the A.I.Ch.E. in Dallas (February 1966).

17. J. H. Duffin, "Improving the Accuracy of the Dynamic Response of a Multistate-Multicomponent Column Model," presented at the 56th National Meeting of the A.I.Ch.E. in San Francisco (May 1965).

18. N. J. Tetlow, D. M. Groves, and C. D. Holland, "Analysis and Control of a Generalized Model of a Distillation Column," presented at the 58th National Meeting of the A.I.Ch.E. in Dallas (February, 1966).

19. C. E. Huckaba, et al., "Experimental Confirmation of a Predictive Model for Dynamic Distillation," *Chemical Engineering Progress Symposium Series,* No. 55, Vol. 61 (1965), pp. 125-135.

20. C. E. Huckaba, F. P. May, and F. R. Franke, "An Analysis of Transient Conditions in Continuous Distillation Operations", presented at the 47th National Meeting of the A.I.Ch.E. in Baltimore (May, 1962).
21. W. L. Luyben, V. C. Verneiuit, and J. A. Gerster, "Transient Response of Ten-Tray Distillation Column: Complete Data and Results," University of Delaware, August 1963.
22. N. A. Lange, *Handbook of Chemistry,* Handbook Publishers, Inc. Sandusky, Ohio (1956), pp. 1424-1438.
23. J. T. Tou, *Modern Control Theory,* McGraw-Hill, New York, 1964.
24. L. C. Tao, "Linear Matrix Differential Equations," *Chemical Engineering,* May 24, 1965, pp. 125-128.
25. G. A. Coulman, "A Computer Oriented Approximate Analytic Solution to a Partial Differential Equation," presented at the 152nd Meeting of the ACS, New York, Sept. 12, 1966.
26. K. R. Westerterp, and H. Kramers, *Elements of Chemical Reactor Design and Operation,* Academic Press, New York, 1963.
27. E. G. del Valle, G. C. April, and R. W. Pike, "Non-Equilibrium Flow and the Kinetics of Chemical Reactions in the Char Zone," NASA-CR-66455 (July 15, 1967).
28. ——, "Non-Equilibrium Flow and the Kinetics of Chemical Reactions in the Char Zone," National ACS Meeting, San Francisco, Calif., April 1968.
29. K. Ogata, *State Space Analysis of Control Systems,* Prentice-Hall Inc., Englewood Cliffs, N.J., 1967.

PROBLEMS

12-1. Multiply matrix **A** times matrix **B** where

$$A = \begin{bmatrix} 1 & 2 \\ 1 & 3 \end{bmatrix}$$

$$B = \begin{bmatrix} 2 & 0 \\ 1 & 4 \end{bmatrix}$$

and the dot product **A** · **B** or **AB** is desired, i.e.,

$$c_{ij} = \sum_k a_{ik} b_{kj}$$

12-2. Form the dot product of **A** and **B**, i.e., **A** · **B** where

$$A = \begin{bmatrix} 1 & 1 \\ 2 & 2 \end{bmatrix}$$

$$B = \begin{bmatrix} -1 & 1 \\ 1 & -1 \end{bmatrix}$$

Now form **B** · **A**.

12-3. The transformation between Cartesian and cylindrical coordinates are $x = r\cos\theta$, $y = r\sin\theta$, and $z = z$. The differential relations are:

$$dx = \cos\theta \ dr - r\sin\theta \ d\theta$$
$$dy = \sin\theta \ dr + r\cos\theta \ d\theta$$
$$dz = dz$$

Express these equations in matrix notation.

Redo this entire problem for the transformation between Cartesian and spherical coordinates.

12-4. Given the basic equation for a second-order lag.

$$\frac{d^2x}{dt^2} + a\frac{dx}{dt} + bx = f(t)$$

Rearrange this into a set of first-order differential equations in matrix format by letting $x = y_1$ and $\dfrac{dx}{dt} = y_2$.

12-5. Find the determinant of **A** if

$$A = \begin{bmatrix} 1 & 3 & 5 \\ 2 & 2 & 1 \\ 1 & 4 & 3 \end{bmatrix}$$

12-6. Find the determinant of **A** if

$$A = \begin{bmatrix} 2 & 2 & 1 \\ 2 & 2 & 1 \\ 1 & 4 & 3 \end{bmatrix}$$

12-7. Given the equations

$$x + 2y + \ z = 4$$
$$3x + 2y + 4z = 10$$
$$5x + \ y + 3z = 7$$

Write these in matrix form of $Ax = b$.

12-8. Calculate the adjA for Prob. 12-7.

12-9. The equations of Prob. 12-7 may be solved:

$$x = A^{-1}b = \frac{\cdot \text{adj } A}{|A|} b$$

Use the results of Prob. 12-8 to find **x**.

12-10. Solve the following set of algebraic equations:

$$4x + \ y - \ z = 5$$
$$x + 3y + \ z = 5$$
$$2x + \ y + 2z = 10$$

12-11. In the nitric acid plant example of Sec. 12-2 the grade of the product (z_1 or z_2) is implied in the elements of **A**. Add a waste stream as a vector element in **y** and show how this introduces a plant efficiency into **A**.

12-12. Section 12-2, illustrated bubble-point calculations in terms of matrices. Take the same approach in the structure of dew-point calculations.

12-13. In Sec. 12-3, a multicomponent distillation model was proposed and the material balance, equilibrium, and summation equations were written. Using the notation of Fig. 12-2 and 12-3 and as given in the text, develop the set of enthalpy balance equations for this column. Use H for vapor enthalpy and h for liquid enthalpy.

12-14. Given the heat exchanger shown, which has a condensing vapor on the shell side and a liquid being heated on the tube side. The exchanger has one tube pass, all fluid properties are assumed to be constant, and the temperature on the shell side is considered constant with respect to position. By using a lumped-parameter model, the flow patterns are assumed to be approximated by a series of perfectly mixed pools. Since the shell-side temperature is independent of position, this entire side can be represented by only one perfectly mixed pool of temperature T_s, the vapor condensation temperature. The flow pattern on the tube side is assumed to be approximated by four perfectly mixed pools in series, as illustrated. Derive the matrix differential equation model for this exchanger in the form

$$\dot{T} = AT + B$$

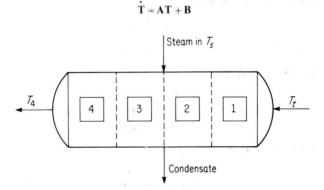

PROB. 12-14.   Description of the vapor-liquid heat exchanger.

12-15. Given the 1–1 liquid-liquid heat exchanger shown. Derive the matrix differential equation in the form

$$\dot{T} = AT + B$$

The variation of physical properties and heat-transfer coefficients with temperature and flow rate will be neglected. The exchanger has two baffles on the shell side, and the exterior is assumed to be perfectly insulated.

On each side of the exchanger the flow pattern is assumed to be approximated by a series of perfectly mixed pools in series as is typically done using a lumped-parameter model. Logically it would be anticipated that more pools would be required to adequately repres-ent the flow pattern on the tube side, although, for simplicity the exchanger in the sketch has the same number of pools on the tube side as on the shell side. Furthermore, the pools on each side coincide in this example, but this is not a necessary condition. For the exchanger shown, there are two pools between each baffle, thus making a total of six pools on each side.

In numbering the pools, pool 1 is the pool at the shell-side inlet. Thereafter, they are numbered consecutively in the direction of the shell-side flow. For convenience in cal-

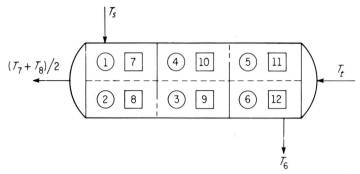

PROB. 12-15.   Liquid-liquid exchanger.   Numbers in circles correspond to shell-side
pools;   numbers in squares correspond to tube-side pools.

culations, the first tube-side pool coincides with the first shell-side pool, and the numbering
continues in the same direction as for the shell side.   To avoid double subscripts, the first
tube-side pool is numbered one higher than the final shell-side pool.

12-16.   Consider the packed-bed distillation column shown.   To use a lumped-parameter model the column is divided into sections, each of which is considered to be a
perfectly mixed pool.   Thus, the vapor and liquid within a pool are of uniform composition
with respect to position, although they may not be in equilibrium.   To keep the analysis
simple, the column is divided into only three pools, and of course, this may be increased
if desired.

The rate of mass transfer is calculated using the overall mass-transfer coefficient
$K_{og}a$, and it is assumed constant throughout the column for simplicity.   The variation
of $K_{og}a$ can be included if desired, or the height of a transfer-unit concept may be used.
Thus in the model $K_{og}a$ can be replaced by V/H_{og}, where V is the molar gas flow rate per
unit are and H_{og} is the height of a transfer unit based on the overall gas phase driving force.

In order to develop the equations to describe this system, the following assumptions
are also made, none of which are inherent restrictions in the method:

1.   Equimolal over flow.
2.   No heat loss to surroundings.
3.   Total condenser.
4.   Feed is a liquid at its bubble point.
5.   Negligible holdup in lines to condenser.
6.   The reboiler acts as an ideal stage.
7.   Binary mixture.

The Antoine equation should be used to calculate the vapor pressure of each component.   Derive the matrix differential equation model in the form

$$\dot{\mathbf{X}} = \mathbf{A}\mathbf{X} + \mathbf{B}$$

12-17.   Given a continuously agitated tank reaction in which the reaction occurring is

$$A_1 \underset{k_1'}{\overset{k_1}{\rightleftarrows}} \quad A_2 \underset{k_2'}{\overset{k_2}{\rightleftarrows}} A_3$$

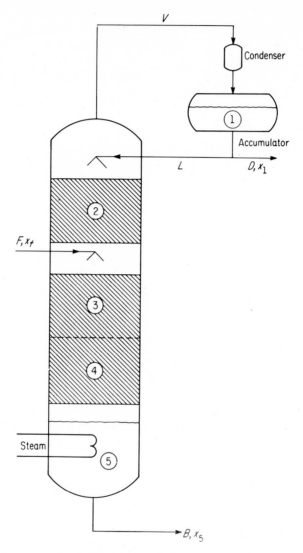

PROB. 12-16.   Continuous contactor.

Also given:

$c_i(n)$ = concentration of ith substance in the nth reactor, i.e., for this problems $n = 1$.

$c_i(0)$ = concentration of the ith substance in the feed entering the ith reactor (there is only 1 reactor in this problem.).

θ = holding time for the reactor.

Write the continuity equation for each component and express them in a single matrix differential equation.

12-18. Redo Prob. 12-17 except assume there are N equal reactors in series. Assume that at $t = 0$ the concentration of all the substances in all the reactors is equal to zero and that the feed entering the first tank is independent of time. What is the matrix model for the nth tank?

12-19. Discuss the solutions for the models of Probs. 12-17 and 12-18.

12-20. Given the air-water cooling tower shown. This is basically a problem in simultaneous heat and mass transport. For steady-state operation a matrix model can be developed of the form

$$\frac{d\mathbf{x}}{dz} = \mathbf{Ax} + \mathbf{b}$$

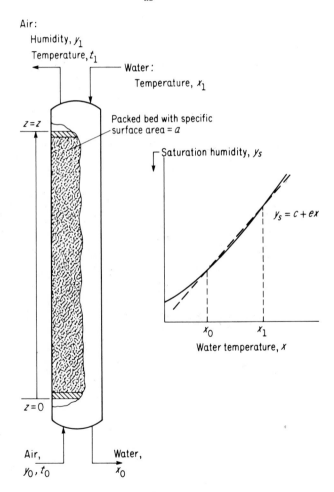

PROB. 12-20

Assume the water flow rate w is constant, and assume the saturation humidity $y_s = c + ex$ (linear) where c and e are constant.   For nomenclature use g as mass-flow rate of dry air, c_g as humid heat capacity, c_w as water heat capacity, x is water temperature, t the air temperature, y the air humidity, and λ the latent heat.

12-21. Consider an empty tubular reactor.   (This has been analyzed by Amundson, Ref. 8.)   Assume plug flow and all diffusive and conductive effects can be neglected.   Heat transfer is characterized by a heat-transfer coefficient at the wall.   Let the length down the axis of the reactor be x.   The reactions which take place in the fluid are given as

$$\sum_{i=i}^{n} a_{ij} A_i = 0, \qquad j = 1, 2, \cdots, m$$

where a_{ij} is positive for a product and negative for a reactant.   There are m reactions and n chemical species.   The nomenclature is as follows:

G = total mass flow rate
T = temperature of reaction mixture
t = ambient temperature
h = heat-transfer coefficient at tube wall
a = cross-sectional area of reactor
p = pressure
P = perimeter of reactor
h_i = partial molar enthalphy of ith species
ΔH_j = heat of reaction of jth reaction
A_i = ith chemical species
C_{pt} = molar heat capacity of kth species
a_{ij} = stoichiometrix coefficient of ith species in jth reaction
f_{ij} = rate of formation of ith species in jth reaction in moles per unit volume per unit time.
τ = shearing stress at wall
f = friction factor
g_i = moles of species i per unit mass of reaction mixure
u = linear velocity along the axis of the tube
w = mass velocity of reaction mixture

Derive the matrix model for the conservation of mass for this reactor.

12-22. For the reactor of Prob. 12-21, derive the matrix model for the enthalpy balance.

12-23. For the reactor of Prob. 12-21, derive the nomentum balance.

12-24. Consider a system of chemical reactions performed in a batch reactor in which all the species are coupled.   If for example there are three chemical species, A_1, A_2, and A_3, (where these symbols stand for both the species and their concentration) and they are related by the reaction scheme

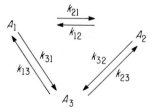

PROB. 12-24

where k_{ij} is the absolute rate constant and $k_{ij} A_j$ is the rate of formation of A_i from A_j in moles per unit volume per unit of time. The equations for the reactor are[8]

$$\frac{dA_1}{dt} = -(k_{21} + k_{31})A_1 + k_{12}A_2 + k_{13}A_3$$

$$\frac{dA_2}{dt} = k_{21}A_1 - (k_{12} + k_{32})A_2 + k_{23}A_3$$

$$\frac{dA_3}{dt} = k_{31}A_1 + k_{32}A_2 - (k_{13} + k_{23})A_3$$

Generalize this to n chemical species, each of which is coupled to every other species by chemical reaction, to give a matrix model of the form

$$\frac{d\mathbf{A}}{dt} = \mathbf{K}\mathbf{A}$$

12-25. Show that Eq. 12-43 is valid.

12-26. Consider the differential equation

$$\frac{d^2c(t)}{dt^2} + \frac{3dc(t)}{dt} + 2c(t) = r(t)$$

(a) Define one state variable as $c(t)$ and the second as $dc(t)/dt$. Develop the state equation. (b) Determine the eigenvalues. (c) How are the eigenvalues related to the roots of the characteristic equation? (d) Calculate two independent eigenvectors. (e) What is the canonical form of the state equation? (f) Calculate $\varphi(t)$.

chapter 13

Systems Theory

In this chapter the procedures and principles for developing state-space models for lumped-parameter systems are discussed. As lumped-parameter or discrete systems are encountered more often in electrical engineering than perhaps in any other engineering discipline, most of the theory has been developed by electrical engineers. However, these principles are directly applicable to systems in all engineering fields. As the development of the state-space model via the linear graph is the primary subject of this chapter, graph theory will be discussed somewhat in several places throughout the chapter.

Perhaps several words of caution should be voiced before the reader continues with this chapter. First, the material presented is of little use in the analysis of small systems, but has a definite advantage in the analysis of large systems requiring machine-aided analysis. Second, it is hoped the reader's patience and perseverance are sufficient so that he reads the entire chapter at least once before judging the method. Several new terms and rules must be mastered, plus the fact that the basic procedure is not consummated until the latter part of the chapter. Such a presentation does not attract its own audience but it seems unavoidable.

13-1. INTRODUCTION

Systems theory is a discipline that treats the analysis of systems composed of discrete physical components, previously denoted as lumped-parameter systems in this text. The systems considered in system theory are all characterized by the fact that they can be viewed as collections of discrete components united at a finite number of interfaces. Examples of such systems are shown in Fig. 13-1, or they can be represented abstractly as in Fig. 13-2, where the closed regions represent the components and the points of contact represent the interfaces. In this chapter the systems-theory approach is presented for developing mathematical models of systems composed of such discrete interacting components.

This mathematical model will be composed of a set of simultaneous algebraic and/or differential equations which describe the behavior of the system and the relationships between the individual components. As evident from Fig. 13-1, the mathematical model is composed of two types of relationships:

1. Relationships describing the behavior of the components, i.e., mathematical models of each of the components.
2. Equations describing their connections or interface arrangement.

Thus the model places the components in the appropriate perspective in regard to the system structure in order to describe the behavior of the complete system. In order to implement this, state-space concepts are very convenient.

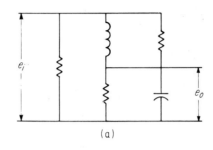

(a)

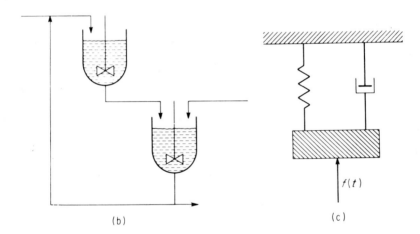

(b)   (c)

FIG. 13-1. Representations of discrete systems. (a) Electrical. (b) Well-mixed tanks. (c) Mechanical.

In order for the above approach to be practical, the mathematical model characterizing the behavior of a component must be independent of the system structure, i.e., how the component is connected to other components of the system. This allows the mathematical models of the various components to be developed independently of the development of the system itself, and in the physical sciences mathematical models of many components are available. In many cases the components may not be physically removed from the system, but

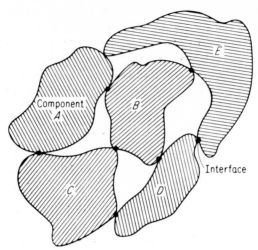

FIG. 13-2.   Abstract representation of a system composed
of discrete components.

system theory applies even when they can be removed only conceptually.   This
allows the investigator to subdivide his system as much as he desires in order to
obtain components for which he can easily obtain a mathematical model.

Generally, systems theory can be divided into two broad aspects:

1.   Modeling theory, or the procedure for developing the mathematical
     model from the component and their interconnections.
2.   Behavior theory, or the process of analyzing the solution characteristics
     of the mathematical model.

In this chapter the emphasis will be placed on modeling theory.

13-2   COMPONENT MODELS [1]

In the abstract representation of a system, recall that each region indicates a
component, which is connected to other components of the system at its interfaces.   Each component is said to have a *terminal* corresponding to each of its interfaces with the other components.   As a component may have practically any
number of terminals, a procedure is needed for determining what constitutes a
mathematical model of the component.   Specifically, consider the three terminal
component abstractly represented in Fig. 13-3a, for which a real counterpart could
be the orifice meter in Fig. 13-3b.

For a given volumetric flow rate Q through an orifice of specified configuration [2], there is a maximum pressure drop ΔP_m, which is normally measured, and
a permanent pressure drop ΔP_p.   Consequently, this is a three terminal compo-

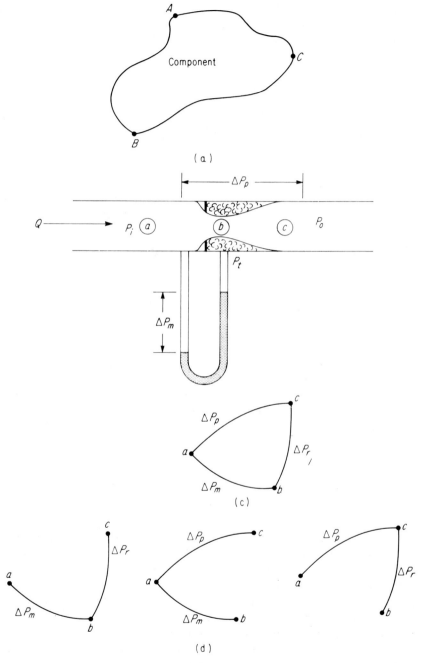

FIG. 13-3.   Application of linear graph concepts to an orifice meter.   (a) Three-terminal component.
(b) Orifice meter.   (c) The edges.   (d) Possible terminal graphs.

nent, i.e., the terminals correspond to the upstream or inlet pressure P_i, the minimum or throat pressure P_t, and the downstream or outlet pressure P_o. The terminals are connected by lines called *edges*, and the three edges connecting these three terminals are shown in Fig. 13-3c.

Two variables, known as *complementary variables*, are associated with the edges of a linear graph. As discussed in Chapter 10, there are two types of variables—*across* and *through* variables. One of the complementary variables is a through variable, typically denoted by y, and the other is an across variable, generally denoted by x. For the orifice meter in Fig. 13-3b, the through variable is the flow rate and the across variable is the pressure. Note that one pair of complementary variables is associated with each edge.

With the above definitions of an edge and a *vertex* (terminal), graph theory has been injected into the discussion. The linear graph forms an important part of the systematic development of a state-space model, and additional concepts will be introduced as the discussion proceeds. The "linear graph", contrary to what the name implies, is equally applicable to linear and nonlinear systems. Another term used in linear graph theory is a *tree*, defined as a set of $n-1$ edges connecting n vertices but forming no closed paths. A *terminal graph* of a component is defined as a tree whose n vertices correspond to an n-terminal component. The tree is not unique, and the three possible trees for the linear graph of the orifice meter are shown in Fig. 13-3d.

With these concepts as a basis, the following postulate of system theory can be stated: The mathematical model of an n-terminal component consists of a set of $n-1$ equations containing $n-1$ pairs of oriented complementary variables x and y identified by an arbitrarily chosen terminal graph. The equations may be algebraic or differential and the variables x and y may be vectors. For the orifice meter, the pressure drop at the throat, the permanent pressure drop, and the pressure drop reclaimed (ΔP_r) can be related to the flow Q. Thus, a model of the orifice meter would be as follows (from the second terminal graph in Fig. 13-3 c):

$$\Delta P_m = f_1(Q) \tag{13-1}$$

$$\Delta P_p = f_2(Q) \tag{13-2}$$

Note that this is a set of $n-1$ equations in $n-1$ pairs of complementary variables (Some confusion may arise because Q is the through variable for both edges in this example.) The above equations relate the pressures at each terminal of the component to the flow rate.

For physical systems, there is typically an orientation of the through variable. For the orifice meter, the fluid flows from vertex a to vertex b to vertex c. This accounts for the orientation of the arrows in Fig. 13-3.

Although a three-terminal component was used for illustrative purposes in this section, analogous models can be developed for the two-terminal components in Table 10-3.

13-3.   THE SYSTEM GRAPH

In the example presented in the above section, note that exactly half the $2(n-1)$ complementary variables were independent variables and the remainder dependent variables. This is true of component models in general. But when components are interconnected to form a system, these independent variables are often no longer independent, i.e., the interconnections place restraints upon them. In general, a system model is developed from two sources—the component model and the interconnection patterns. The constraints arising from the interconnection patterns can be most easily determined from the system graph.

To more clearly present the concept of the system graph, examine the abstract representation of a system composed of several interconnected components as shown in Fig. 13-4a. Note that the terminal graphs of the various components are also shown. The *system graph* is simply the edges and vertices obtained by coalescing the vertices of the component terminal graphs in a one-to-one correspondence with the way in which the terminals of the corresponding components are united to form the system. Thus the graph contains one vertex for each interface and an edge corresponding to each edge in the terminal graphs of the components. However, the system graph may contain closed paths or circuits, which were not allowed in the terminal graphs. One set of constraint equations arises from each circuit. A system graph is shown in Fig. 13-4b for the system in Fig. 13-4a.

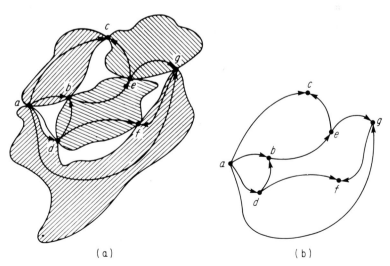

(a)   (b)

FIG. 13-4.   Development of the system graph.   (a) Representation of a system composed of several components.   (b) System graph of this system.

shown in Fig. 13-5a.   The first step in developing the system graph is to specify
the component models.   For the mass, the equation describing its behavior is

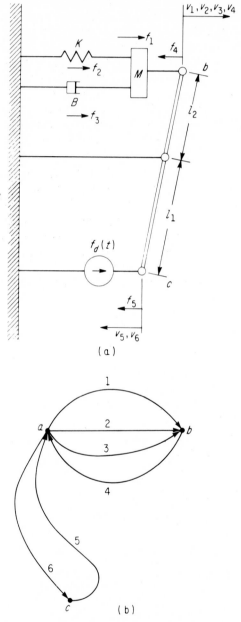

(a)

(b)

FIG. 13-5.   Development of the system graph.   (a)
The system.   (b) The system graph.

$$f_1(t) = M \frac{dv_1(t)}{dt} \tag{13-3}$$

All forces are assumed positive when in the direction indicated in Fig. 13-5a. This is a two-terminal component with the force $f_1(t)$ being the through variable and velocity $v_1(t)$ being the across variable. The spring is described by

$$\frac{df_2(t)}{df} = Kv_2(t) \tag{13-4}$$

where $f_2(t)$ is the force exerted on the spring. The viscous damper is represented by

$$f_3(t) = Bv_3(t) \tag{13-5}$$

where $f_3(t)$ is the force exerted on the damper. All of these components are two-terminal components, and are thus described by one equation with one pair of complementary variables, one of which is a dependent variable and the other an independent variable. These components are represented by edges 1, 2, and 3 in the system graph in Fig. 13-5b.

If the mass of the lever is neglected, it becomes a three-terminal component. The force $f_4(t)$ at the upper end of the lever is related to the force $f_5(t)$ at the lower end by

$$f_4(t) = -\frac{l_1}{l_2} f_5(t) \tag{13-6}$$

where l_1 and l_2 are the respective lengths of the lever. Similarly, the velocity $v_4(t)$ at the upper end is related to the velocity $v_5(t)$ at the lower end by

$$v_5(t) = \frac{l_1}{l_2} v_4(t) \tag{13-7}$$

Note that these two equations can also be expressed by the following vector-matrix equation:

$$\begin{bmatrix} v_5(t) \\ f_4(t) \end{bmatrix} = \begin{bmatrix} 0 & l_2/l_1 \\ -l_2/l_1 & 0 \end{bmatrix} \begin{bmatrix} f_5(t) \\ v_4(t) \end{bmatrix} \tag{13-8}$$

This component is represented by edges 4 and 5 in the system graph.

The force driver is usually included in the system graph. This component is simply represented by

$$f_6(t) = f_d(t) \tag{13-9}$$

This component is represented by edge 6 in the system graph.

The direction of the arrows in the system graph is determined from the direction in which the force is acting if it is in the positive direction. That the forces on the ends of the lever are in the opposing directions is accounted for by the negative sign in Eq. 13-6.

13-4.   LINEAR GRAPH TERMINOLOGY [1]

Since this text is primarily concerned with the mechanics of developing models using linear-graph concepts, the definitions will be stated informally and the proofs will be omitted.   Interested readers should consult the references at the end of this chapter for a more detailed discussion.

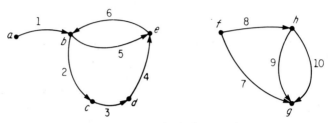

FIG. 13-6.   Example of a linear graph.

The following definitions will be essential to the subsequent development of the technique, and some terms previously defined are repeated.   Figure 13-6 is used to illustrate the important points.

An *edge* consists of an oriented line segment with its end points. Lines 1 through 10 in Fig. 13-6 are edges.   The total number of edges is denoted by e, which equals 10 for the example.

The *vertex* is an end point of an edge.   The vertices in Fig. 13-5 are $a, b, \cdots,$ $h.$ The total number of vertices is denoted by v, which is 8 for the example.

A *finite graph* consists of a finite number of edges and a finite number of vertices.

A subset of the edges of a graph is called a *subgraph* of the graph.   The edges 1, 5, 6, 7, and 9 in Fig. 13-6 form a subset of the edges and a subgraph of the graph.

Edge k is *incident* to vertex l if vertex l is an end point of edge k.   Edge 3 in Fig. 13-6 is incident to vertices c and d only.

An *end vertex* has only one edge incident to it, and the incident edge is called an *end edge*.   Vertex a is an end vertex and edge 1 is an end edge in Fig. 13-6.

The subgraph (1,5,4) in Fig. 13-6 is called a *path*, since one can "proceed" from vertex a to vertex d, traversing the edges 1, 5, and 4 in a sequential manner. Of course, this is not the only path from vertex a to vertex d, and graphs generally contain many paths between any two vertices.   Two paths are said to be *distinct paths* if the only vertices common to both are end vertices.   For example, paths (1,5,4) and (1,2,3) in Fig. 13-6 are distinct paths.

Although a *circuit* has already been defined an intuitive manner, it can be rigorously defined as a set of edges chosen in such a manner that two and only two edges are incident to each vertex in the set.   Even so, it can probably be best visualized as a closed path.

If a graph is a *connected graph,* there is a path between every pair of vertices in the graph. Clearly, the graph in Fig. 13-6 is not a connected graph, since it consists of two connected subgraphs. Such graphs are denoted as *separated graphs,* and arise from such components as hydraulic rams, electric motors, and other types of energy converters.

Previously, a *tree* has been defined for a component graph, and is also defined similarly for the system graph. A *tree* is defined as a subgraph or set of edges which satisfies the following properties:

1. It is connected.
2. It contains all the vertices of the system graph.
3. It contains no circuits.

In Fig. 13-7 one possible tree consists of the edges (1, 2, 4, 7). Of course, the tree of a graph is not unique, and other possible trees of the graph in Fig. 13-7 are (1, 3, 4, 7), (2, 4, 7, 8), (1, 2, 5, 7), etc. Note that there is one and only one path between any pair of vertices in a tree, and a tree with v vertices contains $v - 1$ edges. The *branches* of a tree are the edges of the system graph included in the tree, and those edges of the system graph not in the tree form the *cotree,* or the *complement* of the tree. Note that the cotree is not necessarily connected, and the edges of the cotree are called *chords.* In Fig. 13-7 the cotree of tree (1, 2, 4, 7) is (3, 5, 6, 8). Since the branches of a tree contains $v - 1$ edges, the cotree of a graph containing e edges consists of $e - v + 1$ edges. Note that for graphs such as the one in Fig. 13-6, a tree can be selected in each connected part, but not for the entire graph.

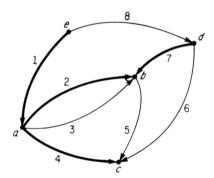

FIG. 13-7. Example of a tree.

Although the tree itself contains no circuits, adding a chord between any pair of vertices forms a circuit. As noted above, a graph with v vertices and e edges contains $e - v + 1$ chords, independent of which tree is selected. Since addition of any one of these chords to the tree forms a circuit, the graph contains exactly $e - v + 1$ distinct circuits. The set of circuits so obtained is defined as the *fundamental circuits.*

A *cut-set* is defined as a set of edges chosen so that

1. removing the cut-set of edges divides the graph into two parts (considering an isolated vertex as a part)
2. no subset of the above set has this property.

The second restriction is necessary in order to insure that no extra edges are included in the cut-set. For example, cut-sets of the graph in Fig. 13-7 are (6,7,8), (8,1), (8,2,3,5,6), etc.

The tree provides a convenient technique for obtaining the cut-sets. Note that if any one branch of a tree is cut, the tree is divided into two unconnected parts. Furthermore, if the restriction is made that *only* one branch of the tree is cut, then there can be only two separate parts of the graph. Thus if a cut-set is chosen such than it contains one and only one branch of a given tree (the other edges in the cut-set being chords), the two properties of a cut-set are automatically satisfied. Since there are $v - 1$ branches in a tree, we will obtain exactly $v - 1$ distinct cut-sets, called the *fundamental cut-sets*. For the graph in Fig. 13-7, the

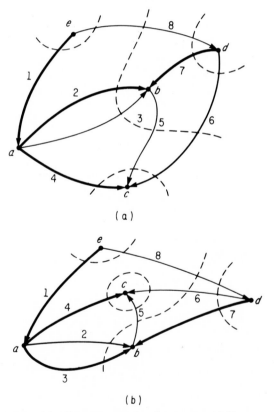

(a)

(b)

FIG. 13-8. Customary notation for cut-sets. (a) The cut-sets. (b) Modification of graph in part (a).

four fundamental cut-sets for the tree shown are (1,8), (6,7,8), (4,5,6), and (2,3,5, 6,8). Each of these are shown by the dashed lines in Fig. 13-8a, which is the customary notation. However, in some cases it is necessary to redraw part of the graph to obtain a different geometrical arrangement. Nonetheless, the edges of the tree remain the same. Such a variation of the system graph is shown in Fig. 13-8b, and a problem illustrating this point is presented in the exercises.

The fundamental circuits and fundamental cut-sets will be subsequently used to reduce the interconnection pattern of the system graph to equation representation. To develop these equations, an *orientation* must be prescribed for both the circuits and the cut-sets. For the circuits the orientation is specified by an arrow, as in Fig. 13-9a, whose direction is the same as that of the chord defining the fundamental circuit. Similarly, the orientation of a fundamental cut-set is synonomous with the direction of the defining branch, as illustrated in Fig. 13-9b.

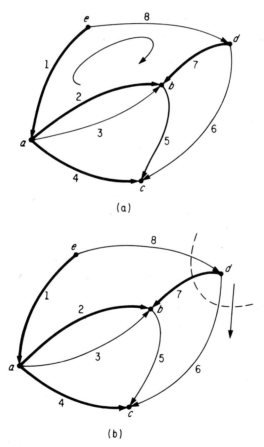

(a)

(b)

FIG. 13-9. Orientation of circuits and cut-sets. (a) Orientation of circuit (1, 2, 4, 8). (b) Orientation of cut-set (6, 7, 8).

Finally, the set of all edges incident to a given vertex is defined as the *incidence set*, and the orientation of an edge is arbitrarily considered positive when it is directed away from the vertex.

This concludes the definitions and basic concepts for the discussions that follow.

13-5. CIRCUIT EQUATIONS

From the orientation of the fundamental circuits the *circuit vectors* can be readily obtained.   For the ith circuit in a system graph, the circuit vector is a row vector denoted as β_i of order $1 \times e$.   The jth element b_{ij} of β_i is defined as follows:

$$b_{ij} = \begin{cases} 1 \text{ if edge } j \text{ is in same direction as circuit orientation} \\ -1 \text{ if edge } j \text{ is directed opposite to circuit orientation} \\ 0 \text{ if circuit } i \text{ does not contain edge } j \end{cases}$$

For example, the four chords in Fig. 13-7 define four circuits whose circuit vectors are

			Edge						
Circuit		1	2	3	4	5	6	7	8
(2, 3)	$\beta_1 = [$	0	−1	1	0	0	0	0	0]
(2, 4, 5)	$\beta_2 = [$	0	1	0	−1	1	0	0	0]
(2, 4, 6, 7)	$\beta_3 = [$	0	1	0	−1	0	1	−1	0]
(1, 2, 7, 8)	$\beta_4 = [$	−1	−1	0	0	0	0	1	1]

One of the basic axioms of system theory is that the product of any circuit vector β_i and a vector $\mathbf{x}$ consisting of the across variables associated with the e edges of the system graph is zero; i.e.,

$$\beta_i \mathbf{x} = 0 \tag{13-10}$$

In essence, this equation is a fundamental circuit equation of the system graph.

With matrix notation, the four circuit vectors $\beta_1, \beta_2, \beta_3,$ and β_4 can be represented by one matrix.   Furthermore, since each fundamental circuit vector must satisfy Eq. 13-10, they can be represented as follows:

$$\begin{bmatrix} 0 & -1 & 0 & 0 & 1 & 0 & 0 & 0 \\ 0 & 1 & -1 & 0 & 0 & 1 & 0 & 0 \\ 0 & 1 & -1 & -1 & 0 & 0 & 1 & 0 \\ -1 & -1 & 0 & 1 & 0 & 0 & 0 & 1 \end{bmatrix} \begin{bmatrix} x_1 \\ x_2 \\ x_4 \\ x_7 \\ \hline x_3 \\ x_5 \\ x_6 \\ x_8 \end{bmatrix} = \mathbf{0} \tag{13-11}$$

Note that in writing this equation the order in which the edges appear has been altered. The first four columns correspond to the branches in the tree, and the last four correspond to chords in the cotree. Note that this same order appears in the vector **x**; i.e., the first four variables are the across variables for the branches and the last four variables are the across variables for the chords. A similar arrangement also occurs for the rows; i.e., row 1 is obtained from the circuit defined by edge 3 (a chord), row 2 from edge 5, etc. This insures that the order of the last four variables (corresponding to the chords) are in ascending order of their subscripts. A similar arrangement will be made in the next section.

In Eq. 13-11 the matrix of the fundamental circuit vectors has $e - v + 1$ rows and e columns. This matrix can be divided into two submatrices, one corresponding to the branches and one to the chords. Since the submatrix corresponding to the chords is the unit matrix **I**, the circuit matrix β can be expressed as

$$\beta = [\,\mathbf{B}\,\vdots\,\mathbf{I}\,]$$

where **B** is a submatrix corresponding to the branches. Thus Eq. 13-11 can be expressed as

$$[\,\mathbf{B}\,\vdots\,\mathbf{I}\,]\left[\frac{\mathbf{x}_b}{\mathbf{x}_c}\right] = \mathbf{0} \qquad (13\text{-}12)$$

where $\mathbf{x}_b$ consists of the $v - 1$ across variables of the branches and $\mathbf{x}_c$ consists of the $e - v + 1$ across variables of the chords.

Expanding Eq. 13-12 yields

$$\mathbf{B}\mathbf{x}_b + \mathbf{I}\mathbf{x}_c = \mathbf{0}$$

or

$$\mathbf{x}_c = -\,\mathbf{B}\mathbf{x}_b \qquad (13\text{-}13)$$

This equation indicates that the across variables for the chords can be readily determined from the across variables for the branches.

13-6. CUT-SET VECTORS

As noted previously, the orientation of a fundamental cut-set corresponds to the direction of the defining branch. For the ith cut-set in a system graph, the *cut-set vector* is a row vector of order $1 \times e$ denoted as α_i. The jth element a_{ij} of α_i is defined as follows:

$$a_{ij} = \begin{cases} 1 & \text{if cut-set } i \text{ contains edge } j \text{ oriented in the same} \\ & \text{direction as the cut-set orientation} \\ -1 & \text{if cut-set } i \text{ contains edge } j \text{ oriented in the opposite} \\ & \text{direction to the cut-set orientation} \\ 0 & \text{if edge } j \text{ is not in cut-set } i \end{cases}$$

For example, the four branches in Fig. 13-7 define the four fundamental cut-sets shown in Fig. 13-8a. The corresponding fundamental cut-set vectors are

		Edge							
Cut-Set		1	2	3	4	5	6	7	8
(1, 8)	α_1 =	[1	0	0	0	0	0	0	1]
(2, 3, 5, 6, 8)	α_2 =	[0	1	1	0	−1	−1	0	1]
(4, 5, 6)	α_3 =	[0	0	0	1	1	1	0	0]
(6, 7, 8)	α_4 =	[0	0	0	0	0	1	1	−1]

Another basic axiom of system theory is that the product of any cut-set vector $\boldsymbol{\alpha}_i$ and a column vector $\mathbf{y}$ consisting of the through variables associated with each of the edges of the system graph is zero; i.e.,

$$\boldsymbol{\alpha}_i \mathbf{y} = 0 \tag{13-14}$$

This equation is a fundamental *cut-set equation* of the system graph.

As for the circuit vectors, the four fundamental cut-set vectors $\boldsymbol{\alpha}_1, \boldsymbol{\alpha}_2, \boldsymbol{\alpha}_3, \boldsymbol{\alpha}_4$ can be represented by one matrix. In addition, each fundamental cut-set equation must satisfy Eq. 13-14. Thus, all four equations may be represented as

$$\left[\begin{array}{cccc:cccc} 1 & 0 & 0 & 0 & 0 & 0 & 0 & 1 \\ 0 & 1 & 0 & 0 & 1 & -1 & -1 & 1 \\ 0 & 0 & 1 & 0 & 0 & 1 & 1 & 0 \\ 0 & 0 & 0 & 1 & 0 & 0 & 1 & -1 \end{array}\right] \left[\begin{array}{c} y_1 \\ y_2 \\ y_4 \\ y_7 \\ \hdashline y_3 \\ y_5 \\ y_6 \\ y_8 \end{array}\right] = \mathbf{0} \tag{13-15}$$

Again, the matrix has been arranged so that the edges corresponding to the branches of the tree appear first, followed by the columns corresponding to the chords. The order of appearance of the through variables in the vector $\mathbf{y}$ is correspondingly rearranged. Also, the rows in the cut-set matrix are arranged so that the through variables for the branches are in ascending order of their subscripts. Note that the order of the subscripts of the across variables in the $\mathbf{x}$ vector is the same as the subscripts of the through variables in the $\mathbf{y}$ vector.

For the cut-set matrix, note that the submatrix consisting of the first four columns is the identity matrix, and corresponds to the branches of the tree. If the submatrix of the last four columns corresponding to the chords is denoted as $\mathbf{A}$, the fundamental cut-set matrix $\boldsymbol{\alpha}$ can be expressed as

$$\boldsymbol{\alpha} = [\mathbf{I} \,\vdots\, \mathbf{A}]$$

The corresponding fundamental cut-set equations are

$$[I \; | \; A] \begin{bmatrix} y_b \\ y_c \end{bmatrix} = 0 \qquad (13\text{-}16)$$

where y_b and y_c correspond to the through variables for the branches and chords, respectively.

Performing the matrix multiplication indicated in Eq. 13-16 yields

$$Iy_b + Ay_c = 0$$

or

$$y_b = - Ay_c \qquad (13\text{-}17)$$

Thus the through variables for the branches are simply related to the through variables for the chords.

13-7. INCIDENCE VECTORS

From the edges incident to the ith vertex in a connected system graph, the incidence vectors γ_i of order $1 \times e$ can be defined. The jth entry g_{ij} of incidence vector γ_i is defined as follows:

$$g_{ij} = \begin{cases} 1 \text{ if edge } j \text{ is incident to vertex } i \text{ and directed outwardly} \\ -1 \text{ if edge } j \text{ is incident to vertex } i \text{ and directed inwardly} \\ 0 \text{ if edge } j \text{ is not incident to vertex } i \end{cases}$$

The incidence vectors for the five vertices in Fig. 13-7 are

Vertex			Edge 1	2	3	4	5	6	7	8
a	γ_1	= [−1	1	1	1	0	0	0	0]
b	γ_2	= [0	−1	−1	0	1	0	−1	0]
c	γ_3	= [0	0	0	−1	−1	−1	0	0]
d	γ_4	= [0	0	0	0	0	1	1	−1]
e	γ_5	= [1	0	0	0	0	0	0	1]

From system theory it can also be shown that each incidence vector must satisfy the relationship

$$\gamma_i y = 0 \qquad (13\text{-}17)$$

If all the incidence vectors are used to define a matrix, its order will be $v \times e$ and it is called the *complete incidence matrix*. Since all the incidence vectors must satisfy Eq. 13-17, the equations developed above for the system graph in Fig. 13-7 can be represented by

$$
\begin{bmatrix}
-1 & 1 & 1 & 1 & 0 & 0 & 0 & 0 \\
0 & -1 & -1 & 0 & 1 & 0 & -1 & 0 \\
0 & 0 & 0 & -1 & -1 & -1 & 0 & 0 \\
0 & 0 & 0 & 0 & 0 & 1 & 1 & -1 \\
1 & 0 & 0 & 0 & 0 & 0 & 0 & 1
\end{bmatrix}
\begin{bmatrix}
y_1 \\ y_2 \\ y_3 \\ y_4 \\ y_5 \\ y_6 \\ y_7 \\ y_8
\end{bmatrix} = \mathbf{0}
\qquad (13\text{-}18)
$$

The incidence equations represented by this equation are called the *complete incidence equations.* Note that the sum of the elements in each column is zero, which arises from the fact the each edge is directed outward from one vertex and inward to one vertex.

Due to this last point, one row (corresponding to one incidence vector) is typically deleted from the incidence matrix, which contains v rows, in Eq. 13-18. Furthermore, it can be shown that the rank of the incidence matrix is $v-1$. Thus deleting a row does not reduce the rank of the matrix. Furthermore, given any $v-1$ rows, the remaining row can be obtained by noting that the elements in all columns must sum to one. Consequently, whenever future reference is made to the incidence matrix, the matrix will be with one row deleted, leaving $v-1$ rows.

As for the circuit and cut-set matrices, the incidence matrix can be arranged so that the columns corresponding to a branch of a tree appear first. Deleting the last row and arranging the incidence matrix in Eq. 13-18 so that columns corresponding to branches appear first yields

$$
\begin{array}{ccccccccc}
\text{Edges} & 1 & 2 & 4 & 7 & 3 & 5 & 6 & 8
\end{array}
$$
$$
\begin{bmatrix}
-1 & 1 & 1 & 0 & 1 & 0 & 0 & 0 \\
0 & -1 & 0 & -1 & -1 & 1 & 0 & 0 \\
0 & 0 & -1 & 0 & 0 & -1 & -1 & 0 \\
0 & 0 & 0 & 1 & 0 & 0 & 1 & -1
\end{bmatrix}
$$

Note that the order of the edges is the same as for the circuit and cut-set matrices. Arranging the incidence matrix in this manner insures that the 4×4 submatrix of the first four columns (corresponding to the branches) is nonsingular. For convenience in notation, this matrix can be represented as

$$
[\pi_b \mid \pi_c]
$$

That π_b must be nonsingular will be useful shortly.

13-8. ORTHOGONALITY RELATIONSHIPS

In the previous three sections, three different matrices were obtained from the system graph. Fortunately, these three matrices can be related quite easily, thus saving considerable work in system analysis.

First we shall examine the relationship between the circuit and cut-set matrices. If the scalar product between a representative circuit vector, say β_i, and a typical cut-set vector, say α_j, is computed, the result will invariably be zero. For example, the scalar product of β_1 and α_1 is

$$\beta_1 \alpha_1{}^T = \alpha_1 \beta_1{}^T = \begin{bmatrix} 1 & 0 & 0 & 0 & 0 & 0 & 0 & 1 \end{bmatrix} \begin{bmatrix} -1 \\ -1 \\ 0 \\ 0 \\ 0 \\ 0 \\ 1 \\ 1 \end{bmatrix} = 0 \tag{13-19}$$

Since this holds for each pair of circuit and cut-set vectors, the circuit and cut-set matrices must satisfy the relationship

$$\beta \alpha^T = \alpha \beta^T = 0 \tag{13-20}$$

where α is the cut-set matrix and β^T is the transpose of the circuit matrix. Substituting the partitioned forms for α and β obtained previously, we obtain

$$\beta \alpha^T = \begin{bmatrix} \mathbf{B} & \vdots & \mathbf{I} \end{bmatrix} \begin{bmatrix} \mathbf{I} \\ \mathbf{A}^T \end{bmatrix} = 0 \tag{13-21}$$

$$\alpha \beta^T = \begin{bmatrix} \mathbf{I} & \vdots & \mathbf{A} \end{bmatrix} \begin{bmatrix} \mathbf{B}^T \\ \mathbf{I} \end{bmatrix} = 0 \tag{13-22}$$

Expanding Eq. 13-21 yields

$$\mathbf{B} + \mathbf{A}^T = 0$$

or

$$\mathbf{B} = -\mathbf{A}^T \tag{13-23}$$

Similarly, expanding Eq. 13-22,

$$\mathbf{B}^T + \mathbf{A} = 0$$

or

$$\mathbf{A} = -\mathbf{B}^T \tag{13-24}$$

Consequently if either the circuit or the cut-set matrix is determined from the system graph, the other can be calculated without even considering the graph.

For example, the circuit matrix developed earlier for the system graph in Fig. 13-7 is

$$\begin{bmatrix} 0 & -1 & 0 & 0 & \vdots & 1 & 0 & 0 & 0 \\ 0 & 1 & -1 & 0 & \vdots & 0 & 1 & 0 & 0 \\ 0 & 1 & -1 & -1 & \vdots & 0 & 0 & 1 & 0 \\ -1 & -1 & 0 & 1 & \vdots & 0 & 0 & 0 & 1 \end{bmatrix} = [\mathbf{B} \vdots \mathbf{I}]$$

Calculating $\mathbf{A}$ from $\mathbf{B}$ by Eq. 13-24 yields the cut-set matrix;

$$[\mathbf{I} \vdots \mathbf{A}] = [\mathbf{I} \vdots -\mathbf{B}^T] = \begin{bmatrix} 1 & 0 & 0 & 0 & \vdots & 0 & 0 & 0 & 1 \\ 0 & 1 & 0 & 0 & \vdots & 1 & -1 & -1 & 1 \\ 0 & 0 & 1 & 0 & \vdots & 0 & 1 & 1 & 0 \\ 0 & 0 & 0 & 1 & \vdots & 0 & 0 & 1 & -1 \end{bmatrix}$$

Note that this is the same cut-set matrix as in Eq. 13-15.

Similarly, the cut-set matrix can be derived from the incidence matrix. In this case, the cut-set matrix is given by

$$[\mathbf{I} \vdots \mathbf{A}] = \pi_b^{-1} [\pi_b \vdots \pi_c] = [\mathbf{I} \vdots (\pi_b^{-1} \pi_c)] \tag{13-25}$$

or

$$\mathbf{A} = \pi_b^{-1} \pi_c \tag{13-26}$$

The property derived in the previous section, that π_b is nonsingular, is essential to insure that π_b has an inverse.

For example, in Sec. 13-7 the matrices π_b and π_c were determined for the system graph in Fig. 13-7 to be

$$\pi_b = \begin{bmatrix} -1 & 1 & 1 & 0 \\ 0 & -1 & 0 & -1 \\ 0 & 0 & -1 & 0 \\ 0 & 0 & 0 & 1 \end{bmatrix}$$

$$\pi_c = \begin{bmatrix} 1 & 0 & 0 & 0 \\ -1 & 1 & 0 & 0 \\ 0 & -1 & -1 & 0 \\ 0 & 0 & 1 & -1 \end{bmatrix}$$

The inverse of π_b is

$$\pi_b^{-1} = \begin{bmatrix} -1 & -1 & -1 & -1 \\ 0 & -1 & 0 & -1 \\ 0 & 0 & -1 & 0 \\ 0 & 0 & 0 & 1 \end{bmatrix}$$

Then **A** can be calculated from Eq. 13-26 to be

$$\mathbf{A} = \pi_b^{-1}\pi_c = \begin{bmatrix} 0 & 0 & 0 & 1 \\ 1 & -1 & -1 & 1 \\ 0 & 1 & 1 & 0 \\ 0 & 0 & 1 & -1 \end{bmatrix}$$

Note that this is the same value of **A** as in Eq. 13-15.

13-9. EXAMPLE

To illustrate the application of these principles to a practical example, consider the mechanical system and its corresponding graph in Fig. 13-5. A possible tree

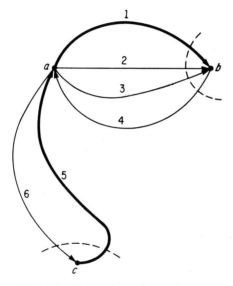

FIG. 13-10. Tree and cut-sets for the system graph in Fig. 13-5b.

for this graph is shown in Fig. 13-10. From this tree, the following circuit matrix can be readily obtained:

Edge

Circuit	1	5	2	3	4	6	
(1, 2)	-1	0	1	0	0	0	
(1', 3)	-1	0	0	1	0	0	= [**B** ⋮ **I**]
(1, 4)	1	0	0	0	1	0	
(5, 6)	0	1	0	0	0	1	

Next, the cut-set matrix can be determined:

Edge

Cut-Set	1	5	2	3	4	6
(1, 2, 3, 4)	1	0	1	1	−1	0
(5, 6)	0	1	0	0	0	−1

$$= [\mathbf{I} \vdots \mathbf{A}]$$

Finally, the incidence matrix can be determined (omitting vertex a):

Edge

Vertex	1	5	2	3	4	6
b	−1	0	−1	−1	1	0
c	0	1	0	0	0	−1

$$= [\boldsymbol{\pi}_b \vdots \boldsymbol{\pi}_c]$$

Alternatively, the cut-set and circuit matrices could have been determined from the incidence matrix.

For the circuit matrix recall the following property:

$$\mathbf{x}_c = - \mathbf{B}\mathbf{x}_b \qquad [13\text{-}13]$$

For this example, this equation becomes (recall that the velocity is the across variable)

$$\begin{bmatrix} v_2 \\ v_3 \\ v_4 \\ v_6 \end{bmatrix} = \begin{bmatrix} 1 & 0 \\ 1 & 0 \\ -1 & 0 \\ 0 & -1 \end{bmatrix} \begin{bmatrix} v_1 \\ v_5 \end{bmatrix} \qquad (13\text{-}27)$$

This equation relates the across variables in the chords to the across variables for the branches.

Similarly, the following relationship can be applied from the cut-set matrix:

$$\mathbf{y}_b = - \mathbf{A}\mathbf{y}_c \qquad [13\text{-}17]$$

Substituting, this equation becomes

$$\begin{bmatrix} f_1 \\ f_5 \end{bmatrix} = \begin{bmatrix} -1 & -1 & 1 & 0 \\ 0 & 0 & 0 & 1 \end{bmatrix} \begin{bmatrix} f_2 \\ f_3 \\ f_4 \\ f_6 \end{bmatrix} \qquad (13\text{-}28)$$

where the force f is the through variable. Thus, the through variables in the branches are related to the through variables for the chords.

Customarily, the through variables associated with the chords in the cotree and the across variables associated with the branches in the tree are denoted as the *primary variables*. The across variables associated with the chords of the cotree and the through variables associated with the branches of the tree are called the *secondary variables*. Note that Eqs. 13-27 and 13-28 give the secondary variables (v_2, v_3, v_4, v_6, f_1, f_5) as functions of the primary variables (v_1, v_5, f_2, f_3, f_4, f_6).

In Sec. 13-3 the following equations were developed for the components of the system:

$$f_1 = M \frac{dv_1}{dt} \qquad [13\text{-}3]$$

$$\frac{df_2}{dt} = K v_2 \qquad [13\text{-}4]$$

$$f_3 = B v_3 \qquad [13\text{-}5]$$

$$f_4 = -\frac{l_1}{l_2} f_5 \qquad [13\text{-}6]$$

$$v_5 = \frac{l_1}{l_2} v_4 \qquad [13\text{-}7]$$

Also recall that f_6 is the driver, and consequently is a specified function of time. These five equations can be expressed as follows:

$$\frac{d}{dt} \begin{bmatrix} v_1 \\ f_2 \end{bmatrix} = \begin{bmatrix} \dfrac{1}{M} & 0 \\ 0 & K \end{bmatrix} \begin{bmatrix} f_1 \\ v_2 \end{bmatrix} \qquad (13\text{-}29)$$

$$\begin{bmatrix} f_3 \\ f_4 \\ v_5 \end{bmatrix} = \begin{bmatrix} B & 0 & 0 \\ 0 & 0 & -l_1/l_2 \\ 0 & l_1/l_1 & 0 \end{bmatrix} \begin{bmatrix} v_3 \\ v_4 \\ f_5 \end{bmatrix} \qquad (13\text{-}30)$$

Note that Eq. 13-29 consists of the terminal equations for the dynamic components and gives the time derivatives of primary variables as explicit functions of secondary variables. Equation 13-30 includes the terminal equations of the components described by algebraic equations, and gives the primary variables as explicit functions of secondary variables.

Next, the circuit and cut-set equations are combined with the component equations. For Eq. 13-29,

$$\frac{d}{dt} \begin{bmatrix} v_1 \\ f_2 \end{bmatrix} = \begin{bmatrix} \dfrac{1}{M} & 0 \\ 0 & K \end{bmatrix} \begin{bmatrix} f_1 \\ v_2 \end{bmatrix}$$

$$= \begin{bmatrix} \dfrac{1}{M} & 0 \\ 0 & K \end{bmatrix} \begin{bmatrix} 0 & 0 & \vdots & -1 & -1 & 1 & 0 \\ 1 & 0 & \vdots & 0 & 0 & 0 & 0 \end{bmatrix} \begin{bmatrix} v_1 \\ v_5 \\ \hline f_2 \\ f_3 \\ f_4 \\ f_6 \end{bmatrix}$$

$$= \begin{bmatrix} 0 & -\dfrac{1}{M} & -\dfrac{1}{M} & \dfrac{1}{M} \\ K & 0 & 0 & 0 \end{bmatrix} \begin{bmatrix} v_1 \\ f_2 \\ f_3 \\ f_4 \end{bmatrix}$$

$$= \begin{bmatrix} 0 & \dfrac{1}{M} \\ K & 0 \end{bmatrix} \begin{bmatrix} v_1 \\ f_2 \end{bmatrix} + \begin{bmatrix} -\dfrac{1}{M} & \dfrac{1}{M} \\ 0 & 0 \end{bmatrix} \begin{bmatrix} f_3 \\ f_4 \end{bmatrix} \tag{13-31}$$

For Eq. 13-30,

$$\begin{bmatrix} f_3 \\ f_4 \\ v_5 \end{bmatrix} = \begin{bmatrix} B & 0 & 0 \\ 0 & 0 & -l_1/l_2 \\ 0 & l_1/l_2 & 0 \end{bmatrix} \begin{bmatrix} v_3 \\ v_4 \\ f_5 \end{bmatrix}$$

$$= \begin{bmatrix} B & 0 & 0 \\ 0 & 0 & -l_1/l_2 \\ 0 & l_1/l_2 & 0 \end{bmatrix} \begin{bmatrix} 1 & 0 & 0 & 0 & 0 & 0 \\ -1 & 0 & 0 & 0 & 0 & 0 \\ 0 & 0 & 0 & 0 & 0 & 1 \end{bmatrix} \begin{bmatrix} v_1 \\ v_5 \\ f_2 \\ f_3 \\ f_4 \\ f_6 \end{bmatrix}$$

$$= \begin{bmatrix} B & 0 & 0 \\ 0 & 0 & -l_1/l_2 \\ 0 & l_1/l_2 & 0 \end{bmatrix} \begin{bmatrix} 1 & 0 \\ -1 & 0 \\ 0 & 1 \end{bmatrix} \begin{bmatrix} v_1 \\ f_6 \end{bmatrix}$$

$$= \begin{bmatrix} B & 0 \\ 0 & -l_1/l_2 \\ -l_1/l_2 & 0 \end{bmatrix} \begin{bmatrix} v_1 \\ f_6 \end{bmatrix} \tag{13-32}$$

Substituting Eq. 13-32 into Eq. 13-31,

$$\frac{d}{dt} \begin{bmatrix} v_1 \\ f_2 \end{bmatrix} = \begin{bmatrix} 0 & -\dfrac{1}{M} \\ K & 0 \end{bmatrix} \begin{bmatrix} v_1 \\ f_2 \end{bmatrix} + \begin{bmatrix} -\dfrac{1}{M} & \dfrac{1}{M} \\ 0 & 0 \end{bmatrix} \begin{bmatrix} B & 0 \\ 0 & -l_1/l_2 \end{bmatrix} \begin{bmatrix} v_1 \\ f_6 \end{bmatrix}$$

$$= \begin{bmatrix} 0 & -\dfrac{1}{M} \\ K & 0 \end{bmatrix} \begin{bmatrix} v_1 \\ f_2 \end{bmatrix} + \begin{bmatrix} -B/M & -Ml_1/l_2 \\ 0 & 0 \end{bmatrix} \begin{bmatrix} v_1 \\ f_6 \end{bmatrix}$$

$$= \begin{bmatrix} -B/M & -1/M \\ K & 0 \end{bmatrix} \begin{bmatrix} v_1 \\ f_2 \end{bmatrix} + \begin{bmatrix} -Ml_1/l_2 \\ 0 \end{bmatrix} f_6 \qquad (13\text{-}33)$$

This differential equation is the *state equation* of the system, and the vector consisting of v_1 and f_2 is the *state vector*. This nomenclature arises because the variables v_1 and f_2 are sufficient to specify the remaining across and through variables in the system, and thus the state of the system. Consequently, the variables in the state vector are called the *state* variables, and the state equation along with the equations necessary to specify the remaining variables comprise the *state model*.

13-10. MAXIMALLY SELECTED TREE

In the above example recall that the mass, a velocity (across variable) driver, was included in the branch of the tree. Furthermore, note that the spring and the force generator, both force (through variable) drivers, were chords in the cotree. In order to simplify the manipulations in developing the state equations, the tree and cotree should be selected in a judicious manner so that they satisfy the following conditions as well as possible:

1. In the equations describing algebraic components, the primary variables are expressed as explicit functions of the secondary variables or time.
2. The derivatives of primary variables are expressed as functions of secondary variables for as many dynamic components as possible. Since the tree and cotree selected specify which variables are primary or secondary, the tree must be selected in a manner so as to comply with the above two conditions. When so selected the tree is called the *maximally selected tree*. Generally, the maximally selected tree will contain as many across drivers as possible in the tree and as many through drivers as possible in the cotree.

For some systems it is not possible to place all the through drivers in the cotree and all the across drivers in the tree.

13-11. DISTILLATION COLUMN

In this section the linear graph is developed for a more complex system, namely, a *distillation column*. To keep the problem relatively simple but yet successfully illustrate the important points, the abbreviated column in Fig. 13-11 is treated. The distilling mixture is binary, equimolal overflow holds, the feed

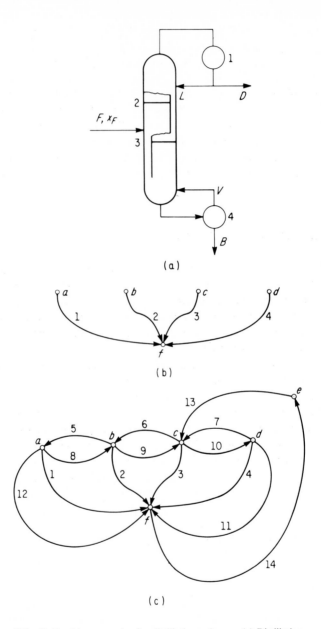

FIG. 13-11.   Linear graph of a distillation column.   (a) Distillation
column.   (b) Section of linear graph corresponding to accumulation
terms.   (c) Complete linear graph.

enters at its bubble point, a total condenser is used, tray efficiencies are unity, and other common simplifying assumptions also hold. It is not necessary to make these assumptions, but they keep the problem simple.

A dynamic model is desired, so each tray will be treated as perfectly mixed with holdup H_t. The holdup of the condenser is H_c, and the holdup in the reboiler is H_b. The linear graph can be initiated by representing the accumulation term corresponding to each of these as shown in Fig. 13-11b. The equations for each of these edges are

$$\text{Edge 1:} \quad m_1 = H_c \frac{dx_1}{dt}$$

$$\text{Edge 2:} \quad m_2 = H_t \frac{dx_2}{dt}$$

$$\text{Edge 3:} \quad m_3 = H_t \frac{dx_3}{dt}$$

$$\text{Edge 4:} \quad m_4 = H_b \frac{dx_4}{dt}$$

where m is the molal rate of accumulation (which can be visualized as a flow), x is the mole fraction light component, and subscripts 1, 2, 3, and 4 refer to the condenser, the top plate, the feed plate, and the reboiler, respectively. Note that x is the across variable for each edge, and m is the through variable.

Now consider the vapor flows in the column. Vapor flows from the top plate to the condenser, which is represented by edge 5 in Fig. 13-11c. The equation for this edge is

$$m_5 = K_2 V x_2$$

where m_5 is the molal rate at which light component enters the condenser in the vapor stream, K_2 is the equilibrium constant for the liquid on the top plate, and V is the vapor flow up the column. Similarly, edges 6 and 7 can be added, and the corresponding equations are

$$m_6 = K_3 V x_3$$
$$m_7 = K_4 V x_4$$

Next, consider the internal liquid flows. Liquid reflux flows from the condenser to the top plate at rate L, which is represented by edge 8 in Fig. 13-11c. The equation for this edge is

$$m_8 = L x_1$$

Similarly, edges 9 and 10 represent the other liquid streams, and the corresponding equations are

$$m_9 = L x_2$$
$$m_{10} = (L + F) x_3$$

Finally, consider the feed stream, the overheads stream, and the bottoms stream.   The bottoms stream is represented by edge 11, whose equation is

$$m_{11} = Bx_4$$

The overheads stream is represented by edge 12, whose equation is

$$m_{12} = Dx_1$$

The feed stream is represented by edge 13, whose equation is

$$m_{13} = Fx_{14}$$

Edge 14 is an across variable driver to give the feed composition x_F, and is represented by the equation

$$x_{14} = x_F$$

This completes the linear graph.

It is interesting to note the significance of the nodes.   Node a corresponds to the condenser, node b corresponds to the top plate, node c corresponds to the feed plate, node d corresponds to the reboiler, node e corresponds to the source of the feed stream, and node f corresponds to the ground or zero value of x.   The equation for this node represents a component material balance for the entire column, which can be verified from node f of the linear graph and the previous equations to be:

$$Fx_F = Lx_1 + H_c \frac{dx_1}{dt} + H_t \frac{dx_2}{dt} + H_t \frac{dx_3}{dt} + H_b \frac{dx_4}{dt} + Bx_4$$

This equation is obtained by summing the through variables at node f.

From the linear graph in Fig. 13-11c and the above equations, it can be shown that

$$\frac{d}{dt} \begin{bmatrix} x_1 \\ x_2 \\ x_3 \\ x_4 \end{bmatrix} =$$

$$\begin{bmatrix} -(L+D)/H_c & VK_2/H_c & 0 & 0 \\ L/H_t & -(VK_2+L)/H_t & VK_3/H_t & 0 \\ 0 & L/H_t & -(VK_3+L+F)/H_t & V_4K_4/H_t \\ 0 & 0 & (L+F)/H_b & -(VK_4+B)/H_b \end{bmatrix} \begin{bmatrix} x_1 \\ x_2 \\ x_3 \\ x_4 \end{bmatrix}$$

$$+ \begin{bmatrix} 0 \\ 0 \\ F/H_t \\ 0 \end{bmatrix} x_F$$

The details are left to the Problems.

13-12. SUMMARY

In this chapter a systematic procedure for developing state-space models of lumped-parameter or discrete systems has been presented. This technique seems to have been developed primarily for electrical systems, and several discussions are available for such systems [1,3,4]. However, it is applicable to any lumped-parameter or discrete system, many of which are found in all branches of engineering. Another similar concept, called the signal-flow graph[5], is also applicable to such systems, but the linear graph seems to be superior, although the two have many points in common.

REFERENCES

1. H. E. Koenig, Y. Tokad, and H. K. Kesevan, *Analysis of Discrete Physical Systems*, McGraw-Hill, New York, 1967.
2. C. O. Bennett and J. E. Myers, *Momentum, Heat, and Mass Transfer*, McGraw-Hill, New York, 1962.
3. S. Seshu and M. B. Reed, *Linear Graphs and Electrical Networks*, Addison-Wesley, Reading, Mass., 1961.
4. W. A. Blackwell and L. L. Grigsby, "A Simple Formulation Procedure for State Space Models of Electric Networks," *Journal of the Franklin Institute*, Vol. 281, No. 6 (June 1966), pp. 486-497.
5. S. J. Mason, "Feedback Theory—Further Properties of Signal Flow Graphs." *Proceedings of the IRE*, Vol. 44, No. 7 (July 1956), pp. 920-926.

PROBLEMS

13-1. Develop a linear graph which describes the system in Prob. 3-1.

13-2. Suppose the mass of water in Prob. 3-1 were W. Develop the linear graph, determine the maximally selected tree, and derive the state equation.

13-3. Draw a terminal graph for the hot-water tank in Prob. 3-2.

13-4. Draw a terminal graph for the thermal bulb in Prob. 3-4.

13-5. Suppose a thermal bulb as considered in the previous exercise is immersed in the hot-water tank in Prob. 3-2. Furthermore, suppose a controller compares the output of the thermal bulb with a desired temperature T_d and manipulates the steam flow S. What is the component graph for the controller? Draw the linear graph for the complete system.

13-6. Draw the linear graph for the series of tanks in Prob. 3-9. Which edges should be in the tree?

13-7. Draw a linear graph for the flash problem in Prob. 3-15.

13-8. Draw a linear graph for the carrier cycle in Prob. 3-19.

13-9. Draw a linear graph for the finite-difference approximation in Prob. 3-24. Use only four elements.

13-10. Develop a finite-difference approximation for the coffee pot in Prob. 7-13 by dividing the bed into only four elements. Draw its linear graph.

13-11. Draw a linear graph for a typical node in the one-dimensional aquifer discussed in Sec. 10-12.

13-12. Develop a linear graph for the hydraulic system in Fig. 10-10.

13-13. Develop the state equation for the column in Sec. 13-11 from the system graph.

chapter 14

Dimensional Analysis

In the previous chapters of this text the emphasis has been placed on a rigorous mathematical treatment of engineering problems. While this is the preferable approach whenever possible, many engineering systems are too complex for a theoretical treatment. For example, turbulent-flow patterns have not been successfully treated mathematically, and an empirical chart is used to calculate pressure drop in conduits when the fluid is in turbulent flow. So many engineering systems are of such a complex nature that some technique is necessary for relating the pertinent variables in these systems. One successful approach has been through the use of dimensional analysis in conjunction with experimentation.

For a physical system, dimensional analysis will indicate which variables or groups of variables are functionally related, but it gives no insight as to the nature of the relationship. Hence, it is usually necessary to determine the actual functional relationship by experimental methods, and perhaps the most important application of dimensional analysis is in the planning of experiments [1]. Applying dimensional analysis to the set of variables describing the system leads to a set of dimensionless groups of the variables which are to be related. As the number of dimensionless groups is always less than the number of original variables, the amount of experimental work required is decreased. In addition, knowing the groups of variables which must be related allows the experimenter to more efficiently plan his approach.

In the discussion that follows, additional advantages and disadvantages of dimensional analysis will become apparent. As the name implies, dimensional analysis is based on the dimensions of the variables, and first an understanding of the concept of dimensions as applied to engineering systems is necessary.

14-1. DISTINCTION BETWEEN DIMENSIONS AND UNITS[1,2]

There is a distinction between "dimensions" and "units," and an understanding of both terms is essential. Examples of dimensions are length and mass, but each of these may be measured in a variety of units. For example, length can be measured in microns, centimeters, feet, meters, miles, or light-years, and each of these is quantitatively related to the others. Although there is nothing fundamental about any dimension, the dimensions of some variables are commonly considered to be combinations of more "fundamental" dimensions. For example, the dimensions of density are mass/(length)^3, and the units may be lb/ft^3, gm/cc, tons/cubic mile, etc.

In order to provide a consistent set of units for measurement purposes, several units have been specified by international agreement. Currently, the units of mass, length, time, temperature, and electrical charge are specified either arbitrarily or by some fundamental physical equation. Hence, it is natural to think of the dimensions of the above entitites as being "fundamental," from which the dimensions of all other variables are "derived." In some cases not all the above "fundamental" dimensions are needed; in other cases additional dimensions are substituted or added to this set. In mechanics the common "fundamental" set consists of mass $[M]$, length $[L]$, and time $[T]$, although the set consisting of force $[F]$, length $[L]$, and time $[T]$ is also frequently used.

The dimensions of many variables may be obtained from an inspection of their definitions or units, whereas the dimensions of some are obtained by physical laws. For example, velocity is defined as the rate of change of position with respect to time; hence its dimensions are [length]/[time]. On the other hand, the dimensions of force may be obtained from Newton's second law of motion:

$$F = ma \qquad (14\text{-}1)$$

where F = force

m = mass

a = acceleration

This relationship indicates that the dimensions of force are the same as the dimension of the product of mass times acceleration, $[ML/T^2]$.

14-2. THE ENGINEERING SYSTEM OF UNITS[3]

As stated in Sec. 1-10, in its customary usage in American engineering circles, the unit of force is the pound force (lb$_f$). This is the force exerted by gravity on a one-pound mass (lb$_m$). Applying Newton's law,

$$F = ma = (1 \text{ lb}_m)\,(32.2 \text{ ft/sec}^2) \qquad (14\text{-}2)$$

In order for the results of the above expression to be in lb$_f$, we must introduce a conversion factor g_c, defined as

$$g_c = \frac{32.2 \text{ ft-lb}_m}{\text{lb}_f\text{-sec}^2} \qquad (14\text{-}3)$$

Inserting this conversion factor, the *gravitational constant*, into Eq. 14-2 yields

$$F = \frac{ma}{g_c} = \frac{(1 \text{ lb}_m)\,(32.2 \text{ ft/sec}^2)}{32.2 \ \dfrac{\text{ft-lb}_m}{\text{lb}_f\text{-sec}^2}} = 1 \text{ lb}_f \qquad (14\text{-}4)$$

Thus the definition of a pound force is satisfied. Although Newton's law in the

Table 14-1

Dimensions of Mechanical Variables [1,4]

Variable	$[M],[L],[T]$ System	$[F],[M],[L],[T]$ System
Length	$[L]$	$[L]$
Mass	$[M]$	$[M]$
Time	$[T]$	$[T]$
Force	$[ML/T^2]$	$[F]$
g_c	$[1]$	$[ML/FT^2]$
Density	$[M/L^3]$	$[M/L^3]$
Pressure (stress)	$[M/LT^2]$	$[F/L^2]$
Velocity	$[L/T]$	$[L/T]$
Acceleration	$[L/T^2]$	$[L/T^2]$
Energy (work)	$[ML^2/T^2]$	$[FL]$
Momentum	$[ML/T]$	$[FT]$
Power	$[ML^2/T^3]$	$[FL/T]$
Viscosity (dynamic)	$[M/LT]$	$[M/LT]$
Viscosity (kinematic)	$[L^2/T]$	$[L^2/T]$
Surface tension	$[M/T^2]$	$[F/L]$
Diffusion coefficient	$[L^2/T]$	$[L^2/T]$
Frequency	$[1/T]$	$[1/T]$

form of Eq. 14-4 must be used instead of the form in Eq. 14-2 when using the engineering set of units, Eq. 14-4 may also be used when the metric units are employed. In this latter case, g_c is simply unity and has no dimensions.

In Table 14-1 are shown the dimensions of several variables in both the $[M]$, $[L]$, $[T]$ and the $[F]$, $[M]$, $[L]$, $[T]$ systems. Note that g_c must be included in the set of variables when using the $[F]$, $[M]$, $[L]$, $[T]$ system, but when using the $[M]$, $[L]$, $[T]$ system, g_c may be omitted. In the exercises at the end of this chapter two examples of fundamental sets consisting of only two dimensions are presented.

In examining systems in which thermal considerations are important, the conversion factor for the mechanical equivalent of heat plays a role similar to g_c. For example, the typical units of heat capacity are Btu/lb°F. The Btu is a measure of energy, and the dimensions of energy in both the $[M]$, $[L]$, $[T]$ and the $[F]$, $[M]$, $[L]$, $[T]$ systems are given in Table 14-1. To use these sets of fundamental dimensions, it would be necessary to convert Btu to its mechanical energy equivalent by the conversion factor J, defined as

$$J = \frac{778 \text{ ft-lb}_f}{\text{Btu}} \tag{14-5}$$

or its counterpart in another set of units.

An alternate technique is to include energy $[Q]$ in the set of fundamental dimensions and J in the set of variables, which is analogous to including g_c in the set of variables when using the $[F]$, $[M]$, $[L]$, $[T]$ system. Table 14-2 gives the dimensions of several additional variables of engineering importance in systems both with and without $[Q]$ in the fundamental set of dimensions.

TABLE 14-2

Dimensions of Thermal Variables

Variable	$[F],[M],[L],[T],[\theta]$ System	$[F],[M],[L],$ $[T],[\theta],[Q]$ System
Temperature	$[\theta]$	$[\theta]$
Energy	$[FL]$	$[Q]$
J	$[1]$	$[FL/Q]$
Heat capacity	$[FL/M\theta]$	$[Q/M\theta]$
Thermal conductivity	$[F/T\theta]$	$[Q/TL\theta]$
Convective heat-transfer coefficient	$[F/TL\theta]$	$[Q/TL^2\theta]$
Thermal diffusivity	$[L^2/T]$	$[L^2/T]$
Heat of vaporization per unit mass	$[FL/M]$	$[Q/M]$
Enthalpy per unit mass	$[FL/M]$	$[Q/M]$
Rate of heat flow	$[FL/T]$	$[Q/T]$

14-3.   AN EXAMPLE USING DIMENSIONAL ANALYSIS[2,5]

To illustrate the ideas of dimensional analysis before discussing the basic principles in detail, consider the pendulum swinging in a vacuum as shown in Fig. 14-1.   Using dimensional analysis, the natural frequency f of oscillation of the pendulum is to be related to the factors influencing f.   Without prior knowledge of the solution of this problem, one may suppose that the variables of importance

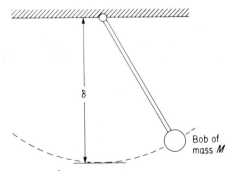

FIG. 14-1.   Swinging pendulum.

are the length δ of the pendulum, the mass m of the bob, and the acceleration g due to gravity. In order to express the relationship between these four variables (f, δ, m and g) in a dimensionless equation, the product of each variable raised to some power must equal a dimensionless constant. Expressing this mathematically,

$$f^a \delta^b m^c g^d = K = \text{dimensionless constant} \tag{14-6}$$

Substituting the dimensions of the four variables into Eq. 14-6,

$$[1/T]^a [L]^b [M]^c [L/T^2]^d = [1] \tag{14-7}$$

or

$$[L]^{b+d} [M]^c [T]^{-a-2d} = 1 \tag{14-8}$$

Requiring the left-hand side to be dimensionless,

$$b + d = 0 \tag{14-9}$$
$$c = 0 \tag{14-10}$$
$$-a - 2d = 0 \tag{14-11}$$

Since c equals zero, the mass of the bob does not affect the frequency of oscillation, as is shown mathematically in elementary physics texts. In addition, Eqs. 14-9 and 14-11 indicate

$$b = -d \tag{14-12}$$
$$a = -2d \tag{14-13}$$

Substituting these values of a, b, and c into Eq. 14-6,

$$f^{-2d} \delta^{-d} g^d = K \tag{14-14}$$

or

$$\left(\frac{g}{\delta f^2}\right)^d = K \tag{14-15}$$

or

$$\frac{g}{\delta f^2} = K^{1/d} = K' \tag{14-16}$$

From physics it is known that

$$f = \frac{1}{2\pi} \sqrt{g/\delta} \tag{14-17}$$

which can be rearranged to a form similar to Eq. 14-16:

$$\frac{g}{\delta f^2} = 4\pi^2 = K' \tag{14-18}$$

Thus dimensional analysis successfully predicts the correct form of the relationship, but does not indicate the value of the constant. If a rigorous mathematical treatment were not possible, the constant could be determined experimentally by only one experiment.

14-4. DIMENSIONAL HOMOGENEITY[1]

If the form of an equation is independent of the fundamental units of measurement, the equation is said to be dimensionally homogeneous. Requiring Newton's law to be dimensionally homogeneous was used earlier to determine the value of g_c from the definition of a pound mass. Equation 14-17 describing the frequency of oscillation of a pendulum is also independent of the units for measuring length and time, as long as a consistent set is used. There is a large number of equations in the engineering literature that are not dimensionally homogeneous, and such equations are usually referred to as *empirical equations*. If Eq. 14-17 is restricted to pendulums on the earth's surface, it reduces to

$$f = 0.9\sqrt{\delta} \qquad\qquad (14\text{-}19)$$

when f is measured in sec^{-1} and δ is in feet. If another set of units is used, the constant 0.9 must be changed.

14-5. THE BUCKINGHAM PI THEOREM[1]

In the pendulum problem there was one dimensionless group, and therefore the value of either a, b, or d could have been specified arbitrarily to simplify the work. Although this would not have been very helpful in this case, it becomes essential in more complex applications. As originally stated by Buckingham [1], the Pi theorem says that the number of dimensionless products in a complete set of dimensionless groups is equal to the total number of variables minus the number of fundamental dimensions. In a complete set of dimensionless groups each original variable appears in at least one dimensionless group and no dimensionless group may be expressed as a combination of other dimensionless groups in the complete set. Although this statement is simple and widely used, it is not infallible, and more rigorous statements are available [1,6].

For the purpose of this text perhaps the best statement of the Pi theorem is as follows: *The number of dimensionless products in a complete set equals the total number of variables minus the rank of the dimensional matrix.*

The rank of the dimensional matrix is equivalent to the maximum number of variables in the original set that will not form a dimensionless product. To illustrate the application of this theorem, a more complex example will be treated.

One of the most fruitful areas in which dimensional analysis has been applied is in mixer design and scale-up.   In order to select the proper drive unit for a mixer, it is necessary to know the power input.   To do this, a small-scale model geometrically similar to the large-scale mixer (prototype) is tested, and the information from the model is used to design the prototype.   From an inspection of the problem the significant variables affecting the power input P are the diameter of the impeller d, the revolutions per minute N, the density of the fluid ρ, the viscosity μ, and the accelleration due to gravity g.   Using the $[F], [M], [L], [T]$ system, the gravitational constant g_c must also be included.   These variables along with their dimensions are tabulated in Table 14-3.   From this table the dimensional matrix is constructed such that each column consists of the exponents in the dimensional expression for the corresponding variable, as shown.

TABLE 14-3

Variables in Mixer Example

Variable	Symbol	Dimensions
Power	P	FL/T
Diameter	d	L
Density	ρ	M/L^3
Rpm	N	$1/T$
Viscosity	μ	M/LT
Acceleration due to gravity	g	L/T^2
Gravitational constant	g_c	ML/FT^2

	P	d	ρ	N	μ	g	g_c
F	1	0	0	0	0	0	-1
M	0	0	1	0	1	0	1
L	1	1	-3	0	-1	1	1
T	-1	0	0	-1	-1	-2	-2

For example, the dimension of P is $[F^1\ M^0\ L^1\ T^{-1}]$.

The rank of the dimensional matrix is the order of the highest order nonzero determinant composed from the dimensional matrix.   In the above case the determinant composed of the first four columns is nonzero:

$$
\begin{bmatrix}
1 & 0 & 0 & 0 \\
0 & 0 & 1 & 0 \\
1 & 1 & -3 & 0 \\
-1 & 0 & 0 & -1
\end{bmatrix} = 1
\tag{14-20}
$$

Hence the rank of the dimensional matrix is four. Since the number of original variables is seven, the number of independent dimensionless groups is $7 - 4$, or 3. By convention, dimensionless groups are commonly denoted by π_1, π_2, and π_3.

To form a dimensionless group π from the original variables, each of the variables is raised to a power as in the following expression:

$$\pi = P^{k_1}\, d^{k_2}\, \rho^{k_3}\, N^{k_4}\, \mu^{k_5}\, g^{k_6}\, g_c^{k_7} \tag{14-21}$$

In terms of the dimensions of the variables, Eq. 14-21 takes the following form:

$$[\pi] = [FL/T]^{k_1}\, [L]^{k_2} [M/L^3]^{k_3}\, [1/T]^{k_4} [M/LT]^{k_5} [L/T^2]^{k_6}\, [ML/FT^2]^{k_7} \tag{14-22}$$

Collecting the exponents of each of the fundamental dimensions, Eq. 14-22 becomes

$$[\pi] = [F]^{(k_1 - k_7)}\, [M]^{(k_3 + k_5 + k_7)}\, [L]^{(k_1 + k_2 - 3k_3 - k_5 + k_6 + k_7)}\ \ [T]^{(k_1 - k_4 - k_5 - 2k_6 - 2k_7)} \tag{14-23}$$

Recalling that $|\pi|$ is dimensionless, the exponent of each of the fundamental dimensions in Eq. 14-23 must be zero, which yields the following four equations:

$$k_1 - k_7 = 0 \tag{14-24}$$

$$k_3 + k_5 + k_7 = 0 \tag{14-25}$$

$$k_1 + k_2 - 3k_3 - k_5 + k_6 + k_7 = 0 \tag{14-26}$$

$$-k_1 - k_4 - k_5 - 2k_6 - 2k_7 = 0 \tag{14-27}$$

Upon inspection, note that the coefficients of the above equation correspond to rows in the dimensional matrix. Hence it is possible to write down directly the equations for the exponents in the dimensionless groups by inspection of the dimensional matrix, thereby simplifying the procedure considerably.

The above set of four equations contains seven unknowns which may be solved for four variables in terms of the remaining three. Three exponents, i.e., the value of three of the k's, may be selected arbitrarily. In order to select the values of these three variables, one must consider how the resulting correlation is to be used. For scale-up of mixers, the power is the variable to be determined, and it would be preferable if it appeared in only one group. Similarly, it is possible to restrict two other variables to appear in only one group, and μ and g are selected in this case.

In selecting the variables to be prescribed, there is one restriction—the determinant composed of the remaining columns of the dimensional matrix must be nonzero. This determinant is

$$\begin{vmatrix} 0 & 0 & 0 & -1 \\ 0 & 1 & 0 & 1 \\ 1 & -3 & 0 & 1 \\ 0 & 0 & -1 & -2 \end{vmatrix}$$

whose value is unity.

To determine the dimensionless group containing the power P, set $k_1 = 1$, $k_5 = 0$, and $k_6 = 0$, thus insuring that μ and g will not appear. Substituting these quantities in Eqs. 14-24 through 14-27.

$$1 - k_7 = 0 \tag{14-28}$$

$$k_3 + k_7 = 0 \tag{14-29}$$

$$1 + k_2 - 3k_3 + k_7 = 0 \tag{14-30}$$

$$-1 - k_4 - 2k_7 = 0 \tag{14-31}$$

Solving these four equations for the individual k's,

$$k_2 = -5 \tag{14-32}$$

$$k_3 = -1 \tag{14-33}$$

$$k_4 = -3 \tag{14-34}$$

$$k_7 = 1 \tag{14-35}$$

Hence from Eq. 14-21,

$$\pi_1 = P^1 d^{-5} \rho^{-1} N^{-3} \mu^0 g^0 g_c^1 = \frac{P g_c}{\rho N^3 d^5} = N_P \tag{14-36}$$

which is commonly called the *power number*.

To determine the second dimensionless group π_2, set $k_1 = 0$, $k_5 = 1$, $k_6 = 0$, and substitute into Eqs. 14-24 through 14-27. Upon solving, $k_2 = -2$, $k_3 = -1$, $k_4 = -1$, and $k_7 = 0$, which indicates

$$\pi_2 = \frac{\mu}{d^2 \rho N} \tag{14-37}$$

However, recall that dimensional analysis indicates nothing about the functional relationship between the individual dimensionless groups. Hence, the reciprocal of the group in Eq. 14-37 could be used instead.

$$\pi_2 = \frac{d^2 \rho N}{\mu} = N_{\text{Re}} \tag{14-38}$$

This group is the *Reynolds number*, N_{Re},

By an analogous procedure the third dimensionless group can be determined. Setting $k_1 = 0$, $k_5 = 0$, and $k_6 = 1$, solution of Eqs. 14-24 through 14-27 yields $k_2 = -1$, $k_3 = 0$, $k_4 = -2$, and $k_7 = 0$. Hence

$$\pi_3 = \frac{g}{dN^2} \qquad (14\text{-}39)$$

If the reciprocal is used instead,

$$\pi_3 = \frac{dN^2}{g} = N_{Fr} \qquad (14\text{-}40)$$

which is the *Froude number, N_{Fr}*.

In this problem, dimensional analysis has reduced the original problem of relating six variables to one of relating three variables, a much easier task. The functional relationship is generally denoted as

$$N_P = \varphi(N_{Re}, N_{Fr}) \qquad (14\text{-}41)$$

where the nature of φ is to be determined experimentally.

To carry this illustration further, consider the design of a mixer in which the effect of gravity, and thus the Froude number, is negligible, which is usually true for mixers with sufficient baffling provided to prevent formation of a vortex. In this case Eq. 14-41 becomes

$$N_P = \varphi \, (N_{Re}) \qquad (14\text{-}42)$$

A geometrically similar model of the large mixer is to be tested, and the information used to design the large version (prototype). Note that if the model is operated at the same N_{Re} as the prototype, then N_P for both mixers is also the same. Hence, it is possible to determine the power input to the large mixer from only one test on the model.

Suppose a mixer 4-ft in diameter for a fluid with the physical properties of water is to be operated at 80 rpm. A 2-ft-diameter model geometrically similar to the prototype is available for testing. To design the mixer, the first step is to determine the rpm to be used when testing the model mixer. Since the model and prototype are to be operated at the same Reynolds number,

$$\left(\frac{d^2 \rho N}{\mu}\right)_m = \left(\frac{d^2 \rho N}{\mu}\right)_p \qquad (14\text{-}43)$$

If water is used in the model, Eq. 14-43 reduces to

$$N_m = N_p \left(\frac{d_p}{d_m}\right)^2 \qquad (14\text{-}44)$$

Since the mixers are geometrically similar, the ratio of the impeller diameters is the same as the ratio of the tank diameters, and

$$N_m = 80 \text{ rpm } (2)^2 = 320 \text{ rpm} \qquad (14\text{-}45)$$

In some cases it is necessary to use a different fluid in the model so that the same Reynolds number may be obtained with a lower rpm.

Assume that when testing the model mixer at the above rpm, the input was 0.1 hp. Since N_p is the same for both the model and the prototype,

$$\left(\frac{Pg_c}{\rho N^3 d^5}\right)_m = \left(\frac{Pg_c}{\rho N^3 d^5}\right)_p \tag{14-46}$$

Solving for the power input to the prototype,

$$P_p = P_m\left(\frac{N_p}{N_m}\right)^3 \left(\frac{d_p}{d_m}\right)^5 \tag{14-47}$$

Substituting numerical quantities into this equation,

$$P_p = 0.4 \text{ hp} \tag{14-48}$$

Thus a simple solution to a seemingly complex problem has been obtained using dimensional analysis.

14-6. SINGULAR DIMENSIONAL MATRIX[1]

In the previous examples the rank of the dimensional matrix has been equal to the number of rows in the matrix, in which case the matrix is said to be *nonsingular*. In some cases the rank is less than the number of rows in the matrix, and the dimensional matrix is said to be *singular*. In such cases it is generally unnecessary to consider all the rows in the dimensional matrix.

For example, consider the case in which the dimensional matrix is as follows:

	A	B	C	D
M	1	− 1	0	1
L	2	3	2	0
T	1	− 1	0	1

All third-order determinants in this matrix are zero, since

$$\begin{vmatrix} 1 & -1 & 0 \\ 2 & 3 & 2 \\ 1 & -1 & 0 \end{vmatrix} = \begin{vmatrix} 1 & -1 & 1 \\ 2 & 3 & 0 \\ 1 & -1 & 1 \end{vmatrix} = \begin{vmatrix} 1 & 0 & 1 \\ 2 & 2 & 0 \\ 1 & 0 & 1 \end{vmatrix} =$$

$$= \begin{vmatrix} -1 & 0 & 1 \\ 3 & 2 & 0 \\ -1 & 0 & 1 \end{vmatrix} = 0 \tag{14-49}$$

However, the dimensional matrix contains at least one nonzero second-order determinant, e.g.,

$$\begin{vmatrix} 0 & 1 \\ 2 & 0 \end{vmatrix}$$

Therefore, the rank of the dimensional matrix is 2. In cases such as this, one of the rows in the original dimensional matrix may be omitted, thus forming a new dimensional matrix whose rank must equal the number of rows. For the above case the original dimensional matrix may be reduced to the following dimensional matrix:

	A	B	C	D
M	1	-1	0	1
L	2	3	2	0

Note that the rank of this matrix is 2. From this point, the dimensionless groups are determined as before.

14-7. THE SIGNIFICANCE OF DIMENSIONLESS GROUPS

In the literature a variety of dimensionless groups have appeared and some 210 of these are tabulated by Catchpole and Fulford [8]. In engineering work it is helpful to attach a physical significance to several common groups, some of which are shown in Table 14-4.

TABLE 14-4

Significance of Dimensionless Groups*

Name	Symbol	Expression	Significance
Brinkman	N_{Br}	$\dfrac{\mu V^2}{k\Delta t}$	Heat generation/heat transferred
Euler	N_{Eu}	$\dfrac{g_c p}{V^2 \rho}$	Pressure force/inertial force
Froude	N_{Fr}	$\dfrac{V^2}{gL}$	Inertial force/gravitational force
Graetz	N_{Gz}	$\dfrac{Wc_p}{kL}$	Thermal capacity of fluid/convective heat transfer
Grashof	N_{Gr}	$\dfrac{L^3 \rho^2 g \beta \Delta t}{\mu^2}$	N_{Re} (buoyancy force/viscous force)
Mach	N_{Ma}	V/V_s	Linear velocity/sonic velocity
Nusselt	N_{Nu}	$\dfrac{hL}{k}$	Total heat transfer/conductive heat transfer

TABLE 14-4 *(continued)*

Name	Symbol	Expression	Significance
Prandtl	N_{pr}	$\dfrac{c_p\,\mu}{k}$	Momentum diffusivity/ thermal diffusivity
Peclet (mass)	N_{Pe_m}	$\dfrac{LV}{D}$	Bulk mass transfer/ diffusive mass transfer
Peclet (heat)	N_{Pe_h}	$\dfrac{LV\rho c_p}{k}$	Bulk heat transfer/ conductive heat transfer
Reynolds	N_{Re}	$\dfrac{LV\rho}{\mu}$	Inertia force/viscous force
Schmidt	N_{Sc}	$\dfrac{\mu}{\rho D}$	Kinematic viscosity/ molecular diffusivity
Sherwood	N_{Sh}	$\dfrac{k_c L}{D}$	Mass diffusivity/ molecular diffusivity
Stanton	N_{St}	$\dfrac{h}{c_p \rho V}$	Heat transferred/ thermal capacity of fluid
Weber	N_{We}	$\dfrac{V^2 \rho L}{\sigma}$	Inertia force/surface tension force

* Reproduced by permission from *Industrial and Engineering Chemistry*, Vol. 58, No. 3 (March 1966), pp. 46-60, copyright 1966 by The American Chemical Society.

14-8. PRECAUTIONS

When presenting experimental data or obtaining correlations with the aid of dimensionless groups, certain subtle factors may enter to make the correlation appear excellent when in fact it is not good at all. In the literature this is evidenced by an example in which hypothetical data actually determined from a table of random numbers were correlated apparently very well by several different functional relationships [9].

Recall at this point that dimensional analysis determines the dimensionless groups based on the variables originally assumed to be significant. Hence there is the possibility that either too many variables were included or some significant variables were omitted, the latter of which could certainly lead to a poor correlation between the dimensionless groups derived. To prevent an apparent correlation of uncorrelated data, several precautions should be taken when using dimensionless groups.

First, log-log paper must be used cautiously when plotting the results. Actually there are only two valid reasons for using log-log graph paper;

1. To determine the constants in an equation of the form

$$y = kx^n$$

2. To present data ranging over several orders of magnitude on one graph. Perhaps the best way to determine if there is really any association between the dimensionless groups is to first plot them on linear graph paper. Second, apparent "stray" points should not be neglected simply because they do not fit the correlation. Third, using a dimensionless group when only one variable in the group was actually varied may "mask" the variation. Another very subtle factor may occur when plotting a dimensionless group or product of dimensionless groups against another group or groups. If the same variable appears in both the groups or combinations of groups being plotted, it will certainly improve the appearance of the correlation. If the variable appears to a high power, this effect is even more pronounced, and moreover it makes the correlation appear good over a wide range of values of the dimensionless groups.

14-9. SUMMARY

In this chapter a review of the use of dimensional analysis to obtain an expression describing a complex engineering system has been presented. Several excellent discussions are available in the literature [1, 2, 3, 4, 7, 8, 10, 11, 12, 13], some of which present proofs of the Pi Theorem [1,6]. Dimensional analysis is a powerful tool, but like all techniques it has its disadvantages and pitfalls. Some additional points are illustrated in the Problems.

REFERENCES

1. H. L. Langhaar, *Dimensional Analysis and Theory of Models,* Wiley, New York, 1951.
2. I. H. Silberberg and J. J. McKetta, "Learning How to Use Dimensional Analysis," *Petroleum Refiner,* Vol. 324, p. 179-183; Vol. 325, pp. 147-150.
3. I. C. Ipsen, *Units, Dimensions, and Dimensionless Numbers,* McGraw-Hill, New York, 1960.
4. Frank Kreith, *Principles of Heat Transfer,* 2d ed., International Textbook, Scranton, Pa., 1965.
5. R. L. Sutherland, *Engineering Systems Analysis,* Addison-Wesley, Reading, Mass., 1958.
6. Van Driest, "On Dimensional Analysis and the Presentation of Data in Fluid Flow Problems," *Journal of Applied Mathematics,* Vol. 13, No. 1 (March 1946), p. A-34.

7. C. O. Bennett and J. E. Myers, *Momentum, Heat, and Mass Transfer*, McGraw-Hill, New York, 1962.

8. J. P. Catchpole and G. Fulford, "Dimensionless Groups," *Industrial and Engineering Chemistry*, Vol. 58, No. 3 (March 1966), pp. 46-60.

9. P. N. Rowe, "The Correlation of Engineering Data," *The Chemical Engineer*, March 1963, pp. 69-76.

10. P. W. Bridgman, *Dimensional Analysis*, Yale U. P., New Haven, 1931.

11. H. E. Huntley, *Dimensional Analysis*, MacDonald & Co., London, 1952.

12. W. J. Duncan, *Physical Similarity and Dimensional Analysis*, Edward Arnold, London, 1953.

13. C. M. Focken, *Dimensional Methods and Their Applications*, Edward Arnold, London, 1953.

PROBLEMS

14-1. When treating astronomical systems, a fundamental set of units consisting of only length and time is very convenient. To define force and mass in this system, Newton's law of motion $F = ma$ and Newton's law of gravitation $F = m\, m'/r^2$ may be used. In this system determine the dimensions of mass, force, density, and energy.

14-2. According to Einstein's law in modern physics, matter is a form of energy. Using this concept a set of fundamental units consisting of only mass and time can be developed if the unit of work is the energy of a gram mass; i.e., work and mass have the same dimensions. Retaining Newton's law of motion $F = ma$, what are the dimensions of length, velocity, force, and density in such a system?

14-3. The drag force on a body moving through a fluid is a function of the velocity, the projected area, the density, and the absolute viscosity (which can be neglected at high velocities). If the quantity to be determined is the force, determine the dimensionless groups required in such a relationship.

14-4. Repeat the mixer problem in the text, but consider the case in which the surface tension is also important.

14-5. Consider a ship moving through the ocean. If we restrict our attention to vessels that are geometrically similar, their shape is completely specified by one dimension L. The drag force on the hull depends upon the characteristic length L, the velocity V with which the ship moves through the water, the density ρ and viscosity μ of the water, and the acceleration due to gravity g. Since energy is dissipated by waves as the ship moves through the water, the term g is important since the energy of the waves depends upon g. Determine a complete set of dimensionless groups that describe this problem. Allow F, μ, and g to appear in only one group each.

14-6. For turbulent flow a pipe, the pressure drop Δp depends upon the diameter D of the pipe, length L of the pipe, the bulk velocity V of the fluid, the density ρ, the viscosity μ, and the roughness height e (the dimension of e is $[L]$). Restricting Δp, μ, L, and e to appearing in only one group each, determine the complete set of dimensionless groups required to express this relationship.

14-7. The height h to which a liquid will rise in a capillary tube depends upon the diameter D of the tube, the surface tension σ, the density ρ of the fluid, and the acceleration due to gravity g. Express this relationship by use of dimensionless groups, using the $[M]$, $[L]$, $[T]$ system. Show that the result is consistent with the equation $\sigma = 1/4\, hDg\rho$ used to determine surface tension.

14-8. In the literature, Benzing (*Ind. Eng. Chem.*, vol. 47, 1955, p. 2087) reports a study on the size of the bubbles formed by a gas issuing from a small orifice beneath the surface of the liquid. The variables affecting the diameter D of the bubbles are the orifice diameter d, the acceleration due to gravity g, the density ρ, the viscosity μ, and the surface tension σ of the liquid. Using the $[F]$, $[M]$, $[L]$, and $[T]$ set of fundamental dimensions, show that a relationship between the above variables can be expressed as

$$\frac{D}{d} = f\left[\frac{g_c\sigma}{\rho\, gd^2}, \frac{\mu^2}{\rho^2 gd^3}\right]$$

14-9. The pressure rise p across a centrifugal pump depends upon the fluid density ρ, the speed (rpm) of revolution N, the impeller diameter D, the volumetric flow rate Q, and the fluid viscosity μ. If it is desired that p, Q, and μ appear in only one group each, determine the dimensionless groups to be related.

14-10. Replacing the pressure rise p by the power input P, repeat exercise 14-9.

14-11. Replacing the pressure rise p by the pump efficiency η, repeat exercise 14-9.

14-12. In many cases, it is possible to combine several variables into only one when making a study of the problem with dimensional analysis. For example, the temperature-time history of a billet cooling in a constant temperature bath is dependent upon the temperature difference $(T - T_b)$ between the billet and the bath at time θ after immersion, the surface conductance (hA_s) between billet and bath, and the heat capacity $(c_p\rho V)$ of the billet. Treating each of these quantities as a single variable, show that the following relationship is consistant with that indicated by dimensional analysis.

$$\frac{T - T_b}{T_0 - T_b} = \exp\left(-\frac{hA_s\theta}{c_p\rho V}\right)$$

14-13. Consider a long homogeneous cylinder with uniform internal heat generation (such as a wire through which current is flowing). The steady state difference between the temperature T_c at the center of the cylinder and the surface temperature T_s in a function of the diameter D, the thermal conductivity k, and the rate of heat generation $\dot{q}$. Using dimensional analysis, suggest a relationship between these variables.

chapter 15

Statistical Techniques

15-1. ACCURACY, PRECISION, AND ERRORS

The end product of an engineering evaluation is frequently a numerical answer. The numerical answer may be a number, an equation, a graphical plot, or a mathematical model. It is desirable to determine how accurate these answers may be and what factors affect their accuracy. An answer in print looks extremely good and the reader seldom questions its accuracy. It is therefore desirable in the first place to define what it meant by *accuracy*. Accuracy is the degree of closeness of an observed value of a variable to the true value of that variable. The observed value is said to be accurate when its value exactly equals the true value. *Accuracy* should not be confused with *precision*. Precision is the degree of reproducibility of an observed value. For example, a weighing machine may indicate a value of 1.12 tons when a 1.0-ton calibrated weight is placed on it. If this value of 1.12 tons is obtained for repeated weighings over a reasonably large interval of time, the machine is obviously precise, but not accurate.

The difference between the true value and the observed value is the error. The error may be positive or negative. If x is any observed quantity and Δx is the error, then an estimate of the true value is given by

$$x \pm \Delta x \qquad (15\text{-}1)$$

Errors in experimental observation can be broadly placed into two categories:

1. Determinate errors
2. Indeterminate errors

Determinate errors are errors that can be detected and corrected for by calibration. For instance, the weighing machine mentioned before could be calibrated by a series of standard weights and a calibration curve could be drawn. The calibration curve will thus give the determinate error.

An *indeterminate error* is an error due to an unassignable cause, although frequently the cause lies in the limitations of the experimental setup and the observer. For instance, a manometer across an orifice meter may have been read very carefully and under ideal conditions will show a small variation in repeated readings for a given value of flow through the meter. The variation may be due to small fluctuations in flow, parallax error in reading the value, or any of a host of plausible reasons, for all of which no calculable corrections can be applied. The error in this case is indeterminate.

Errors can also develop due to calculations. These errors are not due to mistakes in calculation but due to *round-off* or truncation of numerical values. Round-off or truncation of numbers is necessary because of an inability to carry an indefinite number of digits, and also because some of these digits are of no significance. For example, population figures for a country, state, or city are reported in thousands, since these are never arrived at by a head count. It is thus meaningless to talk of a city population of 131,132. The last three digits are not significant unless this figure was arrived at by an actual head count. Only the first three digits are significant. This leads to the concept of significant figures. Consider the numbers 46.007, 15.645, 0.0089206, 1.9873×10^6. All of these numbers contain five significant figures. It should be noted that the zeros after the decimal points are not significant—they merely indicate where these five significant figures should be placed in the number scale. Frequently we wish to carry only a specified number of significant figures. The rules for rounding-off are as follows:

1. For retaining a significant figure increase the digit in the nth place by 1, if the digit in the $(n + 1)$th place is more than 5. *Example:* 46.2586 is rounded to 46.3, and 46.26, correct to three and four significant figures respectively.

2. If the digit in the $(n + 1)$th place is less than 5 the digit in the nth place remains unaltered. *Example*: .00892346 is rounded to .00892, .008923, correct to three and four significant figures.

3. If the digit in the $(n + 1)$th place is exactly 5, then if the digit in the nth place is even, leave it unaltered; if odd, increase by 1. *Example:* 3.425 rounds to 3.42, and 6.775 rounds to 6.78.

It has been indicated earlier that significant figures give rise to round-off errors in arithmetic calculations. This is so because if a number is significant to n digits, then the nth digit can be in error by up to $\pm 1/2$ because of the round-off rules shown above. Arithmetic operations with numbers involving different significant figures will thus be in error. The following rules indicate the error:

1. *Addition and Subtraction.* The answer of a sum of or difference of approximate numbers is correct to the same number of significant digits as the number having the least number of significant digits involved in the addition or subtraction operations. For example, if ten numbers are added, nine of which have five significant digits each and the one remaining three, then the answer is correct to three significant digits.

2. *Multiplication and Division.* In this case the error is not so obvious. Consider the following numbers, 2,436 and 3,442. If these numbers are exact then the answer is not in doubt; if approximate, then because of round-off, the last digit is in error by $\pm 1/2$. In other words the fourth digits could very well have been 5 and 1 respectively. The lower limit of the approximate numbers are thus 2,435 and 3,441. The product of the upper and lower limit of the numbers are respectively 8,384,712 and

8,378,835. If the two products are compared it is seen that the first two significant figures compare exactly, while the third figure is in error. The product of two, numbers of four significant figures each gives an answer which is correct to three significant figures. This example illustrates that the product of approximate numbers will be correct to one significant figure less than the original number. It can thus be appreciated why round-off errors can give rise to meaningless answers in a digital computer. Normally the computer carries eight significant figures; thus eight consecutive multiplications could reduce the answer to a meaningless number. The above analysis is a conservative or rather limiting estimate of the error. In practice, the errors may tend to oppose each other and cancel out. A good rule would be, as in the case of addition and subtraction, to carry as many significant figures in the product or quotient as the factor having the least significant figures entering the computation.

15-2. PROPAGATION OF ERRORS

Variables entering a mathematical relation may be either measured or calculated quantities. It is certainly desirable to know the limits on the error for the various quantities entering an equation. These limits fall into two categories.

1. Given the degree of accuracy (or the error) of the measured quantities, what is the degree of accuracy (or the error) of the calculated quantity?
2. Given the degree of accuracy (or the error) on the calculated quantity, what should be the degree of accuracy (or the error) of the measured quantity?

In order to make such an evaluation, a relationship such as a mathematical model must be available. Consider the equation

$$y = f(x_1, x_2, x_3, \cdots, x_i) \tag{15-2}$$

where the x_i are the measured quantities and y the calculated quantity. Let Δy and Δx_i be the errors in y and x_i respectively. Then

$$y + \Delta y = f(x_1 + \Delta x_1, x_2 + \Delta x_2, x_3 + \Delta x_3, \cdots, x_i + \Delta x_i) \tag{15-3}$$

Expanding y by means of a Taylor series expansion in the neighborhood of a point $x_1, x_2, x_3, \cdots, x_i$:

$$y + \Delta y = f(x_1, x_2, \cdots, x_i) + \frac{\partial f}{\partial x_1} \Delta x_1 + \frac{\partial f}{\partial x_2} \Delta x_2$$
$$+ \cdots \frac{\partial f}{\partial x_i} \Delta x_i + \frac{\partial^2 f}{\partial x_1^2} \frac{(\Delta x_1)^2}{2!} + \cdots \tag{15-4}$$

Subtracting Eq. 15-2 from Eq. 15-4 the result is

$$\Delta y = \frac{\partial f}{\partial x_1} \Delta x_1 + \frac{\partial f}{\partial x_2} \Delta x_2 + \cdots + \frac{\partial f}{\partial x_i} \Delta x_i$$

$$+ \frac{\partial^2 f}{\partial x_i^2} \frac{(\Delta x_1)^2}{2!} + \cdots$$

(15-5)

If the Δ_{x_i} are chosen small enough the second-order and higher terms may be neglected to give

$$\Delta y = \frac{\partial f}{\partial x_i} \Delta x_1 + \frac{\partial f}{\partial x_2} \Delta x_2 + \cdots + \frac{\partial f}{\partial x_i} \Delta x_i \qquad (15\text{-}6)$$

In accordance with Eq. 15-1, the Δ's can be considered the errors and hence Eq. 15-6 gives a relationship between the errors. It should be noted that Eq. 15-6 is correct only when the Δ's are small, otherwise higher-order terms must be considered. The use of Eq. 15-6 will be illustrated with the following example.

Example

Orifice meters are frequently used as flow-measuring devices, and it is necessary to calibrate them. Consider an orifice meter used for measuring the flow rate of water with the pressure drop measured by a mercury manometer. The simplest method of calibration is to obtain the mass-flow rate by collecting and weighing the water flowing through the meter in a definite period of time and noting the pressure drop across the manometer. In a particular run the mass-flow rate W was found to be 19 lb/min, and the corresponding pressure drop across the meter was 9.1 in. Hg. It is desired to know what is the maximum possible error in calculating the orifice coefficient, C_v.

The orifice equation is given by

$$u = C_v \sqrt{2gh_w}$$

where u = velocity of flow, ft/sec through orifice
h_w = pressure loss across meter, ft of water
g = gravitational constant

This can be converted to mass-flow rates by multiplying by the area of the orifice A_o, and the density of water ρ.

$$W \left(\frac{\text{lb}}{\text{min}} \right) = u A_o \rho 60 = C_v A_o \rho 60 \sqrt{2gh_w}$$

Now, putting the orifice pressure drop in terms of the manometer reading gives

$$h_w \rho_w = \rho_m h_m - \rho_w h_m$$

$$h_w = \frac{h_m}{12} \frac{\rho_m - \rho_w}{\rho_w}$$

$$= \frac{(13.6 - 1)}{12} h_m$$

$$= 1.05 h_m$$

where ρ_m = density of manometer liquid
$\quad h_m$ = pressure drop across manometer, in. Hg.
Hence the result is

$$W = C_v A_o \rho_w 60 \sqrt{2g1.05} \sqrt{h_m}$$

Given that the diameter of the orifice is 0.25 in.,

$$W = C_v K \sqrt{h_m}$$

where

$$K = A_o \rho_w 60 \sqrt{2g \, 1.05}$$

The quantities that go to make up K are constants whose values are known to at least the second decimal place, hence they are lumped together.

$$K = \frac{\Pi}{4} (0.25)^2 \text{ in.}^2 \times \frac{1}{144} \frac{\text{ft}^2}{\text{in.}^2} \times 0.623 \frac{\text{lb}}{\text{ft}^3} \sqrt{2 \times 32.2 \times 1.05} \frac{\text{ft}^{1/2}}{\text{sec}} \times 60 \frac{\text{sec}}{\text{min}}$$

$$= 10.5 \frac{\text{lb}}{\text{sec-ft}^{1/2}}$$

Solving for C_v taking the appropriate partial derivatives gives

$$C_v = \frac{1}{K} \frac{W}{\sqrt{h_m}}$$

$$\frac{\partial C_v}{\partial W} = \frac{1}{K \sqrt{h_m}} ; \frac{\partial C_v}{\partial h_m} = \frac{1}{2K} \frac{W}{(h_m)^{3/2}}$$

Substituting in Eq. 15-6, the result is

$$\Delta C_v = \frac{1}{K \sqrt{h_m}} \Delta W - \frac{1}{2K} \frac{W}{h_m^{3/2}} \Delta h_m$$

This equation is valid for small changes in W and h_m. It will be used to find the error in C_v at the point $W = 19$ lb/min $h_w = 9.1$ in. Hg.

$$\Delta C_v = \frac{1}{10.5} \frac{1}{\sqrt{9.1}} \Delta W - \frac{1}{10.5 \times 2} \frac{19}{(9.1)3/2} \Delta h_m$$

$$= 0.0316 \, \Delta W - 0.0332 \, \Delta h_m$$

In order to make an estimate of ΔC_c it is necessary to know ΔW and Δh_m. Let us assume that W can be measured only to 0.1 lb/min and h_m to 0.1 in. Hg. Hence

$$\Delta W = 0.1$$

$$\Delta h_m = 0.1$$

The error ΔW and Δh_m can be either positive or negative, hence the maximum error in C_v is

$$\Delta C_v = 0.0316 \times 0.1 + 0.0332 \times 0.1 = 0.00648$$

also

$$C_v = \frac{1}{K}\frac{W}{\sqrt{h_m}} = \frac{1}{10.5}\frac{19}{\sqrt{9.1}} = 0.6$$

Hence

$$C_v = 0.6 \pm 0.006$$

It should be noted that the error is very small which is verified by experimental results, since it is well known that for turbulent flow, the orifice coefficient is essentially constant. The Reynolds number for this case was 12,000. It should also be noted that the error calculated is the maximum possible, and is valid only for the point in question ($W = 19$lb/min, $h_m = 0.1$ in. Hg). In practice, errors may tend to cancel each other.

15-3. STATISTICAL ANALYSIS

Physical problems seldom have simple solutions. However, this does not mean that it is not possible to define the problem, and generally the important variables are known. What is not known is the functional relationships among the variables. One of the purposes of experimentation is to develop a mathematical model that will approximate the true functional relationship. The data obtained from experimental observations are the values of the dependent variable measured for various values of the independent variables. We wish to determine the following:

1. The functional relationship among the variables.
2. How precisely a variable can be predicted, knowing the value of the remaining variables in the functional relationship—in other words, the magnitude of the error bound on the variables.

These questions can be answered by statistical methods, and it thus becomes desirable to investigate statistical techniques. The question may be asked, what is statistics? The word in normal usage has a broad meaning; and the definition given here is restricted to include definite objectives.

Before these objectives are defined it is necessary to define the terms *distribution* and *population*. The population is the entire object or system under investigation as opposed to the sample which is a fraction or part of the system. Any observation or number obtained from the sample is called a *statistic,* while that from a population is a *parameter.* A statistic (calculated or observed quantity from a sample) due to random error will not be identical to the population parameter, but will be a *random variable,* and will have a relative frequency of occurrence about the parameter. This frequency of occurrence or distribution can be approximated by mathematical functions called *distribution functions.*

Statistics is the science that deals with the assembly, analysis, and interpretation of data obtained by observations and measurements. Three broad regions can be defined.

1. *Mathematical Statistics.* This area deals with the properties of distribution functions that describe various kinds of populations.
2. *Applied Statistics.* This area uses distribution functions as theoretical models for actual populations in order to obtain useful results.
3. *Design of Experiments.* This area employs statistical methods to plan experimental program that will yield data that will give the most information from a statistical point-of view.

15-4. DEFINITIONS OF STATISTICAL PARAMETERS

Given a collection of raw data, one would plot the variables as shown in Fig. 15-1. It is observed that the data scatters about the line drawn. The line may be considered either the true relationship between x and y or a best-fitting line. A precise definition of the best fit will be given later. The equation of the line is given by

$$Y = mX + c \qquad (15\text{-}7)$$

The data is given in the form such that for every x_i there is a corresponding y_i obtained by experimental measurement. The measure of the scatter or the devia-

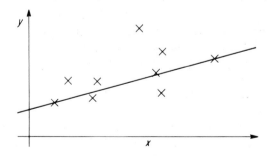

FIG. 15-1.   Relationship between x and y.

tion is then $y_i - Y_i$, where Y_i is calculated from the given value of x_i obtained by substituting into Eq. 15-7.

$$\text{Deviation} \equiv y_i - Y_i \tag{15-8}$$

From Fig. 15-1 it is seen that deviation ($y_i - Y_i$) can be positive or negative, depending on which side of the line the observed point lies. An important quantity used to describe all of the deviations is the variance.

$$\text{Variance} = \sum_{i=1}^{n} (y_i - Y_i)^2/(n-1) \tag{15-9}$$

A quantity such as deviation and variance obtained through calculations performed on experimental data is called a *statistic*. If we knew a physical phenomenon completely, then we could calculate its parameters which, in turn, would give us useful information about the phenomenon. In practice we seldom deal with entire systems; but only a part of it, a presentative sample. Statistical methods are then employed to use sample statistics to predict population parameters. For example, it is desired to evaluate the properties of a certain catalyst. The most reliable evaluation will be made if all of the catalyst is subject to test. After the test is over, we no doubt have "good" data, but no catalyst left. A balance can be achieved by performing tests on a sample. If a proper sample is taken, then the properties of the sample can be used to predict the properties of the catalyst. The above may appear obvious; however, it is very seldom that two samples taken from the same population give identical properties. It is necessary then to test to see if the two samples belong to the same population. This again is a problem of statistics.

It is now appropriate to define *probability*. Probability in a statistical sense can be taken to mean the relative frequency of an event, when the event is observed a large number of times. For example, if one considers the tossing of a coin, then the events are heads and tails. Both events are equally likely, by which is meant that if the coin is unbaised and it is tossed a large number of times the event heads will appear exactly half the number of times. The relative frequency of occurrence of heads is thus 0.5 and by definition above, the probability of heads is 0.5. By a similar reasoning, the probability of tails is also 0.5. It should be observed that the events heads and tails are mutually exclusive, which means that the occurrence of one event excludes the occurence of the other.

In general, probability is a nonnegative number whose value lies between zero and one. A value of zero signifies that the event will not occur and one, that the event is certain to occur.

Consider the general case of n possible events $x_1, x_2, \cdots, x_n$. These events are mutually exclusive and have a probability of occurrence of $p_1, p_2, \cdots, p_n$ respectively. In a given operation at least one event must occur since they are mutually exclusive. Hence the combined probability of all events is unity, or

$$p_1 + p_2 + \cdots + p_n = 1 \tag{15-10}$$

15-5. DISTRIBUTION FUNCTIONS

Variation about a mean value can occur in various ways. In Fig. 15-1 the variation or scatter was with respect to the true or best value in a functional relationship. Another common problem is the small but persistent disagreement between repeated measurements of a variable at a particular point. For example, several trials on a heat exchanger may be made under virtually identical conditions to evaluate the heat transfer coefficient. Despite the utmost care, a small and random variation will be present in the result. Let us assume that the heat-transfer coefficient calculated is good only to two significant figures. The data are hence tabulated in intervals of 10 Btu and the number of times they occur in that interval is recorded as in Table 15-1 and is plotted in Fig. 15-2. The plot in Fig. 15-2 is a frequency-distribution plot called a *histogram*. It gives a measure of

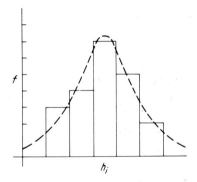

FIG. 15-2. Plot of data in Table 15-1 for frequency of occurrence in an interval as a function of the average value of the interval.

the spread of the data. The dotted line in Fig. 15-2 is the Gauss distribution function which does fit this distribution very well. In general, there are many mathematical distribution functions available. For example, the best known are the Binomial and Poisson (discrete functions), and Maxwell (continuous functions). These functions are obtained by considering infinite sets of data having a random variaton about a mean value and as will be indicated later must satisfy Eq. 15-21.

The Gauss distribution function is by far the most important for the following reasons:

1. Sampling errors seem to fit it very well.
2. It can be used to approximate the discrete distribution functions such as the Binomial and Poisson distributions under certain conditions.
3. It is approached in the limit as described later when the central limit theorem is discussed.

Because of its importance the Gauss distribution function is often referred to

TABLE 15-1

Experimental Data on Heat-Transfer Coefficients

Heat-Transfer Coefficient h[Btu/ (ft)² (hr) (°F)]	Average Value of Interval, h_i	Frequency of Occurrency in Interval, f
190	195	1
200	205	3
210	215	4
220	225	7
230	235	5
250	245	2
		22

as the normal distribution.   The Gauss distribution will be used to illustrate the solutions to problems in the area of applied statistics.

The arithmetic mean is defined as the sum of the observations divided by the total number of observations.   In Table 15-1 the heat-transfer coefficient h_i occurs f_i times in a particular interval of size Δh (in this case Δh equals 10 Btu). Hence total number of observations N is

$$N = \sum_{i=1}^{n} f_i$$

where n = total number of intervals.

The mean heat-transfer coefficient $\overline{h}$ from the definition of mean is

$$\overline{h} = \frac{\sum_{i=1}^{n} f_i h_i}{\sum_{i=1}^{n} f_i}$$

$$= \frac{1}{N} \sum_{i=1}^{n} f_i h_i \tag{15-11}$$

Consider what happens when N tends to infinity:

$$\lim_{N \to \infty} (\overline{h}) = \left[\lim_{N \to \infty} \sum_{i=1}^{n} h_i \frac{f_i}{N} \right]$$

$$= \sum_{i=1}^{n} \overline{h}_i \lim_{N \to \infty} \frac{f_i}{N}$$

The quantity in brackets is the relative frequency of occurrence of h_i and is the probability p_i of h in the range h_i and $h_i + \Delta h$. The quantity $\lim_{n \to \infty} (h)$ is defined as the expected value of h, $E(h)$, hence

$$E(h) = \sum_{i=1}^{n} h_i p_i \qquad (15\text{-}12)$$

As the number of observations N tends to infinity and the interval size Δh tends to zero the histogram of Fig. 15-2 will tend to a smooth curve. p_i will then be $p(h)$, and it is the probability of occurrence of h in the range h and $h + dh$. It is called the *probability density function*. Eq. 15-12 for the continuous case becomes

$$E(h) = \int_{-\infty}^{\infty} h p(h) \; dh \qquad (15\text{-}13)$$

We now define a general quantity called a *moment*. The kth moment about the origin is defined as the expected value of x^k:

$$\mu_k = E(x^k) = \sum_{i=1}^{\infty} x_i^k f(x_i) \qquad \text{for discrete distributions} \qquad (15\text{-}14)$$

$$= \int_{-\infty}^{\infty} x^k f(x) \; dx \qquad \text{for continuous distributions}$$

where $f(x)$ and $f(x_i)$ are probability density functions similar to $p(h)$ of Eq. 15-13.

For $k = 1$, comparing Eq. 15-14 with Eqs. 15-13 and 15-11 it is seen that μ_1 is the mean. Since an infinite number of observations are taken this mean is called μ, the *population mean*.

We can also define the kth moment about the mean μ as the expected value of $(x - \mu)k$ or

$$\mu_k = \sum_{i=1}^{\infty} (x_i - \mu)^k f(x_i) \qquad \text{for discrete distribution} \qquad (15\text{-}14a)$$

$$= \int_{-\infty}^{\infty} (x - \mu)^k f(x) \; dx \qquad \text{for continuous distribution}$$

The second moment μ_2 about the means is usually defined as the variance, σ^2 and is again a population statistic.

The moments are useful in describing the properties of distribution functions. For example the first and second moment are the mean and variance respectively. Higher-order moments are of less general use.

The sample variance $S^2(X)$, is defined in Eq. 15-9, and is considered a best estimate of the population variance, since for large N it approximates Eq. 15-14a:

$$\text{Sample variance} = S^2(x) = \left[\sum_{i=1}^{n} (x_i - \bar{x})^2 \right] \bigg/ (n-1) \qquad (15\text{-}15)$$

The sample variance $S^2(x)$ can be put in more desirable form by expanding square term

$$S^2(x) = 1/(n-1) \left[\sum_{i=1}^{n} (x_i^2 - 2x_i\bar{x} + \bar{x}^2) \right]$$

$$= 1/(n-1) \left[\sum_{i=1}^{n} x_i^2 - \sum_{i=1}^{n} 2x_i\bar{x} + \sum_{i=1}^{n} \bar{x}^2 \right]$$

$$= 1/(n-1) \left[\sum_{i=1}^{n} x_i^2 - 2\bar{x} \sum_{i=1}^{n} x_i + n\bar{x}^2 \right]$$

$$= 1/(n-1) \left[\sum_{i=1}^{n} x_i^2 - 2n\bar{x}\bar{x} + n\bar{x}^2 \right]$$

$$= 1/(n-1) \left[\sum_{i=1}^{n} x_i^2 - n\bar{x}^2 \right]$$

$$S^2(x) = \frac{\sum\limits_{i=1}^{n} x_i^2 - \left(\sum\limits_{i=1}^{n} x_i\right)^2 \bigg/ n}{(n-1)} \qquad (15\text{-}15a)$$

15-6.   GAUSS DISTRIBUTION

The Gauss distribution represents a random variation about the mean μ. The probability density of occurrence of a particular value of x is a continuous distribution given by

$$f = \frac{\exp\left[\frac{1}{2} \frac{(x-\mu)^2}{\sigma} \right]}{2\pi\sigma} \qquad (15\text{-}16)$$

where f is the probability density function of x, x is the value of the variable subject to random error, and μ is the mean value of x.

Equation 15-16 can be normalized or made independent of the units of measurement of x by the following substitution

$$z = \frac{x - \mu}{\sigma} \qquad (15\text{-}17)$$

which gives

$$f = \frac{1}{\sigma \sqrt{2\pi}} \, e^{-1/2 z^2} \tag{15-18}$$

The concept of probability is now applied to the discrete distribution function of Table 15-1, which is shown in Fig. 15-2. Let the frequency of occurrence f_i of a particular observation be in the interval x_i and x_{i+1}. For example, in Table 15-1 for $x_i = 210$, $x_{i+1} = 220$; $f_i = 4$. The total number of observations is $\sum_{i=1}^{n} f_i = N$.

The relative frequency or probability of occurrence in the interval x_i and x_{i+1} will be f_i/N. As x_{i+1} approaches x_i the histogram approaches a continuous curve and the interval shrinks to an infinitesimal dx. f_i thus becomes a continuous function of x, $f(x)$, which is the probability of the variable x in the interval x and $x + dx$. $f(x)$ is also known as the probability density*. If we wish to determine the probability of occurrence in the interval x_1 and x_2 from the extension of the above ideas we have

$$\text{Probability of occurrence between } x_1 \text{ and } x_2 = \frac{\text{frequency of occurrence between } x_1 \text{ and } x_2}{\text{total frequency}} \tag{15-19}$$

which can be expressed as

$$\text{Probability of occurrence between } x_1 \text{ and } x_2 = \frac{\int_{x_1}^{x_2} f(x)\, dx}{\int_{-\infty}^{\infty} f(x)\, dx} \tag{15-20}$$

Equation 15-20 can be evaluated with the Gauss distribution. The event in this case is the value of z, which ranges from minus infinity to plus infinity. For each value of z there is a relative frequency of occurrence of f in the interval z and $z + dz$ given by Eq. 15-18.

It can be shown for the Gauss distribution† that

$$\int_{-\infty}^{\infty} f(x)\, dx = \int_{-\infty}^{\infty} \frac{1}{\sqrt{2\pi}} \, e^{-1/2 z^2}\, dz$$

$$= 1 \tag{15-21}$$

In particular, for the Gauss distribution Eq. 15-20 reduces to

$$\text{Probability of occurrence between } z_1 \text{ and } z_2$$

$$= \int_{z_1}^{z_2} \frac{1}{\sqrt{2\pi}} \, e^{-1/2 z^2}\, dz \tag{15-22}$$

---

* For a more detailed discussion, see, for example, I. S. Sokolnikoff and R. M. Redheffer, *Mathematics of Physics and Engineering*, McGraw-Hill, New York, 1958, p. 632.

† E. G. Craig, *Laplace and Fourier Transforms for Electrical Engineers*, Holt, New York, 1964, p. 366.

Equation 15-21 can be considered to be the total probability of occurrence of z, which is a mutually exclusive event (since for any given event z can have only one value). Equation 15-22 says that the probability of occurrence of z between the limits z_1 and z_2 is the area of the Gauss distribution curve between z_1 and z_2.

It should be noted that if Eq. 15-21 is satisfied by any distribution function $f(x)$ not necessarily the Gauss distribution function, then from Eq. 15-20 $f(x)$ will be a probability density function. It is therefore necessary for a distribution function to satisfy Eq. 15-21.

The following are properties of the Gauss distribution function (Fig. 15-3):
1. The curve is symmetrical about the mean. This is readily seen from Eq. 15-18, since f is an even function of z.

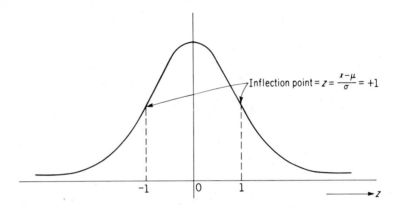

FIG. 15-3. The Gauss distribution.

2. The maximum value of f occurs at $z = 0$. From Eq. 15-17, when $z = 0$, then $x = \mu$, and the maximum occurs at μ, the mean value of x. Also

$$f(0) = \frac{1}{\sigma \sqrt{2\pi}}$$
(15-23)

3. When $z = \pm \infty$, $f(z) = 0$, and the curve is asymptotic to the z in both positive and negative directions.
4. The area under the curve of the normal distribution is unity. It has been shown above that the area between x_1 and x_2 corresponds to the probability of x occurring between x_1 and x_2. This probability may be calculated as follows. From Eq. 15-17 we have

$$\pm x = \mu \pm z\sigma$$
(15-24)

From Eq. 15-24 it is seen that normalizing the Gaussian distribution function results in expressing x in terms of multiples of σ, the standard deviation. Hence, if we wish to determine the probability of a value of x which lies *outside* $x \pm z\sigma$, we have from the previous discussion

$$\int_{-\infty}^{\mu-z\sigma} f\,dx + \int_{\mu+z\sigma}^{\infty} f\,dx = \int_{-\infty}^{z} \frac{e^{-\frac{1}{2}z^2}}{\sqrt{2\Pi}}\,dz + \int_{z}^{\infty} \frac{e^{-\frac{1}{2}z^2}}{\sqrt{2\Pi}}\,dz$$

$$= \int_{-\infty}^{\infty} \frac{e^{-\frac{1}{2}z^2}}{\sqrt{2\Pi}}\,dz - \int_{-z}^{z} \frac{e^{-\frac{1}{2}z^2}}{\sqrt{2\Pi}}\,dz \qquad (15\text{-}25)$$

$$= 1 - 2\int_{0}^{z} \frac{e^{-\frac{1}{2}z^2}}{\sqrt{2\Pi}}\,dz \qquad (15\text{-}26)$$

The first integral in Eq. 15-25 reduces to unity from property 4. The second integral follows from property 1. The second integral can be evaluated from tabulations in reference books, for example, R. H. Perry, C. H. Chilton, and S. D. Kirkpatrick, *Handbook for Chemical Engineers,* 4th ed., McGraw-Hill, New York, 1963, pp. 1-39.

5. The normal distribution is characterized by two parameters, the mean and the standard deviation σ. In Fig. 15-4 the effect of the parameters is shown for two similar distributions. These distributions are not normalized as in Fig. 15-3. In Fig. 15-4a two normal distributions having the same standard deviation but different means are shown. Thus there are two identical bell-shaped curves with the maximum value occurring at the respective mean. In Fig. 1-4b maximum value of the two curves occur at-the same point. Here the means are the same, but the shape is different. The shape of the curve having the smaller value of σ is more pinched, and has a large $f(\mu)$ value. This is in accordance with Eq. 15-24.

It should be remembered that the Gauss distribution is a theoretical distribution function that is used to approximate real populations. This is the reason for using σ and μ, the population statistics, and not $\bar{x}$ and $S(x)$ the sample statistics, hence the need to differentiate between them.

6. In Table 15-2 the area bounded by $\pm z$ for the normal distribution is given. From Eq. 15-17 we see that when $z = 1$, x is one standard deviation away from the mean, hence the term "sigma limits." Table 15-2 gives the sigma limits. It is seen from Table 15-2 that the one sigma limit encloses about 68% of the area, and hence the probability of finding x between $\mu + \sigma$ is about 0.68. Similarly, the probability of finding x between $\mu \pm 3\sigma$ is 0.9973. This means for a normal distribution it is very unlikely to find a value of x larger or smaller than three standard deviations. The three sigma limits are thus frequently chosen for control charts.

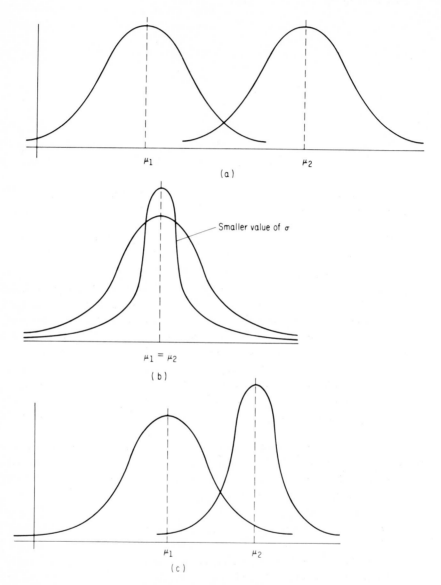

FIG. 15-4.   Effect of σ and μ on the normal distribution.   (a) Same σ, different μ.   (b) Same μ, different σ.   (c) Different μ and σ.

TABLE 15-2

z	Area Enclosed Between $\pm z$ of the Normal Distribution
0.6745	0.50
1.00	0.682
1.65	0.90
1.96	0.95
2.00	0.9545
3.00	0.9973

15-7. APPLICATION OF THE GAUSS DISTRIBUTION

In order to use the Gauss distribution, two population parameters must be specified.   These are the *mean* and the *standard deviation,* and they completely specify the distribution.   One way of obtaining this information is to use a large-size sample which makes Eq. 15-15 a close approximation of Eq. 15-12a. The following example illustrates the use of the Gauss distribution.

Example
Electric light bulbs have an average life of 1,000 hr with a standard deviation of 50 hr.   What is the upper and lower limit for the life of a bulb with 5% error?
Since we are to determine the limits on the life of an electric bulb with a 5% probability of error, the area enclosed between A and B of Fig. 15-5 must be 0.95 and the cross-hatched area 0.05.   From property 1 of the normal curve it follows that the area from $-\infty$ to B is 0.975.   From a table of the area under the normal curve we find that from $-\infty$ to $z = 1.96$ the area is 0.975.
Using Eq. 15-24, we have

$$A = \mu - \sigma z = 1,000 - 50 \times 1.96 = 902$$
$$B = \mu + \sigma z = 1,000 + 50 \times 1.96 = 1,098$$

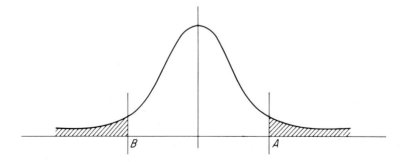

FIG. 15-5.   Probability limits on a normal distribution.

With 95% probability, the life of a bulb will be not less than 902 hours and not more than 1,098 hours. These lower and upper limits are known as the *confidence limits*. The confidence limits define an interval of certainty in which a given percentage of the observations *n* tends to infinity. This percentage depends on the level of probability specified.

If a confidence limit with 99% probability is desired, then, referring to Table 15-2, $z \pm 2.326$ and 99% of the area is enclosed under the normal distribution. Hence the confidence limits are $1,000 \pm 2.236 \times 50$, which is equal to $1,000 \pm 116$. Thus as the probability or the confidence with which we wish to predict a value increases, the confidence limits also increase. From the above discussion we have

$$\text{Confidence limit} = \mu \pm z_p \sigma \qquad (15\text{-}27)$$

where the subscript *p* is the desired probability.

Consider Fig. 15-6, which is the normal distribution for the electric light-bulb example. The 95% confidence limits are also shown. Now, if a particular bulb had a life of, say, 1,020 hours, what statement can be made about the bulb? Since it lies within the confidence limits, it can be said that least 95% probability that this bulb is part of the population. On the other hand, if the life of the bulb was found to be 1,100 hours, since it lies outside of the confidence limits, we can say with 5% chance of being wrong that this does not belong to the population. This suggests a method to test individual sample data as to whether they belong to a given population or not. Note in the above light-bulb example that with 1% probability (confidence limits are $1,000 \pm 116$) that the bulb does belong to the population.

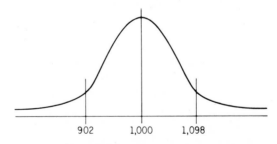

FIG. 15-6.   Normal distribution of electric light bulb problem.

When a distribution function is used to make a test to determine whether a given sample statistic belongs to a given population, a statement is made (called a *hypothesis*) concerning a population statistic. If the test shows that the hypothesis to be improbable, the hypothesis is rejected; otherwise it is accepted. Because the acceptance or rejection of a hypothesis is based on probability not certainty, two types of errors are involved:

Type I error—rejecting a hypothesis when it is true.

Type II error—accepting a hypothesis when it is false.

A 5% chance of a type I error should be interpreted as meaning that this type of error rejecting a hypothesis which is true, will be committed five times out of every hundred, when this test is made on a large number of similar observations.

To illustrate the use of the above test consider the following example. Suppose a customer observed that a certain bulb had a life of 900 hours. The customer wishes to test whether this bulb belongs to the population of bulbs of previous example having a mean of 1,000 hours and standard deviation of 50 hours

We set up a hypothesis $H_0 : \mu = 1000$. Where H_0 stands for the null hypothesis or the statement that there is no difference (hence the term *null*) between the stated mean (in this case 1,000 hours) and the mean of the population. The alternate hypothesis is called H_1 and states that there is a difference between the population mean and the stated mean. We also need to set a significance level or probability of error called the α value for the test. Let us choose an α level or error of 5%.

From the previous example for $\alpha = 5\%$ the mean will lie between the limits 902 and 1,098 hours. Since the given life of the bulb is 900 hours and is located out of these limits, with a 5% chance of being wrong the hypothesis is rejected. The given bulb is considered as not a part of the population.

A little reflection will show that decreasing the probability of being wrong increases the confidence limits and the type II error—the probability of accepting a false hypothesis. It is therefore important in a statistical investigation to first define the hypothesis and then perform the analysis so as not to bias the results.

The following example illustrates how Type I and Type II errors occur.

Example

Catalyst A, whose average yield for a particular reaction is 0.8 has a standard deviation of 0.04. Catalyst B gives and average yield for the same reaction of 0.9 has a standard deviation of 0.04. A normal distribution will be assumed. In Fig. 15-7 the distribution for the two catalysts is shown. The 95% confidence limit for the two catalysts from Eq. 15-27 is $0.8 \pm 1.96 \times 0.4 = 0.8 \pm .0784$.

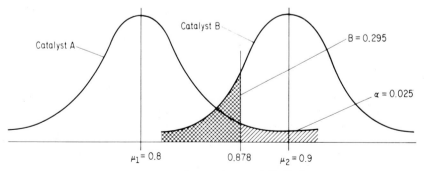

FIG. 15-7. α and β values for normal distribution.

If a sample of a catalyst on a particular test gave a yield value of 0.86, then referring to Fig. 15-7, it can be said that with 2.5% chance of being wrong (called the α value) that it is catalyst A. Note, however, this sample may really have been catalyst B. From Fig. 15-7, we see that the probability of confusing catalyst B for A is 29.5% (called the β value). The α refers to Type II error and depends on the value of the probability or α and μ_2.

15-8. THE STUDENT-FISHER t-DISTRIBUTION

In practice the population statistics are not known, and hence the normal distribution is of limited value. This is because before confidence limits, can be calculated the standard deviation of the population has to be known. This is overcome by using the Student-Fisher t-distribution. The t-distribution is valid for populations that are normal and is given by

$$f = K\left(1 + \frac{t^2}{n}\right)^{-n+\frac{1}{2}} \tag{15-28}$$

where $\quad t = \dfrac{\bar{x} - \mu}{S(\bar{x})}$

n = degrees of freedom

$S(\bar{x})$ = estimate of standard deviation and

K = a constant that depends on n.

The degrees of freedom are quantities over which there is no control. In this example it is one less than the number of sample data points. This is because one degree of freedom has been used to calculate the mean, x. This is true, since if the mean and any $N-1$ data points are given, then the Nth data point can be calculated from Eq. 15-11 and is uniquely determined.

In order to use this results an estimate $S(x)$ of σ, the standard deviation of the population, is needed. $S(x)$, the estimate of the standard deviation, is obtained from

$$S^2(x) = \frac{\Sigma(x - \bar{x})^2}{n - 1}$$

or

$$S(x) = \sqrt{\frac{\Sigma(x - \bar{x})^2}{n - 1}} \tag{15-29}$$

When n is large the correction is negligible and in fact reduces to Eq. 15-14a. The central limit theorem is used to obtain the standard deviation of the mean of the sample whose estimated standard deviation is $S^2(x)$.

The theorem will be stated here without proof* and applies to population which

---

* For a simple proof, see I. S. Sokolinkoff and R. M. Redheffer, *Mathematics of Physics and Modern Engineering*, McGraw-Hill, New York, 1958, p. 667.

are discrete or continuous and described by any distribution function. The theorem states that *for a population having a finite variance $S^2(x)$, the variance of means of samples of size N drawn from this population will have a variance $S^2(\bar{x})$ given by*

$$S^2(\bar{x}) = \frac{S^2(x)}{N} \qquad (15\text{-}30)$$

If instead of the estimate variance of the population we had the actual population parameter σ^2 them Eq. 15-30 will become

$$\sigma_{\bar{x}}^2 = \frac{\sigma^2}{N}$$

From Eq. 15-30 we have

$$S(\bar{x}) = \frac{S(x)}{\sqrt{n}} \qquad (15\text{-}31)$$

Figure 15-8 is a plot of a *t*-distribution, which is seen to be bell-shaped. The exact shape of the curve depends on *n,* the degrees of freedom, and approaches the normal distribution for large *n.* Note for large *n,* $S(\bar{x}) \to \sigma_x$ and $t \to z$. From the figure it is seen that, except for the additional parameter *n,* the distribution is similar to the Gauss distribution.

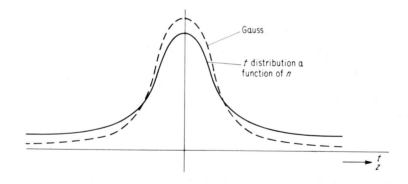

FIG. 15-8. Comparison of Gauss and *t* distributions.

Table 15-4 is a table of *t* values with α (the probability of risk) and the degrees of freedom as parameters. For example, for 98% probability and 10 degrees of freedom, the *t* value is 2.764. The following example will illustrate its use.

Example
The data given are the analysis of several samples of a crude reported to have

TABLE 15-4
Table of t Values

D. F.	0.5	0.1	0.05	0.02	0.01	0.001
1	1.000	6.314	12.706	31.821	63.657	636.619
2	.816	2.920	4.303	6.965	9.925	31.598
3	.765	2.353	3.182	4.541	5.841	12.941
4	.741	2.132	2.776	3.757	4.604	8.610
5	.727	2.015	2.571	3.365	4.032	6.859
6	.718	1.943	2.477	3.143	3.707	5.959
7	.711	1.895	2.365	2.998	3.499	5.405
8	.706	1.860	2.306	2.896	3.355	5.041
9	.703	1.833	2.262	2.821	3.250	4.781
10	.700	1.812	2.228	2.764	3.169	4.578
11	.697	1.796	2.201	2.718	3.106	4.437
12	.695	1.782	2.179	2.681	3.055	4.318
13	.694	1.771	2.160	2.650	3.012	4.221
14	.692	1.761	2.145	2.624	2.977	4.140
15	.691	1.753	2.131	2.602	2.947	.4.073
16	.690	1.746	2.120	2.583	2.921	4.015
17	.689	1.740	2.110	2.567	2.898	3.965
18	.688	1.734	2.101	2.552	2.878	3.922
19	.688	1.729	2.093	2.539	2.861	3.883
20	.687	1.725	2.086	2.528	2.845	3.850

a sulfur content of less than 2.5%. It is desired to evaluate if the samples do represent an oil containing a sulfur content of less than 2.5%.

$$\% S = 2.1, 1.9, 2.4, 2.3, 2.5, 2.4, 2.6, 2.5, 2.7, 2.3, 2.4, 2.3$$

In order to test the data, it shall be assumed that it does belong to a normal distribution with a population mean of 2.5% sulfur. In other words, we have the null hypothesis

$$H_0 : \mu = 2.5$$

A significance level must be chosen–i.e., a value of $\alpha = 0.05$ is set, i.e., a 5% risk that a true value* will be rejected.

Since a standard deviation is not given, the t test is used:

$$N = 12$$
$$\Sigma x = 28.4$$
$$\Sigma x^2 = 67.72$$
$$\bar{x} = \frac{28.4}{12} = 2.37$$

* The α level, the probability of rejecting the hypothesis when true, is generally chosen at 5% in absence of specific data on β level (the probability of not rejecting the hypothesis H_0 when some other totally different hypothesis is true.)

The estimate of the standard deviation is then

$$S^2(x) = \frac{\Sigma(x)^2 - \bar{x}\Sigma(x)}{n-1} \qquad \text{(from. Eq. 15-15a)}$$

$$S^2(x) = \frac{67.72 - (2.37) \times (28.4)}{11} = 0.0375$$

From Eq. 15-31, the standard mean deviation is the sample

$$S^2(\bar{x}) = \frac{0.0375}{12} = 0.003121$$

$$S(\bar{x}) = 0.0557$$

We wish to determine whether these crude samples belong to a population having a sulfur content less than 2.5% sulfur. Hence, we are interested in examining only those values of sulfur content which are larger than 2.5% due to random error. From Fig. 15-9 it is seen that this is a one-tailed test. For 95% con-

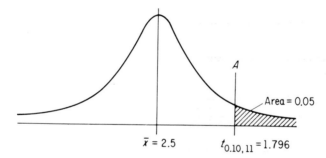

FIG. 15-9. One-tailed "*t*" text.

fidence level and 11 degrees of freedom, 95% of the area must lie to the left of line A. From Table 15-4 $t_{0, 10, 11}$ is equal to 1.796. Note, the t table gives areas corresponding to $\pm z$, hence a z-value of 0.10 corresponds to an area of .05 to the right of A in Fig. 15-9. From Eq. 15-28, the result is

$$t = \left|\frac{\bar{x} - \mu}{S(\bar{x})}\right| = \left|\frac{2.37 - 2.5}{0.0557}\right| = 2.334$$

and from Table 15-4, $t_{0.10,11} = 1.796$.

Since the calculate t is larger than the tabulated value of t, the mean of the data sample falls to the right of the line A, and hence the hypothesis is rejected. The data sample is not from a population which has a sulfur content less than 2.5%, with a chance of being in error of 5%.

15-9. CORRELATION AND REGRESSION

So far statistical methods to test if a sample mean belongs to a given population has been discussed. It was shown that in the absence of the population statistics, the t test could be used. In this section, the analysis of experimental data will be discussed. Regression is the technique of fitting a mathematical model to observed data. Correlation tests will indicate how good the model fits the data. It is important to note that a good fit does not necessarily imply a correct model. For example, a reaction may be first-order in a given range, while actually being second-order. Thus in the narrow range for which the data was taken, a first-order seems a very good fit. Statistics is thus a tool or device and should be recognized for its value.

Figure 15-1 showed a scatter diagram as a plot between two variables. A straight line was drawn through the data. The question arises concerning the criteria for a good fit. It was indicated that the independent variable x is generally under control and can be chosen as exactly as desired. The variable Y, the dependent variable, is subject to random error. The mathematic model chosen to represent this data was a straight line:

$$\hat{Y} = ax + b \tag{15-32}$$

At this stage it is necessary to distinguish between the quantities $\hat{Y}$, Y, and y. $\hat{Y}$ is the value obtained from Eq. 15-32, Y is the true value and y the value obtained by experiment. This differentiation is made because these three quantities in general are not identical. As the number of data points increase, $\hat{Y}$ will tend to approach Y, the true value.

The deviation between experimental and predicted value is given by Eq. 15-8. It can be seen that a good fit would mean that all of the experimental values would lie on the line, and the deviation would be zero. Thus, the sum of the deviations being minimized appears to be a good criterion for a fit. However, since the deviation could be positive or negative, their sum may cancel to give a small value, even when there is a large scatter. The sum of the squares of the deviation does not suffer from this disadvantage. The most widely used criterion for a good fit would therefore be to minimize the sum of the squares*. This is called the *method of least squares*. The sum of the square of error is given by

$$Q = \sum_{i=1}^{N} (y_i - \hat{Y}_i)^2 \tag{15-33}$$

For the present, consider $\hat{Y}_i$ given by Eq. 15-32. In the general case, $\hat{Y}_i$ would be represented by the mathematical model for which the least-squares fit is desired. From Eq. 15-32, it is seen that $\hat{Y}_i$ will depend on the values of a and b

---

* For conditions under which least-squares analysis is valid, see, for example, Normal L. Johnson, and Fred C. Leone, *Statistics and Experimental Design in Engineering and Physical Science,* Wiley, New York, 1964, p. 382.

selected. Thus in order that Q in Eq. 15-33 be a minimum, we have from maxima and minima theory of calculus that

$$\frac{\partial Q}{\partial a} = \frac{\partial}{\partial a} \left[\sum_{i=1}^{N} (y_i - \hat{Y}_i)^2 \right] = 0 \tag{15-34}$$

$$\frac{\partial Q}{\partial b} = \frac{\partial}{\partial b} \left[\sum_{i=1}^{N} (y_i - \hat{Y}_i)^2 \right] = 0 \tag{15-35}$$

Substituting Eq. 15-32 for $\hat{Y}_i$ in 15-34 simplifying, the result is

$$\frac{\partial}{\partial a} \left[\sum_{n=1}^{N} (y_i - Y_i)^2 \right] = 0$$

$$\sum_{n=1}^{N} 2(y_i - Y_i) \left(\frac{\partial Y_i}{\partial a} \right) = 0$$

$$- 2 \sum_{n=1}^{N} (y_i - ax_i - b) x_i = 0$$

$$\Sigma y_i x_i - a\Sigma x_i^2 - b\Sigma x_i = 0$$

$$\Sigma y_i x_i = a\Sigma x_i + b\Sigma x_i \tag{15-36}$$

Similarly, it can be shown that Eq. 15-35 can be reduced to

$$\Sigma y_i = a\Sigma x_i + Nb \tag{15-37}$$

Solving Eq. 15-36 and 15-37 simultaneously, the value of a is

$$a = \frac{\Sigma y_i x_i - (\Sigma x_i)(\Sigma y_i)/N}{\Sigma x_i^2 - (\Sigma x_i)^2/N} \tag{15-38}$$

The value of b is obtained from Eq. 15-37 and is given by

$$b = \frac{\Sigma y_i}{N} - a\frac{\Sigma x_i}{N}$$

$$= \bar{y} - a\bar{x} \tag{15-39}$$

A linear model is generally more desirable for its simplicity. Also through a given set of points a number of curves may be drawn; and it is more difficult to determine which curve best fits the data.

Table 15-5 lists some typical mathematical models for $\hat{Y}$. Figure 15-10 shows how the shape of the curves could vary for different values of the parameters indicated in Table 15-5.* It should be noted that for Equation B the curve is very sensitive to small changes in values of the parameters.

---

* For a more detailed discussion see L. Bryce Anderson, "Regression Analysis Correlates Relationships between Variables," *Chemical Engineering,* May 1963, p. 175.

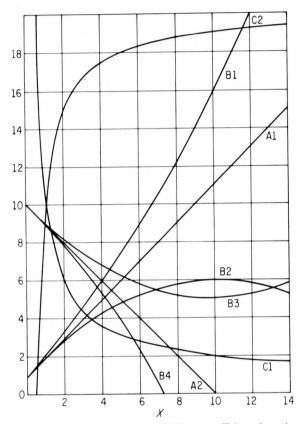

FIG. 15-10. Numerical values of linear coefficients determine shape of plotted equations. *Source:* L. Bryce Andersen, "Regression Analysis Correlates Relationship Between Variables," *Chemical Engineering,* May 1963, p. 175, with permission.

TABLE 15-5

Useful Forms for Empirical Equations
(Coefficients are linear)

Equation	General Form	Specific Constants for Curves			Curve
		a	b_1	b_2	
A	$Y = a + b_1 X$	1	1	......	A1
		10	−1	......	A2
B	$Y = a + b_1 X + b_2 X^2$	1	1	0.05	B1
		1	1	−0.05	B2
		10	−1	0.05	B3
		10	−1	−0.05	B4
C	$Y = a + \dfrac{b_1}{X}$	1	10	......	C1
		20	−10	......	C2

15-10. POLYNOMIAL REGRESSION

When the function Y is in the form of a polynomial such as:

$$Y_i = a_0 + a_i X + a_2 X^2 + a_3 X^3 + \cdots + a_n x_n$$

$$= \sum_{k=0}^{N} a_k X^k \tag{15-40}$$

This equations lends itself very readily to least-squares analysis.   The value of X, for a given set of X_i, is dependent on the constants a_k.   Equation 15-34 will therefore be of the form

$$\frac{\partial Q}{\partial a_k} = 0 \qquad k = 0, 1, 2, \cdots, n \tag{15-41}$$

Equation 15-41 will give rise to n equations in the n unknowns, which are the constants a_k.   For example, for Y being a third-order equation the result is

$$\frac{\partial Q}{\partial a_k} = \Sigma \left[\frac{\partial \, (y - a_0 - a_1 x - a_2 x^2 - a_3 x^3)^2}{\partial a_k} \right] \tag{15-42}$$

where $k = 0, 1, 2, 3$.

$$- 2\Sigma \ (y - a_0 - a_1 x - a_2 x^2 - a_3 x^3) \quad \left[\ \frac{\partial}{\partial a_k} \, (a_k x^k) \right] = 0$$

In Eq. 15-42 the range over which the summation is carried is not indicated, but implied, to be similar to that of Eq. 15-40.   This is tacitly implied in all future summation, unless otherwise indicated.

For the $k = 0$, 1, 2, 3, after simplifying the results are

$$\Sigma a_0 + a_1 \Sigma x + a_2 \Sigma x^2 + a_3 \Sigma x^3 = \Sigma y$$

$$a_0 \Sigma x + a_1 \Sigma x^2 + a_2 \Sigma x^3 + a_3 \Sigma x^4 = \Sigma yx$$

$$a_0 \Sigma x^2 + a_1 \Sigma x^3 + a_2 \Sigma x^4 + a_3 \Sigma x^5 = \Sigma yx^2 \tag{15-43}$$

$$a_0 \Sigma x^3 + a_1 \Sigma x^4 + a_2 \Sigma x^5 + a_3 \Sigma x^6 = \Sigma yx^3$$

In matrix form this reduces to

$$[A] [S] = [P] \tag{15-44}$$

where the symmetric matrix $[A]$ is the coefficients of Eq. 15-43 and is made up of the sum of the product of y and x.   Eq. 15-44 can be solved by inverting the matrix or by using Gauss reduction.

When the polynomial of Eq. 15-40 is of kth degree, the $[A]$ and $[P]$ matrices can be very easily deduced from Eq. 15-43 and are

$$[A] = \begin{bmatrix} N & \Sigma x & \cdots & \Sigma x^k \\ \Sigma x & \Sigma x^2 & \cdots & \Sigma x^{k+1} \\ \cdot & & & \\ \cdot & & & \\ \cdot & & & \\ \Sigma x^k & \Sigma x^{k+1} \cdots & & \Sigma x^{k+k} \end{bmatrix} \tag{15-45}$$

$$[P] = \begin{bmatrix} \Sigma y \\ \Sigma yx \\ \cdot \\ \cdot \\ \cdot \\ \Sigma yx^k \end{bmatrix} \qquad S = \begin{bmatrix} a_0 \\ a_1 \\ \cdot \\ \cdot \\ \cdot \\ a_k \end{bmatrix} \tag{15-46}$$

In general, the more constants used to fit the data (i.e., the higher the degree of the polynomial) the better the fit. However, this improvement reaches a point of diminishing returns. A common procedure is to use a computer program for the general case of the polynomial of the kth degree, and solve Eq. 15-44. The sum of the square of error Q_k is calculated from Eq. 15-33. The best fit will be given by the smallest value of Q_k. The sum of the squares of the deviation O_k will never reach a minimum, but the difference as k increases grows smaller and smaller. This is because the curve tries to pass through every point. A real danger in this method is that for large number of parameters (degree of the polynomial) the curve will "wiggle" or cycle in between data points and hence is not suitable for interpolation. Regression models should be used with caution for extrapolation. In general, one arbitrarily stops when the improvement in Q_k from the higher-order polynomial is not worth the additional complexity introduced by using this higher-order polynomial. For example, heat capacity data for gases are reported in most cases with $k = 2$ and sometimes $k = 3$.

Least-squares analysis can also be applied for linear equations with multiple variables. This is known as *linear multiple regression*. The mathematical model in this case is

$$Y = a_0 + a_1 x_1 + a_2 x_2 + \cdots \tag{15-47}$$

$$Y = a_0 + \sum_{i=1}^{n} a_i x_i$$

The least-squares analysis will give an equation in the form of Eq. 15-44 where the matrices $[A]$ and $[P]$ are given by

$$[A] = \begin{bmatrix} \Sigma x_1 & \Sigma x_1^2 & \Sigma x_2 x_1 \cdots & & & \Sigma x_n x_1 \\ \Sigma x_2 & \Sigma x_1 x_2 & \Sigma x_2^2 & \cdots & & \Sigma x_n x_2 \\ \cdot & & & & & \\ \cdot & & & & & \\ \cdot & & & & & \\ \Sigma x_i & \Sigma x_1 x_i & \Sigma x_2 x_1 \cdots & \Sigma x_i^2 \cdots & & \Sigma x_n x_i \\ \cdot & & & & & \\ \cdot & & & & & \\ \cdot & & & & & \\ \Sigma x_n & \Sigma x_1 x_n & \Sigma x_2 x_n \cdots & & & \Sigma x_n^2 \\ & & & & & \cdot \end{bmatrix}$$

$$[P] = \begin{bmatrix} \Sigma y \\ \Sigma x_1 y \\ \Sigma x_2 y \\ \cdot \\ \cdot \\ \cdot \\ \Sigma x_n y \end{bmatrix} \qquad S = \begin{bmatrix} a_0 \\ a_1 \\ \cdot \\ \cdot \\ \cdot \\ \cdot \\ a_n \end{bmatrix}$$

15-11. NONLINEAR REGRESSION

When the mathematical model is nonlinear, least-squares analysis cannot in general be done by hand calculations, and a digital computer is necessary. The following example will illustrate the complexity of the problem.

Let the mathematical model be of the form

$$Y = \frac{a}{b + x} \tag{15-48}$$

$$Q = \Sigma (y - Y)^2$$

$$\frac{\partial Q}{\partial a} = 2 \frac{\partial}{\partial a} \left[\left(y - \frac{a}{b + x} \right) \left(- \frac{1}{b + x} \right) \right] = 0$$

$$\frac{\partial Q}{\partial b} = 2 \Sigma \left[\left(y - \frac{a}{b + x} \right) \left(\frac{a}{b + x^2} \right) \right] = 0$$

Hence this gives

$$\Sigma \left[\left(y - \frac{a}{b + x} \right) \left(\frac{1}{b + x} \right) \right] = 0 \tag{15-49}$$

$$\Sigma \left[\left(y - \frac{a}{b + x} \right) \left(\frac{1}{(b + x)^2} \right) \right] = 0 \tag{15-50}$$

Equations 15-49 and 15-50 must be solved for a and b. It is obvious that the equations are nonlinear and therefore must be solved by trial and error. The Newton-Raphson method could be used to obtain a solution. The procedure is as follows. The Taylor series expansion for two variables is

$$f(a + \Delta a, b + \Delta b) = f(a,b) + \frac{\partial f(a,b)}{\partial a} da + \frac{\partial f(a,b)t}{\partial b} db + \cdots \qquad (15\text{-}51)$$

If (a,b) are the roots of $f(a,b)$ then $a + \Delta a$, $b + \Delta b$ are better approximations of the roots, hence $f(a \pm \Delta a, b + \Delta b) = 0$. Omitting the higher derivatives, Eq. 15-51 can be put in the following form:

$$f(a_k, b_k) + (a_{k+1} - a_k)\frac{\partial f(a_k, b_k)}{\partial a} + (b_{k+1} - b_k)\frac{\partial f(a_k, b_k)}{\partial b} = 0 \qquad (15\text{-}52)$$

$$g(a_k, b_k) + (a_{k+1} - a_k)\frac{\partial g(a_k, b_k)}{\partial a} + (b_{k+1} - b_k)\frac{\partial g(a_k, b_k)}{\partial b} = 0 \qquad (15\text{-}53)$$

where

$$f(a,b) = \sum \left[\left(y - \frac{a}{b+x} \right) \left(\frac{1}{(b+x)} \right) \right] \qquad (15\text{-}54)$$

$$g(a,b) = \sum \left[\left(y - \frac{a}{b+x} \right) \left(\frac{1}{b+x^2} \right) \right] \qquad (15\text{-}55)$$

$$\Delta a = (a_{k+1} - a_k) \qquad (15\text{-}56)$$

$$\Delta b = (b_{k+1} - b_k) \qquad (15\text{-}57)$$

The subscript k stands for the kth iteration.

The computation procedure is as follows:

1. Assume a value of a_b, b_k, any value will serve; for example, $(0,0)$ or $(1,1)$.

2. Calculate $f(a,b)$, $g(a,b)$, $\frac{\partial f(a,b)}{\partial a}$, $\frac{\partial f(a,b)}{\partial b}$, $\frac{\partial g(a,b)}{\partial a}$, $\frac{\partial g(a,b)}{\partial b}$ for the point (a_k, b_k). These values can be calculated from Eq. 15-52 and 15-53.

3. All quantities except Δa and Δb are known in equation and since these equations are linear in Δa and Δb, solve for Δa and Δb by Cramer's rule.

4. From Eq. 15-56 and 15-57, calculate a_{k+1} and b_{k+1}. The iterative procedure is thus complete, with $a_{k+1} = a_k$ and $b_{k+1} = b_k$ steps 1 to 3 can be repeated to generate better estimates of a_k and b_k. The computation is stopped when two sucessive iteration differ by the desired degree of precision.

To find the parameters in the equation $Q = \Sigma (y - Y)^2$ subject to the condition that Q is a minimum falls in the general area of optimization discussed in Chapter 16. The treatment outlined above is also called a *root-finding technique*, and is by no means the only method to be used. Depending on the type of nonlinear function to be fitted one or more of the more powerful optimization techniques will give a more rapid convergence to the solution'

15-12.   LINEARIZATION OF DATA

From the preceding section it is seen that nonlinear mathematical models lead to complex procedures for evaluating parameters of the mathematical model. Frequently it becomes questionable if one should go to the trouble of a nonlinear analysis, especially when the experimental data have questionable accuracy.   A simpler procedure would be to linearize the model and then use least-squares analysis to arrive at the constants.

For example, if the mathematical model for Y is given by

$$\hat{Y} = ae^{bx} \tag{15-58}$$

then this equation could be linearized by taking logarithims of both sides of the equation.

$$\ln \hat{Y} = \ln a + bx \tag{15-59}$$

Making a change of variables, we have

$$\zeta = m\gamma + c \tag{15-60}$$

where

$$\begin{aligned}
\zeta &= \ln \hat{Y} \\
\gamma &= x \\
m &= b \\
c &\div= \ln a
\end{aligned} \tag{15-61}$$

Equation 15-60 is of the form of Eq. 15-32, hence, m and c can be obtained by using Eqs. 15-38 and 15-39.   In terms of the new variables, they are

$$m = \frac{\Sigma\zeta\gamma - (\Sigma\zeta)(\Sigma\gamma)/N}{\Sigma\gamma^2 - (\Sigma\gamma)^2/N} \tag{15-62}$$

$$c = \zeta - m\gamma \tag{15-63}$$

The values a and b are then calculated from Eq. 15-61.   It should be noted that the parameters b and c obtained in this manner do not give the least Q.   This is because the least-squares analysis was used to minimize the deviations of ζ and not Y.   However, for most engineering analysis this is sufficient.   Table 15-6 gives linear transformations for some of the common nonlinear functions often encountered.

Equations 6, 7, and 8 of Table 15-5 were included for completeness.   It is not recommended that a least-squares analysis be performed.   The method of averages of the next section is recommended here.   This is because d in Equation 6 is obtained by a plot of the data on rectangular paper, then d is the intercept, when $X = 0$, and hence $Y = d$.   In Equations 7 and 8 $Y_n,(X_n)$ is obtained by plotting

TABLE 15-6

Linearization of Typical Equations

Y	ς	γ	m	c
1. $Y = a + bX$	Y	X	b	a
2. $Y = aX^b$	$\log Y$	$\log X$	b	$\log a$
3. $Y = ab^x$	$\log Y$	X	$\log b$	$\log a$
4. $Y = a + b/X$	Y	$1/X$	a	a
5. $Y = \dfrac{X}{a + bX}$	$1/Y$	$1/X$	a	b
6. $Y = d + aX^b$	$\log(Y - d)$	$\log X$	b	$\log a$
7. $y = a + bX + dX^2$	$\dfrac{Y - Y_n}{X - X_n}$	X		
8. $y = \dfrac{X}{a + bX} + c$	$\dfrac{X - X_n}{Y - Y_n}$	X		

the data points on rectangular paper and drawing a smooth curve by eye alone through the points. $Y_n, (X_n)$ is then any point on the smooth curve.

Method of Averages

The least-squares analysis for linear equations is accurate; however, it is rather tedious to do by hand, especially when the number of data points is large. The method of averages is approximate but is a better procedure to use rather than eye in a straight line. It is useful only for linear equations and is recommended for Equations 6, 7, and 8 of Table 15-5. An example will best illustrate the method.

Example

The data below represent a linear relationship between x and y. It is required to obtain the best-fitting line by the method of averages and compare by the method of least squares.

	y	x	
1.	1.0	0.313	$\Sigma x = 14.169$
2.	1.3	0.5	$\Sigma y = 27.00$
3.	2.0	0.937	$\Sigma x^2 = 29.3848$
4.	2.4	1.189	$\Sigma xy = 53.8326$
5.	2.9	1.50	
6.	3.1	1.75	
7.	4.2	2.31	
8.	4.9	2.73	
9.	5.2	2.94	

Solution: Mathematical model is

$$\hat{Y} = ax + b \qquad\qquad [15\text{-}32]$$

In the method of averages, the data are arranged in increasing or decreasing order and divide into two groups. In this case there are nine data points. There is a choice of having either four or five in the first group. The first four data points are taken in group one, the last five in group two. The data are substituted in Eq. 15-32 and added as shown below.

$$
\begin{aligned}
1.0 &= a & 0.313 + b \\
1.3 &= a & 0.5 + b \\
2.0 &= a & 0.937 + b \\
2.4 &= a & \underline{1.189 + b} \\
6.7 &= a & 2.939 + 4b
\end{aligned} \quad \text{I}
$$

$$
\begin{aligned}
2.9 &= a & 1.5 + b \\
3.1 &= a & 1.75 + b \\
4.2 &= a & 2.31 + b \\
4.9 &= a & 2.73 + b \\
5.2 &= a & \underline{2.94 + b} \\
20.30 &= a & 11.23 + 5b
\end{aligned} \quad \text{II}
$$

Solving Equations I and II for a and b, the result is

$$a = 1.578$$
$$b = 0.516$$

A slightly different answer would have been obtained if the split was made such that the first group consisted of five points. This is a drawback of the method of averages. For an even set of data points, the method of averages gives a unique answer.

For the method of least squares, from Eq. 15-38 the result is

$$a = \frac{53.8326 - 14.169 \times 27.0 \times 1/9}{29.3848 - (14.169)^2 \times 1/9} = 1.600$$

From Eq. 15-39 the result is

$$b = \frac{27}{9} - 1.600 \cdot \frac{14.169}{9} = 0.481$$

Least-squares method gives $a = 1.60$ and $b = 0.481$.

Table 15-7 gives a comparison of the two methods.

$$Q_{avg} = 0.0415$$
$$Q_{least\ sq} = 0.0378$$

$$\Sigma|y - \hat{Y}|_{avg.} = 0.3834$$
$$\Sigma|y - \hat{Y}|_{least\ sq.} = 0.3630$$

TABLE 15-7

Comparison of Least Squares and Method of Averages

		Method of Averages		Least Squares	
x	y	$\hat{Y}$	$y - \hat{Y}$	$\hat{Y}$	$y - \hat{Y}$
0.313	1.0	1.01	−.01	0.981	.019
0.5	1.3	1.305	−.005	1.281	.019
0.937	2.0	1.9946	.0054	1.980	.02
1.189	2.4	2.392	.008	2.383	.017
1.50	2.9	2.883	.017	2.881	.019
1.75	3.1	3.278	−.178	3.281	−.181
2.31	4.2	4.161	.039	4.178	.022
2.73	4.9	4.824	.976	4.849	.051
2.94	5.2	5.155	.045	5.185	.015

From the above results, it is seen that the method of averages compares favorably with least squares. In this case, the two methods gave almost the same results because there was very little spread in the original data. From Eq. 15-39 it will be noticed that for linear regression, the average of the data points is a point on the least-squares line. Hence to plot the least-squares line, all one needs to do is to locate the average point and use the slope or the intercept to draw the line through the average point (x,y).

When the scatter in the data is large, it is not obvious whether a linear fit is satisfactory. From the preceeding section it is seen that the best fitting line can be obtained regardless of whether the data are linear or not. All that is required to obtain the linear parameters a and b is the necessary summation terms indicated in Eqs. 15-38 and 15-39.

If the data were obtained from a population that did behave in a linear manner, then the scatter of the observed data will be due to random error. Correlation used the t test to indicate if the scatter was due to random variation alone and measures the probability of acceptance of a linear fit.

The sum of the squares of the deviation is given $\Sigma(y - \hat{Y})^2$ and as indicated earlier, is a measure of goodness of the correlation—the larger its value, the poorer the fit; the smaller its value, the better the fit. By itself this tells us little. The following manipulations bring it into a more desirable form. Substituting Eq. 15-31, we have

$$\Sigma(y - \hat{Y})^2 = \Sigma(y - ax - b)^2$$

Substituting for b from Eq. 15-39 and arranging,

$$\Sigma(y - \hat{Y})^2 = \Sigma[(y - \bar{y}) - a(x - \bar{x})]^2 \qquad (15\text{-}64)$$

In order to make the algebra more compact, the following terms are defined:

$$\begin{aligned}
\Sigma'x &= \Sigma(x - \bar{x}) \\
\Sigma'y &= \Sigma(y - \bar{y}) \\
\Sigma'x^2 &= \Sigma(x - \bar{x})^2 \\
\Sigma'y^2 &= \Sigma(y - \bar{y})^2 \\
\Sigma'xy &= \Sigma(x - \bar{x})(y - \bar{y})
\end{aligned} \qquad (15\text{-}65)$$

Simplifying Eq. 15-64 and using the nomenclature defined by Eq. 15-65, we have

$$\Sigma(y - Y)^2 = \Sigma'y^2 - 2a\Sigma'xy + a^2\Sigma'x^2 \qquad (15\text{-}66)$$

Using the same method as that for Eq. 15-15a, it can be shown that

$$\Sigma'x^2 = \Sigma x^2 - \bar{x}\Sigma x \qquad (15\text{-}67)$$

$$\Sigma'xy = \Sigma xy - \bar{x}\Sigma y \qquad (15\text{-}68)$$

Substituting Eqs. 15-67 and 15-68 in Eq. 15-38 gives

$$a = \frac{\Sigma'xy}{\Sigma'x^2} \qquad (15\text{-}69)$$

and

$$a\Sigma'xy = \frac{(\Sigma'xy)^2}{\Sigma'x^2} = \frac{(\Sigma'xy)^2}{(\Sigma'x^2)^2} \cdot \Sigma'x^2 = a^2\Sigma'x^2$$

$$a\Sigma'xy = a^2\Sigma'x^2 \qquad (15\text{-}70)$$

Substituting Eq. 15-70 in Eq. 15-66 and simplifying gives

$$\Sigma(y - \hat{Y})^2 = \Sigma'y^2 - a^2\Sigma'x^2$$

or

$$\Sigma(y - \hat{Y})^2 = \Sigma'y^2 - a\Sigma'xy \qquad (15\text{-}71)$$

From Eq. 15-71 it is seen that the sum of the squares of the deviation is a difference of two terms. The first term is the square of the deviation about the mean value. The second term is subtracted from the first and since a was obtained by least-squares analysis, it can be considered as the square of the deviation of the observed values removed by correlation. Let C be the sum of the squares removed by correlation. Then

$$C = a^2\Sigma'x^2 = a\Sigma'xy$$

The *correlation coefficient* r^2 can be likened to an efficiency factor and is defined as the ratio of the sum of the square of the deviation removed by correlation, to the total deviation originally present.

$$r^2 = \frac{C}{\Sigma'y^2} \tag{15-72}$$

$$r^2 = 1 - \frac{\Sigma(y - \hat{Y})^2}{\Sigma'y^2}$$

$$= \frac{a^2 \Sigma'x^2}{\Sigma'y^2} \quad \text{(follows from Eq. 15-71)}$$

$$r = a\sqrt{\frac{\Sigma'x^2}{\Sigma'y^2}} \tag{15-73}$$

From Eq. 15-72 it is seen that for perfect correlation $\Sigma(y - Y)^2 = 0$ and hence $r = 1$. It should be noted that when the slope is negative, perfect correlation will give $r = 1$, hence the limits on r are

$$-1 \leq r \leq 1 \tag{15-74}$$

A value of $r = 0$ means no correlation.

Fisher has related the correlation coefficient to the t test. Table 15-8 gives a tabulation of r values for a given degree of freedom and probability level. The degrees of freedom in this case is N - 2. Two degrees of freedom have been removed by the slope and intercept, in other words, given any N - 2 points, the slope and intercept, the remaining two points can be calculated. By calculating the slope and intercept, the arbitrariness of the problem has been reduced by 2. For example, with 98% certainty and 10 degrees of freedom from Table 15-7, it can be found that $r = 0.658$. Hence, due to random variation alone, an r value of 0.658 or higher is expected, and the risk of rejecting the data, when they truly belonged to the population, is 2%.

The t test can be used to estimate the limits of accuracy of the various parameters that enter the linear regression. As has been seen in the case of the t test, an estimate the variance is needed and then, by using the central limit theorem, an estimate the standard deviation is made. The variance for various parameters entering the linear regression is stated here without proof, as the proof is beyond the scope of elementary statistics. It will suffice to say that they are easy to use and indicate useful results. The example that follows illustrates their use.

Estimate of the variance of regression:

$$S^2(Y) = \frac{\Sigma(y - \hat{Y})^2}{N - 2} \tag{15-73}$$

Estimate of variance of $\bar{y}$,

$$S^2(\bar{y}) = \frac{S^2(Y)}{n} \tag{15-74}$$

Estimate of variance of slope a,

TABLE 15-8

Table of Correlation Coefficients r.*

D.F.	Probability of a larger value of r				
	0.1	0.05	0.02	0.01	0.001
1	0.988	0.997	1.0	1.0	1.0
2	0.900	0.950	0.980	0.990	1.0
3	0.805	0.878	0.934	0.959	0.991
4	0.729	0.811	0.882	0.917	0.974
5	0.669	0.754	0.833	0.874	0.951
6	0.622	0.707	0.789	0.834	0.925
7	0.582	0.666	0.750	0.780	0.898
8	0.549	0.632	0.716	0.765	0.872
9	0.521	0.602	0.685	0.735	0.847
10	0.497	0.576	0.658	0.708	0.823
12	0.458	0.532	0.612	0.661	0.780
14	0.426	0.497	0.574	0.623	0.742
16	0.400	0.468	0.542	0.590	0.708
18	0.378	0.444	0.516	0.561	0.679
20	0.360	0.423	0.492	0.537	0.652
25	0.323	0.381	0.445	0.487	0.597
30	0.296	0.349	0.409	0.449	0.554
35	0.275	0.325	0.381	0.418	0.519
40	0.257	0.304	0.358	0.393	0.490
45	0.243	0.288	0.338	0.372	0.465
50	0.231	0.273	0.322	0.354	0.443

* Abridged from Table VI of Statistical Tables for Biological, Agricultural and Medical Research by R. A. Fisher and Frank Yates, Oliver & Boyd, Ltd., Edinburgh and London, 1953, reprinted with permission of publishers.
Source: William Volk, *Applied Statistics for Engineers* (McGraw-Hill, New York, 1968), p. 231, with permission of the publisher.

$$S^2(a) = \frac{S^2(Y)}{\Sigma' x^2} \qquad (15\text{-}75)$$

Estimate of variance of any estimated mean y_i for a given x_i:

$$S^2(\bar{y}_i) = S^2(Y)\left[\frac{1}{N} + \frac{(x_i - \bar{x})^2}{\Sigma' x^2}\right] \qquad (15\text{-}76)$$

Estimate of variance of any single estimated value of $\bar{y}_i$:

$$S^2(y_i) = S^2(Y)\left[1 + \frac{1}{N} + \frac{(x_i - \bar{x})^2}{\Sigma' x^2}\right] \qquad (15\text{-}77)$$

Estimate of variance of intercept b:

$$S^2(b) = S^2(Y)\left[\frac{1}{N} + \frac{\bar{x}^2}{\Sigma'x^2}\right] \qquad (15\text{-}78)$$

Example

The following experimental data were taken on aqueous solution of sodium chloride at 50°C. y is the density in grams per cc and x is the percent concentration by weight.   It is desired to obtain a linear correlation.   Figure 15-11 is a plot of

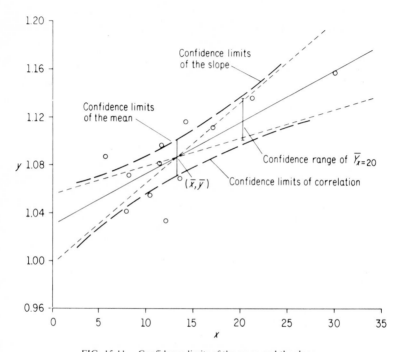

FIG. 15-11.   Confidence limits of the mean and the slope.

the experimental points, from which it appears that a linear correlation is doubtful.   A linear least-squares analysis will, however, be performed.   The calculations are indicated below.

Data

x wt.,%	y, gm/cc
5.5	1.082
7.75	1.038
8.0	1.068
10.22	1.051
11.1	1.077

Data

x wt.,%	y, gm/cc
11.3	1.093
11.9	1.031
13.7	1.068
14.0	1.112
17.0	1.109
20.8	1.132
29.5	1.152

$N = 12$
$\Sigma x = 160.75$
$\Sigma x^2 = 2626.44$
$\Sigma y = 13.013$
$\Sigma y^2 = 14.1265$
$\Sigma xy = 176.42$

$$\Sigma' x^2 = \Sigma x^2 - \frac{(\Sigma x)^2}{N} = 2,626.44 - \frac{(160.75)^2}{12} = 473.06 \qquad (15\text{-}67)$$

$$\Sigma' y^2 = \Sigma y^2 - \frac{(\Sigma y)^2}{N} = 14.1265 - \frac{(13.013)^2}{12} = 0.015$$

$$\Sigma' xy = \Sigma xy - \frac{\Sigma x \Sigma y}{N} = 176.42 - \frac{(160.75)(13.013)}{12} = 2.10 \qquad [15\text{-}68]$$

$$r = \frac{\Sigma' xy}{(\Sigma' x^2 \Sigma' y^2)^{1/2}} \qquad \text{(combining Eqs. 15-73 and 15-69)}$$

$$= \frac{2.1}{[(473.06)(.015)]^{1/2}} = 0.787$$

Degrees of freedom $= N - 2 = 10$

Referring to Table 15-8, it is seen that for 10 degrees of freedom, $r = 0.708$ for a probability level of 0.01 and $r = 0.823$ for a probability level of 0.001. Hence with 99% confidence it can be said that X and Y correlate, but not with 99.9% confidence.

$$\bar{x} = \frac{160.75}{12} = 13.40$$

$$\bar{y} = -\frac{13.013}{12} = 1.0844$$

$$a = \frac{\Sigma' xy}{\Sigma' x^2} = \frac{2.1}{473.06} = 0.00444 \qquad \text{(from Eq. 15-69)}$$

The least-squares line is thus

$$Y = [1.0844 - (0.00444)(13.40)] + 0,00444X$$
$$Y = 0.00444X + 1.0249 \tag{15-79}$$

The least-squares line is drawn in Fig. 15-11, using the point (x,y) and the intercept.

The use of Eqs. 15-73 through 15-78 will now be demonstrated.

$$S^2(Y) = \frac{(1 - r^2)\,\Sigma' y^2}{N - 2} \tag{15-80}$$

The above equation is obtained by combining Eqs. 15-73 and 15-72.

$$S^2(Y) = \frac{(1 - .787)(.015)}{10} = 0.00032$$

$$S^2(a) = \frac{S^2(Y)}{\Sigma' x^2} = \frac{0.00032}{473.06} \quad \text{(from Eq. 15-75)}$$

$$= 6.746 \times 10^{-7}$$

$$S(a) = 0.000821$$

Note that $S(a)$ is the estimated standard deviation:

$$S^2(y) = \frac{S^2(Y)}{N} = \frac{0.00032}{12} = 0.0000267 \quad \text{(from Eq. 15-74)}$$

$$S(y) = 0.00517$$

From Table 15-4, for a 95% confidence level ($\alpha = 0.05$) and ten degrees of freedom,

$$t_{.05,10} = 2.228$$

The 95% confidence range of $\bar{y}$ is thus

$$\bar{y} = 1.0844 \pm (2.228)(0.00517)$$
$$= 1.0844 \pm 0.0115$$
$$\bar{y} \text{ ranges from } 1.08977 \text{ to } 1.0711$$

Similarly, the 95% confidence range of the slope a is given by

$$a = 0.00444 \pm (2.228)(0.000821)$$
$$= 0.00444 \pm 0.00183$$

The slope thus ranges from 0.00627 to 0.00261.   This range of the slope has been sketched in Fig. 15-11 by means of dotted lines.

To estimate Y at $X = 20$, for example.   The value of an average estimate Y is calculated from Eq. 15-79, and the confidence range is obtained from Eq. 15-76.

$$\bar{Y}_{x=20} \quad = 1.0249 + (0.00444)20 = 1.1137$$

$$S(\bar{Y}_{x=20}) = \left[S^2(y) \left[\frac{1}{N} + \frac{(x-\bar{x})^2}{\Sigma' x^2} \right] \right]^{1/2}$$

$$= \left[.00032 \left(\frac{1}{12} + \frac{(20-13.4)^2}{473.06} \right) \right]^{1/2} = 0.00633$$

$$\bar{Y}_{x=20} \quad = 1.1137 \pm (2.228)(.00633)$$

$$= 1.1137 \pm 0.0141$$

$\bar{Y}_{x=20}$ ranges from 1.1278 to 1.0996

The above is an average value of the density at $x = 20$.

If it is desired to estimate the confidence of a single value at, say, $x = 20$, then Eq. 15-77 is used to find the confidence range:

$$S(Y_{x=20}) = \left[S^2(r) \left[1 + \frac{1}{N} + \frac{(x-\bar{x})^2}{\Sigma' x^2} \right] \right]^{1/2}$$

$$= \left[0.00032 \left(1 + \frac{1}{12} \frac{(20-13.4)^2}{473.06} \right) \right]^{1/2}$$

$$= 0.0104$$

$$Y_{x=20} \quad = 1.1137 \pm (2.228)(0.0104)$$

$$= 1.1137 \pm 0.0232$$

It should be noted that the confidence for single estimate is much less accurate than the average.

Finally, Eq. 15-78 is employed to obtained as estimate for b, the intercept.

$$S^2(b) = S^2(r) \left[\frac{1}{N} + \frac{\bar{x}^2}{\Sigma' x^2} \right]$$

$$= 0.00032 \left[\frac{1}{12} + \frac{(13.40)^2}{473.06} \right] = 0.001321$$

$$S(b) = 0.0364$$

$$b = 1.0249 \pm (2.228)(0.0364)$$

$$= 1.0249 \pm 0.0811$$

15-13. SUMMARY

In this chapter the modeling of experiments by means of statistical techniques was covered. First the types of error were discussed. Next, if the error can be estimated and a mathematical model of the process was available, then from

Taylor's theorem it was possible to obtain a relationship between the error of the dependent and independent variables. Many times neither the model nor an estimate of the error is available. Since the error can be due to unassignable causes and can be randomly distributed, an estimate of the error can be obtained by fitting a distribution function and using mathematics of probability. This led to the discussion of the normal distribution. It was next shown that the normal distribution was of limited value but served to illustrate the principles of the t distribution. The next topic discussed was regression or curve fitting. It was shown that the mathematics is much simpler for a linear model and hence one reason for linearization of experimental data. The t distribution, besides giving an estimate of the error of the variables defining the process, can be used to obtain an estimate the error of the various parameters entering the regression model. In this brief introduction the objective was to orient the reader to the language of statistics and to indicate the use of statistical analysis in engineering problems

REFERENCES

The references given here are for further reading and are in increasing order of complexity.

1. Elmer B. Mode, *Elements of Statistics,* 3d ed. Prentice-Hall, Englewood Cliffs, N. J., 1961.
2. William Volk, *Applied Statistics for Engineers,* McGraw-Hill, New York, 1958.
3. Bernard Ostle, *Statistics in Research,* Iowa State U. P., Ames, 1963.

PROBLEMS

15-1. Each of the following variables have an error as indicated.

$$w = 100 \pm 10$$
$$x = 4.60 \pm 0.02$$
$$y = 0.00137 \pm 0.00005$$
$$z = 200 \pm 5$$

Determine the maximum absolute error and the relative error in p for the mathematical models shown below.

(a) $p = x + y - z$

(b) $p = x \log (z - y)$

(c) $p = xyz$

(d) $p = \int_{r=w}^{r=z} \dfrac{dr}{x - \ln r}$

(e) $\quad xp^2 + zp + y = 0$

Hint: The relative error is given by $\Delta p/p$, where p is the approximate value obtained from the model and Δp is the absolute error.

15-2. It is desired to measure the average velocity in a pipe by weighing a quantity of water W issuing from the pipe in a given time t. Calculate the maximum absolute error in measuring the average velocity given the following additional data.

Variable	Approximate value	Degree of precision
W	75 lb	± 2 lb
t	52 sec	± 1.0 sec
D	3/4 in.	± 0.005 in.

15-3. In Prob. 15-2, if it is desired that the average velocity is to be measured such that maximum allowable error is ± 1.0%. What should be the allowable errors in the measured quantities W, t, D?

Hint: If $W = W_a^a, W_b^b, \cdots, W_n^n$, then from Eq. 15-6, $\hspace{2cm}$ (1)

$$\Delta W = (W_b^b, \cdots, W_n^n) a W_a^{n-1} \Delta W_a + (W_a^a W_c^c, \cdots, W_n^n) \, b W_{b \Delta W_b}^{n-1} + \cdots \hspace{0.5cm} (2)$$

Combining (1) and (2), we have

$$\frac{\Delta W}{W} = a \frac{\Delta W_a}{W_a} + b \frac{\Delta W_b}{W_b} + \cdots$$

where $\dfrac{\Delta W_i}{W_i}$ is the relative error. Assume that the relative error is the same for all the measured quantities in Prob. 15-3.

15-4. In the contact process for producing sulfuric acid from elemental sulfur it was found that the efficiency of the process average 98% with a standard deviation of 2.5. If the plant uses 1,000 tons per day of sulfur what will be the maximum amount of sulfur released with the stack gases per hour?

15-5. In Prob. 15-4, it was observed on a particular day that the efficiency of the plant was 96.5% probability, is the plant running satisfactorily?

15-6. Pilot plant yield determinations were run for a new process and found to be (10.7, 8.4, 9.6, 11.5, 10.2). (a) What is the mean yield, what is the 95% confidence limit on the mean? (b) We have obtained 6 more data points from the new process at new operating conditions. The observed yields are 10.3, 10.1, 11.5, 10.4, 11.7, 10.2 What are the chances (β value) that these yields are significantly different from the yields under the original conditions?

15-7. Ten samples of a monomer were divided in half, one-half were polyermized under existing conditions, the other half were polymerized with some additive which would change its hardness. Does the additive make any difference in the hardness?

Data

Hardness Number

Sample No.	Untreated	Treated
1	62	65
2	71	67
3	70	71
4	54	62
5	65	63
6	68	64
7	68	62
8	69	71
9	73	75
10	61	69

Hint: The t test when used to find the difference between two means is as follows:

$$t = \frac{|\bar{x}_1 - \bar{x}_2|}{\bar{s}(x)\sqrt{\dfrac{1}{n_1} + \dfrac{1}{n_2}}}$$

where n_1, n_2 are the number of samples and $\bar{s}(x)$ is the pooled estimate of the standard deviation from both sets of data which is given by

$$\bar{s}(x) = \sqrt{\frac{\Sigma' x_1^2 + \Sigma' x_2^2}{n + n_2 - 2}}$$

15-8. Two different feeds were used in a test run on a new catalyst. The results of the test are shown in the table.

Yield, Volume %.

Feed A	Feed B
80.8	70.5
81.7	77.0
78.9	78.1
79.9	75.6
84.1	76.9
80.3	73.8

It is claimed that with the new catalyst both feeds can be treated in the same unit. (a) Is the precision of the unit significantly different when running with feed A than with feed B? (b) Is the yield with feed A higher than with feed B? (c) If the yields are significantly different, write the 95% confidence interval for the true difference.

15-9. The following data were obtained on the enthalpy of methane at 1 atm pressure.

Temperature, °F	Enthalpy, Btu/lb
− 200	630

Temperature, °F	Enthalphy, Btu/lb
− 100	650
0	824
100	851
200	875
300	1,110
400	1,050
500	1,200

Calculate (a) The linear regression equation, (b) the correlation coefficient (is it significant); (c) the residual variance (the fraction of the sum of squares not accounted for by correlation) and (d) the enthalpy of methane at 340°F and − 10°F. Give both the point an interval estimate.

15-10. In a spray-type column used as a heat exchanger, the following data was obtained for the heat transfer from the continuous phase to oil drops (James W. McCaskill, Ph.D. dissertation, Louisiana State University, 1967). A preliminary plot indicated that the data fits Equation 2 of Table 15-6. (a) Write a computer flow diagram for nonlinear least-squares analysis. (b) Find the least-squares equation of the linearized data. (c) Linearize the data and obtain the equation by using the method of averages. Compare the result with part (b).

HTU/D	U_D/U_C
181	14.15
92.1	5.22
85.4	2.93
92.8	1.77
82.6	1.21
188.5	23.95
153.5	11.50
145.5	7.18
127.5	5.12
119.5	3.81
195.0	25.3
160.0	11.0
144.0	7.23
133.0	5.17
123.0	3.60

In order to make the arithmetic simpler, code the data by subtacting the mean from the data. Example;

$$X = (\text{HTU}/D) - (\text{HTU}/D) \quad \text{mean}$$

$$Y = (U_{D/U_C}) - (U_{D/U_C}) \quad \text{mean}$$

Show that the slope obtained for the new variable X, $Y \cdots$ unaffected by coding the data.

chapter 16

Analytical and Numerical Optimum-Seeking Methods

16-1. OPTIMIZATION OF MODELS

One of the main reasons for obtaining a mathematical model of a chemical or refinery process is to be able to predict the optimum conditions that maximize the profit or minimize the cost of operation of the process. Up to this point the development of the mathematical model only has been discussed, and it is now in order to study some of the methods of obtaining the values of the independent process variables that give the maximun profit or minimum cost. However, before this can be done an additional function is required that expresses the profit per unit time as a function of the independent variables. In many cases this is a simple linear equation made up of the sum of the product of the unit cost or sales value and the flow rates of the entering and leaving process streams. Also, a constant is usually included to cover overhead costs as labor and utilities. This equation is referred to by several names. It is called the *profit* or *cost function,* the *response surface,* the *criterion,* the *return function* (dynamic programming) and the *objective function* (linear programming). This function can also be a more complicated one, since operating cost can vary as the reciprocal of the independent variable (*e.g.,* compressor operations) and can be a function of time (*e.g.,* depreciation of the capital equipment or deactivation of catalyst.)

Of the various optimization techniques, each is best suited to operate on a particular type of mathematical model. Consequently when formulating the model it behooves us to have strong and weak points of various optimization techniques in mind. For example, if a process can be adequately described with equations that are linear the optimum operating conditions can be obtained by the powerful techniques of linear programming. If a process can be formulated as a series of stages with branches and loops, the no less powerful techniques of dynamic programming can be applied to obtain the optimum. However, if the process is very complex as is the case with many chemical processes one is forced to use numerical search techniques to grope around on the response surface seeking a peak or valley. These search techniques can be combined with the dynamic programming algorithm to give very effective combined optimization procedures. In fact, other optimization techniques can be used for suboptimization also, depending on the nature of the model.

Search techniques or optimum-seeking methods as they are sometimes referred are very effectively combined with other methods as previously mentioned. The basis of these methods are the classical theory of maxima and minima. Consequently our approach to the study of the optimization will first be a review of the classical theory of maxima and minima which leads directly to the search techniques. This will be followed with a chapter on dynamic programming and then one on linear programming. In this way the reader will be prepared to not only formulate the mathematical model of the process but apply the three important optimization techniques to obtain the best operating conditions for the process.

16-2. ANALYTICAL METHODS

Classical theory of maxima and minima (analytical methods) is concerned with finding the maximum or minimum (extreme points) of a function. In particular, we seek to determine the values of n independent variables $x_1, x_2, \cdots, x_n$ of a function that reaches an extreme point. Before starting with the development of the mathematics to locate extreme points consider a geometrical interpretation of the shape of a surface which is a function if two independent variables. This is shown in Fig. 16-1a, and a cross section through the points A and B is shown in Fig. 16-1b. Also shown is the first derivation of the curve through points A and B.

In this example point A is an absolute maximum which occurs at the top of a sharp ridge. Here the first derivation is discontinuous. A second but lower maximum is located at point B (a local maximum). At point B the first derivatives of $f(x_1, x_2)$ are zero, and B is called a *stationary point*. It is not necessary for stationary points to be maxima or minima as illustrated by stationary point C, a saddle point. In this example the minimum does not occur in the interior of the region but on the boundary at points D and E (local minima). To determine the absolute minima it is necessary to compare the value of the function at these points.

In essence, then, the basic problem of determining the maximum profit or minimum cost for process using the classical theory becomes one of locating all of the local maxima or minima and then comparing the individual values to determine the absolute maximum or minimum. The example has illustrated the places to look which are

1. At stationary points
2. On the boundaries
3. At discontinuities in the first derivative

When the function and its derivatives are continuous, the local extreme points will occur at stationary points. However, it is not necessary that all stationary points be local extreme points, as saddle points can occur also.

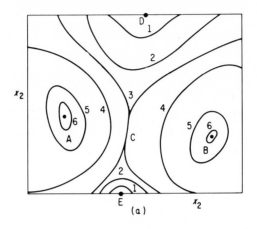

(a)

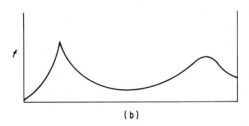

(b)

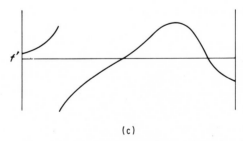

(c)

FIG. 16-1. Geometric interpretation of function $f(x_1, x_2)$. (a) Topological map. (b) Cross section through points A and B. (c) First derivative of curve through points A and B.

Locating Stationary Points

Based on our geometric intuition, we can readily understand the famous Weierstrass' theorem, which guarantees the existence of extreme points. It says: *Every function which is continuous in a closed domain possesses a maximum and minimum value either in the interior or on the boundary of the domain.*

There is another theorem which tells us how to locate extreme points in the interior of a region of a continuous function. It says: *A continuous function of n variables attains a maximum or a minimum in the interior of a region, only at those values of the variables for which the n partial derivatives either vanish simultaneously (stationary points) or at which one or more of these derivatives cease to exist [i.e., are discontinuous].*

Thus the problem becomes one of locating points where the partial derivatives are zero or where some of them are discontinuous. Analytically the stationary points can be located by simultaneously solving the n algebraic equations which result in setting the n partial derivatives equal to zero. The algebraic equations can be examined for points of discontinuities or use a numerical search to locate them.

Locating Local Maxima and Minima

As we have seen, it is not necessary for all stationary points to be local maxima and minima since there is a possibility of saddle or inflection points. We now need to develop conditions to permit us to evaluate stationary points as maxima or minima. These sufficient conditions will be developed for the cases of 1, 2, and n independent variables. Once the local maxima and minima are located, it is necessary to compare the individual points to locate the absolute maximum and minimum.

One Independent Variable

To develop criteria establishing whether a stationary point is a local maximum or minimum, we begin by performing a Taylor series expansion about stationary point x_0.

$$f(x) = f(x_0) + f'(x_0)(x - x_0) + f''(x_0)(x - x_0)^2 + \text{higher-order terms} \quad (16\text{-}1)$$

We select x sufficiently close to x_0 such that the higher-order times are negligible compared to the second-order terms. Since the first derivative vanishes at the stationary point, the above equation becomes

$$f(x) = f(x_0) + \tfrac{1}{2}f''(x_0)(x - x_0)^2 \quad (16\text{-}2)$$

We can say whether x_0 is a local maximum or minimum by examining the value of $f''(x_0)$, since $(x - x_0)^2$ is always positive.   This is summarized in the following:

If $f''(x_0) > 0$ then	$f(x_0)$ is a minimum
$f''(x_0) < 0$	$f(x_0)$ is a maximum
$f''(x_0) = 0$	examine $f'''(x_0)$

If the second derivative is zero, it is necessary to examine higher-order derivatives.   In general, if $f'(x_0) = f''(x_0) = \cdots = f^{(n-1)}(x_0) = 0$, the Taylor's series expansion becomes

$$f(x) = f(x_0) + \frac{1}{n!} f^{(n)}(x_0)(x - x_0)^n \tag{16-3}$$

If n is even, then $(x - x_0)^n$ is always positive, and the result is

If $f^{(n)}(x_0) > 0$ then	$f(x_0)$ is a minimum
$f^{(n)}(x_0) > 0$	$f(x_0)$ is a maximum

If n is odd, then $(x-x_0)^n$ changes sign as x moves from $x < x_0$ to $x > x_0$, and thus there is an inflection point.

These results can be summarized in the following theorem.

If at a stationary point the first and possibly some of the higher derivatives vanish, then the point is or is not an extreme point according as the first nonvanishing derivative is of even or odd order.   If it is even, there is a maximum or minimum according as the derivative is negative or positive.

Sign of a Quadratic Form

To perform a similar analysis for a function with more than one independent variable, it is necessary to determine what is called the *sign of the quadratic form*. This is usually written as

$$Q(\mathbf{A}, \mathbf{x}) = \sum_{i=1}^{n} \sum_{j=1}^{n} a_{ij} x_i x_j \tag{16-4}$$

where a_{ij} are the components of symmetric matrix $\mathbf{A}$ and $a_{ij} = a_{ji}$.

It turns out [1] that we can determine if Q is always positive or negative for all values of x_i and x_j by evaluating the sign of the minor determinates of $\mathbf{A}$, D_i.

$$D_i \equiv \begin{vmatrix} a_{11} & a_{12} \cdots & a_{1i} \\ a_{21} & & \\ a_{i1} & a_{i2} \cdots & a_{ii} \end{vmatrix} \tag{16-5}$$

The important results that will subsequently be used are

> If $D_i > 0$ for $i = 1, 2, \cdots n$ then $Q > 0$ (positive definite)
> If $D_i < 0$ for $i = 1, 3, \cdots$ then $Q < 0$ (negative definite)
> $\quad D_i > 0$ for $i = 2, 4 \cdots$
> If D_i is neither of these no statement can be made.

Two Independent Variables

To develop the criteria for a local maximum or minimum for a stationary point (c_1, c_2) of a function of two variables, a Taylor series expansion is made about the point.

$$
\begin{aligned}
f(x_1, x_2) = f(c_1, c_2) &+ f_{x_1} \cdot (x_1 - c_1) + f_{x_2} \cdot (x_2 - c_2) \\
&+ \tfrac{1}{2}[f_{x_1 x_1} \cdot (x_1 - c_1)^2 + 2f_{x_1 x_2} \cdot (x_1 - c_1)(x_2 - c_2) \\
&+ f_{x_2 x_2} \cdot (x_2 - c_2)^2] + \text{higher-order terms}
\end{aligned} \tag{16-6}
$$

where the subscripts x_1 and x_2 indicate partial differentiation with respect to those variables and evaluation at the stationary point.

The term in the bracket of the above equation is a quadratic form and $f(c_1, c_2)$ will be a minimum or a maximum accordingly as the bracket term is positive or negative. These results can be summarized by the following.

> $f(c_1, c_2)$ is a minimum if $f_{x_1 x_1} > 0$ and $\begin{vmatrix} f_{x_1 x_1} & f_{x_1 x_2} \\ f_{x_2 x_1} & f_{x_2 x_2} \end{vmatrix} > 0$
>
> $f(c_1, c_2)$ is a maximum if $f_{x_1 x_1} < 0$ and $\begin{vmatrix} f_{x_1 x_1} & f_{x_1 x_2} \\ f_{x_2 x_1} & f_{x_2 x_2} \end{vmatrix} > 0$

N Independent Variables

The results of the previous section can be extended to the case of n independent variables by defining the following determinant:

$$
D_i \equiv \begin{vmatrix}
f_{x_1 x_1} & f_{x_1 x_2} & \cdots & f_{x_1 x_i} \\
f_{x_2 x_1} & \cdots & & \vdots \\
\cdot & & & \\
\cdot & & & \vdots \\
f_{x_i x_1} & \cdots & & f_{x_i x_i}
\end{vmatrix} \tag{16-7}
$$

Then for a stationary point to be a local minimum,

$$
D_i > 0 \text{ for } i = 1, 2, ..., n
$$

For the stationary point to be a local maximum,

$$D_i < 0 \text{ for } i = 1, 3, 5, \dots$$
$$D_i > 0 \qquad i = 2, 4, 6, \dots$$

If neither of these fit the situation, refer to Hancock[1] for an extensive discussion of other cases.   We will not consider any further cases, as these results are usually sufficient.

Example 16-1[2]

Determine the optimum operating pressure and recycle ratio for a process where the hydrocarbon feed is mixed with recycle and compressed before being passed into a catalytic reactor.   The product and unreacted material are separated by distillation, and the unreacted material is recycled.   The pressure, P and recycle ratio R must be selected to minimize the total annual cost for the required production rate of 10^7 pounds per year.   The feed is brought up to pressure at an annual cost of $1,000P$, mixed with the recycle stream, and fed to the reactor at an annual cost of $4 \times 10^9 / PR$.   The product is removed in a separator at a cost of $10^5 R$ per year and the unreacted material is recycle in a recirculating compressor which consumes $1.5 \times 10^5 R$ annually.   Also determine the annual cost and show that it is a minimum.

Solution: The equation giving the operating cost is

$$C(\$/\text{yr}) = 1,000P + 4 \times 10^9/PR + 2.5 \times 10^5 R \qquad (1)$$

Equating the partial derivatives of C with respect P and R to zero gives

$$\partial C/\partial P = 1,000 - 4 \times 10^9/RP^2 = 0 \qquad (2)$$
$$\partial C/\partial R = 2.5 \times 10^5 - 4 \times 10^9/P^2 R = 0 \qquad (3)$$

Solving simultaneous gives

$$P = 1,000 \text{ psi and } R = 4$$

Substituting to determine the corresponding cost gives

$$C = 3 \times 10^6 \ (\$/\text{yr})$$

$C(P,R)$ is a minimum if

$$\frac{\partial^2 C}{\partial P^2} > 0 \quad \text{and} \quad \begin{vmatrix} \dfrac{\partial^2 C}{\partial P^2} & \dfrac{\partial^2 C}{\partial P \partial R} \\[2ex] \dfrac{\partial^2 C}{\partial R \partial P} & \dfrac{\partial^2 C}{\partial R^2} \end{vmatrix} > 0$$

Where each derivative is evaluated at $(P = 1,000, R = 4)$.   Performing the appropriate partial differentiation and evaluation at the stationary point gives

$$\partial^2 C / \partial P^2 = 2 > 0 \qquad \text{and} \qquad \begin{vmatrix} 2 & 10^3/4 \\ 10^3/4 & 10^6/8 \end{vmatrix} = 3 \times 10^6/16 > 0$$

Thus the stationary point is a minimum.

Analytical Techniques Applicable to Constraints

To this point the situation has been considered where the independent variables could take on any value; positive or negative, finite or undefined. In actuality the values of the independent variables are limited since they are things such as flow rates, temperatures, and pressures. Consequently there are constraints on the variables if nothing more than the fact that they must be nonnegative. In many cases they are bounded within limits as dictated by the process equipment.

First techniques to locate the stationary points of functions subject to equality constraints will be developed, and an example illustrating the techniques will be given. It turns out that the sufficient conditions for establishing that a stationary point is a maximum or minimum is directly applicable to the situation where constraints are the same in which the inequality is converted to an equality constraint as discussed in Chapter 18.

There are three methods of locating the extreme points of the function $f(x_1, x_2, ..., x_n)$ of n independent variables subject to m constraint equations $g_i(x_1, x_2, ..., x_n) = 0$. Here $i = 1, 2, ..., m$ and $n > m$. These are direct substitution, solution by constrained variation and method of Lagrange multipliers.

Direct Substitution

This simply means to substitute the constraint equations directly into the function to be extremized. This will result in an equation with $(n - m)$ unknowns, and the previous techniques for unconstrained optimization are applicable.

Unfortunately it is not always possible to perform these substitutions when the constraint equations, and the return function are somewhat complicated. It is then necessary to resort to other means.

Solution by Constrained Variation

This method is seldom used but furnishes the theoretical basis for the third technique. It is best illustrated for the case of two independent variables by considering Fig. 16-2. Here it can be seen that there is a local minimum of the constrained system at point A and a local maximum at point B. (The maximum of the unconstrained system is at C).

At point A the curve $f(x_1, x_2) = \text{constant}$ and the curve $g(x_1, x_2) = 0$ are tangent and have the same slope. This means that differential changes, dx_1 and dx_2 produce the same change in the dependent variables $f(x_1, x_2)$ and $g(x_1, x_2)$. This can be expressed as

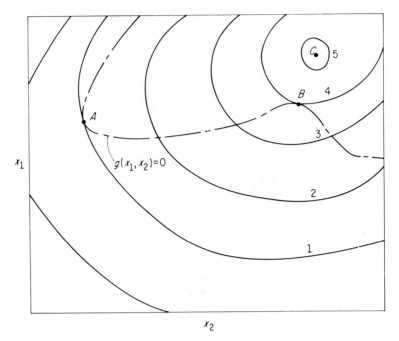

FIG. 16-2.   Sketch of return function $f(x_1, x_2)$ and constraint equation $g(x_1, x_2) = 0$.

$$\left(\frac{dx_1}{dx_2}\right)_f = \left(\frac{dx_1}{dx_2}\right)_g \tag{16-8}$$

Taking the total derivatives of both f and g gives

$$df = \frac{\partial f}{\partial x_1}\, dx_1 + \frac{\partial f}{\partial x_2}\, dx_2 = 0 \qquad dg = \frac{\partial g}{\partial x_1}\, dx_1 + \frac{\partial g}{\partial x_2}\, dx_2 = 0 \tag{16-9}$$

since $f =$ constant at points A or B and $g = 0$.

It should be remembered in this case that $\partial f/\partial x_1$ and $\partial f/\partial x_2$ are not zero. They were zero in the unconstrained case.

Combining the above Eqs. 16-8 and 16-9 gives

$$\frac{\partial f}{\partial x_1} \cdot \frac{\partial g}{\partial x_2} - \frac{\partial f}{\partial x_2} \cdot \frac{\partial g}{\partial x_1} = 0 \tag{16-10}$$

This is the equation to be solved in combination with the constraint equation to locate the stationary points.   This technique will be illustrated with an example subsequently.   First let us consider the extension to the general case.

General Case of n Independent Variables and m Constraint Equations

In general we are interested in finding the stationary points of a function $f(x_1, x_2, ..., x_n)$ subject to m constraint equations $g_i(x_1, x_2, ..., x_n) = 0$ where $i = 1, ... m$ and $n > m$.

The same reasoning applies in $n + 1$ dimensional space as applied to the three-dimensional space above, and this results in the following equations.

$$df = \frac{\partial f}{\partial x_1} dx_1 + \frac{\partial f}{\partial x_2} dx_2 + \cdots + \frac{\partial f}{\partial x_n} dx_n = 0$$

$$rg_1 = \frac{\partial g_1}{\partial x_1} dx_1 + \frac{\partial g_1}{\partial x_2} dx_2 + \cdots + \frac{\partial g_1}{\partial x_n} dx_n = 0$$

$$\tag{16-11}$$

$$dg_m = \frac{\partial g_m}{\partial x_1} dx_1 + \frac{\partial g_m}{\partial x_2} dx_2 + \cdots + \frac{\partial g_m}{\partial x_n} dx_n = 0$$

The $(n - m)$ equations corresponding to Eq. 16-10 of the two dimensional case can be written in terms of Jacobian determinants which are

$$J\left(\frac{f, g_1, g_2, ..., g_m}{x_1, x_2, ..., x_m, x_{m+1}}\right) = 0$$

$$J\left(\frac{f, g_1, g_2, ..., g_m}{x_1, x_2, ..., x_m, x_n}\right) = 0 \tag{16-12}$$

The Jacobian determinant for the first above is

$$J\left(\frac{f, g_1, g_2, ..., g_m}{x_1, x_2, ..., x_{m+1}}\right) = \begin{vmatrix} \dfrac{\partial f}{\partial x_1} & \dfrac{\partial f}{\partial x_2} & \cdots & \dfrac{\partial f}{\partial x_{m+1}} \\[2ex] \dfrac{\partial g_1}{\partial x_1} & \dfrac{\partial g_1}{\partial x_2} & \cdots & \dfrac{\partial g_1}{\partial x_{m+1}} \\[2ex] \dfrac{\partial g_m}{\partial x_1} & \dfrac{\partial g_m}{\partial g_m} & \cdots & \dfrac{\partial g_m}{\partial x_{m+1}} \end{vmatrix} \tag{16-13}$$

To illustrate the complexity of this method and give a basis for the results presented by the Jacobian, consider the following example.

Optimize

$$f(x_1, x_2, x_3)$$

subject to

$$g_1\ (x_1,\ x_2,\ x_3) = 0$$
$$g_2\ (x_1,\ x_2,\ x_3) = 0$$

At a stationary point

$$\frac{\partial f}{\partial x_1}\ dx_1 + \frac{\partial f}{\partial x_2}\ dx_2 + \frac{\partial f}{\partial x_3}\ dx_3 = 0 \qquad (16\text{-}14a)$$

$$\frac{\partial g_1}{\partial x_1}\ dx_1 + \frac{\partial g_1}{\partial x_2}\ dx_2 + \frac{\partial g_1}{\partial x_3}\ dx_3 = 0 \qquad (16\text{-}14b)$$

$$\frac{\partial g_2}{\partial x_1}\ dx_1 + \frac{\partial g_2}{\partial x_2}\ dx_2 + \frac{\partial g_2}{\partial x_3}\ dx_3 = 0 \qquad (16\text{-}14c)$$

To eliminate the differential dx_3, multiply Eq. 16-14a by $\partial g_1/\partial x_3$, Eq. 16-14b by $\partial f/\partial x_3$, and subtract.   This gives

$$\left(\frac{\partial f}{\partial x_1}\frac{\partial g_1}{\partial x_3} - \frac{\partial f}{\partial x_3}\frac{\partial g_1}{\partial x_1}\right)dx_1 + \left(\frac{\partial f}{\partial x_2}\frac{\partial g_1}{\partial x_3} - \frac{\partial f}{\partial x_3}\frac{\partial g_1}{\partial x_3}\right)dx_2 = 0 \ (16\text{-}15a)$$

Performing a similar manipulation with Eqs. 16-14b and 16-14c gives

$$\left(\frac{\partial g_1}{\partial x_1}\frac{\partial g_2}{\partial x_3} - \frac{\partial g_2}{\partial_1}\frac{\partial g_1}{\partial x_3}\right)dx_1 + \left(\frac{\partial g_1}{\partial x_2}\frac{\partial g_2}{\partial x_3} - \frac{\partial g_2}{\partial x_2}\frac{\partial g_1}{\partial x_3}\right)dx_2 = 0 \quad (16\text{-}15b)$$

Eliminating x_2 by a similar manipulation gives

$$\left[\left(\frac{\partial f}{\partial x_1}\frac{\partial g_1}{\partial x_3} - \frac{\partial f}{\partial x_3}\frac{\partial g_1}{\partial x_1}\right)\left(\frac{\partial g_1}{\partial x_2}\frac{\partial g_2}{\partial x_3} - \frac{\partial g_2}{\partial x_1}\frac{\partial g_1}{\partial x_3}\right)\right.$$

$$\left. - \left(\frac{\partial g_1}{\partial x_1}\frac{\partial g_2}{\partial x_3} - \frac{\partial g_2}{\partial x_1}\frac{\partial g_1}{\partial x_3}\right)\left(\frac{\partial f}{\partial x_2}\frac{\partial g_1}{\partial x_3} - \frac{\partial f}{\partial x_3}\frac{\partial g_1}{\partial x_3}\right)\right]\ dx_1 = 0$$

$$(16\text{-}16)$$

This is the equation to solve with the two constraint equations for the stationary point values of x_1, x_2, and x_3.   It will be left to the student to verify if this equation is the same result obtained from the Jacobian.

The reason the term in the bracket must be zero, since the equation is zero is that dx_1 is an arbitrarily selected and finite quantity.

Method of Lagrange Multipliers

The most common method to handle constraints is to employ Lagrange multipliers.   To illustrate this technique, consider multiplying Eq. 16-14b by λ_1 and Eq. 16-14c by λ_2; then add these to Eq. 16-14a.   The result is

$$\left(\frac{\partial f}{\partial x_1} + \lambda_1 \frac{\partial g_1}{\partial x_1} + \lambda_2 \frac{\partial g_2}{\partial x_1}\right) dx_1 + \left(\frac{\partial f}{\partial x_2} + \lambda_1 \frac{\partial g_1}{\partial x_2} + \lambda_2 \frac{\partial g_2}{\partial x_2}\right) dx_2$$

$$+ \left(\frac{\partial f}{\partial x_3} + \lambda_1 \frac{\partial g_1}{\partial x_3} + \lambda_2 \frac{\partial g_2}{\partial x_3}\right) dx_3 = 0 \tag{16-17}$$

The constants λ_1 and λ_2, the Lagrange multipliers, are to be determined such that each of the terms in the bracket are zero. This guarantees that stationary points will be obtained of the unconstrained system $L = f + \lambda_1 g_1 + \lambda_2 g_2$, since Eq. 16-17 can be written as

$$\frac{\partial}{\partial x_1}(f + \lambda_1 g_1 + \lambda_2 g_2) dx_1 + \frac{\partial}{\partial x_1}(f + \lambda_1 g_1 + \partial^2 g_2) dx_2$$

$$+ \frac{\partial}{\partial x_3}(f + \lambda_1 g_1 + \lambda_2 g_2) = 0 \tag{16-18}$$

L is called the *augmented function*.

Consequently to locate the stationary points of the following system five equations are to be solved for x_1, x_2, x_3, λ_1, and λ_2.

$$\frac{\partial f}{\partial x_1} + \lambda_1 \frac{\partial g_1}{\partial x_1} + \lambda_2 \frac{\partial g_2}{\partial x_1} = 0 \tag{16-19a}$$

$$\frac{\partial f}{\partial x_2} + \lambda_1 \frac{\partial g_1}{\partial x_2} + \lambda_2 \frac{\partial g_2}{\partial x_2} = 0 \tag{16-19b}$$

$$\frac{\partial f}{\partial x_3} + \lambda_1 \frac{\partial g_1}{\partial x_3} + \lambda_2 \frac{\partial g_2}{\partial x_3} = 0 \tag{16-19c}$$

$$g_1(x_1, x_2, x_3) = 0 \tag{16-19d}$$

$$g_2(x_1, x_2, x_3) = 0 \tag{16-19e}$$

The above equations result if one considers determining the stationary points of the augmented function $L = f + \lambda_1 g_1 + \lambda_2 g_2$ and considers the Lagrangian multipliers as independent variables. In general, with n independent variables and m constraint equations one must solve a system of $n + m$ equations to locate the stationary points. Although this sounds formidable, it is generally easier than obtaining the solution by constrained variation or by direct substitution unless the system is very simple. It can also be used with the numerical techniques of the next section when equality constraints are involved.

To illustrate these techniques consider an extension of Example 16-1.

Example 16-2

For the process in Example 16-1 it is necessary to maintain the product of the pressure and recycle ratio equal to 9,000 psi. Determine the optimal values of the pressure and recycle ratio and the minimum cost within this constraint by direct substitution, constrained variation and Lagrange multipliers. Again we wish to minimize C:

$$C = 1,000P + 4 \times 10^9/PR + 2.5R \tag{1}$$

but subject to the constraint

$$PR = 9000 \tag{2}$$

   A. Direct Substitution. Solving (2) above for P and substituting into (1) gives

$$C = 9 \times 10^6/R + 4 \times 10^6 + 2.5 \times 10^5 R \tag{3}$$

Setting $dC/dR = 0$ and solving gives

$$R = 6 \text{ and } P = 1,500 \text{ psi}$$

The corresponding cost is

$$C = 3.33 \times 10^6$$

which is greater than the unconstrained system as would be expected.

   B. Constrained Variation. The equations to be solved for this case are

$$\frac{\partial C}{\partial P} \frac{\partial}{\partial R} (9,000 - PR) - \frac{\partial C}{\partial R} \frac{\partial}{\partial P} (9,000 - PR) = 0 \tag{4}$$

$$9,000 - PR = 0 \tag{2}$$

Equation (4) simplifies to

$$P = 2,500 R \tag{5}$$

which when solved simultaneously with Equation (2) gives the same results as part *A*.

   C. Lagrange Multipliers. The augment function is formed

$$L = 1,000P + 4 \times 10^9/PR + 2.5 \times 10^5 R + \lambda (9,000 - PR) \tag{6}$$

Setting partial derivatives of L with respect to P, R, and λ equal to zero gives

$$1,000 - 4 \times 10^9/P^2 R - \lambda R = 0 \tag{7}$$

$$2.5 \times 10^5 - 4 \times 10^9/PR^2 - \lambda P = 0$$
$$9,000 - P \cdot R = 0$$

Solving the above simultaneously gives

$$P = 1,500, \quad R = 6, \lambda = 117.3$$

Method of Steepest Ascent

A further application of the method of Lagrange multipliers is the derivation of steepest ascent (descent) on the surface of a function to be maximized (minimized). This result will be extremely valuable when numerical methods are discussed.

To illustrate the direction of steepest ascent, a geometric representation is shown in Fig. 16-3. We wish to obtain the direction of steepest ascent, that is we wish to obtain the maximum value of df/ds where $f(x_1, x_2, ..., x_n)$ is a function of

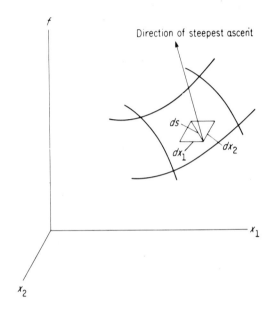

FIG. 16-3. Geometric representation of the direction of steepest ascent.

n variables. There is the constraint equation relating dx_1, dx_2, ..., dx_n and ds. The system is

Maximize

$$\left[\frac{df}{ds} = \sum_{i=1}^{n} \left(\frac{\partial f_i}{\partial x_i} \right) \left(\frac{dx_i}{ds} \right) \right] \tag{16-20}$$

subject to

$$(ds)^2 = \sum_{i=1}^{n} (dx_i)^2 \tag{16-21}$$

To obtain the maximum value of df/ds the augmented function is formed which is

$$L = \sum_{i=1}^{n} \left(\frac{\partial f}{\partial x_i} \right) \left(\frac{dx_i}{ds} \right) + \lambda \left[1 - \sum_{i=1}^{n} \left(\frac{dx_i}{ds} \right)^2 \right] \tag{16-22}$$

where the bracketed term is a rearrangement of Eq. 16-21.

Differentiating L with respect to the independent variables (dx_i/ds) and equating to zero gives

$$\frac{\partial f}{\partial x_i} - 2\lambda \left(\frac{dx_i}{ds} \right) = 0 \qquad \text{for } i = 1, 2, \cdots, n \tag{16-23}$$

where these n equations are solved simultaneously with the constraint equation for the values of dx_i/ds and λ. Combining Eqs. 16-21 and 16-23 and solving for λ gives

$$\lambda = \pm \, 1/2 \left[\sum_{i=1}^{n} \left(\frac{\partial f}{\partial x_i} \right)^2 \right]^{\frac{1}{2}} \tag{16-24}$$

and solving for dx_i/ds gives

$$\frac{dx_i}{ds} = \pm \frac{\partial f/\partial x_i}{\left[\sum_{k=1}^{n} \left(\frac{\partial f}{\partial x_k} \right)^2 \right]^{\frac{1}{2}}} \qquad \text{for } i = 1, 2, \cdots, n \tag{16-25}$$

The denominator of the above expression is not a function of i, and dx_i/ds is proportional to $\partial f/\partial x_i$. The plus sign indicates the direction of steepest ascent and the negative sign the direction of steepest descent. If k is the constant of proportionality, Eq. 16-25 becomes

$$\frac{dx_i}{ds} = \pm \, k \left(\frac{\partial f}{\partial x_i} \right) \tag{16-26}$$

and

$$\frac{df}{ds} = \pm \, k \sum_{i=1}^{n} \left(\frac{\partial f}{\partial x_i} \right)^2 \tag{16-27}$$

Consequently the direction of steepest ascent can be determined by evaluating the partial derivatives of f. This result is invaluable in specifing in which direction to search for a peak or valley, and is the basis for numerical techniques as Steep Ascent and Steep Ascent Partan. It should be noted that when dealing with physical systems, the direction of steepest ascent (descent) may be only a direction of steep ascent (descent) depending on the scales used to represent the independent variables. This is discussed and illustrated by Wilde [3] and Wilde and Beightler [12].

16-3. NUMERICAL METHODS

To now methods to locate extreme points have been discussed that have required an analytical function.   We are now ready to discuss techniques applicable to situations that do not require an equation to maximize or minimize.   This is generally the rule, since many times we are trying to optimize an operating chemical process or a complicated mathematical model of a process where the partial derivatives of the independent variables can not be readily formed. Other situations might require the previous specification of experiments such as the location of thermocouples in a fixed-bed reactor to locate the maximum temperature in the catalyst bed.

Search Problems and Search Plans

We are solving what are called *search problems*.   This is any investigation that seeks the optimal value of an unknown function.   Search problems can be classified by the number of independent variables—one or more than one.   For the solution of one-dimensional problems there are powerful techniques based on the *minimax principle*.   Unfortunately, this is not available for multidimensional problems.

Search problems can also be classified depending on the presence or not of random factors.   Deterministic problems are considered to have no experimental error or random factors present.   An example here is a mathematical model of a chemical process.   With a stochastic process, random factors particularly experimental errors are present.   An example here is an operating process where the measurement of the variables involve the usual random errors in measurement. To analyze stochastic processes, one considers them as deterministic with noise superimposed.   The effect is to slow the search for the optimum.

To perform the search, a search plan is needed.   This is a set of instruction for performing n experiments, x_1, x_2, ..., x_n, (values of the independent variables). An experiment can be the measurement of the dependent variable (e.g., temperature, pressure and space velocity) of an operating chemical process; or performing the calculation of the dependent variable by a mathematical model of a process having specified the independent variables.

Search plans can be classified as either simultaneous or sequential.   A simultaneous search plan has the location of each experiment specified before any result is known.   An example here is the previous one of the location of the thermocouples in the fixed-bed reactor.   A sequential search plan permits the experimenter to base future experiments on past outcomes.   The advantage of sequential search plans increase exponentially with the number of experiments over simultaneous search plans.

Our approach will be to examine first one-dimensional simultaneous and sequential search plans.   This will be followed by a discussion of some of the more widely used multidimensional search techniques.   The material discussed up to now on numerical methods and to be subsequently discussed is paraphrased from parts of the books by Wilde [3] and Wilde and Beightler [12].   Refer to these excellent books for amplification of the material discussed here and other methods to analyze stochastic processes, which are not covered in the present text.

Single Variable Search

The simplest case to optimize is that for a one-independent variable, unimodal function with no experimental error.   Optimal policies will be developed for both sequential and simultaneous searches.   We will search for the maximum, for convenience of discussion, as a minimization problem can be transposed to a maximization problem by seeking the greatest value of $-y$, rather than the least values of y.   Psychologically, management prefers to maximize profits rather than minimize costs.

A unimodal function must have one peak or valley, crudely speaking.   Examples of unimodal functions are shown in Fig. 16-4.   They can be convex, continuous, arbitrary, constrained, or discrete.   The most general definition does not require the use of derivatives of the function, but just uses the results, y_1 and y_2 from two experiments x_1 and x_2 ($x_2 > x_1$) which are placed on the closed interval $a \le x \le b$.   Then y is said to be unimodal on the interval from a to b if $y_2 > y_1$ for $x^* > x_2$ and if $y_1 > y_2$ for $x_1 > x^*$ (where x^* is the value specifing the location of the maximum.)

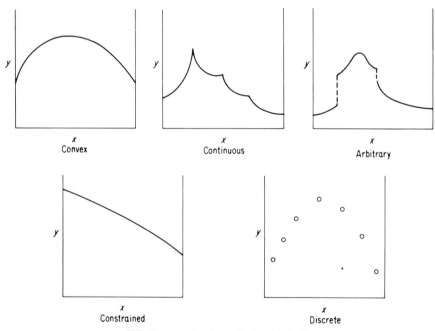

FIG. 16-4.   Various types of unimodal functions.

Measuring the Effectiveness of a Search Plan

We are interested in not only searching for the optimum but performing the search in an optimal manner. We wish to use the optimal search plan. To determine this optimal search plan, let us first compare the two search plans shown in Fig. 16-5 to determine the most effective. To do this, we need a criterion to measure the effectiveness of a search plan. The criterion to be used is the size of the interval of uncertainty remaining after placing of the experiments. This does not depend on the outcome of the experiments, which is very important in removing lucky outcomes from consideration. To illustrate this point, if we based our comparison of the search plans on the results of experiments when x_1 attained the largest value we would conclude that Search Plan A was more effective than Search Plan B, since the interval of uncertainty is 0.1 less for A than B. However, if the highest outcome occurred at x_2 or x_3 we would have concluded B was the more effective. Therefore to eliminate chance from affecting the comparison of search plans they must be compared based on the longest interval of uncertainty that would result. Then the longest interval of uncertainty depends only on the search plan and not on the outcome of experiments.

This can be expressed by the following:

$$L_n(\mathbf{x}_k) = \max_{1 \leq K \leq n} [1_n(\mathbf{x}_k, K)] \qquad (16\text{-}28)$$

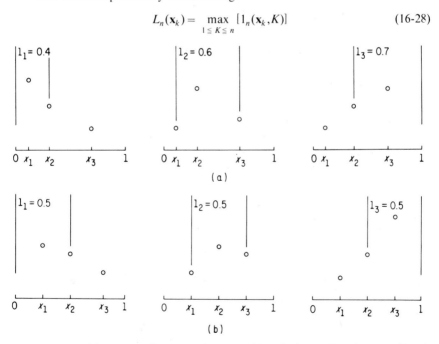

FIG. 16-5. Possible outcomes for two search plans. (a) Search plan A. Experiments are located at $x_1 = 0.1$, $x_2 = 0.3$, and $x_4 = 0.7$. (b) Search plan B. Experiments are located at $x_1 = 0.25$, $x_2 = 0.50$, and $x_3 = 0.75$.

where x_k represents the $k = 1, 2, ..., n$ experiments x_k of the search plan and l_n (x_k, K) represents all of the possible final intervals of uncertainty. K is the index of the greatest outcome which also varies from 1 to n, depending on all possible experimental results. To illustrate the use of Eq. 16-28, apply it to search plans A and B.

Search Plan A

$$L_3(x_1 = 0.1, \quad x_2 = 0.3, \quad x_3 = 0.7) = \max_{1 \le K \le 3} [l_3(x_k, 1), l_3(x_k, 2), l_3(x_k, 3)]$$

$$L_3 = \max_{1 \le K \le 3} [0.4, 0.6, 0.7] = 0.7$$

Search Plan B

$$L_3(x_1 = 0.25, \quad x_2 = 0.50, \quad x_3 = 0.75) = \max_{1 \le K \le 3} [l_3(x_k, 1), l_3(x_k, 2), l_3(x_k, 3)]$$

$$L_3 = \max_{1 \le K \le 3} [0.5, 0.5, 0.5] = 0.5$$

Based on this criterion of effectiveness, Search Plan B is the best.

A general expression for l_n after n experiments can be written as

$$l_n = x_{k+1} - x_{k-1}$$

where

$$x_{k-1} \le x^* \le x_{k+1} \quad \text{and} \quad y(x_k) = \max_{\mathbf{x}} [y(x_1), y(x_2), ..., y(x_n)]$$

Minimax Principle

A completely conservative measure of effectiveness is now available, and no matter how the experiments turn out, the final interval will be no greater than L_n, and if we are lucky it will be less. Now it is optimal to use the search plan that has the smallest final interval of uncertainty, L_n^*. All of the possible search plans, x_k, must be examined to obtain L_n^*.

$$L_n^* = \min_{x_k} \{L_n(x_k)\} = L_n(x_k^*) \tag{16-29}$$

Combining Eqs. 16-28 and 16-29 gives the minimax scheme:

$$L_n^* = \min_{x_k} \left\{ \max_{1 \le K \le n} [l_n(x_k, K)] \right\} \tag{16-30}$$

This is completely conservative where chance *only* determines the final position but *not* the length of the final interval of uncertainty. Based on this principle optimal simultaneous and sequential search plans will be developed.

Simultaneous Search

Simultaneous searches are not as effective as sequential searches, but have to be faced as in the case of the monitoring of the hot spot in a fixed-bed reactor. To narrow the interval of uncertainty, at least two experiments must be used. Using the unit interval for convenience but with no loss in generality, the following can be written from Eq. 16-28 as

$$L_2 = \max[x_2, 1 - x_1] \tag{16-31}$$

The smallest interval of uncertainty would be $x_1 = 0.5$, $x_2 = 0.5$ but this is the same point. If ϵ is the minimum separation between experiments such that a difference in the outcome can be detected, then L_2^*, the optimum is

$$L_2^* = \max[0.5 + \epsilon/2, \quad [1 - (0.5 - \epsilon/2]]$$
$$L_2^* = 0.5 + \epsilon/2 \tag{16-32}$$

The interval is only reduced by an amount ϵ if an additional experiment is used:

$$L_3 = \max[x_2, (x_3 - x_1), (1 - x_2)]$$

The optimal final interval of uncertainty is

$$L_3^* = 0.5 \tag{16-33}$$

Uniform Pairs

As can be seen, it is only efficient to perform an even number of experiments. This is called *search by uniform pairs*. The results for $n = 4$ and $n = 6$ have been illustrated by Wilde, [3] and this is show isn Fig. 16-6. The general expression to locate the points for an n (even) search is given by

$$x_k = \frac{(1 + \epsilon)[(k + 1)/2]}{(n/2) + 1} - \left\{ \left[\frac{k + 1}{2}\right] - \left[\frac{k}{2}\right] \right\} \epsilon \tag{16-34}$$

where the bracket specifies using the greatest integer value of the number e.g., $[3.5] = 3$. The corresponding ϵ minimax length is

$$L_n^* = \frac{1 + \epsilon}{(n/2) + 1} \tag{16-35}$$

which is obtained by taking $(x_{k+1} - x_{k-1})$.

Sequential Search

With a sequential search it is possible to make use of previous information to locate subsequent experiments. The result is a much larger reduction in the final interval of uncertainty for the same number of experiments.

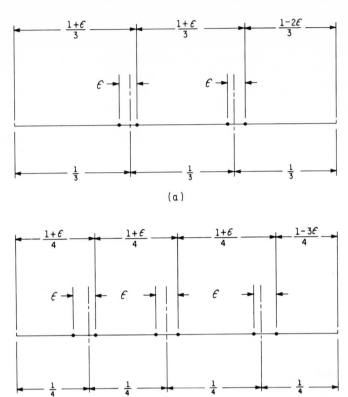

FIG. 16-6. Simultaneous search by uniform pairs. (a) Four simultaneous experiments ($n = 4$). (b) Six simultaneous experiments ($n = 6$).

Dichotomous Search

If there are only two experiments, a good thing to do is place them symmetrically in the center of the interval, a distance ϵ apart. On a unit initial interval, the final interval after two experiments is

$$L_2 = 1/2 + \epsilon/2$$

Placing the third and fourth in the interval remaining after two experiments gives

$$L_4 = 1/2[1/2 + \epsilon/2] + \epsilon/2 = 1/4 + 3\epsilon/4$$

Repeating,

$$L_6 = 1/2[1/4 + 3\epsilon/2] + \epsilon/2 = 1/8 + 7\epsilon/8$$

For n (even) experiments the optimum is located within the interval

$$L_n = 1/2^{n/2} + (1 - 1/2^{n/2})\epsilon \tag{16-36}$$

and the star or asterisk superscript has not been written, since there is the search plan which is more efficient which is called the Fibonacci Search Plan. However, before discussing the Fibonacci Search Plan the final interval of uncertainty for this plan and for the simultaneous one should be compared. As can be seen by proceeding simultaneously, the final interval of uncertainty decreases linearly with n (even), Eq. 16-35. However, using the sequential search plan, the final interval of uncertainty decreases exponentially with n. To reduce this final interval to one percent of its original length, 198 simultaneous experiments are required, whereas only 14 dichotomous experiments are required.

Fibonacci Search Plan

This is the most efficient of the one-dimensional search plans. The method will be described, and the reader is referred to Wilde[3]and Wilde and Beightler [12] for the development of the procedure and other minimax searches.

Having specified the number of experiments to be performed, the location of the first experiment is given by

$$x_1 = L_2^* = \frac{F_{n-1}}{F_n} + \frac{(-1)^n \epsilon}{F_n} \tag{16-37}$$

F_n is the nth Fibonacci number. This famous sequence of numbers is generated by the following relation:

$$F_n = F_{n-1} + F_{n-2}$$
$$F_0 = F_1 \equiv 1 \tag{16-38}$$

The second experiment is placed symmetrically from the other end of the initial interval and gives two experiments with which to eliminate a portion of the interval. The search is continued by placing experiments symmetrically within the remaining interval of uncertainty. This is illustrated in Fig. 16-7 for the case of $n = 4$ with a function whose maximum is near the right-head side of the interval.

The final interval of uncertainty after n experiments is given by

$$L_n^* = 1/F_n + F_{n-2} \epsilon / F_n \tag{16-39}$$

For $n = 4$, $\epsilon = 0.05$, $F_4 = 5$, and $F_2 = 2$ of Fig. 16-6, the result is

$$L_4^* = 0.220$$

In Fig. 16-7, we have shown the final interval of uncertainty located to the right-hand side of the initial interval. The location of this interval is given by chance alone, and the size is determined by the number of experiments alone.

The disadvantage of this search technique is that the number of experiments must be specified. The following technique does not have this disadvantage but sacrifices some on the reduction on the final interval of uncertainty.

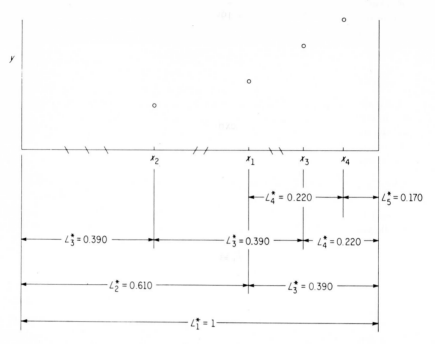

FIG. 16-7.   Fibonacci search for $n = 4$ and $\epsilon = 0.05$ with maximum near right-hand side of the initial interval.

Golden Section Search

Referring to Fig. 16-6, an equation that relates the intervals of uncertainty is $L_{j-1} = L_j + L_{j+1}$. This search technique holds the ratio of succesive intervals constant, i.e., $L_j/L_{j-1} = \tau$. Noting that $(L_{j-1}/L_j)(L_j/L_{j+1}) = \tau^2$, we can combine the relations to obtain

$$\tau^2 = \tau + 1 \qquad (16\text{-}40)$$

This equation has a positive root $\tau = 1.1618....$

Thus the experiments are started at L_2 where

$$x_1 = L_2 = L_1/1.618$$

and continued with

$$L_3 = L_2/1.618$$

to the final interval at

$$L_n = L_1/\tau^{n-1} \qquad (16\text{-}41)$$

It can be shown that the final interval of uncertainty for the Golden Section search is 17% larger than a Fibonacci search for large n. The advantage being that the search can be started without specifying the number of experiments for this relatively small penalty. Also, a search can be started with Golden Section and the a switch to a Fibonacci search can be made.

In the searches up to now, a final interval of uncertainty has been specified. More often the optimum is thought of as a point which is a point where we know to a certain accuracy, a *precision point*. The next search, a modification of the Fibonacci search, is useful here.

Lattice Search

Knowing the accuracy required, the number of precision points on the interval of uncertainty can be determined. Let us now distribute these points uniformly on an interval from 1 to $K + 1$. A Fibonacci search can now be used to reduce this interval. The first point of a Fibonacci search with no experimental error on a unit interval is given by

$$L_2^* = \frac{F_{n-1}}{F_n} \qquad (16\text{-}42)$$

It is necessary to have L_2^* fall on a lattice point, that is the product of L_2^* and the length of the 1 to $k + 1$ interval L_{k+1} must be an integer. The way this can occur is to have L_{k+1} equal to F_n, and from Eq. 16-42

$$L_{k+1} L_2^* = F_{n-1} \qquad (16\text{-}43)$$

Consequently all that is necessary is to add sufficient fictitious experiments at one end of the interval such that the above equation holds. Once having placed the first experiment, all others will fall on the lattice points. It turns out that one less experiment will be required than with other search techniques, since they require a point to be in the final interval of uncertainty. This is not necessary for lattice search, as discrete points are used and their position is already specified.

Comparisons of One-Dimensional Searches

As can be seen in Table 16-1, the Lattice search has the largest reduction ratio, since one experiment is saved using discrete points. For optimizing other than discrete functions, the Fibonacci search is best, and this is followed by the Golden Section, then the Dichotomous search. This reduction ratio for the Fibonacci search given in the table is also the Fibonacci numbers. One-dimensional searches have been extended to cover the situation were a set of two or more experiments are placed sequentially, and these are referred to as Odd Block and Even Block searches [12]. The Fibonacci search is a special case of the Odd Block search, and all of these have been shown to be optimal in the minimax sense. The reader is referred to Wilde and Beightler [12] for the details and minimax proofs of these one-dimensional searches.

TABLE 16-1

Reduction Ratio for Various Sequential Search Plans[3]

Number of Experiments	Fibonacci Search	Golden Section Search	Lattice Search	Dichotomous Search
0	1	1	1	1
1	1	1	1	1
2	2	1.62	2	2
3	3	2.62	4	2
4	5	4.24	7	4
5	8	6.85	12	4
6	13	11.09	20	8
7	21	17.94	33	8
8	34	29.0	54	16
9	55	47.0	88	16
10	89	76.0	143	32
11	144	123	232	32
12	233	199	376	64
13	377	322	609	64
14	610	521	986	128
15	987	843	1,596	128
20	10,946	9,349	17,710	1,024
24	75,025	64,078	121,392	4,096

Multidimensional Searches

Unfortunately for searches involving more than one independent variable the powerful minimax concept is not applicable even if the function to be optimized is unimodal except for the very special case of rectangular unimodality[12]. One-dimensional searches could be performed on all of the parallel lines in the experimental region, and then the optimum points of each line could be compared to find the optimum of the region. This is not practical, since a large number of lines would have to be examined, meaning essentially the performing of an exhaustive search on one of the variables. In essence the problem is one of eliminating areas or volumes to locate the small region containing the optimum. Therefore, multidimensional searches must eliminate areas that do not contain the optimum or climb toward the optimum in rapid steps.

Multidimensional search techniques can be considered to have three parts—an opening gambit, a middle game, and an end game. The opening gambit is the selection of a starting point. Usually there is sufficient technical knowledge of the process to select a satisfactory starting point. If there is not, then the midpoint, for example, can be used.

The objective of the middle game is to climb toward the peak as rapidly as possible. The response surface is explored only when necessary to guide successive moves. Then when it appears that the optimum (maximum or minimum) has been reached, extensive exploration is to be performed using a quadratic fit of the response surface at the apparent optimum (end game). This will permit a check to insure the point is in actuality an extreme point rather than a saddle point.

It will be assumed that sufficient technical information is known about the process to select a starting point (e.g., the current operating conditions). Refer to Wilde [5] for the location of other starting points as the midpoint or the median.

There have been many middle-game procedures developed but generally involve contour tangent elimination or gradient and ascent [12]. For contour tangent elimination to locale the optimum, a strongly unimodal function is required which is very restrictive and because of this, refer to Wilde [6] for a discussion of this method. However, the method of Steep Ascent and Steep Ascent Partan, the basis of many successful middle-game algorithms [7], will be discussed.

A discussion will follow on the end-game procedures. This involves a quadratic fit of the surface to determine if the apparent optimum is an actual optimum rather than a saddle point. Finally, the Kiefer-Wolfowitz stochastic approximation procedure will be presented as a method to optimize in the presence of experimental error.

Method of Steep Ascent

The method of steep ascent involves determining the gradient line at the starting point. One then moves an arbitrary distance along this line and determines a new and hopefully higher value of the criterion function. A more effective way is to perform a one-dimensional search along the gradient line to locate the high point. Then the procedure is repeated with a new gradient line determined at the new point until an optimum is reached.

The gradient line was given by Eq. 16-26 as

$$\frac{dx_i}{ds} = \pm k \left(\frac{\partial f}{\partial x_i} \right) \qquad [16\text{-}26]$$

To implement this equation numerically, approximate the derivatives by differences as given here:

$$\frac{dx_i}{ds} \doteq \Delta x_i \qquad \frac{\partial f}{\partial x_i} \doteq \frac{f_i - f_o}{\Delta x_i} = M_i \qquad (16\text{-}44)$$

where f_i is the value of f measured by perturbing x_i only. The gradient line is

$$\Delta x_i = \pm k M_i \qquad (16\text{-}45)$$

where the plus sign gives the direction of steep ascent, the minus sign the direction of steep descent. To illustrate the calculation of the gradient line, consider the following example.

Example 16-3

The function f depends on four independent variables x_1, x_2, x_3, x_4 and data about point $f(\mathbf{x}_0)$ is given below. Determine the gradient line in the direction of steep ascent.

Calculation of Partial Derivatives

Data on Function f Center point

x_1	x_2	x_3	x_4	f
0	1	-1	3	$5 = f_0$
1	1	-1	3	$7 = f_1$
0	-1	-1	3	$6 = f_2$
0	1	1	3	$8 = f_3$
0	1	-1	2	$5 = f_4$

$$M_1 = (7\text{-}5)/(1) \quad = \quad 0$$
$$M_2 = (6\text{-}5)/(-2) = -0.5$$
$$M_3 = (8\text{-}5)/(2) \quad = \quad 1.5$$
$$M_4 = (5\text{-}5)/(-1) = \quad 2$$

Gradient line through point $(0, 1, -1, 3)$ is

$$x_1 = 2k$$
$$x_2 = 1 - 0.5k$$
$$x_3 = -1 + 1.5k$$
$$x_4 = 3$$

A one-dimensional search could now be performed on k to locate the high (low) point along the gradient.

The procedure of measuring the gradient and performing a one-dimensional search to locate the optimum is shown in Fig. 16-8 for successive steps of a function of x_1 and x_2 with elliptical contours.

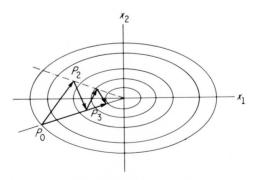

FIG. 16-8. Search by steep ascent on a function with
elliptical contours.

Steep-Ascent Partan

This procedure involves an acceleration step which makes it very effective in climbing functions with ridges. The acceleration step is shown in Fig. 16-8 with a one-dimensional search along a line through points P_0 and P_3. For this function of two independent variables with elliptical contours, the optimum is located in exactly two climbs along the gradient and one acceleration step.

This can be extended to a function with n independent variables and the optimum will be uniquely determined if the contours are ellipsoidal. The accompanying table outlines the procedures and the numbering of the points such that a climb along the gradient results in an odd-numbered point and an acceleration in an even-numbered point.

Steep-Ascent Partan Moves

Number of Independent Variables

Move	2	3	4	n
Start	P_0			
Climb	$P_0 \to P_2$			
Climb	$P_2 \to P_3$	$P_4 \to P_5$	$P_6 \to P_7 \cdots$	$P_{2n-2} \to P_{2n-1}$
Accelerate	$P_0 - P_3 \to P_4$	$P_2 - P_5 \to P_6$	$P_4 - P_7 \to P_8$	$P_{2n-4} - P_{2n-1} \to P_{2n}$

The search terminates at the optimum after $2n$ experiments if the contours of the return function are ellipsoidal. As can be seen two additional searches are required for each additional independent variable.

Nonellipsoidal Contours

The methods of Steep Ascent and Steep-Ascent Partan are applicable to any function. Of course, the optimum would not be reached in exactly $2n$ steps. In fact it can seem from Fig. 16-8 that steep-ascent partan will arrive at the optimum by the forth point for a function of two independent variables with radially similar contours. Further, one would expect excellent performance on any function from Steep Ascent Partan, as it will seek and follow ridges. This is born out in a recent comparison of search techniques by Leon [8]. One should refer to Leon [8] and Wilde and Beightler [12] for a discussion of popular optimization techniques currently in use.

End-Game Procedure

A quadratic fit to the response surface should be made to insure the middle-game procedure is not confounded by a resolution ridge or a saddle point. The procedure is conveniently conducted in two steps. First, a noninteracting approximation is performed, and examined as being a suitable fit to the surface. If not, the second step is to construct the interacting approximation for an improved fit to the surface. The one that best describes the surface can then be examined to determine if the result is an extreme point or a saddle point. The reason for this order is that usually the middle-game procedures have developed the necessary points for the noninteracting approximation. It can be determined immediately, and if it is not satisfactory, one additional point need be measured to construct the interacting approximation. The mathematical tools are already available, and it is only necessary to indicate the numerical technique.

Noninteracting Approximation

A quadractic approximation to a surface can be determined by the following Taylor series expansions for n independent variables $\bar{x} = (x_1, x_2, \cdots, x_n)$.

$$f(\mathbf{x}) = f(\mathbf{x_0}) + \sum_{i=1}^{n} \left(\frac{\partial f}{\partial x_i}\right)_{x_0} dx_i + 1/2 \sum_{i=1}^{n} \sum_{j=1}^{m} \left(\frac{\partial^2 f}{\partial x_i \partial x_j}\right)_{x_0} dx_i dx_j \quad (16\text{-}46)$$

A noninteracting approximation has the second cross-partial derivatives equal to zero, and using the following:

$$M_i = \left(\frac{\partial f}{\partial x_i}\right)_{x_0} \quad \text{and} \quad M_{ii} = \left(\frac{\partial^2 f}{\partial x_i^2}\right)_{x_0} \quad (16\text{-}47)$$

The result is

$$\Delta f = \sum_{i=1}^{n} M_i \Delta x_i + \frac{1}{2} \sum_{i=1}^{n} M_{ii} (\Delta x_i)^2 \quad (16\text{-}48)$$

The M_is and M_{ii}s can be evaluated by the following finite-difference relations:

$$M_i = \frac{f(x_1, x_2, \cdots, x_i + \Delta x_i, \cdots, x_n) - f(x_1, x_2, \cdots, x_i, \cdots, x_n)}{\Delta x_i} \quad (16\text{-}49)$$

$$M_{ii} = \frac{f[x_1, x_2, \cdots, x_i + \Delta x_i, \cdots, x_n] + f[x_1, x_2, \cdots, x_i - \Delta x_i, \cdots, x_n] - 2f[x_1, x_2, \cdots, x_n]}{(\Delta x_i)^2}$$

$$(16\text{-}50)$$

Having obtained this approximation, the next step is to predict the location of the optimum by the approximation. Setting the appropriate partial derivative equal to zero gives the following result to predict the location of the optimum:

$$\Delta x_i^* = x_i^* - x_i = -M_i/M_{ii} \quad \text{for} \quad i = 1, 2, ..., n \quad (16\text{-}51)$$

If the value of Δx_i^* is not within the resolution of Δx_i, i.e., $(\Delta x_i^* < \Delta x_i)$, the new coordinates x_i^* can be used as a point to continue with the middle game search.

If the value of Δx_i^* is within this resolution, the accuracy of the noninteracting approximation can be determined. This is accomplished by comparing the value of f at $(\mathbf{x} + \Delta \mathbf{x})$ and that predicted by the quadratic approximation, Eq. 16-48. If comparable results are obtained, the sign of the quadratic form is determined.

Interacting Quadratic Approximation

However, if it is necessary to include the interacting terms, the following equation can be used.

$$M_{ij} = [(f_{ij} - f_0) - M_i \Delta x_i - M_j \Delta x_j - 1/2 \, (M_{ii} \Delta x_i^2 + M_{jj} \Delta x_j^2)] / \Delta x_i \Delta x_j \qquad (16\text{-}52)$$

Having now obtained the noninteracting approximation which is

$$\Delta f = \sum_{i=1}^{n} M_i \Delta x_i + \sum_{i=1}^{n} \sum_{j=1}^{n} M_{ij} \Delta x_i \Delta x_j \qquad (16\text{-}53)$$

it is necessary to determine the sign of the quadratic approximation. Determining the sign of a quadratic form has been discussed in the previous section; here the reader may refer to the following summary of results:

f is a maximum	if Δf is negative definite
minimum	positive definite
saddle point	indefinite

If the fit of the surface does predict a saddle point, there is a higher (lower) point on the response surface. This point can be predicted by determining the extreme points of the interacting quadratic approximation.

Stochastic Approximation Procedures

Before leaving this topic, search techniques will be briefly discussed that converge to an optimum in the face of random error. Random (e.g., experimental) error clouds the perception of what is happening and greatly hampers the search for the optimum. *Stochastic approximation procedures* (as they are called) deal with random error as noise superimposed on a deterministic process. We are forced to consider convergence for these procedures before we consider efficiency. The work of Dvoretzky, and Kiefer and Wolfowitz in this area has been summarized in an excellent manner by Wilde [9]. We will only outline the procedure for using the Kiefer-Wolfowitz stochastic approximation procedure for k independent variables.

Kiefer-Wolfowitz Multidimensional Stochastic Approximation Procedure

With noise present we are forced to creep to prevent being confounded by random error. However, for unimodal functions, it can be shown that the procedure converges in mean square and with probability one.

Having selected a starting point, the procedure continues according to the following equation for k independent variables:

$$\begin{bmatrix} x_1, n+1 \\ x_2, n+1 \\ \cdot \\ \cdot \\ \cdot \\ x_{k,n+1} \end{bmatrix} = \begin{bmatrix} x_1, n \\ x_2, n \\ \cdot \\ \cdot \\ \cdot \\ x_{k,n} \end{bmatrix}$$

$$+ \frac{a_n}{c_n} \begin{bmatrix} z(x_{1,n}+c_n, x_{2,n}, \cdots, x_{k,n}) - z(x_{1,n}-c_n, x_{2,n}, \cdots, x_{k,n}) \\ z(x_{1,n}, x_{2,n}+c_n, \cdots, x_{k,n}) - z(x_{1,n}, x_{2,n}, -c_n, \cdots, x_{k,n}) \\ \cdot \\ \cdot \\ \cdot \\ z(x_{1,n}, x_{2,n}, \cdots, x_{k,n}+c_n) - z(x_{1,n}, x_{2,n}, \cdots, x_{k,n}-c_n) \end{bmatrix}$$

$$[16-46]$$

For convergence, the parameters a_n and c_n must meet the following criteria:

$$\lim_{n \to \infty} a_n = 0$$

$$\lim_{n \to \infty} c_n = 0$$

$$[16-47]$$

$$\sum_{n=1}^{\infty} a_n = \infty$$

$$\sum_{n=1}^{\infty} \left(\frac{a_n}{c_n}\right)^2 < \infty$$

Consider the following example illustrating the use of the Kiefer-Wolfowitz procedure:

Example 16-4

Develop the procedure to obtain the minimum of a function of the form $[A x_1 (x_1 - x_1^*)^2 + B x_2 (x_2 - x_2^*)^2]$ which is affected by experimental error. The value of the minimum is somewhere on the interval $1 \leq x_1^* \leq 3$, $1 \leq x_2^* \leq 3$. Starting with the midpoint of the interval, give the equations for the second, third and last of twenty trials.

Solution: $a_n = 1/n$, $c_n = 1/n^{\frac{1}{4}}$ satisfies the criterion of Eq. 16-47.

For $\mathbf{x}_2 = (x_{1,2}, x_{22})$, $n = 1$:

$$\begin{bmatrix} x_{1,2} \\ x_{2,2} \end{bmatrix} = \begin{bmatrix} 2 \\ 2 \end{bmatrix} + \begin{bmatrix} z(3,2) - z(1,2) \\ z(2,3) - z(2,1) \end{bmatrix}$$

For $\mathbf{x}_3 = (x_{1,3}, x_{2,3})$, $n = 2$:

$$\begin{bmatrix} x_{1,3} \\ x_{2,3} \end{bmatrix} = \begin{bmatrix} x_{1,2} \\ x_{2,2} \end{bmatrix} + \frac{1}{2^{3/4}} \begin{bmatrix} z(x_{1,2} + 2^{-1/4}, x_{2,2}) - z(x_{1,2} - 2^{-1/4}, x_{2,2}) \\ z(x_{1,2}, x_{2,2} + 2^{-1/4}) - z(x_{1,2}, x_{2,2} - 2^{-1/4}) \end{bmatrix}$$

For $\mathbf{x}_{20} = (x_{1,20}, x_{2,20})$, $n = 19$

$$\begin{bmatrix} x_{1,20} \\ x_{2,20} \end{bmatrix} = \begin{bmatrix} x_{1,19} \\ x_{2,19} \end{bmatrix} + \frac{1}{19^{3/4}} \begin{bmatrix} z(x_{1,19} + 19^{-1/4}, x_{2,19}) - z(x_{1,19} - 19^{-1/4}, x_{2,19}) \\ z(x_{1,19}, x_{2,19} + 19^{-1/4}) - z(x_{1,19}, x_{2,19} - 19^{-1/4}) \end{bmatrix}$$

There are variations of the above procedure and one uses only the sign of the approximation to the derivatives. This can be employed effectively when there is difficulty with convergence, due to the shape of the curve on either side of the optimum. Also, a base point can be utilized in evaluating the derivative. However, this procedure is said not to converge as rapidly as the above method.

16-4. SUMMARY

In this chapter we have reviewed the analytical methods for locating extreme points of functions with and without constraints. This presentation was an elaboration on material by Law [13] and Leitmann [14]. These methods formed the basis for numerical search techniques that were subsequently discussed. This included the important minimax principle, which is the basis for one-dimensional searches. Two important middle-game procedures, Steep Ascent and Steep-Ascent Partan, were then discussed for multidimensional searches, and the end-game procedure was outlined. The end-game procedure is used to prevent middle-game procedures from being confounded by resolution ridges or saddle points. The chapter was closed with the Keifer-Wolfowitz stochastic approximation procedure that will locate the optimum in the face of random error.

REFERENCES

1. H. Hancock, *Theory of Maxima and Minima,* Dover, New York, 1960, p. 103.
2. J. D. Wilde, *Ind. Eng. Chem.,* Vol. 57, (August 1965), No. 8, p. 18.
3. D. J. Wilde, *Optimum Seeking Methods,* Prentice-Hall, Englewood Cliffs, N.J., 1965.
4. *Ibid.,* p. 29.
5. *Ibid.,* p. 99.
6. *Ibid.,* p. 93.
7. A. Lavel, and T. P. Vogel, *Recent Advances in Optimization Techniques,* Wiley, New York, 1966, p. 23.
8. *Ibid.,* p. 28.
9. D. J. Wilde, *Optimum Seeking Methods,* Ch. 6.
10. D. J. Wilde, *Ind. Eng. Chem.,* Vol. 57, No. 8 (August 1965), p. 18.
11. D. J. Wilde, *Optimum Seeking Methods,* p. 145f.
12. D. J. Wilde, and C. S. Beightler, *Foundations of Optimization,* Prentice-Hall, Englewood Cliffs, N.J.

13. V. J. Law, *Notes on Optimization Theory,* Chemical Engineering Dept., Tulane University, 1965.

14. T. N. Edelbaum, "Theory of Maxima and Minima," in G. Leitmann, *Optimization Techniques with Application to Aerospace Systems,* Academic Press, New York, 1962.

PROBLEMS

16-1. In a chemical process, hydrocarbon feed is mixed with recycle and heated before being passed into a catalytic reactor. The product and unreacted material are separated by distillation, and the unreacted material is recycled. The optimum temperature T, and recycle ratio R must be selected to minimize the total annual cost, both direct operating expenses and amortized capital expenditures, for a required production rate of $F = 50,000$ bbl/day. The feed is brought up to temperature in a furnace at a cost of 25c/bbl per 100° F above 70°F, mixed with recycle stream and fed to the reactor at cost of $5 \times 10^8/F \cdot R\$$ per day for a 75% conversion. Product is removed by destillation at a cost of 50c/bbl of product, and the unreacted material is recycled in a recirculating compressor which consumes 15c/bbl of recycle annually. (a) Determine the optimal values of the temperature and recycle ratio, and the corresponding minimum operating cost. Show that this is a minimum. (b) For technical reasons the ratio of the temperature and recycle ratio must be equal to 5000°F. Determine the optimal values of the temperature and recycle ratio, and the corresponding minimum operating cost within the constraint. Do this by direct substitution, constrained variation, and Lagrange multipliers.

16-2. A five-experiment search plan and the corresponding outcomes are given below on the interval $5 \leq x \leq 18$. Find the possible final intervals of uncertainty and indicate which of these is the actual one and the maximum one.

Independent variable . . .	5.5	7.8	10.2	14.6	17.1
Dependent variable	2.7	3.1	7.2	5.4	3.8

16-3. Design a five-experiment Fibonacci search for the preceding problem and compare the final intervals of uncertainty for both search plans.

16-4. Conduct a search for the maximum on the function $y = 1 + 9x - 2x^2$ on the interval $0 \leq x \leq 3$, using six Golden Section points. What is the maximum value of y attained and what is the length of the final interval of uncertainty?

16-5. Develop a computer program flow diagram for Golden Section search. Show the necessary logic, such that a program could be readily written for this search.

16-6. Discuss the procedure for applying a Fibonacci search to a unimodal function of two independent variables.

16-7. Use the method of steep ascent and Steep-Ascent Partan to locate the minimum of the function $y = x_1^2 + 3x_2 + 5x_3^2$, starting at the point (2, 1, 3). Why is Steep-Ascent Partan more effective?

16-8. Evaluate the noninteracting and interacting quadratic approximation of the function $y = 3x_1^2 + x_2^2 + 2x_3^2 - 12x_3 + 31$ at the point (2, 1, 3), and determine if the point is a maximum, or inflection point.

16-9. There are search methods that use logical procedures for searching for the optimum. Develop a logical procedure than selects a base point and perturbs about the base point. However, when the perturbed point is better than the base point, use it as the new base point. Next accelerate through the point obtained after the perturbations from the base point, and proceed with the search. Discuss success and failures in their affect on acceleration steps. This type of search has been named *Pattern Search* [11].

16-10. Find the minimum of the function given in Prob. 16-7, starting at the same point. However this time experimental error is involved and the Kiefer-Wolfowitz procedure must be used, employing $a_n = 1/n$ and $c_n = 1/4$ and $n = 1, 2, ..., 10$. Simulate experimental error by flipping a coin and add (subtract) a constant $k = 0.1$ from y if the coin turns up heads (tails).

chapter 17

Dynamic Programming

17-1. BASIC THEORY

Dynamic programming reduces the effort required for optimization by breaking a multivariable problem into a series of interrelated problems, each containing only a few variables. Therefore, it can be applied to any process that can be broken down into stages. The stages may be units of time, process components, or any other suitable idea.

Richard Bellman[1] first presented the basic theory of dynamic programming in the early 1950's and postulated the "Principle of Optimality," which states:

> An optimal policy has the property that whatever the initial state and initial decisions, the remaining decisions must constitute an optimal policy with regard to the state resulting from the first decision.

The Principle of Optimality is stated mathematically as

$$f_n(S_n) = \max_{d_n} \ [R_n(S_n, d_n) + f_{n-1}(S_{n-1})] \tag{17-1}$$

where $n =$ number of stages remaining in the process
$S_n =$ input to nth stage,
$d_n =$ decision at nth stage
$f_n(S_n) =$ maximum return from an n-stage process with input S_n to nth stage
$R_n(S_n, d_n) =$ return from stage n with input S_n and decision d_n
$S_{n-1} =$ output from stage n and input to stage $n - 1$
$f_{n-1}(S_{n-1}) =$ maximum return from stages 1 through $n - 1$.

This means simply that since every component in a serial structure influences every downstream component; and only the last component is independent. The last component can be suboptimized independently for each possible state of the input it receives. Once this is done, the last two components are grouped together and suboptimized independently for each possible state of input. This process continues until the entire structure is included in the optimization.

17-2. GENERAL ANALYSIS

Each stage of a serial structure is characterized by four factors (Fig. 17-1).

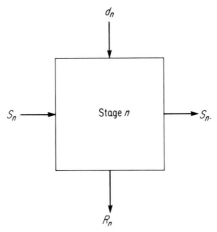

FIG. 17-1. Stage of any process.

These are the input, s_n; a decision, d_n; the output, s_{n-1}; and the return R_n. The output and return are dependent on the input and decisions. For a multistage serial process the output from stage n is the input to stage $n-1$ (Fig. 17-2). The set of variables that are outputs from the stages of the process are referred to as *state (dependent) variables* and the set of variables that are inputs to the stages are the *decision (independent) variables*.

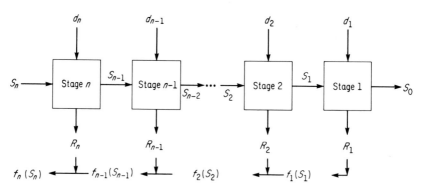

FIG. 17-2. Multistage process.

It is clear that the optimum for an n-stage system is the maximum of the sum of the return from stage n and the optimal return from stages 1 to $n-1$. Thus, for every value of the input, there is an optimal decision which causes the combined n-stage return to be a maximum. Therefore one must determine the optimal value of the decision for each possible state of the input.

A dynamic programming analysis usually begins with the last stage and ends

at the first stage.   The last stage of a serial process has an output that affects no
other unit in the system.   Therefore the decision that yields the optimum for every
possible input can be determined for the last stage (Fig. 17-3).

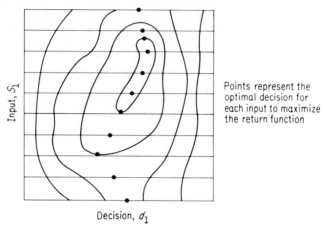

Points represent the
optimal decision for
each input to maximize
the return function

FIG. 17-3.   Contours of a return function.

For a two-stage process, the optimum decision that yields the maximum combined return is desired.   Numbering backward, stage 1 must operate with the
input received from stage 2.   The decision of stage 2 is varied.   The maximum
return is obtained from the sum of the return from stage 2 and the optimum
return from stage 1.   This is repeated for various inputs to stage 2.   As given by
Aris [2], the maximum profit from two stages is

$$
\begin{array}{l}
\text{Maximum profit} \\
\text{from two stages}
\end{array}
=
\text{maximum of}
\left(
\begin{array}{l}
\text{profit from} \\
\text{second stage}
\end{array}
+
\begin{array}{l}
\text{maximum profit from} \\
\text{the first stage with} \\
\text{feed produced by} \\
\text{the second stage}
\end{array}
\right)
$$

(17-2)

For a multistage system, first find $f_1(S_1)$.   With $f_1(S_1)$ known, find $f_2(S_2)$, then
$f_3(S_3)$.   This is continued until all stages are included.

The effort expended for this computation is just that of considering every
decision for every possible input for every stage.   It is ideally suited for machine
computation.

17-3.   NETWORK SYSTEMS

Some problems can be formulated in terms of networks.   The application
of dynamic programming can best be illustrated by the following example.

Example 17-1

A tank truck of an expensive chemical produced in San Francisco is to be delivered to any major port on the east coast for shipment to Europe. The cost for shipment across the Atlantic is essentially the same from any major port shown. Select the optimum route (lowest mileage) from San Francisco to the east coast. The relative distances between cities along possible routes are shown on the network diagram in Fig. 17-4.

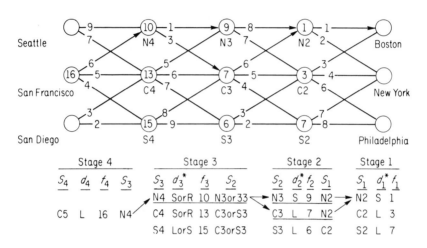

FIG. 17-4. Dynamic programming solution of a network problem.

Solution: To solve the problem one essentially works backward. In the circles are placed the optimal decision that would be made if the truck would have arrived at that city. For example, if at stage 1 the optimal route led to the northern city (N2), the optimal decision, d_1^*, would be to drive to Boston (S—straight, not L—left or R—right) which is closer than New York. Therefore, for each value of the state variable, the optimal decision, d_2^*, is tabulated at each stage. At stage 2 the return (mileage) for each state variable (city N3, C3, S3) is added to the optimal return (mileage) from stage 1 to determine the optimal decision (L, S, or R) at stage 2 as shown in the table. The output from stage 2, S_2, is the input to stage 1. The process is repeated for the four stages of the process. At the fourth stage the state variable has only one value, C5, and it is only necessary to determine the optimal decision corresponding to C5 which is N4.

The optimal return (minimum relative milage) is determined as 16, and the optimal policy is determined the underlined input-output-input sequence back through the table. The optimal policy is start at C5 (San Francisco) left to N4; then straight to N3, straight to N2, and straight to N1 (Boston) or right to C3, left to N2, and straight to N1 (Boston). As can be seen, there is not a unique optimal path.

17-4. APPLICATION TO A REFINERY PROCESS

Example 17-2

The refinery process for producing isooctane by reacting butene with isobutane in the presence of sulfuric acid catalyst (alkylation) is an excellent one to illustrate the dynamic-programming algorithm. In a typical alkylation process, four stirred reactors are used in series as shown in Fig. 17-5 where the catalyst flows from one reactor to another as it is degraded (decreases in H_2SO_4 concentration) Feed is introduced into each reactor and product is removed from the settler associated with each reactor. The product is collected from each reactor and sent to a large distillation column for fractionation. The controllable variables for each reactor are the temperature ($40 - 60°F$), d_1; feed space velocity (0.1 to 0.5 bbl butenes/lb catalyst/hr), d_2; butene feed concentration ($10 - 40\%$), d_3; and catalyst concentration (entering at 98% H_2SO_4), S_4. Usually the exit catalyst concentration is not less than 85% as competing polymerization reactions become important below this concentration.

Mathematical models are available that relate the profit with the independent variables. The profit is determined by the yield and quality of the product which in turn are affected by the independent variables. For example, the product quality is greater at the lower reactor temperatures but the cost of maintaining the lower temperatures is greater. The yield decreases with higher space velocity (feed rate), but more total product is produced. The yield and quality are better with lower butene feed concentrations, but more separation of unreacted isobutane is required which increases the recovery cost. A better product is produced at higher catalyst concentrations, but more catalyst is required as the discard composition increases. Thus it can be seen that a balance of the variables will be needed at each stage to maximize the profit. Also shown in Fig. 17-5 are some relative values of the profit for the optimum decisions.

For each stage the state variable is the catalyst concentration to the stage. Having specified the catalyst concentration (state variable), we must perform a three-dimensional search on temperature, space velocity, and butene concentration (decision variables) to locate the optimum profit. We can now make effective use of search techniques for the three-dimensional optimization. Some results are indicated in Fig. 17-5 in terms of a relative profit at each stage for a range of values on the catalyst concentration (state variable). As would logically be expected, the catalyst concentration is maintained as high as possible through out the process.

The following illustrates the application of the dynamic programming algorithm to determine the profit at each stage. At stage 1 the profit is just that

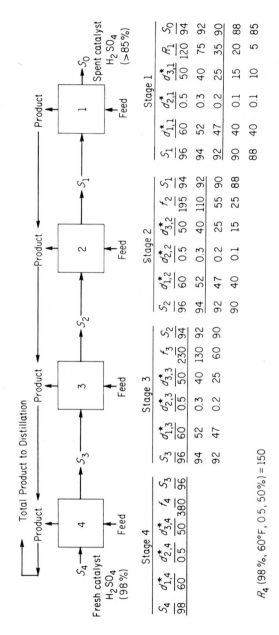

Stage 1

S_1	$d^*_{1,1}$	$d^*_{2,1}$	$d^*_{3,1}$	R_1	S_0
96	60	0.5	50	120	94
94	52	0.3	40	75	92
92	47	0.2	25	35	90
90	40	0.1	15	20	88
88	40	0.1	10	5	85

Stage 2

S_2	$d^*_{1,2}$	$d^*_{2,2}$	$d^*_{3,2}$	f_2	S_1
96	60	0.5	50	195	94
94	52	0.3	40	110	92
92	47	0.2	25	55	90
90	40	0.1	15	25	88

Stage 3

S_3	$d^*_{1,3}$	$d^*_{2,3}$	$d^*_{3,3}$	f_3	S_2
96	60	0.5	50	230	94
94	52	0.3	40	130	92
92	47	0.2	25	60	90

Stage 4

S_4	$d^*_{1,4}$	$d^*_{2,4}$	$d^*_{3,4}$	f_4	S_3
98	60	0.5	50	380	96

R_4 (98%, 60°F, 0.5, 50%) = 150

FIG. 17-5. Diagram of a typical sulfuric acid alkylation process and dynamic programming optimization.

calculated for that stage for various values of the catalyst concentration (state variable S_2) making the optimum decisions for the temperature, $d_{1,1}$; space velocity, $d_{2,1}$; and butene feed concentration, $d_{3,1}$. Now the maximum profit at stage 2 for a given value of the entering catalyst concentration, S_3 is obtained by making the optimal decisions at stage 2 to maximize the sum the profit from stage 2 and the optimal profit from stage 1. That is,

$$f_2(S_2) = \max_{d_{1,2}, d_{2,2}, d_{3,2}} \{R_2 + f_1\} \tag{17-3}$$

and for

$$S_2 = 96\% \ H_2SO_4$$
$$f_2(96\%) = 120 + 75 = 195$$

and

$$d_{1,2}^* = 60°F$$
$$d_{2,2}^* = 0.5 \text{bbl/bbl/hr}$$
$$d_{3,2}^* = 50\%$$

The other results are determined similarly. Continuing for stage 3,

$$f_3(S_3) = \max_{d_{1,3}, d_{2,3}, d_{3,3}} \{R_3 + f_2\} \tag{17-4}$$

and for

$$S_3 = 96\% \ H_2SO_4$$
$$f_3(96\%) = 120 + 110 = 230$$

For stage 4, the entering concentration is fixed at 98% H_2SO_4, consequently we only have to determine the optimum values for one value of the state variable. This is referred to as an *initial value problem*. The result is

$$f_4(S_4) = \max_{d_{1,4}, d_{2,4}, d_{3,4}} \{R_4 + f_3\} = 150 + 230 = 380 \tag{17-5}$$

and the maximum profit is 380. The optimal decisions can be obtained by moving back through the table, and this specifies how each reactor should be operated.

Before leaving the problem there are two points to be noted. First is that a family of optimal solutions can be developed with only slightly more work. If the table at stage 4 is completed for more values of the state variable (and in a finer interval), the optimal policy and maximum profit can be obtained without further work if the fresh catalyst concentration should change. If the problem was solved by search techniques alone, the solution would have to be repeated by varying all 13 independent variables. Thus dynamic programming makes available a family of solutions.

The second point is that the problem has been solved as an initial value problem. There are times when it would be desirable to fix the outlet composition and determine the inlet composition. This is referred to as a *final-value problem*. The method to solve this type of problem is discussed in the next section, along with procedures for handling loops and branches. First, however, it is necessary to formalize and expand the notation to be able to do this.

17-5. VARIABLES, TRANSFORMS AND STAGES

Although it was adequate in the preceding problem, the process flow diagram is generally not sufficient for developing an optimization plan because information flow rather than material flow must be depicted. Consequently functional equations and the corresponding functional diagrams of the stages of the process are developed to describe this information flow. Let us modify the definitions of state and decision variables in characterizing a stage of the process. This is necessary to distinguish between inputs to one stage that may have been outputs from the preceding stage. However, when loops and branches are involved the following nomenclature will be most helpful.

All outputs from a stage which are not returns (values predicted by the criterion function) are called *state variables* and written $\tilde{S}_i$, where i is the index of the input generating the state. The transformation equation having S_i as the independent variable is called the *transition function*, T_i, and is a function of the inputs. All inputs that are not *state variables*, S_i, are called *decision variables*, d_i. The functional diagram is a representation of the transaction equations, and this

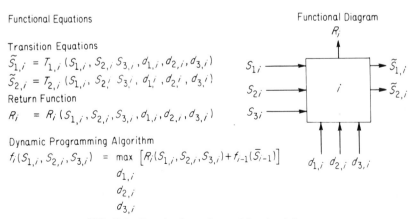

Functional Equations

Transition Equations
$$\tilde{S}_{1,i} = T_{1,i}(S_{1,i}, S_{2,i}, S_{3,i}, d_{1,i}, d_{2,i}, d_{3,i})$$
$$\tilde{S}_{2,i} = T_{2,i}(S_{1,i}, S_{2,i}, S_{3,i}, d_{1,i}, d_{2,i}, d_{3,i})$$

Return Function
$$R_i = R_i(S_{1,i}, S_{2,i}, S_{3,i}, d_{1,i}, d_{2,i}, d_{3,i})$$

Dynamic Programming Algorithm
$$f_i(S_{1,i}, S_{2,i}, S_{3,i}) = \max_{\substack{d_{1,i} \\ d_{2,i} \\ d_{3,i}}} \left[R_i(S_{1,i}, S_{2,i}, S_{3,i}) + f_{i-1}(\tilde{S}_{i-1}) \right]$$

Functional Diagram

FIG. 17-6. Functional equations and functional diagram.

is illustrated in Fig. 17-6. Thus a more general definition of *a stage* is a collection of one or more functions all depending on the same input variables. By using these functional equations and diagrams we can decompose systems and develop the necessary dynamic programming algorithms. This will include simple serial systems as just discussed previously and the more complex ones with branches and loops. Thus when a process is to be optimized by dynamic programming, the functional equations and diagrams can be developed describing the information flow. Then the appropriate material and energy balance equations, etc., can be developed which are the transition functions describing the outputs in terms of the inputs according to the functional diagrams and equations. Then results are available to generate the numbers from which the optimal solution is selected.

17-6. SERIAL OPTIMIZATION [4]

A serial system is the one discussed up to now, with the output of one stage as the input to the following stage. This is illustrated in Fig. 17-7 and is the same as the multistage process of Fig. 17-2. The functional equations are also given and include the transition equations, incident identity and return function. The incident identity gives the interaction between the stages. The return function specifies that the total profit R from the process is the sum of the profits from the individual stages.

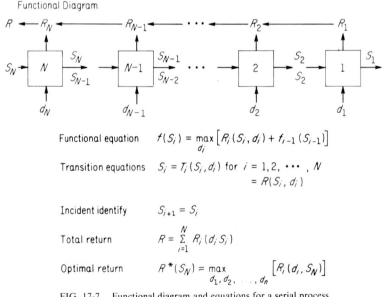

Functional Diagram

Functional equation $f(S_i) = \max\limits_{d_i} \left[R_i(S_i, d_i) + f_{i-1}(S_{i-1}) \right]$

Transition equations $S_i = T_i(S_i, d_i)$ for $i = 1, 2, \cdots, N$
$$= R(S_i, d_i)$$

Incident identify $S_{i+1} = S_i$

Total return $R = \sum\limits_{i=1}^{N} R_i(d_i, S_i)$

Optimal return $R^*(S_N) = \max\limits_{d_1, d_2, \ldots, d_n} \left[R_i(d_i, S_N) \right]$

FIG. 17-7. Functional diagram and equations for a serial process.

Serial optimization can be subdivided into initial value, final value, and two-point boundary-value problems.

Initial Value Problem

The dynamic programming algorithm for the ith stage of the initial-value problem is simply

$$f_i(S_i) = \max_{d_i} [R_i(S_i,d_i) + f_{i-1}(S_{i-1})] \tag{17-6}$$

Now to show that the optimal return at stage i is only a function of S_i, the stage variables at stage i, the incident relations and transition functions can be used as follows:

$$S_{i-1} = \tilde{S}_i = T_i(S_i,d_i) \tag{17-7}$$

Substituting the above into Eq. 17-6 gives

$$f_i(S_i) = \max_{d_i}\left\{ R_i(S_i,d_i) + f_{i-1}[T_i(S_i,d_i)] \right\} \tag{17-8}$$

For stage 1 there is one decision involved:

$$f_1(S_1) = \max_{d_1} R_1(S_2,d_1) \tag{17-9}$$

and there is one decision for each stage. At the last stage, stage N, the functional equation is -

$$f_N(S_N) = \max_{d_N} \{R_N(S_N,d_N) + f_{N-1}[T_N(S_N,d_N)]\} \tag{17-10}$$

If the value of S_N is selected that gives the maximum return $f_N(S_N)$, the problem is referred to as an N-decision, one-state, optimization problem. (There are $N+1$ independent variables and at the last stage a two decision, no state suboptimization is performed. However, if S_N is specified (a "choice" variable) as was the case in the alkylation process it is referred to as an N-decision, no-state optimization problem, and there are N independent variables. The set of values of the decisions, d_i, and state, S_N, that maximize the return function is called the *optimal policy*. N partial optimizations have been used to obtain an optimal return and optimal policy for the whole process.

Final-Value Problem

For this situation the output $\tilde{S}_1$ from the first stage is specified (choice variable). There are two approaches to solve this problem—*state inversion* and *decision inversion*.

State inversion simply means to transform the final-value problem into an initial-value problem by obtaining the N inverse transition functions. That is solve the transition equations for S_i in terms of $\tilde{S}_i$, i.e.,

$$S_i = \tilde{T}_i(\tilde{S}_i,d_i) \quad \text{for} \quad i = 1,2,\cdots,N \tag{17-11}$$

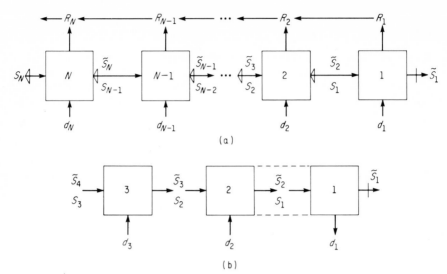

(a)

(b)

FIG. 17-8.   Functional diagrams for (a) state inversion (b) decision inversion.

This results in reversing the arrows in Fig. 17-7 as shown in Fig. 17-8a.   Reversing the numbering on the stages and the problem becomes an initial-value one.   The problem is an N-decision, one-state optimization problem.

In some cases inverting the transfer functions is an impossible task, and the technique of decision inversion is employed.   Here the roles of d_1 and S_1 are interchanged.   The stage 1 transfer function is

$$\check{S}_1 = T_1(S_1,d_1) = \text{const.} \tag{17-12}$$

This equation can be put in form

$$d_1 = \hat{T}_1(S_1,\check{S}_1) \tag{17-13}$$

and d_1 is uniquely determined by specifying S_1, since $\check{S}_1$ is a constant for this case. Hence stage 1 is decisionless and must be combined with stage 2.   (This is one of the rules to simplify dynamic-programming problems that will be subsequently discussed.)   This is shown diagrammatically in Fig. 17-8b with the arrow reversed on d_1 indicating that it is no longer a decision, and the arrow crossed on $\check{S}_1$ indicating that it is fixed.

The functional equation for stages 1 and 2 is now

$$f(S_2) = \max_{d_2} [R_2(S_2,d_2) + R_1(S_1,d_1)] \tag{17-14}$$

which can be combined with Eq. 17-18 and

$$S_1 = \check{S}_2 = T_2(S_2,d_2) \tag{17-15}$$

$$f(S_2) = \max_{d_2} \{R_2(S_2,d_2) + R_1\{T_2(S_2,d_2), \hat{T}_1[T_2(S_2,d_2),\tilde{S}_1]\} \qquad (17\text{-}16)$$

which shows that $f(S_2)$ results from a one-decision optimization.

The usual initial-value problem applies to the rest of the problem. The overall optimization involves $N-1$ total stages with $N-2$ one-decision, one-state partial optimization and at the Nth stage a two-decision, no-state partial optimization.

To illustrate state inversion we need only to consider a complete table developed at stage 4 of the alkylation process example (Fig. 17-5). Correspondingly, there would only be one entry in the table at stage 1 if the choice variable was taken as 98% H_2SO_4 for S_4.

To illustrate decision inversion Table 17-1 was developed to show the combination of stages 1 and 2. Only decisions $d_{1,2}^*$, $d_{2,2}^*$, and $d_{3,2}^*$ are required, since specifying S_2 automatically determine $d_{1,1}$, $d_{2,1}$, and $d_{3,1}$ of stage one as $\tilde{S}$ is a constant.

$$d_{11} = \hat{T}_{1,1}(S_2,\tilde{S}_1)$$
$$d_{2,1} = \hat{T}_{2,1}(S_2,\tilde{S}_1)$$
$$d_{3,1} = \hat{T}_{3,1}(S_2,S_1) \qquad (17\text{-}17)$$

TABLE 17-1

Decision Inversion for the Alkylation Process Example

Stage 2						Stage 1					
S_2	$d_{1,2}^*$	$d_{2,2}^*$	$d_{3,2}^*$	f_2	S_1	S_1	$d_{1,1}$	$d_{2,1}$	$d_{3,1}$	R_1	S_0
96	60	0.5	50	160	94						
94	52	0.3	40	100	92	94	52	0.3	40	40	88
92	47	0.2	25	55	90	92	47	0.2	25	25	88
						90	40	0.1	15	20	88

Two-Point Boundary-Value Problem

This type of problem arises when both the initial value and the final value are specified. The recommended way to handle this problem is to perform decision inversion, condense stages 1 and 2, and proceed as an initial-value problem. The functional equations for this case are:

Stage 1 and 2

$$f_2(S_2,\tilde{S}_1) = \max_{d_2} \{R_2(S_2,d_2) + f_1[T_2(S_2,d_2), \tilde{S}_1]\} \qquad (17\text{-}18)$$

This is a one-state, one-decision optimization at one stage, since $\tilde{S}_1$ is a specified constant. The optimization continues in the usual fashion:

$$f_3(S_3) = \max_{d_3} \{R_3(S_3,d_3) + f_2[T_3(d_3,S_3)]\} \qquad (17\text{-}19)$$

At the Nth stage, the functional equation is

$$f_n(S_n) = \max_{d_n} [R_n(S_n, d_n) + f_{n-1}(S_{n-1})] \tag{17-20}$$

This is a no-state, one-decision, one-stage partial optimization, since S_n is a specified constant.

Therefore, to solve the two-point boundary-value problem, we first perform a decision inversion followed by the usual solution for an initial-value problem. This involves an $N-2$ one-state, one-decision and one no-state, one-decision suboptimization. Two-point boundary-value problems always require decision inversion.

17-7. CYCLIC OPTIMIZATION

This is a special case of the two-point boundary-value problem where $S_N = \check{S}_1$. The functional diagram is shown in Fig. 17-9. The approach to solve this problem is to pick a value of $\check{S}_1 = S_N = C_1$ and proceed as a two-point boundary-value problem to determine the optimum return:

$$f_N(C_1) = \max_{d_1, d_2, \cdots, d_N} \left\{ \sum_{i=1}^{N} R_i(d_i, C_1) \right\} \tag{17-21}$$

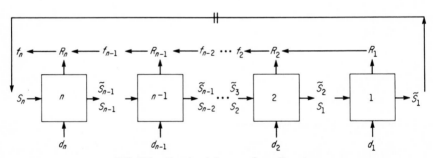

FIG. 17-9. Functional diagram of a cyclic system.

Then perform a one-dimensional search by varing C_1 until the maximum return is located. A Fibonacci search or Golden Section search can be used effectively here for the search on the cut-state values. Fixing the value of a state variable is referred to as a *cut-state,* and is indicated on a functional diagram by two slashes.

17-8. BRANCHED SYSTEMS [4]

Branched systems consist of two types. These are ones with converging branches and diverging branches. A feedforward loop (bypass) is a special case of a diverging branch and a feedback loop (recycle) is a special case of a converging branch. The discussion will begin with diverging branches, the simpler of the two cases.

Diverging Branches

The functional diagram of a diverging branch is given in Fig. 17-10. The branch consists of stages $1'$ through M' and has the following transition functions, incident identities, and return function:

$$\text{Transition functions: } \tilde{S}_{i'} = T_{i'}(S_{i'}, d_{i'}) \qquad (17\text{-}22)$$

$$\text{Incident identities: } \tilde{S}_{i'} = S_{i'-1} \qquad (17\text{-}23)$$

$$\text{Return function: } R_{i'} = R[S_{i'}, d_{i'}] \qquad (17\text{-}24)$$

$$\text{for } i' = 1', 2', \cdots, M'$$

The maximum return for the diverging branch is

$$f_{M'}(S_{M'}) = \max_{d_{1'}, d_{2'}, \cdots, d_{M'}} \left[\sum_{i=1'}^{M'} R(d_{i'}, S_{M'}) \right] \qquad (17\text{-}25)$$

To find this return requires the solution of an initial value serial problem which is "easily" done.

To connect the branch to the main system at stage k, the following are the transition functions and dynamic programming algorithm (functional equation):

$$\text{Transition functions: } \tilde{S}_{1k} = T_{1k}(S_k, d_k) = S_{k-1} \qquad (17\text{-}26)$$

$$\tilde{S}_{2k} = T_{2k}(S_k, d_k) = S_{M'} \qquad (17\text{-}27)$$

$$\text{Incident equations: } \tilde{S}_{1k} = S_{k-1}, \tilde{S}_{2k} = S_{M'} \qquad (17\text{-}28)$$

Functional equation:

$$f_k(S_k) = \max_{d_k} [R_k(S_k, d_k) + f_{M'}(S_{M'}) + f_{k-1}(S_{k-1})] \qquad (17\text{-}29)$$

This can be combined with the transition functions to give an algorithm in d_k and S_k only:

$$f_k(S_k) = \max_{d_k} \{R_k(S_k, d_k) + f_{M'}[T_{2k}(S_k, d_k)] + f_{k-1}[T_{1k}(S_k, d_k)]\} \qquad (17\text{-}30)$$

This is referred to as *absorption of a diverging branch.*

If the final value of the diverging branch is specified, decision inversion is performed and the final stage $(1')$ is condensed with the previous. Then one-decision optimization is continued. If the final value is a choice value, decision inversion is performed and the final stage $(1')$ is condensed with the previous. Then it is necessary to proceed with two-state $(S_{i'}, S_{1'})$, one-decision optimization.

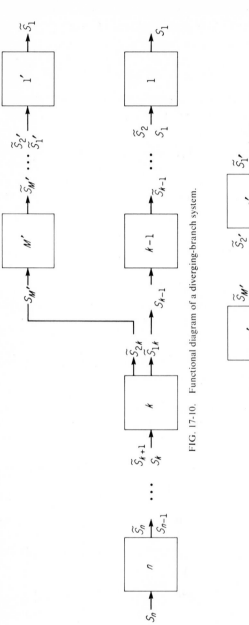

FIG. 17-10.   Functional diagram of a diverging-branch system.

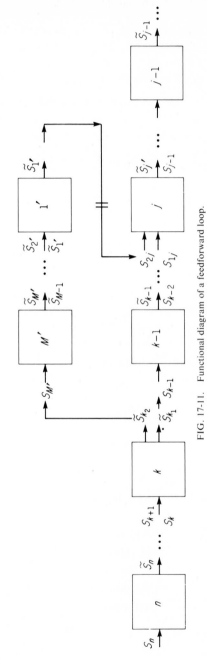

FIG. 17-11.   Functional diagram of a feedforward loop.

A special case of a diverging loop is a feedforward loop as shown schematically in Fig. 17-11. The approach to decompose this structure is to first write the transfer and return functions for stage j, which are

$$\tilde{S}_j = T_j(S_{j1}, S_{j2}, d_j) \tag{17-31}$$

$$R_j = R(S_{j1}, S_{j2}, d_j) \tag{17-32}$$

Now if $S_{j2} = \tilde{S}_{1'}$ (= const.) is treated as a cut-state, the result is a diverging-branch problem, having a fixed output. To stage $j-1$ on the main system, it is the usual initial-value problem. At stage j the functional equation in terms of S_{1j}, S_{2j}, and d_j is

$$f_j(S_{1j}, S_{2j}) = \max_{d_j} \{R_j[S_{1j}, S_{2j}, d_j] + f_{j-1}[T_j(S_{1j}, S_{2j}, d_j)]\} \tag{17-33}$$

The computations are continued to stage k and the diverging branch is absorbed:

$$f_k(S_k) = \max_{d_k} \{R_k(S_k, d_k) + f_{k-1}(S_{k-1}) + f_{M'}(S_{M'})\} \tag{17-34}$$

At this point the process is repeated for new values of $S_{1'}$ (using one dimensional search techniques) to locate the best values of $S_{1'} = S_1^*$. This best value is the one that gives the best value of $f_k(S_k)$. Then the partial optimizations are continued as usual.

Converging Branches

The functional diagram of a converging branch is given in Fig. 17-12. The branch consists of stages $1'$ through M' and in this case we wish to optimize the return from the branch and the main system. The obvious approach is to perform state inversion on the branch and convert the system to one with a diverging branch. Unfortunately this is usually difficult to accomplish.

The next approach is slightly more complicated to indicate, but is easier to implement. The functional equation at stage k is

$$f_{M+k}(S_{1k}) = \max_{S_{2k}, d_k} \{R_k(S_{1k}, S_{2k}, d_k) + f_{k-1}(S_{k-1}) + f_{M'}(S_{2k}, S_{M'})\} \tag{17-35}$$

The above indicates that state variable S_{2k}, should be treated as a choice variable. A two-dimensional search is used to maximize the right-hand side of the equation which includes a term for the return at stage k, the maximum return from the main system and the maximum return from the branch.

To obtain the maximum return from the branch, first cut the state at the branch, $\tilde{S}_{1'}$. This is then a final-value problem with $S_{M'}$ a choice variable or a two-point boundary-value problem if $S_{M'}$ is fixed. The procedure here calls for decision inversion, stage combination, and proceeding with one-state, one-decision optimization.

It is not necessary to carry $S_{M'}$ forward as a state variable, and the partial

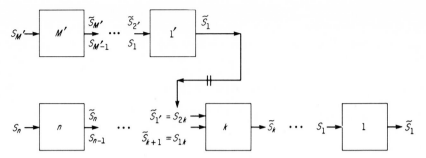

FIG. 17-12.   Functional diagram of a converging-branch system.

optimization is continued forward as a serial problem past 'stage k.   To illustrate these techniques consider the following example.

Example 17-3 *

A manufacturing process is arranged as shown in Fig. 17-13, along with the operating conditions and cost for each unit.   Raw material which comes in two variations, A and B, is successively heated and purified before it is sent to a reactor where it undergoes transformation into impure products.   The impure product is cleaned up in a "finisher" and the final product is sold.   A new heating-purifying train has just been built, and it is operated in parallel with the old one.   The new unit processes the same amount of material as the old unit, which is then mixed and pumped to the reactor.   The impurity content is averaged here.   After the reaction and finishing steps, two grades of product can be marketed—"colossal" and "stupendous."

We wish to find the maximum profit and the corresponding optimum policy for the plant operations.   It is *not* necessary for both plants to use the same raw material.

Solution: The problem involves a converging branch with choice variables as an initial value on the branch, and on the main system.   The functional diagram for the branch and main system is show in Fig. 17-14.   As in a serial problem, the optimal decisions are obtained for the various values of the state variables at stage 1.   At stage 2 the functional equation is

$$f_{2+4}(S_{12}) = \max_{S_{22}, d_2} \quad \{R_2(S_{12}, S_{22}, d_2) + f_1(S_1) + f_4 \cdot (S_{22}, S_4 \cdot)\}$$

This requires a two-dimensional search on the branch to locate the best value of S_{22} and on the main system to locate d_2^*, that maximizes the return from the branch and the main system for the various values of the state variable S_{12}.

* This example was developed by Prof. D. J. Wilde as a quiz problem for his optimization course at Stanford University.

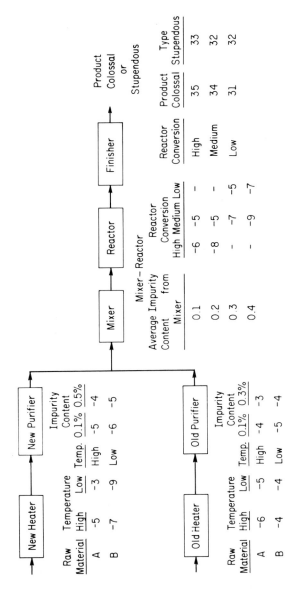

New Heater → New Purifier → (Mixer)

Raw Material	Temperature High	Low
A	-5	-3
B	-7	-9

Impurity Content Temp.	0.1%	0.5%
High	-5	-4
Low	-6	-5

Old Heater → Old Purifier → (Mixer)

Raw Material	Temperature High	Low
A	-6	-5
B	-4	-4

Impurity Content Temp.	0.1%	0.3%
High	-4	-3
Low	-5	-4

Mixer–Reactor

Average Impurity Content from Mixer	Reactor Conversion High	Medium	Low
0.1	-6	-5	-
0.2	-8	-5	-
0.3	-	-7	-5
0.4	-	-9	-7

Finisher → Product Colossal or Stupendous

Reactor Conversion	Product Colossal	Product Type Stupendous
High	35	33
Medium	34	32
Low	31	32

FIG. 17-13. Flow diagram of a manufacturing process to produce products colossal and stupendous with operating cost and sales prices tabulated.

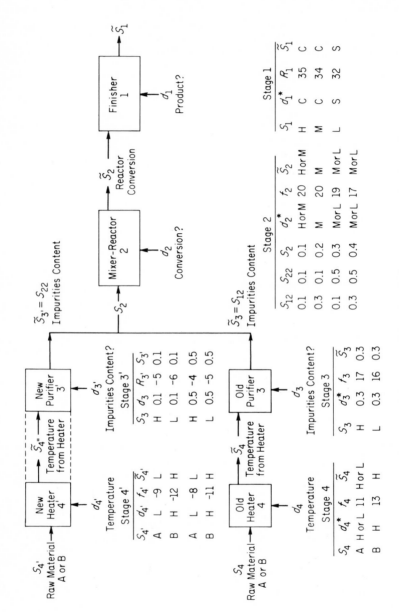

FIG. 17-14. Functional diagram with optimal policy for the manufacturing process operations.

Decision inversion is required for the branch and the results of this are shown under stages 3′ and 4′ for cut state values of 0.1% and 0.5%. The minimum cost to operate the branch is − 9 for an impurities content of 0.1% and − 8 for 0.5%. We now have the information to complete stage 2.

To illustrate the procedure at stage 2, consider the case for a value of S_{12} = 0.3% impurities content and a cut-state value of $S_{22} = 0.1\%$:

		High	Medium	Low
	R_2	− 8	− 5	–
	f_1	35	34	32
	f_4	− 9	− 9	− 9
$f_2 (S_{12} = 0.3, S_{22} = 0.1) = \max \{$		18	20	– $\}$
$= 20$				

performing a search on S_{22} and d_2 the results shown at stage 2 are obtained.

The partial optimization is then continued to stage 3 with the value of S_{12} corresponding to the largest value of f_2. The maximum profit of 13 is obtained using B as feed to the old heater and A to the new heater. The optimal policy is indicated on Fig. 17-14. Another optimal solution is also shown for the case of both units having to receive the same feed. This is A and the maximum profit is 11 for this case.

17-9. PROCEDURES AND SIMPLIFYING RULES

The previous discussion on applying dynamic programming to systems with loops and branches was developed by Aris, Newhauser, and Wilde [4]. A previous article by Mitten and Newhauser [5] gives a stepwise procedure for applying dynamic programming. Although the procedure is almost obvious, it is worth repeating for reenforcement.

1. Break the process down into stages.
2. Choose the units for measuring the return from each stage and from the entire process.
3. For each stage, determine:
   (a) the possible inputs, decisions, and outputs.
   (b) the output and return for every input-decision combination.
4. Apply the Principle of Optimality recursively, to find the optimal return from the process and the optimal decisions at each stage.

Based on the results of the article with Aris and Nemhauser, Wilde [6] formulated six important rules for simplifying a system to make a dynamic programming optimization plan more efficient. These are:

Rule 1. Irrelevant Stages

If a stage has no return and if its outputs are not inputs to other stages of the system, then the stage and its decisions may be eliminated.

It is not necessary to consider a stage that does not affect the return function.

Rule 2. Order of Partial Optimization

Partial optimization must proceed in the direction opposite the state flow (or the recursive application of the Principle of Optimality).

This is a guide to the choice of transformations to put a system under study into the required form.

Rule 3. Stage Combination

If a stage has as many or more output variables than it has input variables, then elimination of the output variables by combination with the succeeding stage should be considered.

Since an exhaustive search is required on the state variables, an overall savings of effort is obtained when state variables are eliminated, even at the expense of obtaining more decision variables. Multidimensional search techniques can be applied to decision variables. This leads to the following corollary.

Corollary to Rule 3—Decisionless Stages.

Any decisionless stage should be combined with an adjacent stage. Choose the one that eliminates the most state variables.

Rule 4. Fixed Output Constraints

Any fixed output should be transformed into an output by inverting either the state or decision variables. This transforms a final-value problem into an initial-value problem.

The result of applying this rule reduces the apparent number of decision variables. This is the case for the final-value problem, where decision inversion transforms a decision into an output which is completely specified by the input state variable.

Rule 5. Small Loops

Any loop with less than four decision variables should be optimized with respect to all decision variables simultaneously.

A loop with three decision variables has one of these eliminated by the procedure of cutting the state. This calls for decision inversion at stage 1 and combination of stages 1 and 2. Thus a one-state, one-decision partial optimization is required at stages 1 and 2 and a no-state, one decision optimization at stage 3. This is repeated, searching on S_3 for the maximum return. Rather than go through this procedure, Wilde suggests that it is easier to just perform a three-dimensional search.

Rule 6. Cut-State Location

Cut-states should always be inputs to multiple input stages.

This rule converts loops into diverging branches, which are easier to manipulate than are converging branches. Further, it replaces a state variable with one on which a one-dimensional search can be performed.

17-10. ADVANTAGES AND DISADVANTAGES

It is clear that the Principle of Optimality is a relatively simple concept. Difficulty arises in the formulation of a particular problem, as there is no clearly defined method to follow for a specific case.   Each case requires an individual analysis, and a proper formulation can mean the difference between success and failure.   The primary limitation, however, is in the number of input vectors (state variables) to each stage; a limit of three or possibly four is within the range of computational feasibility.   An exhaustive search of more than this is essentially impossible, even with the larger memory available in current digital computers.

The chief advantage of dynamic programming is its reduction of effort required to find an optimum.   Dynamic programming can be compared to an exhaustive search, since there is no basis for comparison with optimum-seeking methods presently.   For this comparison, consider an n-stage process with k decisions at each stage.   An exhaustive search requires k^n calculations, whereas dynamic programming requires nk.   The savings from using dynamic programming are evident.   For a process of $k = 3$ and $n = 100$, an exhaustive search requires 5.15×10^{46} calculations, whereas dynamic programming requires only 300.

17-11. APPLICATION OF DYNAMIC PROGRAMMING TO THE CONTACT PROCESS—A CASE STUDY[8]

Brief Description of the Process

The contact process plant for the production of sulfuric used in this study was designed to produce 98% sulfuric acid as a primary product and high pressure steam as a secondary product.   Sulfur input has been arbitrarily fixed at 10,000 lb/hr, which corresponds to a medium-sized plant.   This input will yield a theoretical maximum of 31,210 lb/hr of 98% sulfuric acid.

The process is shown schematically in Fig. 17-15.   Dry air and molten sulfur enter the burner and react to form sulfur dioxide.   The reaction is exothermic and goes to completion.

In the oxidation of sulfur dioxide to sulfur trioxide, the reaction rate increases with temperature, and the equilibrium conversion decreases with temperature. Therefore two converters are used, with partial conversion occurring in each. The temperature input to each is controlled by a waste-heat boiler.

The hot gas from the burner enters waste-heat boiler 1, is exchanged with water, and enters converter A after having been cooled to a suitable temperature. Partial oxidation to sulfur trioxide occurs in the presence of vanadium catalyst. This reaction is exothermic.

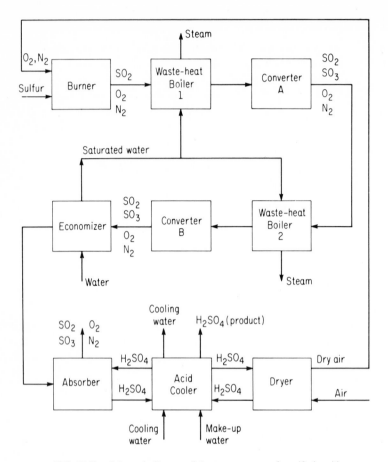

FIG. 17-15.   Schematic diagram of the contact process for sulfuric acid.

The stream next enters waste-heat boiler 2, is exchanged with water, and enters converter B after having been cooled to a suitable temperature. The water from waste-heat boilers 1 and 2 is taken off as high-pressure steam. Further oxidation occurs in converter B, again with a vanadium catalyst.

The gas goes next to the economizer, where it is exchanged with water. The water used is condensate that is preheated in the economizer before going to the two waste-heat boilers.

The cooled gas, containing sulfur trioxide, sulfur dioxide, oxygen, and nitrogen, is passed in countercurrent flow with 98% sulfuric acid, which absorbs sulfur trioxide. The unabsorbed gas is vented to the atmosphere.

Air is drawn in from the atmosphere through the dryer, where it is passed in countercurrent flow with 98% sulfuric acid, which absorbs the moisture from the air. The dry air flows to the burner, and the dilute acid is mixed with the concentrated acid from the absorber.

Since there is not enough moisture in the air to supply all of the water required, make-up water is added to hold the concentration at 98%. A sensing element regulates the amount of water required for make-up.

Heat of absorption and heat of dilution have raised the temperature of the acid. It is cooled in the acid cooler by water flowing over exposed pipes. The water is recirculated through a cooling tower. The cooled acid provides the overhead for the dryer and absorber and 98% acid, which is drawn off as product.

Dynamic-Programming Analysis

The dynamic programming analysis begins with the overall flow plan of nine stages (see Fig. 17-15). Each stage corresponds to an actual unit of the process. It is readily apparent that the scheme must be revised to permit an efficient optimization.

In Fig. 17-16 a diagram shows the nine stages with only input and decision variables depicted. The system output, W_O, is the rate of flow of product acid and is the dependent variable. Decision variables are W_{W_1}, the rate of flow of water to the economizer; W_{W_2} or W_{W_3}, the rate of flow of water to the waste-heat boilers 1 and 2 respectively; W_{W_O}, the rate of flow of water to the acid cooler; and W_A, the rate of air input to the process. The flow rate of sulfur to the burner, W_F, is fixed; and its input arrow is shown crossed, indicating this fact.

There are two recycle streams: one from the dryer to the burner and one from the economizer to each of the waste-heater boilers. These must be eliminated to permit a dynamic programming solution.

The interior recycle stream can be eliminated if the three flow rates, W_{W_1}, W_{W_2}, and W_{W_3}, are allowed to be independent. Specifically, the flow from the economizer is not considered to be the input to the boilers. An interim cost can be assigned to the input and output streams of each; the economizer in producing a stream of value per unit mass, and each of the boilers is receiving a stream of the same value per unit mass. The questionable feature of doing this is the possibility that the optimum determined may require a flow through the economizer that is far different from that required through the boilers. It was decided, however, that this was not probable. Nevertheless, a check was made to determine the extent of this difference. At the optimum the sum of the inputs to the boilers was essentially equal to the output from the economizer.

Additional simplification is possible. Decisionless stages must be combined with adjacent stages according to the rules previously discussed.

1. The burner is decisionless and is combined with waste-heat boiler 1.
2. Converter A is decisionless and is combined with waste-heat boiler 1.
3. Converter B is decisionless and is combined with waste-heat boiler 2.

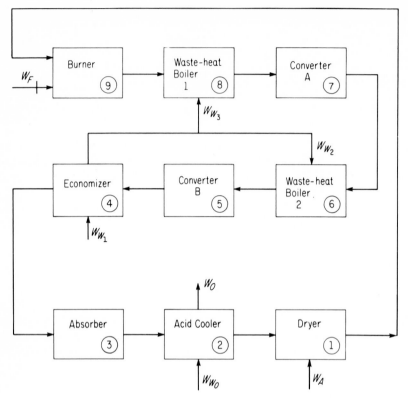

FIG. 17-16.   Decision variables affecting the contact process.

4. The absorber is decisionless and is combined with the acid cooler.
The resulting scheme is shown in Fig. 17-17.   There are now five stages, each
with one input and one decision.   A loop between the burner and the dryer makes
the system cyclic.

The approach to handle cyclic processes eliminates the loop by cutting the
state between the dryer and the burner.   The system is now a serial one. · By
cutting the state, the input of the dryer is fixed, and this makes the stage
decisionless, since this fixes the input of wet air.   It is then combined with the
acid cooler-absorber.

A closer analysis reveals that cutting the state requires fixing two state variables: temperature, T_1 and flow rate W_1 of the dry-air stream.   Having the
dryer, acid cooler, and absorber combined, the value of W_{W_O}, the acid cooler
water flow rate, is uniquely determined for a specified input to the stage.   Thus
the decision on W_{W_O} is removed, and this stage must be combined with the econo-

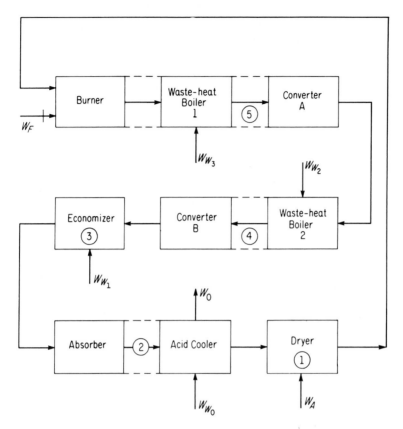

FIG. 17-17. Intermediate system, five stages.

mizer.   The resulting scheme is shown in Fig. 17-18.   It is a three-stage, initial-value, serial problem.

Dynamic programming is readily applied to this scheme.   There is one input and one decision at each stage.   It is necessary only to determine the optimal policy of the serial system for each fixed value of the cut-state.   A sequence of these serial problems is to be solved.   The value of the cut-state giving the best results is the optimum.

The stepwise procedure presented earlier is now applied to the three-stage system.   Step 1, breaking the process into stages, is already completed.   Step 2, the unit of return measurement, is chosen to be the net profit per hour of operation.

Step 3, the determination of inputs, outputs, and functional equations, will now be outlined.   The specific transition and return functions are given by Lowry [9].

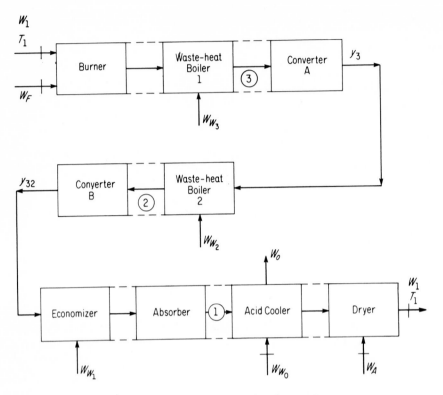

FIG. 17-18.   Final optimization plan, three stages.

As shown in Fig. 17-18, the input to stage three is the cut state values of W_1 and T_1.   The decision variable is W_{W_3}, the flow rate to waste-heat boiler 1, which controls T_3, the output temperature of the stage which determines the conversion y_3.   The return, R_3, is the sum of the cost of sulfur, the cost of the stream input to the boiler, and the profit from the steam produced.   The functional equation for stage 3 is

$$f_3 = \max_{W_{W_3}} \; [R_3(W_{W_3}) + f_2(y_3)]$$

The input to stage 2 is the ouput from stage 3, y_3 which is the conversion of SO_3 in converter A.   The decision variable is W_{W_2}, the flow rate to waste-heater boiler 2, which controls T_2, the output temperature of the stage which determines y_2, the conversion at stage 2.   The output is the total conversion $y_{32}(= y_2 + y_3)$ of the SO_2 to SO_3.   The return R_2, is the sum of the cost of the input stream to the boiler and the profit from the steam produced.   The functional equation for stage 2 is

$$f_2(y_3) = \max_{W_{W_2}} \; [R_2(y_3, W_{W_2}) + f_1(y_{32})]$$

Step 4, the recursive application of the functional equations is performed. The calculation begins with stage 1. $R_1 (y_{32}, W_{W_1})$ is calculated for every y_{32}, W_{W_1} combination. The optimal decision, $W_{W_1}^*$, is designated for each input, yielding $f_1 (y_{32})$.

For the two-stage process, the combined two-stage return is calculated for every stage 2 input-decision combination and the optimal decision, $W_{W_2}^*$, designated, yielding $f_2 (y_3)$. Next, the combined three-stage return is calculated for the cut-state value and the optimal decision, $W_{W_3}^*$, designated. This yields the overall optimum f_3. This calculation is repeated for each cut-state value. The value of the cut-state giving the highest value is the optimum for the process. This also specifies the optimum operating policy.

The strategy just described is the one that resulted after considerable trial-and-error attempts to formulate the problem. No all-purpose algorithm is available, and a strategy must be developed by experience, intuition, and luck.

The first problem was that of eliminating the interior loop. It encloses only two variables, and unless it can be eliminated by some unconventional method, Wilde's Rule No. 5 precludes using a dynamic programming approach, since less than four decision variables are involved. The solution was to allow each stream to be independent and, while not entirely correct, does permit an effective dynamic programming analysis.

Decisionless stages should be combined with adjacent stages, but when a choice of adjacent stages exists, it should be made to eliminate the most state variables. However, it may be necessary to make an arbitrary choice. This choice can affect the complexity of the succeeding analysis, and only an actual trial can determine whether or not the effect of a particular choice is adverse. The present case afforded a choice of combinations for each decisionless stage. Numerous plans were devised and rejected before the final plan emerged.

The location of the cut-state did not necessarily have to be, and although other choices were considered, no other appeared to afford any special advantage.

Adherence to Wilde's rules can simplify somewhat the task of problem formulation. In addition, it is suggested that a problem formulator be thoroughly familiar with the process under study.

Results: A discussion of the computer implementation of the model is given by Lowry [10]. This was performed on an IBM 7040 digital computer and average execution required 12 minutes. The results were that an overall optimum was selected and the operating policy necessary to achieve that particular value was specified. The best return was $230.57 per hour and was obtained when the cut-state value of W_1 was 135,000 lb/hr and T_1 was 430°K. In Table 17-2 the operating policy necessary to achieve this return is shown. This was determined by varying the values of the cut-state variables over a realistic range.

The entire schedule of optimal decisions for each stage input for the optimum return is shown in Table 17-3. As indicated by the table, the overall optimum re-

TABLE 17-2

Overall Optimum Operating Policy

$$W_1 = 135{,}000 \text{ lb}/hr \qquad T_1 = 430°\text{K } (157°\text{C})$$

Stage 3

Optimal Return f_3 ($ /hr)	Output, $\check{S}_3$ y_3^*	Converter A Temperature $T_3^*(°\text{K})$	Decision, d_3^* $W_{W_3}^*$ (lb/hr)
230.57	0.36226	1,000.0	25,513

Stage 2

Output $\check{S}_2$ y_{32}^*	Decision, d_2^* $W_{W_2}^*$ (lb/hr)	Converter B Temperature T_2^* (°K)
0.9859	22,472	700

Stage 1

Absorber Gas Temperature T_i^* (°K)	Decision, d_1^* $W_{W_1}^*$ (lb/hr)	Output, $\check{S}_1$ W_O^* (lb/hr)	Water Flow Rate to Acid Cooler $W_{W_0}^*$ (lb/hr)
325	45,214	30,768 Production rate òf H_2SO_4	207,501

quires an output from stage 3 of $y_3^* = 0.36226$. The computer was programmed to interpolate for this value as an input to stage 2, obtained the output, y_{32}^*, and the optimal operating policy, T_2^*, $W_{W_2}^*$, for stage 2. The stage 2 output is the required input to stage 1. The computer interpolates for this value and obtains the optimal output, W_0^*, and the optimal operating policy, $W_{W_1}^*$, $W_{W_0}^*$, T_i^*, for stage 1.

A close examination reveals that each of the functions is essentially linear in the region of importance; therefore the linear interpolation used was accurate. Also, the sum of $W_{W_2}^*$ and $W_{W_3}^*$ (47,985) approaches that of $W_{W_1}^*$ (45,214) with a deviation of less than 6%. This is sufficient to justify the method used to break the loop in the dynamic programming analysis.

Operating policies for selected cut-state values of W_1 are presented in Table 17-4 with the corresponding optimal value of T_1. Total conversion y_{32} increases as W_1 increases, but the total return f_3 reaches a peak and then decreases. The optimal policy occurs at $W_1 = 135{,}000$ lb/hr. In addition to the optimal policy, these seven alternative operating policies are generated and are all within 1% of the maximum profit attainable. This is possible because the parameters are adjusted for each cut-state to seek the greatest return.

Each optimal policy selected the highest values of T_1, the air to the burner, and T_3, the gas temperature from converter A; and the lowest value of T_2, the gas

TABLE 17-3

Stage Results for Overall Optimum

Stage 3:

Return, f_3	Output, S_3		Decision, d_3
f_3 ($/hr)	y_3^*	T_3^* (°K)	$W_{W_3}^*$ (lb/hr)
230.57	0.36226	1000	25,513

Stage 2:

Input, S_2	Decision, d_2		Return, f_2	Output, S_2
y_3	$W_{W_2}^*$ (lb/hr)	T_2^* (°K)	f_2 ($/hr)	y_{32}^*
0.30	24,523	700	353.83	0.9859
→ 0.35	→ 22,828	→ 700	→ 353.49	→ 0.9859
0.40	21,215	700	353.16	0.9859
0.45	19,657	700	352.85	0.9859
0.50	18,136	700	352.54	0.9859
0.55	16,637	700	352.23	0.9859
0.60	15,147	700	351.93	0.9859
0.65	13,653	700	351.63	0.9859
0.70	12,141	700	351.33	0.9859
0.75	10,593	700	351.01	0.9859
0.80	8,983	700	350.68	0.9859
0.85	7,264	700	350.34	0.9859
0.90	5,342	700	349.95	0.9859
0.95	2.931	700	349.46	0.9859

Stage 1:

Input, S_1	Decision, d_1			Return, R_1	Output, S_1	
y_{32}	$W_{W_1}^*$ (lb/hr)	T_2^* (°K)	$W_{W_0}^*$ (lb/hr)	R_1 ($/hr)	W_O^* (lb/hr)	T_2^*
0.80	81,512	325	152,736	284.54	24,967	
0.82	80,167	325	158,630	291.57	25,591	
0.84	78,733	325	164,523	298.59	26,215	
0.86	77,183	325	170,416	305.61	26,840	
0.88	75,481	325	176,309	312.62	27,464	
0.90	73,551	325	182,201	319.63	28,088	
0.92	71,364	325	188,093	326.62	28,712	
0.94	68,695	325	193,986	333.59	29,336	
0.96	65,208	325	199,877	340.53	29,960	
→ 0.98	→ 59,827	→ 325	→ 205,769	→ 347.39	→ 30,585	
→ 1.0	→ 10,135	→ 325	→ 211,660	→ 352.41	→ 31,209	

TABLE 17-4

Optimum Operating Policy for Various Cut-State Values

	W_1 (lb/hr)	T_1 (°K)	
A	85,000	430	
B	105,000	430	
C	115,000	430	
D	135,000*	430*	* Overall optimum
E	155,000	430	
F	175,000	430	
G	195,000	430	
H	215,000	430	

Stage 1

	T_i^*(°K)	$W_{W_1}^*$ (lb/hr)	W_0^*(lb/hr)	$W_{W_0}^*$ (lb/hr)
A	350	34,445	30,497	251,983
B	350	37,314	30,697	239,593
C	350	38,383	30,738	232,830
D	325*	45,214*	30,768*	207,501*
E	325	49,582	30,795	191,815
F	325	54,241	30,813	176,033
G	325	59,061	30,825	160,196
H	325	63,979	30,832	159,134

Stage 2

	y_{32}^*	$W_{W_2}^*$ (lb/hr)	T_2^* (°K)
A	0.9766	17,960	700
B	0.9830	20,003	700
C	0.9844	20,959	700
D	0.9859*	22,472*	700*
E	0.9867	22,945	700
F	0.9873	24,657	700
G	0.9877	26,360	700
H	0.9880	28,056	700

Stage 3

	f_3 ($/hr)	y_3^*	T_3^*(ªK)	$W_{W_3}^*$ (lb/hr)
A	228.51	0.3070	1,000	32,342
B	230.29	0.3379	1,000	29,727
C	230.49	0.3479	1,000	28,347
D	230.57*	0.3623*	1,000*	25,513*
E	230.46	0.3721	1,000	22,618
F	230.25	0.3792	1,000	19,688
G	230.00	0.3847	1,000	16,734
H	229.49	0.3890	1,000	13,223

temperature from converter B. The equilibrium conversion increases with decreasing temperature, and high temperature permits more steam to be made. It appears that the optimal policy must always choose the lowest temperature possible in converter B and the highest temperature possible in converter A. This combination will always yield the greatest return when steam and acid production are considered as contributing jointly to the profit. If steam profit is zero, or if its importance is reduced, the results would approach the case where the by-product steam is not used and maximum acid production is desired. The degree of conversion achieved in converter A is about one-half of that currently realized in industrial cases where there is no use for the steam produced. A plant designed to operate for maximum profit, however, must utilize the value of steam production if steam is required for other processes. It should be realized, moreover, that dynamic programming would choose an optimal policy for a plant requiring maximum acid production.

17-12. SUMMARY

The general principles of dynamic programming have been presented. This included the development of the functional equations and diagrams for serial and branching systems. Rules have been presented for formulating a process problem into a dynamic programming system and further rules have been given to simplify the system. A case study of the application of dynamic programming to an actual chemical process has been presented which showed that the technique is a very effective optimization procedure for these types of processes.

REFERENCES

1. R. Bellman, *Dynamic Programming,* Princeton U. P., Princeton, N. J., 1957, p. 83.
2. R. Aris, *The Optimal Design of Chemical Reactors,* Academic Press, New York, 1961, p. 4.
3. D. J. Wilde, private communication (1964).
4. R. Aris, G. L. Nemhauser, and D. J. Wilde, "Optimization of Multistage Cyclic and Branching Systems by Serial Procedures," *A.I.Ch.E. Journal,* Vol. 10, No. 3, (November 1964), p. 913.
5. L. G. Mitten, and G. L. Nemhauser, "Multistage Optimization," *Chem. Engr. Prog.,* Vol. 54, No. 1, (January 1963), p. 53.
6. D. J. Wilde, "Strategies for Optimizing Macrosystems", *Chem. Engr. Prog.,* Vol. 61, No. 3, (March 1965), p. 86.
7. S. M. Roberts, *Dynamic Programming in Chemical Engineering and Process Control,* Academic Press, New York, 1964, pp. 9-10.

8. Ivan Lowry, *A Dynamic Programming Study of the Contact Process,* M.S. thesis, Louisiana State University, Baton Rouge, La., 1965.

9. *Ibid,* Appendix A.

10. *Ibid,* p. 33.

PROBLEMS

17-1. Shown in the sketch is a partially completed functional diagram of a process that involves a feedforward loop and a converging branch. Complete the functional diagram by labeling it with the appropriate subscripts on the state and decision variables. Then give the appropriate functional equations, transition functions and incident identities for each stage in the space indicated. Also give the type of partial optimization at each stage, and describe how the feedforward loop and converging branch are evaluated and included in the main branch.

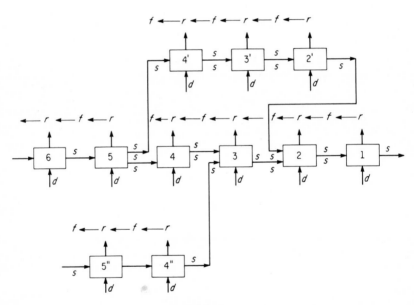

PROB. 17-1

17-2. In the flow diagram below, a simplified version of a catalytic cracking unit and associated separation facilities is shown.

Develop the dynamic programming functional equations and diagram for this case. Define each decision, state, and return variable, and describe how to calculate the return at each stage. Be sure to simplify the functional diagram where required by Wilde's rules, and indicate the steps to obtain the optional return and policy.

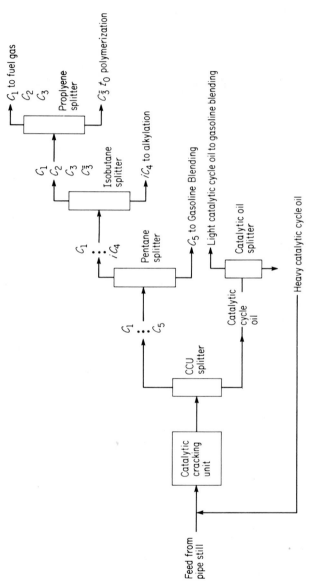

PROB. 17-2

17-3. Develop the dynamic programming functional equation and diagrams for the allocation of feed to three chemical reactors operated in parallel. The product quality from each reactor depends on the catalyst's age and activity and is different for each reactor.

17-4. A unit of time may be used as a dynamic programming stage. This is illustrated by the problem of determining when to replace a piece of equipment. In order to maximize the overall profit made from operating a piece of equipment over a period of years the net profit per year, salvage value and purchase price of new equipment need to be known. The following gives the net profit, in thousands of dollars, per year, r_i starting with a new piece of equipment at $t = 0$.

t	0	1	2	3	4	5	6	7	8	9	10	11	12
r_i	15	14	13	12	11	9	7	5	3	1	0	0	0

Inflation affects the purchase price p of new equipment as shown.

t	0	1	2	3	4	5	6	7	8	9	10	11	12
p		12	12	13	13	14	14	15	15	15	16	16	16

The salvage value s varies as follows:

t	0	1	2	3	4	5	6	7	8	9	10	11	12
s	5	4	3	2	1	0	0	0	0	0	0	0	0

Determine the optimum equipment replacement policy for a 12-year period, using dynamic programming.

17-5. Referring to Example 17-1, determine the shortest distance between the major cities on the East and West Coast.

17-6. Solve Example 17-2 as a network problem.

17-7. Give three other final optimization plans for the sulfuric acid contact process that would correspond to the one of Fig. 17-18.

chapter 18

Linear Programming

18-1. INTRODUCTION

The term *programming* of linear programming does not refer to computer programming but to scheduling. Linear programming was developed about 1947 before the advent of the computer when George B. Dantzig [1] recognized a generalization in the mathematics of scheduling and planning problems. Linear programming developed as the computer developed and now problems involving several thousand independent variables and constraints equations can be solved. The technique has been applied to refining and chemical-plant optimization, livestock-feed blending, routing of aircraft and utilization of crews, transportation and allocation problems, and optimization of corporate strategy–to name only a few.

The application of linear programming has been successful when a large number of interrelated choices exist and where the best choice is far from obvious. Often such problems involve vast sums of money for a small improvement. A typical example is a large oil refinery where the stream flow rates are very high, and a small improvement is multiplied by a large number to obtain considerable amount of savings on an annual basis.

18-2. CONCEPTS AND GEOMETRIC INTERPRETATION

As the name indicates, all equations that are used in linear programming analysis must be linear. Although this appears to be a severe restriction, there are many problems that can be cast in this context. In a linear programming formulation, the equation that determines the profit or cost of operation is referred to as the *objective function*. It must have the form of the sum of linear terms. The equations which describe the limitations under which the system must operate are called the *constraints*. The variables must be nonnegative, *i.e.,* positive or zero only.

The best way to introduce the subject is with an example to illustrate the method and give some geometric intuition about the problem.
Example

A small chemical company makes two types of small solid-fuel rocket motors for testing; for motor A the profit is $ 3 per motor and motor B the profit is $ 4 per motor.   A total processing time of 80 hours per week is available to produce the motors and an average of 4 hours per motor is required for A, while only 2 hours per motor is required for B.   However, due to the hazardous nature in the material in B, 5 hours preparation and cleanup time are required per motor, while only two hours per motor are required for A.   The total preparation time available is 120 hours per week.   Determine the number of each motor that should be produced to maximize the profit.

   Solution: The objective function and constraint equations for this case are

   Maximize

$$3A + 4B \qquad \text{Profit}$$

subject to

$$4A + 2B \leq 80 \qquad \text{Processing time}$$
$$2A + 5B \leq 120 \qquad \text{Preparation time}$$

It would be tempting to make all B motors using the preparation time limitation $120/5 = 24$ for a profit of \$96.   If all A motors were made, there is a processing time limitation $80/4 = 20$ for a profit of \$60.   However, there is a best solution; and this can be seen from Fig. 18-1.   Here the constraint equations are shown, and the small arrows show the region that are allowable for the variables.   For the processing time and preparation time, any values of the variables lying above the lines violate the constraint equations.   Consequently allowable values must

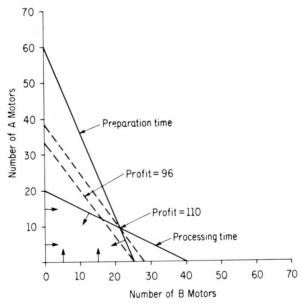

FIG. 18-1.   Constraints and objective function for maximizing rocket-motor profit.

lie on or inside the lines and the A and B axes (since A and B must be nonnegative). This is called the *feasible region.* The objective function is shown for $P = 96$, and this is one of the family of lines:

$$3A + 4B = P$$

where P can increase as long as the values of the variables stay in the feasible region. Increasing P it can be seen that the maximum value will be at the vertex $A = 10$, $B = 20$, which gives $P = \$110$.

The usual response surface—experimental region plot is shown in Fig. 18-2. As can be seen we have a plane as the response surface and the highest point is the vertex $A = 10$, $B = 20$. The intersection of the response surface and planes of $P = $ constant give a line in the response surface as shown for $P = 96$. The projection of this line on the experimental region is the same line shown in Fig. 18-1 for $P = 96$.

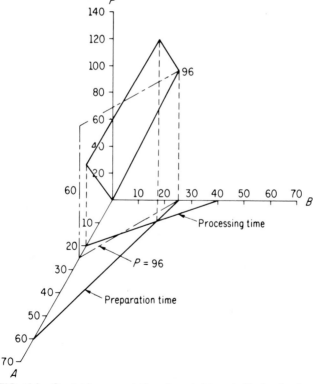

FIG. 18-2. Geometric representation of constraints and objective function for maximizing rocket-motor profit.

18-3. GENERAL STATEMENT OF THE LINEAR PROGRAMMING PROBLEM

There are several ways to write the general mathematical relations for linear programming.   First, in the usual algebraic notation,

Objective Function

Maximize

$$c_1 x_1 + c_2 x_2 + \cdots + c_n x_n \tag{18-1a}$$

Constraint Equations

subject to

$$x_i \geq 0 \text{ for } i = 1, 2, \cdots, n \tag{18-1b}$$

$$
\begin{aligned}
a_{11} x_1 + a_{12} x_2 + \cdots + a_{1n} x_n &\geq b_1 \\
a_{21} x_1 + a_{22} x_2 + \cdots + a_{2n} x_n &\geq b_2
\end{aligned} \tag{18-1c}
$$

$$\cdot \qquad\qquad\qquad\qquad \cdot$$
$$\cdot \qquad\qquad\qquad\qquad \cdot$$
$$\cdot \qquad\qquad\qquad\qquad \cdot$$

$$a_{m_1} x_1 + a_{m_2} x_2 + \cdots + a_{m_n} x_n \geq b_n$$

We seek the values of the x_i's that maximize the objective function.   The coefficients, c_i's, of the x_i's are referred to as *cost coefficients*.   The values of the x_i's must satisfy the constraint equations.   Usually there are more unknowns than equations, i.e., $n > m$ and some of the values of the x_i's are arbitrarily specified (usually zero).   The independent variables can be flow rates in a chemical process, and the constraint equations can be material and energy balances of the processes and plant involved.

The general formulation can also be written as

Maximize

$$\sum_{i=1}^{n} c_i x_i \tag{18-2a}$$

subject to

$$x_i \geq 0 \qquad j = 1, 2, \cdots, n \tag{18-2b}$$

$$\sum_{i=1}^{n} a_{ij} x_i \geq b_j \qquad j = 1, 2, \cdots, m \tag{18-2c}$$

Vector notation is another convenient method of writing the above.

Maximize

$$\mathbf{c} \cdot \mathbf{x} \qquad (18\text{-}3a)$$

subject to

$$\mathbf{x} \geq 0 \qquad (18\text{-}3b)$$

$$\mathbf{Ax} \geq \mathbf{b} \qquad (18\text{-}3c)$$

To obtain an optimal solution of the above equation set the inequalities are converted to equalities by introducing slack variables as is shown in the next section. This gives a set of equations to be solved which has more independent variables than equations, and some must be arbitrarily specified as has been indicated. To fit into the mathematical framework of linear programming, these "extra" variables must be set equal to zero, and the reason will be obvious by the subsequent discussion. A solution of the constraint equations, as equalities, where as many variables are nonzero (positive or negative) as there are constraint equations is a *basic solution*. All of the other variables are zero. A *basic feasible solution* is a solution of the constraint equations where the nonzero variables are positive. A *basis* is a collection of the nonzero variables. It will be seen subsequently that the optimum (maximum or minimum) of the objective function is a basic feasible solution. This is determined using the *simplex method* of linear programming. However it is best to illustrate the simplex method with an example before stating the simplex method theorems which establish the global optimality.

Slack Variables

The illustration will involve inequalities, and these did not affect the solution in the previous example since it was solved graphically. It is necessary to have equality constraints, but many times the constraints are inequalities. However, inequalities can be converted to equalities by introducing another variable, a slack variable. Consider the inequality

$$x_1 + x_2 \leq b$$

If a positive x_3, of magnitude to be determined, is added to the left-hand side of the above, the equation is converted to the equality

$$x_1 + x_2 + x_3 = b$$

The physical interpretation of a *slack variable*, as x_3, is that it represents the difference between the largest value of the sum of the variables $(x_1 + x_2)$ and the smallest. If the slack variable is zero, as it is in some cases, this is the situation where the largest value of the other variables are optimum. The slack variable would be subtracted from the left-hand side if that side were greater than the right-hand side.

18-4. DEMONSTRATION OF THE SIMPLEX METHOD

The simplex method is an algorithm that steps from one vertex of the intersections of the constraint equations to another vertex in a manner such that the objective function always increases. Without attempting to show a model associated with the following set of equations [2], let us see how the algorithm operates!

Maximize

$$x_1 + 2x_2$$

subject to

$$2x_1 + x_2 \leq 10$$
$$x_1 + x_2 \leq 6$$
$$-x_1 + x_2 \leq 2$$
$$-2x_1 + x_2 \leq 1$$

Inserting the slack variables gives the following set:

Maximize

$$x_1 + 2x_2 \qquad\qquad\qquad = z$$

subject to

$$2x_1 + x_2 + x_3 \qquad\qquad\qquad = 10$$
$$x_1 + x_2 \qquad + x_4 \qquad\qquad = 6$$
$$-x_1 + x_2 \qquad\qquad + x_5 \qquad = 2$$
$$-2x_1 + x_2 \qquad\qquad\qquad + x_6 = 1$$

where z is the value of the objective function. There are six independent variables, and there will be four nonzero variables in the basis. All of the basic feasible solutions are shown in Table 18-1 and correspond to the vertices of the convex polygon as shown in Fig. 18-3.

TABLE 18-1

Basic Feasible Solutions of the Equation Set

Vertex	x_1	x_2	x_3	x_4	x_5	x_6
A	0	0	10	6	2	1
B	0	1	9	5	1	0
C	1	3	5	2	0	0
D	2	4	2	0	0	1
E	4	2	0	0	4	7
F	5	0	0	1	7	11

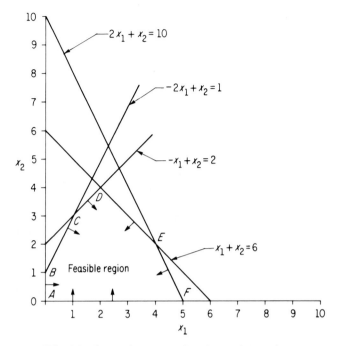

FIG. 18-3. Geometric representation of constraint equations.

One of the basic solutions for the above equations set is

$$x_1 = 0, \quad x_2 = 6, \quad x_3 = 4, \quad x_4 = 0, \quad \underline{x_5 = -4} \text{ and } \underline{x_6 = -5}.$$

Comparing the variables that constitute a basis at one vertex with one at an adjacent vertex, it is seen that each have all but one variable in common. Therefore, to obtain basis B from basis A it is necessary to remove x_6 from the basis (i.e. set x_6 = zero and bring x_2 into the basis (i.e., solve for x_2 = zero). The simplex method moves from vertex to vertex, and each time it is in a direction of an improved value of the objective function. This is the key to the simplex algorithm. The technique moves from vertex to vertex with Gaussian elimination.

The procedure to be followed in solving a linear programming problem is:

1. Place the problem in a linear programming format with constraint equations and objective function.
2. Introduce any slack variables and have positive right-hand sides on the constraint equations.
3. Select an initially feasible basis. If all constraint equations were inequalities, the slack variables can be used.
4. Perform algebraic manipulations to express the objective function in terms of the nonbasic variables only. This determines the value of the objective function for the variables in the basis.

5. Inspect the objective function and select the variable with the largest positive coefficient to bring into the basis for maximization. If there are no positive coefficients the optimum has been reached.

6. Inspect the constraint equations to select one to be used to eliminate the variable to be brought into the basis. The selection is made such that none of the right-hand sides become negative. This is necessary so none of the variables in the new basis will become negative.

7. Perform the same elimination on the objective function to remove the new variable in the basis. Have only variables *not* in the basis in the objective function. This determines the new value of the objective function.

8. Repeat the procedure of steps 5 through 7 until all coefficients are negative in the objective function for maximization. Thus if any of the nonbasic variables are increased this would result in a decrease in the objective function.

Applying the above procedure to the preceding equation set, the first two steps are already completed. The slack variables will be used as the initial feasible basis (step 3). The set can be written with the variables not in the basis in the objective function and is

Maximize

$$x_1 + 2x_2 \qquad\qquad = z \qquad z = 0$$

subject to

$$
\begin{aligned}
2x_1 + x_2 + x_3 &= 10 & x_3 &= 10 \\
x_1 + x_2 \quad + x_4 &= 6 & x_4 &= 6 \\
-x_1 + x_2 \quad\quad + x_5 &= 2 & x_5 &= 2 \\
-2x_1 + x_2 \quad\quad\quad + x_6 &= 1 & x_6 &= 1
\end{aligned}
$$

Continuing with the procedure, the variable with the largest positive coefficient is associated with x_2. Thus increasing x_2 will increase the objective function (step 5). The fourth constraint equation will be used to eliminate x_2 from the objective function (step 6). x_2 is said to enter the basis and x_6 to leave. Proceeding with the elimination gives

Maximize

$$5x_1 \qquad\qquad\qquad - 2x_6 = z - 2 \qquad z = 2$$

subject to

$$
\begin{aligned}
4x_1 \quad + x_3 \quad\quad - x_6 &= 9 & x_3 &= 9 \\
3x_1 \quad\quad + x_4 \quad - x_6 &= 5 & x_4 &= 5 \\
x_1 \quad\quad\quad + x_5 - x_6 &= 1 & x_5 &= 1 \\
-2x_1 + x_2 \quad\quad\quad + x_6 &= 1 & x_2 &= 1
\end{aligned}
$$

The nonzero variables in the basis are x_2, x_3, x_4, and x_5 and the objective function has increased from $z =$ zero to $z = 2$.

The procedure is repeated (step 8) selecting x_1 to enter the basis. The third constraint equation will be used, and x_5 will leave the basis. Performing the manipulations gives

Maximize

$$-5x_5 + 3x_6 = z - 7 \qquad z = 7$$

subject to

$$
\begin{aligned}
x_3 - 4x_5 + 3x_6 &= 5 & x_3 &= 5 \\
x_4 - 3x_5 + 2x_6 &= 2 & x_4 &= 2 \\
x_1 \quad x_5 \quad - x_6 &= 1 & x_1 &= 1 \\
x_2 + 2x_5 \quad - x_6 &= 3 & x_2 &= 3
\end{aligned}
$$

The procedure is repeated and x_6 is selected to enter the basic. The second constraint equation is used and x_4 leaves the basis. The manipulations result in:

Maximize

$$-3/2x_4 - 1/2x_5 = z - 10 \qquad z = 10$$

subject to

$$
\begin{aligned}
x_3 - 3/2x_4 + 1/2x_5 &= 2 & x_3 &= 2 \\
1/2x_4 - 3/2x_5 + x_6 &= 1 & x_6 &= 1 \\
x_1 \quad + 1/2x_4 - 1/2x_5 &= 2 & x_1 &= 2 \\
x_2 \quad + 1/2x_4 + 1/2x_5 &= 4 & x_2 &= 4
\end{aligned}
$$

All of the coefficients of the objective function are negative, and increasing either x_4 or x_5 will only decrease the objective function. The maximum is reached, and the optimal basic feasible solution is obtained.

Referring to Table 18-1 the set of basic feasible solutions, it is seen that the simplex method was started at vertex A. The first application of the procedure stepped to the adjacent vertex, B, with an increase in the objective function to 2. Proceeding the simplex method, then moved the solution to vertex C, where the objective function was increased to 7 at the next application of the algorithm the optimum is reached for $z = 10$ at vertex D. At this point the application of the simplex method is stopped since the maximum was reached.

18-5. THE MATHEMATICS OF LINEAR PROGRAMMING

To prove the theorems that establish the previous procedure and locate the optimal solution of the linear programming problem, the mathematics of convex sets and linear inequalities has to be developed. This is done in any of the standard texts devoted to the subject and is beyond the scope of this brief discourse. However, the appropriate theorems with words of explanation will be given such that the concepts are conveyed, and the more interested can refer to the standard works such as Garvin [3].

A feasible solution, as was previously discussed, is any solution to Eqs. 18-1b and 18-1c, and a convex set consists of points which correspond to the feasible solutions of the constraint equations.   A convex set is illustrated in Fig. 18-4a for two dimensions and made up of any of the points lying inside or on the boundaries of the feasible region.   An analogous definition of a convex set is a collection of points such that it contains any two points A and B it also contains the straight line $\overline{AB}$ between the points.   A nonconvex set is shown in Fig. 18-4b.   An extreme point or vertex of a convex set is a point that does not lie on any segment joining two other points in the set.   The important theorem relating convex sets with feasible and basic feasible solutions is

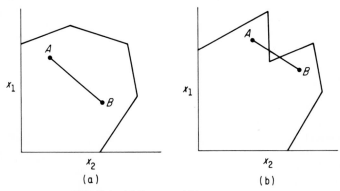

FIG. 18-4.   (a) Convex and (b) nonconvex sets.

The collection of feasible solutions constitute a convex set whose extreme points correspond to a basic feasible solution [4].

In the proof of the above theorem, it is shown that a linear combination of any two feasible solutions is a feasible solution and hence lies on a straight line between the two, thus constituting a convex set.   To prove that a basic feasible solution is an extreme point, it is assumed that a basic feasible solution can be expressed as a linear combination of feasible solutions and shown by contradiction that this is impossible.   Thus it must be an extreme point.

The next important theorem is:

If a feasible solution exists, then a basic feasible solution exists [5].

This is proved by showing that a basic feasible solution can be constructed from a feasible solution.

The next theorem relates the value of the objective function with the basic feasible solution:

If the objective function possesses a finite maximum (or minimum), then at least one optimal solution is a basic feasible solution [6].

This theorem can be proved by writing any point P as a convex combination of extreme points. Then since the objective function is a linear function it can be written as a sum of linear terms obtained by substituting the convex combination for the point P. Then the value of the objective function evaluated at P is less that obtained by substituting the maximum value of the objective function for each value of the objective function in the linear combination. Thus there is an extreme point at which the objective function assumes its maximum value. Also, one can be located where the objective function obtains its minimum value by a similar technique.

To formalize the simplex computational procedure consider the set of equations with a basic feasible solution $\mathbf{x} = (x_4, x_5, x_6)$.

Maximize

$$c_1 x_1 + c_2 x_2 + c_3 x_3 \qquad\qquad = z_0 \qquad (18\text{-}4a)$$

subject to

$$
\begin{aligned}
a_{11} x_1 + a_{12} x_2 + a_{13} x_3 + x_4 & = b_1 \\
a_{21} x_1 + a_{22} x_2 + a_{23} x_3 + x_5 & = b_2 \qquad (18\text{-}4b) \\
a_{31} x_1 + a_{32} x_2 + a_{33} x_3 + x_6 &= b_3
\end{aligned}
$$

If c_1 is the largest positive coefficient and b_1/a_{11} is the smallest positive ratio, then x_1 enters the basis and x_5 leaves the basis. Performing the elimination the result is

Maximize

$$(c_2 - c_1 a_{12}/a_{11})x_2 + (c_3 - c_1 a_{13}/a_{11})x_3 - (c_1 a_{14}/a_{11})x_4 \qquad = z_0 - c_1 b_1/a_{11} = z_1$$

$$(18\text{-}5a)$$

subject to

$$
\begin{aligned}
x_1 + a_{12}/a_{11} x_2 + a_{13}/a_{11} x_3 + a_{14}/a_{11} x_4 &= b_1/a_{11} \\
(a_{22} - a_{21} a_{12}/a_{11})x_2 + (a_{23} - a_{21} a_{13}/a_{11})x_3 - a_{21} a_{14}/a_{11} x_4 + x_5 &= b_2 - a_{21} b_1/a_{11} \quad (18\text{-}5b) \\
(a_{32} - a_{31} a_{12}/a_{11})x_2 + (a_{33} - a_{31} a_{13}/a_{11})x_3 - a_{31} a_{14}/a_{11} x_4 + x_6 &= b_3 - a_{31} b_1/a_{11}
\end{aligned}
$$

If $z_1 > z_0$, then there is an improvement in the objective function and the solution is continued. If $z_1 < z_0$ then no improvement in the objective function is obtained, and $\mathbf{x}$ is basic feasible solution that maximizes the objective function. The following theorem is:

If for any basic feasible solution $\mathbf{x}_k = (x_1, x_2, \cdots, x_m)$ the condition $z(\mathbf{x}_k) > z(\mathbf{x}_j)$ for all $j = 1, 2, \cdots n (j \neq k)$ then $\mathbf{x}_k$ is a basic feasible solution that maximizes the objective function.

A corresponding result can be obtained for the basic feasible solution that minimizes the objective function.

18-6. ARTIFICIAL VARIABLES

To start a linear programming problem it is necessary to have the equations in canonical form as Eq. 18-4a and 18-4b are.   Many times it is convenient to just add as many new variables (artificial variables) as there are constraint equations to give an initial basis to start the solution. This is permissible, and it can be shown that the feasible solutions to original problems are feasible solutions to the augmented problem.   However, it is necessary to modify the objective function to insure that all artificial variables leave the basis.   This is accomplished by adding terms to the objective functions which consist of products of the artificial variable, and a negative coefficient that can be made arbitrarily large for the case of maximizing the objective function.   Thus it is insured that the artificial variables leave the basis immediately.

At this point it is reasonable to wonder if this is not a significant amount of computations for convenience only.   The answer would be yes if one were solving a linear programming problem for the first time.   However, this is not usually the case, since the state of the art for linear programming computations is quite advanced, and there are sophisticated and fast LP codes available that only require the values of the coefficients in the equation set to obtain an optimal solution.   In fact, there are companies that specialize in furnishing efficient LP programs, and large-capacity programs can be obtained from the IBM General Share Library.   Thus, developing a linear model of the particular process is the main effort required, and then one of the available general LP codes can be used to obtain the optimal solution.   Also, most major companies have a group that are expert in implementation of the LP algorithms.   Therefore, in this space available it is best to discuss with an illustration describing the development of equations for a process rather than computer implementation of the algorithm.

18-7. LINEAR PROGRAMMING OPTIMIZATION
OF A SIMPLE REFINERY

The flow diagram of an extremely simple petroleum refinery is shown in Fig. 18-5.   The refinery produces premium gasoline which sells for $ 5/ bbl and has an octane number of ≥ 95, regular gas for $ 4.50/ bbl and an octane number of ≥ 89, and fuel oil for $ 2.50/ bbl and a contamination number no more than 55. To meet market demands the refinery must produce at least 25,000 bbl/day of premium gasoline, 10,000 bbl/day of regular gasoline and 30,000 bbl/day of fuel oil.   The current cost of crude is $ 2.75/ bbl.

Operating cost for separation in the crude oil still is $ 0.25/ bbl for each of the products produced, light virgin naphtha, heavy virgin naphtha, and virgin distillate.   The virgin distillate and part of the heavy virgin naphtha (HVNCC) are sent to the catalyctic cracking unit.   The cost of operation depends on the feed

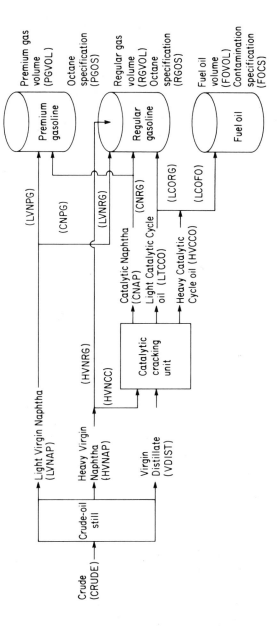

FIG. 18-5.   Flow diagram of a simple refinery.

TABLE 18-2

Refinery Objective Function and Constraints.

		Crude Oil Still Activities				Catalytic Cracking Activities		Premium Gasoline Blending		Regular Gasoline Blending				Fuel Oil Blending		Right Hand Side
Row	Name	CRUDE	LVNAP	HVNAP	VDIST	HVNCC	VDIST	LVNPG	CNPG	LVNRG	HVNRG	CNRG	LCORG	LCOFO	HVCCO	
Object Function	Profit	−2.75	−0.25	−0.25	−0.25	−0.10	−0.15	5.00	5.00	4.50	4.50	4.50	4.50	2.50	2.50	=
Premium Gas blending Constraints	PGVOL							1	1							≥ 25,000
	PGOS							−3	2							≥ 0
Regular Gas blending Constraints	RGVOL									1.	1	1	1			≥ 10,000
	RGOS									3	−5	8	−1			≥ 0
Fuel oil Blending Constraints	FOVOL													1	1	≥ 30,000
	FOCS													−5	4	≤ 0
Catalytic cracker Capacity Constraints	CCCC					1	1									≤ 50,000
Crude still Capacity	CSCC	1														≤ 100,000
Material Balance Constraints	LVNAP	−0.2						1		1						= 0
	HVNAP	−0.5									1					= 0
	VDIST	−0.3					1									= 0
	CNAP					−0.7	−0.1		1			1				= 0
	LTCCO					−0.4	−0.3						1	1		= 0
	HVCCO					−0.2	−0.7								1	= 0

to the unit and is $0.10/bbl for the heavy virgin naphtha and $0.15/bbl for the virgin distillate.

It is convenient to present the above information in a matrix form as Table 18-2. Thus one can see at a glance the cost coefficients and constraint equations. The first line gives the objective function where the cost are shown as negative and sales prices positive. The second line gives the premium gasoline volume-blending constraint and says that at least 25,000 bbl/day of premium must be produced. The third line gives the octane number requirement of premium. The octane number of the light virgin naphtha is 92 and the catalytic naphtha is 97. Thus

$$92LVNPG + 97 \ CNPG \geq 95 \ (LVNPG + CNPG)$$

Rearranging gives:

$$-3LVNPG + 2CNPG \geq 0$$

This is the equation shown in the third line. Similar relations are shown for regular gasoline and fuel-oil blending. The inequality is reversed in the case of the contamination number. The accompanying table gives the specifications on each blending component.

Component	Octane Number	Contamination Number
Heavy catalytic oil	—	59
Light catalytic cicle oil	88	50
Catalytic naphtha	97	—
Heavy virgin naphtha	84	—
Light virgin naphtha	92	—

Next the catalytic cracking unit is capacity (CCCC) must not exceed 50,000 bbl/day. Correspondingly, the crude-oil still capacity is limited to 100,000 bbl/day.

Following in the table are the material balance constraints. For example, the line LVNAP says that the sum of the light virgin naphtha to premium and regular gasoline must equal the total from the crude still. Outputs are taken as positive and inputs negative. As shown, the crude is separated into three volume fractions—0.2 light virgin naphtha, 0.5 heavy virgin naphtha, and 0.3 virgin distillate. Blank spaces in the matrix are zero.

For the catalytic cracking unit, the operating cost for one barrel of heavy virgin naphtha is $ 0.10 and the product distribution is 0.7 bbl of catalytic naphtha, 0.4 bbl of light catalytic cycle oil, and 0.2 bbl of heavy catalytic cycle oil. The increase in volume is due to the cracking of large molecules to small ones. These material balances are shown at lines CNAP, LTCCO, and HVCCO. Correspondingly, the virgin distillate product distribution from the catalytic cracking unit is 0.1 bbl of catalytic naphtha, 0.3 bbl of light catalytic cycle oil, and 0.7 bbl of heavy catalytic cycle oil per barrel of virgin distillate. The numbers are shown under the appropriate columns completing the material balances.

For this simple refinery model a total of 14 constraint equations were involved. For an actual refinery there would be hundreds of constraint equations, but all would be developed in a similar fashion. As can be seen, the model (constraint equations) are simple, and it was possible to consider only one set of operating conditions for the catalytic cracking unit and crude still. When different operating conditions are included, it is necessary to consider discrete values.

A search can be performed by repeating the problem for changes in operating conditions. The linear model can be used to approximate the actual nonlinear models of each process unit. Searching in this fashion has been referred to as *sectionally linearized linear programming*. A program is available in the IBM Share Library that takes a nonlinear model, makes a linear approximation, calculates the optimum, compares with a best previous result, and continues until no improvement is obtained. Thus a nonlinear search is attempted with linear programming.

18-8.   SENSITIVITY ANALYSIS

Having obtained the optimal solution for one case, it would be desirable to know how much the cost coefficients could change, for example, before it is necessary to resolve the problem. In fact there are five areas that should be examined for their effect on the optimal solution. These are:

1. Changes in the right-hand side of the constraint equations, b_i.
2. Changes in the coefficients of the objective function, c_i.
3. Changes in the coefficients of the constraint equations, a_{ij}.
4. Addition of new variables.
5. Addition of more constraint equations.

Changes in the right-hand side of the constraint equations correspond to changes in the maximum capacity of a process unit or the availability of a raw material, for example. Changes in the coefficients of the objective function correspond to changes of the cost or the sale price of the products, for example. Changes in the coefficients of the constraint equations correspond, for example, to changes in flow rate or quality of process streams. Addition of new variables and constraint equations correspond to the addition of new process units in the overall plant. Knowing how these various factors vary without changing the optimal solution supplies valuable information and reduce the number of times the linear programming problem must be solved.

Some preliminary mathematical expressions must be developed for the analysis of the effect of the above five areas on the optimal solution. These are the inverse of the optimal basis and the Lagrange multipliers. To obtain the inverse of the optimal basis, $\mathbf{A}^{-1}$, consider that the optimal basis $x^* (x_1, x_2, \cdots, x_m, 0, 0, \cdots,$

0) has been found by the previously described simplex method to optimize the objective function. There are m constraint equations and n variables as given by Eqs. 18-1a, 18-1b, and 18-1c. For convenience the variables in the optimal basis have been rearranged to go from 1 to m, and there are $(n - m)$ variables not in the basis. The optimal solution to this linear program problem is

$$\mathbf{c} \cdot \mathbf{x}^* = \max_{\mathbf{x}} \mathbf{c} \cdot \mathbf{x} = z^* \tag{18-6a}$$

and

$$\mathbf{Ax}^* = \mathbf{b} \tag{18-6b}$$

To solve for $\mathbf{x}^*$, multiply both sides of the above equation by the inverse of the optimal basis, $\mathbf{A}^{-1}$ whose elements are β_{ij}:

$$\mathbf{x}^* = \mathbf{A}^{-1}\mathbf{b} \tag{18-7}$$

It should be noted that $\mathbf{A}^{-1}$ is not obtained from the last step of the simplex method, but it has to be obtained from the original formulation of the problem but employing the optimal basis found from the simplex method.

The linear programming problem could be solved by the classical method of Lagrange multipliers. However the simplex method gives a systematic procedure of locating the optimal basis. Having located the optimal basis by the simplex method, the Lagrange multiplier formulation, and the inverse of the optimal basis will be used to determine the effect of change in the right-hand side on the optimal solution. Multiplying each constraint, Eq. 18-6b by the Lagrange multiplier λ_i and adding to the objective function gives, after some rearrangement,

$$\left[c_1 + \sum_{i=1}^{m} a_{i1} \lambda_i \right] x_1 + \left[c_2 + \sum_{i=1}^{m} a_{i2} \lambda_i \right] x_2 + \cdots + \left[c_m + \sum_{i=1}^{m} a_{im} \lambda_i \right] x_m$$

$$+ \left[c_{m+1} + \sum_{i=1}^{m} a_{im+1} \lambda_i \right] x_{m+1} + \cdots + \left[c_n + \sum_{i=1}^{m} a_{in} \lambda_i \right] x_n = z^* + \sum_{i=1}^{m} b_i \lambda_i$$

$$\tag{18-8}$$

where x_1 to x_m are positive numbers and x_{m+1} to x_n are zero.

To solve this by classical means, the partial derivatives of z with respect to the independent variables and the Lagrange multipliers would be set equal to zero. Taking the partial derivatives of z with respect to the Lagrange multipliers just give the constraint equations, and taking the partial derivatives with respect to the independent variables, x_j ($j = 1, 2 \ldots$),

$$\frac{\partial z}{\partial x_j} = \left[c_j + \sum_{i=1}^{m} a_{ij} \lambda_i \right] = 0 \qquad j = 1, 2, \cdots, m \tag{18-9}$$

and x_j for $j = m + 1$, ..., n is zero, since $\mathbf{x}^*$ is the Lagrange multipliers. Solving for λ_i in the above the result is given below in matrix notation

$$\lambda = - \left[\mathbf{A}^{-1} \right]^T \mathbf{C} \tag{18-10}$$

or the ith Lagrange multiplier λ_i is computed, since the transpose of the inverse of the optimal basis is known, since $\mathbf{A}^T = [\mathbf{A}^{-1}]^T$.

$$\lambda_i = - \sum_{k=1}^{m} \beta_{ki} c_k \qquad i = 1, 2, \cdots, m \tag{18-11}$$

where β_{ki} are the elements of the inverse of the optimal basis. With this as background, the effect of the five changes on the optimal solution can be determined.

Changes in the Right-Hand Side

The original optimal solution will remain optimal as long as it remains a basic feasible solution, since the values of the c_i's are not affected by changes in the b_i's. Consequently the values of the variables in the basis must remain greater than zero for changes in the b_i's. To determine how changes in the b_i's affect the values of the independent variables in the basis, the ith component of Eq. 18-7 is used. This is

$$x_i^* = \sum_{k=1}^{M} \beta_{ik} b_k \qquad i = 1, 2, \cdots, m \tag{18-12}$$

For a change in b_k an amount Δb_k, the change in the ith variable in the basis is

$$x_{i,\text{ new}}^* = x_{i,\text{ old}}^* + \sum_{k=1}^{m} \beta_{ik} \Delta b_k > 0 \tag{18-13}$$

and $x_{i,\text{new}}^*$ remain be positive for the solution to remain a basic feasible solution. The problem must be resolved if x_i^* becomes negative.

To determine change in value of the objective function for changes in the b_i's, it is noted that

$$z^* = - \sum_{i=1}^{m} b_i \lambda_i \tag{18-14}$$

from Eq. 18-8, since the left-hand side is zero the terms in the brackets are zero for the variables in the basis and the variables that are not in the basis are zero. For the change Δb_k the change in the value of the objective function is

$$z_{\text{new}}^* = z_{\text{old}}^* - \sum_{i=1}^{M} \lambda_i \Delta b_i \tag{18-15}$$

Generally large linear programming computer programs have included as part of the computations, the calculation of $x^*_{i,\text{new}}$ and $z^*_{i,\text{new}}$ for specified values of the Δb_i's, one at a time. Also the values of the Δb_i's are computed that will cause a change in basis ($x^*_{i,\text{new}} - x^*_{i,\text{old}} = 0$).

Changes in the Coefficients of the Objective Function

It is necessary to consider changes in the cost coefficients in both of the variables in the basis and those not in the basis. Referring to Eq. 18-8, the coefficients on the variables that are not in the basis must remain positive for minimization, and if the coefficient becomes negative it would be profitable to have that variable enter the basis.

$$\left[c_j + \sum_{i=1}^{M} a_{ij}\lambda_i \right] > 0 \qquad j = m+1, \cdots, n \qquad (18\text{-}16)$$

The values of the Lagrange multipliers are affected by changes in the cost coefficients in the basis, and they are related by Eq. 18-11.

If c'_j is defined as the following:

$$c'_j = \left[c_j + \sum_{i=1}^{M} a_{ij}\lambda_i \right] \qquad (18\text{-}17)$$

and c'_j must remain greater than zero, by Eq. 18-16.

Let c_k be the cost coefficient of the variables in the basis. Then substituting for c_k from Eq. 18-11 gives

$$c'_j = c_j - \sum_{i=1}^{M} a_{ij} \sum_{k=1}^{M} \beta_{ki} c_k \quad \text{for } j = m+1, \cdots, n$$

or

$$c'_j = c_j - \sum_{k=1}^{M} \sum_{i=1}^{M} a_{ij}\beta_{ki} c_k \quad \text{for } j = m+1, \cdots, n \qquad (18\text{-}18)$$

For a change, Δc_j, in the nonbasic variable cost coefficient, c_j, and for a change, Δc_k, in the basic variable cost coefficient c_k, it can be shown that the following equation holds:

$$c'_{j,\text{new}} = c'_{j,\text{old}} + \Delta c_j - \sum_{k=1}^{M} \Delta c_k \sum_{i=1}^{M} a_{ij}\beta_{ki} \quad \text{for } j = m+1, \cdots, n \qquad (18\text{-}19)$$

For the optimal solution to remain optimal,

$$c'_{j,\text{new}} > 0 \qquad (18\text{-}20)$$

and if Eq. 18-20 does not hold, a new optimal solution must be obtained. The objective function will change according to the following if the basis remains the same:

$$z^*_{new} = z^*_{old} + \sum_{k=1}^{M} x_k \Delta c_k \tag{18-21}$$

If $c'_{j,new} < 0$, then the problem must be resolved, it is usually convenient to introduce an artificial variable and proceed from this point to the new optimal solution. Large linear programming codes usually have this provision incorporated in them.

Changes in Coefficients of the Constraint Equations

Referring to Eq. 18-17, it is seen that for changes in the a_{ij}'s for the nonbasic variables will cause changes in c'_j. For the optimal solution to remain optimal, $c'_j > 0$, and if not the problem must be resolved. The use of artificial variable is usually incorporated in large linear programming computer programs. Changes in the coefficients of the constraint equations, a_{ij}, of the basic variables, are beyond the scope of this discussion, but are treated in detail by Garvin [3].

Addition of New Variables

The effect of adding new variables can be determined by considering the results of the addition of new variables to Eq. 18-8. If k new variables are added, then k additional terms will be added to Eq. 18-8, and the coefficient of the kth term is

$$\left[c_{n+k} + \sum_{i=1}^{M} a_{i,n+k} \lambda_i \right] > 0 \tag{18-22}$$

which must be greater than zero for the original optimal solution to remain optimal. If the term in the brackets above is less than zero, the solution can be improved and the problem can be resolved. The use of artificial variables is usually the procedure in handling additional variables that cause a change in the optimal solution.

Addition of More Constraint Equations

For the addition of more constraint equations, the procedure is to add artifical variables and proceed with the solution to the optimum. The artifical variables supply the canonical form to proceed with the solution. The addition of more constraint equations correspond, for example, to the addition of more process units in a petroleum refinery or chemical plant. This can be seen from the previous section.

18-9. SUMMARY

The optimization procedure of linear programming was introduced with a geometric interpretation. Then the simplex algorithm to find the optimum was illustrated, and theorems were given verifying that the optimum was actually attained. The chapter concluded with an illustration for developing the objective function and constraint equations for a simple petroleum refinery and a discussion of sensitivity analysis.

REFERENCES

1. G. B. Dantiz, *Linear Programming and Extensions,* Princeton U. P., Princeton, N.J. 1963.
2. *An Introduction to Linear Programming,* IBM Data Processing Application Manual E20-8171, IBM Corporation, White Plains, N.Y., 1964.
3. W. W. Garvin, *Introduction to Linear Programming,* McGraw-Hill, New York, 1966.
4. *Ibid.,* p. 10.
5. , *Ibid.,* p. 12.
6. *Ibid.,* p. 21.

PROBLEMS

18-1. A company makes two levels of purity of a product which is sold in gallon containers. Product A is of higher purity than product B with profits of $0.40/gal made on A and $0.30/gal made on B. Product A requires twice the processing time of B; and if all the company produced was B, it could make 1,000 gal/day. However, the raw material supply is sufficient for only 800 gal/day of both A and B combined. Product A requires a container of which only 400 gal/day are available while there are 700 gal/day available for B. Assuming all of the product can be sold of both A and B, what volumes of each should be produced?

18-2. Obtain the linear programming solution of the following problem:

Maximize

$$x_1 + 2x_2$$

subject to

$$x_1 + x_1 \leq 6$$
$$x_1 + 3x_2 \geq 6$$
$$x_1 - x_2 \leq 2$$
$$-x_1 + 3x_2 \leq 10$$

18-3. For the simplex method to converge to an optimal solution, the objective function must change for the better each time. However, there are cases when the objective function does not change when the new variable enters the basis. This is referred to as *degeneracy.* Explain what happens by solving the following linear programming problem.

Maximize

$$2x_1 + x_2$$

subject to

$$x_1 + 2x_2 \leq 10$$
$$x_1 + x_2 \leq 6$$
$$x_1 - x_2 \leq 2$$
$$x_1 - 2\omega x_2 \leq 1$$
$$2x_1 - 3x_2 \leq 3$$

18-4. Discuss how linear programming might be applied to a nonlinear system of objective functions and constraints. The name *sectionally linearized linear programming* has been given this procedure.

18-5. Give a comparison of linear programming, dynamic programming, and search techniques. Discuss the advantages and disadvantages of each and how each might be applied to chemical refinery processes.

18-6. To illustrate the use of artificial variables, solve the following linear programming problem employing artificial variables.

Maximize

$$x_1 + 3x_2$$

subject to

$$x_1 + 4x_2 \geq 24$$
$$5x_1 + x_2 \geq 25$$
$$x_1 \geq 0, x_2 \geq 0$$

Use p as the cost coefficient associated with the artificial variables in the objective function. It is only necessary to carry the problems to the point where the artificial variables have left the basis. Be sure to state the reasoning used at each step in solving the problem. Why can the artificial variables be dropped from further consideration when they have left the basis?

18-7. Derive Eq. 18-19 from Eq. 18-18 which relates the changes in the basic and nonbasic variable cost coefficients.

18-8. If one attempts to solve Prob. 18-6 there would be some difficulty. What is this difficulty? Could the set of equations of Prob. 18-6 represent the profit function and material balances of a chemical process?

18-9. Having solved Prob. 18-3, what are the largest changes in the b_i's for the optimal solution to remain optimal?

18-10. Having solved Prob. 18-3, what are the largest changes in the cost coefficients that can be made and still have the optimal solution remain optimal?

Index